Corn

Its Origin
Evolution
and Improvement

Corn

*Its Origin
Evolution
and
Improvement*

Paul C. Mangelsdorf

1974

The Belknap Press of

Harvard University Press

Cambridge, Massachusetts

To my wife

Who for almost fifty years
has shared my interest
in *Zea Mays*

Preface

I hope that I may be excused for entering on these personal details, as I give them to show that I have not been hasty in coming to a decision. Charles Darwin, 1859

This book is the culmination of forty-eight years of research on the corn plant, but my interest in corn goes back even further. Indeed I can scarcely remember a time when I did not have a curiosity about the corn plant. Long before I understood its function, I admired and was intrigued by the graceful fascicle of silks, often delicately colored, that emerged from the tender lateral shoots. Then soon after the silks had withered the shoot began to swell, becoming in several weeks a roasting ear and if not then picked and consumed, in several months a full ear with ripe kernels arranged in orderly rows. Even as a boy I knew that this plant was different from any of the others that I saw or grew. And since my father was a commercial seedsman and florist and my mother also a plant lover, it was only natural that during my boyhood I should have seen or grown virtually all of the cultivated plants of garden, field, or greenhouse that were adapted to northeastern Kansas where I grew up.

I also became quite familiar at an early age with the natural vegetation of the region. This was the result of frequent, long, Sunday afternoon walks with my father and my older brother, Albert, who later became a renowned sugar cane breeder. One of the things that especially impressed me was the difference between certain cultivated plants and their wild counterparts. The wild strawberries were so much smaller but more flavorful than their garden-grown relatives. The wild gooseberries were tiny compared to the cultivated fruits. The flowers of violets and columbines were smaller and more dainty than those of horticultural varieties grown in the greenhouse or garden. I speculated on the reasons for the differences and supposed that they were due to differing environments. Yet when I transplanted wild columbines or violets they continued to bear the same dainty, delicate flowers that they had produced in their native habitats. It was only natural that I should have been puzzled and intrigued by these circumstances.

It was only natural, too, that when ready for college I should have decided to study scientific agriculture, and to enroll at Kansas State Agricultural College for this purpose. There it was my good fortune to come in contact almost immediately with John H. Parker, who was in charge of the improvement work on corn and small grains and who gave me a job as one of his student assistants. I served in that capacity for four years and was an apprentice in the full sense of the word. Observing that I was much more interested in corn than in the other cereals in his charge, Parker encouraged me to read all of the early Kansas publications on corn and in a genetics seminar during my senior year, assigned me the task of reading and reviewing the recently published book, *Inbreeding and Outbreeding*, by Edward M. East and Donald F. Jones, in which the authors described the research that formed the basis for the development of hybrid corn. Also, during my senior year, Parker practically turned over to me the experimental work on corn. I am still appalled at some of the naïve things I did that year at the expense of the State of Kansas. But I now know that education is usually expensive no matter what form it takes or who pays for it. There was one product of this experience that proved to be especially valuable to me. When Dr. Donald F. Jones of the Connecticut Agricultural Experiment Station, one of the inventors of hybrid corn and at that time its chief promoter, asked Parker for recommendations for a position as his assistant, because of the interest I had displayed in corn and the opportunity that had been given to me to experiment with it, I was, among Parker's three student assistants, considered the most likely candidate for this position and was the one especially recommended.

So it was that in June 1921, shortly after receiving my diploma, I found myself headed for New Haven to become graduate assistant to Dr. Jones. One of the perquisites of my position was the privilege of spending the six winter months at Harvard University in pursuing graduate work under the supervision of Professor Edward M. East, who had been mentor to both H. K. Hayes and D. F. Jones of hybrid-corn fame as well as to another famous corn geneticist, R. A. Emerson. Professor East had, still earlier, initiated the corn-breeding program at the Connecticut Station after serving as a chemist in the famous Illinois experiments on selection for chemical composition in corn.

The five and a half years that I spent at the Connecticut Station and as a graduate student at Harvard were undoubtedly among the most fruitful of my life. At the Connecticut Station I learned all that was then known about breeding and producing hybrid corn. At both the station and Harvard I learned a great deal about theoretical genetics and other aspects of biology. It was at Harvard, too, where a certain professor raised a question in my mind that was later to stimulate a line of research which I have followed for many years. In the early twenties many biologists were quite skeptical about the conclusions of geneticists, especially those concerned with the chromosomes as carriers of heredity. Among these was Professor W. J. V. Osterhout, who in his course in plant physiology liked to chide any graduate students in genetics who were enrolled in his class about the highly speculative conclusions of geneticists, especially those of the Morgan-Sturtevant-Bridges-Muller school. At one period, looking straight at me, he remarked, "You geneticists may know something about the hereditary mechanisms that distinguish a red-eyed from a white-eyed fruit fly but you haven't the slightest inkling about the hereditary mechanisms that distinguish fruit flies from elephants."

This remark disturbed me because, as a student of genetics, I assumed, as I supposed did most geneticists, that the hereditary mechanisms responsible for differences between species and genera were the same as those concerned with more superficial characteristics such as eye color in fruit flies. Nevertheless I had to admit that there was then little or no tangible evidence to support such an assumption. Indeed some biologists asserted catagorically that the hereditary mechanisms responsible for differences between genera and species and other taxa resided not in the nucleus but in the cytoplasm, and there was some evidence from embryology to support their views. The question that Professor Osterhout raised continued to plague me for several years.

At both the Connecticut Station and at Harvard I read widely on the literature pertaining to corn and was especially intrigued by articles concerned with the hybridization of corn with its relatives. It was obvious from the published articles that the closest relative of corn is the Mexican plant teosinte, and it was also obvious from the illustrations that accompanied the articles that teosinte, although then classified as a separate genus, is in many of its characteristics quite similar to corn. An attempt by Collins and Kempton (1916) to cross a more distant relative of corn, *Tripsacum*, by teosinte produced a plant exactly like the male parent. This result, which the authors called "patrogenesis," was

a unique phenomenon and it aroused my curiosity about *Tripsacum dactyloides*, a a species which was reported to be widely distributed in the eastern United States including parts of Kansas but one that I had never encountered in my frequent excursions into the countryside.

When I told Dr. Jones of my curiosity about *Tripsacum* I found that he, too, had never seen the plant, so together we went to the small herbarium maintained by the Department of Plant Pathology at the Connecticut Station and there found a specimen which had been collected near New Haven some years earlier by Professor Alexander Evans of Yale University. I was impressed by the resemblance of the relatively long, hairy styles of this plant to the silks of corn and wondered if taxonomists had not been in error in considering *Tripsacum* and corn to be only distantly related: separate genera in the same botanical tribe. I persuaded Dr. Jones that *Tripsacum* perhaps deserved more study than it had yet received, and he in turn asked Professor Evans to accompany us on a field trip to show where he had collected *Tripsacum*. The plants were still there exactly where Professor Evans had found them some years previously. We collected both plants and seeds for our proposed experiments.

This was in the fall of 1926. In January 1927, I moved to Texas to take charge of the experimental work on corn and small grains at the Texas Experiment Station at College Station. My first planting of experimental corn in Texas included various genetic stocks with which I had been working in Connecticut and had brought with me to Texas. Planted in March, these emerged promptly and grew more rapidly than any of the Texas varieties, but suddenly in May, when only about knee high, they came into flower, many of the plants too small to produce more than nubbins which were almost useless for my purposes. I realized at once that genetic stocks from New England would not be adapted to Texas conditions, and I began to think about other kinds of genetic research that I might carry on in Texas to supplement the practical breeding work which was my principal assignment.

Shortly after this I made my first trip to several of the Texas substations where experiments on corn were being conducted, and

it was at one of these, the Angleton Station in the Gulf Coast region, that I saw two things that were to affect the nature of my research for the next forty years. Among the grasses being tested at the station for forage production was perennial teosinte, which at that time had already survived several winters. In the right-of-way of the railroad which ran through the station and which was well fenced against trespassing livestock were numerous plants of *Tripsacum* of a type quite different from that which Dr. Jones and I had collected in New Haven a few months earlier.

Riding back on the train to College Station that evening and thinking over the many things I had seen on this trip, it suddenly occurred to me that perhaps I had a unique opportunity in Texas to study the relatives of corn and their hybrids with corn and with each other. Perhaps I might even bring some evidence to bear on the question that Professor Osterhout had raised in his class: are the characteristics which distinguish species and genera inherited in the same way as those which distinguish more superficial characters? Corn, *Zea mays*, was known to hybridize easily with its close relative, teosinte, at that time considered by botanists to be a distinct genus, *Euchlaena*. Because corn had been extensively used as a subject of genetic research, marker genes were known for the majority of its chromosomes. Even then Dr. R. A. Emerson at Cornell and his graduate student, George W. Beadle, later a Nobel Laureate, were crossing teosinte with stocks of corn carrying two marker genes on each of various chromosomes. Their purpose was to determine whether the crossing over between loci is the same in maize-teosinte hybrids as it is in maize. In planning to make similar crosses, my purpose was to find out to what extent characteristics which distinguish the two genera are determined by the chromosomes. I planned, also, to attempt crossing corn with *Tripsacum* in the hope of stimulating apomictic reproduction in corn, thereby obtaining in one generation completely homozygous strains of corn to be used in the production of hybrid corn.

Later this same year Dr. Robert G. Reeves, who had worked on the cytology of corn at Iowa State College, joined the staff of the Department of Biology at Texas A & M College. Since he wished to continue research on some aspects of corn cytology, I asked him to join me in a project on the cytogenetics of maize and its relatives. We asked the Texas Station for modest financial support, initially $300 per year if I remember correctly, for our joint project. The partnership between Bob Reeves and me proved to be congenial and fruitful beyond our fondest hopes. Our research contributions complemented each other to produce results which neither of us could have achieved alone.

Our crosses of corn carrying marker genes on various chromosomes with teosinte were completely successful, and populations resulting from backcrosses to double-recessive marker stocks showed that the genes responsible for the characteristics distinguishing these two genera were unquestionably carried on the chromosomes. One of these crosses involved two marker genes on chromosome 4, *su*, which is responsible for sugary endosperm, a characteristic of sweet corn, and *Tu*, the locus responsible for pod corn, a peculiar type in which the kernels are enclosed in floral bracts as they are in other cereals and the majority of grasses. In classifying our populations we became thoroughly familiar with this character and the nature of its variation, a circumstance which was to play an important part in our later speculations on the origin of corn and its relatives.

The pollinations of corn and *Tripsacum* were initially somewhat disappointing, since they produced no apomictic reproduction in the corn. Examination of the pollinated silks showed that the pollen grains of *Tripsacum* had germinated and their pollen tubes had entered the styles but apparently had not reached the ovules. Since the styles of *Tripsacum* are much shorter than those of corn and their pollen grains much smaller, it occurred to me that perhaps apomixis could be induced by shortening the styles of corn to about the same length as those of *Tripsacum*. Pollinations made on such pruned silks in the summer of 1929 produced a few small seeds which were sacrificed for cytological determination and which Reeves found to be not parthenogenetic corn but true hybrids of corn and *Tripsacum*.

The following year we made pollinations of corn and *Tripsacum* on an extensive scale and produced thousands of hybrid seeds, from which we obtained about thirty living hybrid plants, the first maize-*Tripsacum* hybrids ever to be reported. These proved to be partially fertile when pollinated by corn, and we were able to carry the hybrids into the second, third, and subsequent generations.

Neither Reeves nor I were able to spend a great amount of time on our joint project, he because of a heavy teaching load and I because of the extensive breeding program on corn in which I was engaged and on wheat, barley, and oats, which I supervised. But in the summer of 1937 I found time to lay out all of the "ears"—pistillate spikes—of the segregating generations of our hybrids of corn with both teosinte and *Tripsacum*. One thing that impressed me especially was the occurrence of closely similar types in the two hybrids. About this same time I received in the mail a reprint of a paper by Kempton and Popenoe reporting the rediscovery of teosinte in Guatemala, where it had been collected in the nineteenth century but had since been unknown to botanists. In reading this paper I was impressed, among other things, by the authors' report that teosinte sometimes occurred in the vicinity of extensive populations of *Tripsacum*.

One evening, pondering what I had been seeing and reading, it suddenly occurred to me that if teosinte were not the progenitor of corn, as some botanists had supposed, but instead a hybrid of corn and *Tripsacum* then all of the facts would fall into place. The ancestral form of cultivated corn could then be assumed to have been a wild form of pod corn. Repeated natural hybridization between corn and teosinte would account for some of the characteristics of modern races of corn. Before finally dropping off to sleep that night, I had framed what we later called the tripartite hypothesis and outlined in my own mind the treatise that Reeves and I soon began to write and which we entitled "The Origin of Indian Corn and Its Relatives."

Reeves accepted the hypothesis with enthusiasm and pointed out facts about the previously collected cytological data which seemed to support it. It was not until several days later that I suddenly remembered that Edgar Anderson, then of the Missouri Botanical Garden, on one of his collecting trips in Texas had seen our teosinte and *Tripsacum* plants and their hybrids with corn and had once asked me "Is there any possibility that teosinte might be a hybrid of corn and *Tripsacum*?" In reply I ridiculed the idea so sharply that there was no further conversation about it. Now I wrote Anderson to inquire whether he remembered asking this question and if so had he asked it seriously. He replied that he remembered the incident well and that the question was quite serious since he had been impressed by the fact that teosinte appeared to be intermediate between corn and *Tripsacum* in certain characteristics. I shall never know, and for all practical purposes it makes little difference, whether the idea that teosinte is a hybrid of corn and *Tripsacum* came to me wholly as a result of our experiments or whether it emerged from the subconscious to which it had been relegated when I so sharply rejected it on an earlier occasion.* I do know that I have consistently been reluctant to engage in speculation that has no experimental or other factual support. Once when Dr. R. A. Emerson visited College Station and saw our work on hybrids of corn and its relatives, he asked what ideas I then had about the origin of corn. My reply was, "None. At this point I am interested only in assembling the necessary facts." And this was true. For this reason I take some measure of satisfaction in the remark attributed to Horace, "The man who makes the experiment deservedly claims the honor and the reward." The important fact is that this part of the hypothesis—the hybrid origin of teosinte—whether inspired by Anderson's sentient observations or by our own extensive experiments, or by both, made it immediately possible to develop the other two parts.

Now I needed library facilities beyond those available at College Station to consult the literature on the history of corn and time to write my part of the projected treatise. Since Texas colleges were then not authorized to grant sabbatical leaves, the station director, Mr. A. B. Conner, for whose sympathetic understanding and appreciation of the need

*Recent electron-microscope studies of the pollen of corn, teosinte, *Tripsacum* and maize-*Tripsacum* hybrids by my colleague, Elso Barghoorn, his graduate student Umesh C. Banergee, and my former associate Walton C. Galinat have shown convincingly that teosinte is not a hybrid of maize and *Tripsacum*. The theory is, however, still of historical interest because of the extensive research that it has stimulated.

of basic research I shall always be grateful, arranged to have me officially assigned to work for six months at Harvard University, to which I had been invited by my former mentor, Professor East. In those six months I completed the draft of my part of the book. Reeves, in the meantime, had completed the draft of his part. We then spent almost a year putting the two drafts together and revising certain parts.

The book, a volume of 315 pages, was published in September 1939 in an edition of 8,000 copies at a cost to the Experiment Station, through a state printing contract, of 28 cents per copy. One of my minor satisfactions, some years later, was to see copies listed by booksellers at $2.50 each.

It was the publication of this monograph, I feel sure, that was mainly responsible for my being invited to come to Harvard as Professor of Botany and Assistant Director of the Botanical Museum. I accepted the invitation eagerly for, although my years in Texas had been rewarding, I now needed opportunities that Texas could not provide: for exploration —to seek for wild corn—and for the study of archaeological materials which I considered to hold the key to corn's evolutionary history.

During my first year at Harvard I raised funds, primarily from the Massachusetts Society for Promoting Agriculture, to send Dr. Hugh C. Cutler and his wife, Marian, to South America to search for wild corn, especially in Paraguay, a little-known country from which had come a number of reports on pod corn. The Cutlers found no wild corn but learned a great deal about corn and other plants of South America. After Pearl Harbor, their experience was put to good use in searching for supplies of natural rubber in Brazil.

Shortly after moving to Cambridge, I became acquainted with the late Dr. A. V. Kidder, then dean of American archaeologists. He complimented me on the recently published work on the origin of corn and added "I believe that the botanical research in which you and your colleagues are engaged will someday give us the key to the origin of American agriculture and the rise of the prehistoric American cultures and civilizations." Thanking him for his generous remark, I countered, "On the contrary, I think that future archaeological excavations will tell the story of the evolution of corn." It turns out that we were both right but that I was more nearly right than he.

I did describe one collection of archaeological corn during my first year in Cambridge, but prehistoric corn did not really begin to make sense to me until I had first participated in identifying and describing the living races of maize of Mexico and other countries. This project had its beginnings in 1941 when the Rockefeller Foundation invited three agricultural scientists, Dr. E. C. Stakman, a plant pathologist from Minnesota University, Dr. Richard Bradfield, a soils expert from Cornell University, and me to make a survey of Mexican agriculture to determine whether it was feasible and advisable for the foundation to undertake a program aimed at helping the Mexican people to improve their agriculture. In our report to the foundation we recommended that such a program be initiated and that special attention be given to the improvement of corn, which in Mexico was—and still is—the basic food plant. The project was subsequently initiated in 1943 and under the direction of Dr. J. George Harrar, later the foundation's president, was quickly an outstanding success. In 1967, the three original consultants, Stakman, Bradfield, and I, published a history of the program and its subsequent ramifications in a book entitled *Campaigns against Hunger*.

Chosen to head the corn-improvement work of the Mexican Program was Dr. E. J. Wellhausen, but since he could not leave his position immediately I was asked to come to Mexico in August 1943 to begin cooperation with Mexican agronomists. In our travels over the country, Harrar and I collected corn at every opportunity, and when Wellhausen arrived in Mexico he not only continued the practice but later employed a Mexican *agronomo*, Sr. Ephraim Hernandez X., to collect systematically throughout the country. By 1948 the collection included almost 2,000 entries, many of which had been grown and observed in field plantings at the experiment station at Chapingo. Wellhausen, who had by that time been joined by one of my former Texas students, Lewis M. Roberts, invited me to come to Mexico and participate in classifying and describing the Mexican races of corn. The results of our joint studies were published in a Mexican edition in 1951 and an English edition in 1952 of a book entitled *Races of Maize in Mexico*.

It was this same year, 1948, in which Herbert Dick, a graduate student in archaeology in the Peabody Museum at Harvard,

excavated a once-inhabited rock shelter, Bat Cave, in New Mexico. Working with him was a graduate student in botany, C. Earle Smith, whose expenses, including a sleeping bag, were paid by a grant from the Botanical Museum in anticipation of receiving in exchange any prehistoric corn and other vegetal material which the excavation might turn up. Although Harvard's then comptroller objected to this expenditure as improper—paying for something not yet received—he finally reluctantly allowed it.

So far as prehistoric corn is concerned the expedition hit a "jackpot." Cobs of corn were found by Dick and Smith at every level of the cave, and at the bottom of the accumulated debris, which was some six feet in depth, there occurred tiny cobs dated by radiocarbon determinations of associated charcoal at 5,600 years. This proved to be both a popcorn and a form of pod corn as Reeves and I had postulated that primitive corn, if ever discovered, probably would be. The corn from higher levels showed a distinct evolutionary sequence that included evidence of hybridization with teosinte.

The paper which we published describing this prehistoric corn attracted wide attention among archaeologists and resulted in a flow to the Botanical Museum of archaeological material from a number of sites in Mexico, among them the cobs from a cave in northeastern Mexica, La Perra Cave, excavated by Dr. Richard S. MacNeish, then on the staff of the National Museum of Canada.

Having failed to find living wild corn in any part of the hemisphere and convinced by the earliest archaeological specimens that primitive corn was both a popcorn and a form of pod corn, I and my associate, Dr. Galinat, who had joined me in 1954, undertook to produce a genetically reconstructed ancestral form by combining the principal characteristics of popcorn and pod corn. We succeeded in developing a pod-popcorn that had a means of dispersing its seeds and which might have been capable of surviving in the wild in a suitable environment. Indeed in a simulated wild habitat competing with aggressive weeds and weedy grasses, it more than held its own.

In the meantime MacNeish, convinced that prehistoric wild corn might be found in once-inhabited caves of Mexico, had after several disappointments discovered caves in the Tehuacán Valley in Mexico in which the earliest cobs appeared to be those of a wild corn.

Galinat and I made an intensive study of the specimens, comprising almost 25,000, turned up by MacNeish in five caves in Tehuacán. These show that the earliest corn dated at ca. 5000 B.C. was probably wild corn, that it was a popcorn and a form of pod corn; that this corn was domesticated, and after crossing with teosinte or *Tripsacum*, suddenly evolved in an explosive manner to create productive varieties which formed the nutritional basis of the advanced cultures and civilizations which the Spaniards encountered when they conquered Mexico in the sixteenth century.

Now the time has come for me to summarize the research on the corn plant in which I have been engaged for forty-eight years. Some readers, I suppose, will regard spending a lifetime in research on a single plant as a form of monomania. Perhaps by some standards it is, but this is really the only way to get this kind of a job done. And spending a lifetime on a single important food plant is not very different from spending it on a single human disease, as Hans Zinsser did on plague (and wrote about it in a fascinating book, *Rats, Lice, and History*) or spending a lifetime on a single organ, as Paul Dudley White has done on the heart. Perhaps this kind of specialization is not essentially different from that practiced by scholars in the humanities spending a lifetime in scholarly research on a single great literary figure such as Shakespeare or a great historical figure such as Lincoln or even a great war such as the Civil War to which more than one historian has devoted a lifetime of research and study.

In recent years when it has become apparent that the pressure of an explosively expanding world population is creating worldwide food problems, I have sometimes argued that botanists should know as much about the world's principal food plants, of which corn is one, as we know about some of the destructive agents of the world, as the medical profession, for example, knows about the principal human diseases, and as aeronautical engineers know about the world's principal bombers and guided missiles.

This argument, although it undoubtedly has some validity, also contains an element of rationalization on my part. Scientists like to believe, of course, that their research is important and in some manner useful, but for the most part scientists do what they do because they love doing it. Their motivation is more a matter of self-expression than a conscious desire to save the world from hunger, disease, or other vicissitudes of life, although these desires are by no means lacking.

So it has been with me in my lifelong study of the corn plant. Corn is a mystery, and mysteries are there to be solved as surely as mountains are there to be scaled. And spending a lifetime on a single important plant does not necessarily mean a dull life. On the contrary, I can testify that it can be an interesting and at times exciting life. One week, for example, may have found me making pollinations in the experimental cornfield, for the purpose of developing a genetically reconstructed ancestral form; the next week in the museum studying prehistoric corncobs from some once-inhabited dry caves; and the third week I might have been on a plane headed for Mexico, Colombia, or Peru to study the living races of corn of this hemisphere.

The important thing is that as a result of my researches in botany, genetics, archaeology, and history, and those of my graduate students, associates, and contemporaries, we now have at least partial solutions to the mystery of corn. Indeed so much progress has been made in the past twenty-five years that I have on occasion given a talk on the subject with the bold title, "A Botanist's Dream Come True."

One of the most important factors contributing to the solution of these problems has been cooperation between botanists and archaeologists. My own studies have been greatly enriched by the cooperation and specimens that I have received from graduate students and former graduate students of the Peabody Museum of Anthropology and Archaeology. These include Herbert Dick, Robert Lister, Reynold Ruppe, David Kelley, Michael Coe, and Cynthia and Henry Irwin. Especially important and valuable has been my collaboration with Dr. Richard S. MacNeish, who has turned up more prehistoric corn—including prehistoric wild corn—and other early remains of American cultivated plants than any other single archaeologist and who is now recognized as the leading authority on the beginnings of New World agriculture. Also quite fruitful has been the collaboration with two long-time associates, Robert G. Reeves and Walton C. Galinat, whose researches have been complementary to mine.

Adding significantly to our present understanding of the corn plant have been the researches of my graduate students: James Cameron, John Rogers, John Edwardson, William Hatheway, Alexander Grobman, Surinder Sehgal, Raju Chaganti, Gordon Johnston, H. Garrison Wilkes, and Ramana Tantravahi. All chose problems related in some way to the origin and evolution of corn and its relatives, and their published works, considered as a whole, represent a monumental contribution. I have relied heavily upon them. Two former postdoctoral fellows, Y. C. Ting and Angelo Bianchi, also have made valuable contributions. I am deeply indebted to the Rockefeller Foundation for the opportunity, acting as their consultant in agriculture, to become familiar with maize in indigenous American cultures; to participate in the published classification and description of the races of maize of Latin America; and for two generous grants to support my research from 1959 to 1969.

This book might never have been written had I not received a cordial invitation from Sir Joseph Hutchinson, Director of the School of Agriculture, University of Cambridge, to spend a sabbatical leave at his institution. There my wife and I spent a pleasant and fruitful six months in 1962 with Sir Joseph as our gracious and thoughtful host. The unique facilities of the school's library were made conveniently available to us through the good offices of Mr. F. A. Buttress, Librarian. We enjoyed frequent discussions with Dr. Geoffrey Bushnell of the Museum of Archaeology, who introduced me to the extensive literature on early maize in Africa.

Since my retirement from Harvard University I have had the privilege of being a member of the Department of Botany at the University of North Carolina. Here Mrs. Betty Zouk, Librarian, has been exceptionally helpful in assembling needed publications from other departments and institutions.

Various chapters of the book have been read by Walton C. Galinat, George W. Beadle, H. Garrison Wilkes, Hugh C. Cutler, David H. Timothy, Major M. Goodman, Elso S. Barghoorn, William L. Brown, and Sherret S. Chase. Although several of the readers do not agree with my principal conclusions, all have made helpful criticisms and comments. I am especially grateful to Marcus M. Rhoades, who read the entire manuscript and made many invaluable criticisms and suggestions.

Special Acknowledgment

My colleagues and I are deeply indebted to the Instituto Nacional De Anthropologia E Historia, Mexico, and to Professor Jose Luis Lorenzo, Head of its Department of Prehistory for arranging that the archaeological remains of corn from sites in the Tehuacán Valley be loaned for botanical studies to the Botanical Museum of Harvard University. Without this generous cooperation the studies reported in Chapter 15 could not have been made.

Contents

Corn

*Its Origin
Evolution
and Improvement*

Introduction

History celebrates the battlefields whereon
we meet our death, but scorns to speak of
the ploughed fields whereby we live; it
knows the names of the kings' bastards, but
cannot tell us the origin of wheat.

Henry Fabre

As it is with wheat so it is also with corn. History can tell us almost nothing about its origin except in a negative way. The very absence of any Old World historical references to maize before 1492 is evidence of its American origin. The earliest historical reference to corn tells us that it was on November 5, 1492, when two Spaniards, whom Christopher Columbus had delegated to explore the interior of Cuba, returned to him with a report of "a sort of grain they called *maiz* which was well tasted, bak'd, dry'd, and made into flour."

Actually Columbus may have encountered maize several weeks earlier than this. He landed on the Bahamian island of San Salvador on October 12, 1492, the date now celebrated as Columbus Day. On Sunday, October 14, he sailed along the island seeing several villages and many natives who, he recorded in his journal, brought "various foodstuffs." Since corn was and still is one of the staple foods of San Salvador, Columbus probably saw it there, but his journal does not show that it made any impression on him. A few days later, however, visiting the Bahamian island which he named Fernandino, now known as Long Island, he wrote in his journal: "This island is very green and flat and very fertile, and I have no doubt that all the year they sow and reap panizo."

Panizo is the Italian word for a millet, *Panicum miliaceum* L., that is commonly cultivated in Italy and other parts of the Old World. Since it is quite unlikely that it grew in the Bahamas, Landstrom (1967) may be correct in concluding that Columbus probably meant Indian corn, which was grown on

these islands but was then unknown in Europe. Whether Columbus first encountered corn in the middle of October or the first week of November is primarily an academic question. There is, however, no doubt about the year, 1492.

Later explorers to America found corn being grown by the Indians in all parts of America where agriculture was practiced from Canada to Chile. We now know that it was the basic food plant of all of the advanced cultures and civilizations of the New World. The seminomadic hunting and fishing Indians in both North and South America augmented their diet of fish and game with corn from cultivated fields. The more advanced mound builders of the Mississippi Valley and the cliff dwellers of the southwest were corn-growing and corn-eating people. The highly civilized Maya of Central America, the energetic and warlike Aztecs of Mexico, and the fabulous Incas of Peru and adjoining countries, all looked to corn for their daily bread. The abundant harvest that this cereal yielded gave these ancient people a measure of leisure which the constant quest for food in a hunting and food-gathering culture could not provide —leisure for weaving beautiful fabrics, for molding exquisite pottery, for building magnificent highways and towering pyramids; leisure to study the sun and the moon and the stars and the seasons, to invent a system of arithmetic, and to perfect a calendar more accurate than the Old World calendar of the same period. Along with these developments there grew sophisticated religions which included among the numerous deities, maize gods, and which involved solemn ceremonies

associated with the planting, growing, and harvesting of corn. These were corn-fed civilizations, and corn was indeed "the grain that built a hemisphere."

Later, corn became the bridge over which European civilization traveled to a foothold in the New World. The European colonist not only adopted the corn plant as the principal source of his sustenance but also embraced an elaborate cultural complex which the American Indian had developed through centuries of trial and error and which included companion crops, methods of culture, harvesting, storage, and utilization.

The planting of corn in hills, its interplanting with beans and squashes, the use of husking pegs in harvesting, the storing of the ears in ventilated cribs, the use of green corn for roasting ears, the removal of the hull with lime to make hominy are only a part of the Indian inventions adopted with slight changes by the early settlers. One of the most remarkable aspects of this complex is the fact that the combination of corn, beans, and squashes is now recognized as furnishing an adequate, indeed an excellent diet. Corn supplies carbohydrates, small amounts of protein, and fat; the beans represent the principal source of protein, but more important still, they contain adequate amounts of some of the "dietary essential," amino acids—the building blocks of proteins—in which corn is deficient, especially tryptophane and lysine. Beans can also remedy corn's notorious deficiency in two vitamins, riboflavin and nicotinic acid. Squashes are valuable in supplying additional calories as well as vitamin A, and their seeds furnish an increment of wholesome fat in which a diet of corn and beans alone is barely adequate.

How the American Indians with no knowledge of chemistry discovered that corn, beans, and squashes together provide an adequate diet is still a mystery, but somehow they did and they did this thousands of years ago. It was a highly successful complex not only because it was adequate in both quantity and quality but also because it made efficient use of the land, especially for a people who lacked draft animals to perform the various tillage operations. The beans climbed and twined on the stalks of corn, exposing their leaves to the sun without drastically shading the leaves of the corn plant, and the squash vines spread out over the ground between the hills of corn, choking out the weeds. The growing of corn, beans, and squashes together in the same fields is still practiced on small farms in New England and in the hill country of Kentucky and Tennessee.

Spreading westward from New England and the South as the railroads opened the great fertile lands of the Midwest, corn found almost the ideal conditions for its culture, and two distinct kinds, the New England Flint and the Southern Dent, brought into the region by settlers from North and South, hybridized to produce a new race which helped to create what has since become the world's most important agricultural region, the Corn Belt of the United States.

Corn has now become the basic food plant of our modern American civilization. It is the most efficient plant that we Americans have for capturing the energy of the sun and converting it into food. True, we consume directly only small amounts of corn: roasting ears, breakfast cereals, Indian pudding, and, for a somewhat different purpose, a beverage invented by a Kentucky minister of the Gospel, Bourbon whiskey. But transformed, as three-fourths of it is, into meat, milk, eggs, and other animal products, it is our basic food plant, as it was of the people who preceded us in this hemisphere.

Columbus carried the grain called "maiz" to Spain when he returned. Within one generation it had spread through Southern Europe. Within two generations it had spread around the world. Only two other New World products, tobacco and syphilis, spread with equal rapidity.*

Today, corn or maize, as it is known in many parts of the world, is grown in every state in the United States and every suitable agricultural region of the globe. A crop of corn is maturing somewhere in the world every month of the year. It grows from north latitude 58° in Canada and Russia to south latitude 40° in the southern hemisphere. Fields of corn are growing below sea level in the Caspian plain and at altitudes of more than 12,000 feet in the Peruvian Andes. Corn is grown in regions with less than ten inches of annual rainfall in the semiarid plains of Russia and in regions of more than 400 inches of rainfall on the Pacific coast of Colombia. It thrives almost equally well in the short summers of Canada and the perennial summers of the tropical equatorial regions of Ecuador and Colombia. No other crop is distributed over so large an area, and only one other, wheat, occupies a larger acreage. Today corn is grown on more than 270 million acres of land, and it produces an annual grain crop of nearly 9 billion bushels. Corn is one of perhaps not more than twelve or fifteen species of cultivated plants—each one the principal source of food of millions of people —which quite literally stand between mankind and starvation.

To thrive or even to exist under such a wide variety of environmental conditions, a species usually must possess a great diversity of forms. This characteristic the corn plant exhibits to a degree probably not found in any other crop plant. Corn varieties differ from each other in many characteristics and by wide extremes. Commercially speaking there are five main types—dent, flint, flour, sweet, pop—which are distinguished by differences in the nature of the storage material in the grain. Any one of these types may come in a variety of colors.

Superimposed upon the diversity of kernel types and colors is a tremendous variability in numerous vegetative and other morphological characteristics. There are early maturing varieties, such as the Gaspé flint from the Gaspé Peninsula in Canada or Cinquantino from the Pyrenees Mountains of Spain which mature in sixty to seventy days, and very late varieties in the American tropics that require ten or eleven months to reach maturity. The number of leaves varies from eight to forty-eight, the height of stalk from less than two feet to more than twenty, the number of stalks produced by a single seed ranges from one to fifteen. The size of the ear varies from the tiny ears of some of the popcorn varieties which are no larger than a man's thumb to the gigantic corn grown in the Jala Valley of Mexico, which produces ears up to two feet in length borne on stalks so tall that the ears may be conveniently harvested from horseback and so stiff and strong that they are sometimes used for pickets in enclosures for domestic animals.

In spite of this overwhelming diversity of form, almost all of the principal types of corn known today, dent, flint, sweet, flour, and pop, were already being grown by the Indians when America was discovered, and all of them are classified by botanists within a single species, *Zea Mays* L. All the varieties of corn known today are readily hybridized with each other, and the hybrids are, almost without exception, completely fertile.

What was the nature of the wild or primitive corn from which our multitude of present varieties has developed? Where, when, and how was a species, so hardy that it could survive in the wild, converted to a cultivated plant so specialized and so dependent upon man's ministrations that it would soon disappear from the face of the earth if deprived of man's protection? These questions are all parts of one of the most intriguing mysteries of modern times; a mystery which has occupied the attention of archaeologists, who know that the ancient civilizations of Mexico, Central America, and Peru were based upon the culture of corn; ethnologists, who know that corn played a prominent part in the religious beliefs and ceremonies of these ancient peoples and was one of the chief motifs in their decorations; and botanists, geneticists, and cytologists, who realize that a solution to this problem may throw revealing light upon the processes of evolution under domestication, expand our understanding of the corn plant, and enable us to improve it still further. And as the mystery of corn has elicited the interest of various kinds of scholars, so research aimed at the solution of the mystery has employed various kinds of evidence and techniques.

Alphonse De Candolle, the eminent nineteenth-century student of the origins of cultivated plants, employed four kinds of evidence in seeking origins: (1) history; (2) philology or linguistics; (3) botany, with particular reference to geographic distribution of wild relatives; (4) archaeology. He emphasized the fact that all possible sources of evidence should be employed since any one alone would seldom provide a complete and satisfactory answer.

History, it turns out, and as Fabre long ago complained, tells us virtually nothing about the origin of corn, since there is no historical evidence bearing on the problem before 1511. Maize is not mentioned in the Bible, and in fact there is no ancient Hebrew word for it. The word "corn," which does appear in the Bible frequently, is used there as a synonym for "grain." The Greek and Roman writers, Theophrastus and Pliny, who discoursed intelligently and at length on other crop plants, make no mention of maize. Evidently the Greeks had no word for it, and the Romans were no better off. The Vedas of India—poems or hymns in Sanskrit which record the folklore of their period and mention many plants used for food, in medicine, and in various religious ceremonies—say nothing about maize. Extensive search of the pre-Columbian Chinese literature reveals no evidence that the Chinese scholars were acquainted with it. The records left by those ancient Americans, the Aztecs, Mayas, and the Incas, tell us nothing of the origin of maize except in the form of various highly implausible myths. They do, however, show how important it was in their economic and religious life and so provide a part of the considerable body of convincing evidence that maize was an American plant.

Linguistics tells little about the origin of maize except that the absence of a word for maize in Old World languages points to its New World origin. Linguistics has, however, been employed by some students in attempting to prove that maize had reached parts of the Old World, especially Africa, before 1492.

Within the past forty years the remaining two methods of De Candolle, botany and archaeology, augmented by the twentieth century methods of genetics and cytology, have made phenomenal progress—more, perhaps, than was made in the previous four hundred years—toward the solution of the mystery of corn. Especially effective has been the cooperation between botanists and archaeologists. Another important factor has been the airplane, which has made it possible to travel to remote parts of the hemisphere which previously were almost inaccessible and to collect wherever they occur the living races of corn. The result of these diverse studies and of collaboration between various disciplines has been that we now have definitive answers to two of the questions posed earlier in this chapter: "where?" and "when?" and we have at least partial answers to the third "how?" It is the purpose of this book to summarize the present state of the evidence bearing on these questions.

1 The Modern Corn Plant

This Corne is a marvellous strange plant.

Lyte, 1619

Of all of the New World plants which excited the curiosity and wonder of the herbalists of the immediate post-Columbian period, none proved to be more intriguing than Indian corn or maize. The ears of this plant and to only a slightly lesser extent, the tassels, were quite baffling to the students of that period and continued to perplex botanists for more than four centuries. Nowhere has the recognition of the peculiarities of the corn plant been more vividly expressed than by Lyte in his *New Herbal*, a translation of Dodoens, 1619.* The description is as follows: "This Corne is a marvellous strange plant, nothing resembling any other kind of grayne; for it bringeth forth his seede cleane contrarie from the place whereas the Floures grow, which is against the nature and kinds of all other plants, which bring forth their fruit there, whereas they have borne their Floure . . . at the highest of the stalkes, grow idle and barren eares, which bring forth nothing but the floures or blossomes."

Readers familiar with the corn plant will recognize Lyte's "idle and barren eares" as the staminate inflorescences commonly known as the tassels. Botanists recognize the description as that of a monoecious (from the Greek "one house") plant which bears the male organs, the stamens, and the female organs, the pistils, in separate flowers on the same plant. The majority of grasses are "perfect flowered," bearing the stamens and pistils within the same flower. But monoecism in grasses is by no means rare. It is characteristic

*This description in a slightly different form first appeared in an edition entitled *A Nievve Herbal* in 1578.

of all the species of the botanical tribe, *Maydeae*, to which corn belongs. Indeed, it is this characteristic more than any other that has been the basis for the establishment of this particular tribe by taxonomists.

The botanical characteristics of the modern mature corn plant which are involved in its reproduction are illustrated in Figure 1.1. These and other characteristics, which are especially important in the research on corn's origin and evolution reported in later chapters, are described below in general terms. A more detailed description of the structure, growth, and reproduction of corn is found in Kiesselbach (1949), Sass (1955), and Weatherwax (1955).

Like all other cereals, corn is a grass. It is somewhat more robust than wheat, rice, or barley but quite similar in this respect to the grain sorghums, which in general appearance it superficially resembles. Like all other cereals, corn bears its flowers in "spikelets," the units of grass inflorescences, and which are themselves small inflorescences comprising two or more flowers. Corn's uniqueness among the important cereal grasses of the world lies in the nature of its inflorescences. The terminal inflorescence, the tassel (Fig. 1*A*, *C*) usually bears only male spikelets. These occur in pairs, one member of each pair being sessile, without a stalk, and the other pedicellate, borne on a pedicel, a slender stalk supporting a single flower. Each of these spikelets contains two flowers, and each flower bears three pollen sacs or anthers (Fig. 1.1*D*) packed tightly with pollen grains estimated at about 2,500 in number by Sturtevant (1881), who also calculated that the total

number of pollen grains produced per plant is 18 million, a ratio of at least 9,000 pollen grains for each ovule or potential seed. Once while waiting for a bus, I counted the number of spikelets on an average tassel of New England corn, multiplied this figure by two to obtain the number of flowers, by three to get the total number of anthers, and by 2,500 to obtain a figure for total number of pollen grains. My figure, 15 million, came quite close to Sturtevant's estimate, but, like his, it was based on the assumption of 2,500 pollen grains per anther, a figure which seems never to have been adequately confirmed. Other estimates of the pollen production of vigorous plants have ranged from 30 to 60 million pollen grains (Arber, 1934) and the estimates of the number of pollen grains produced for each kernel being as high as 25,000 to 50,000. Kiesselbach (1949) has estimated that in a Nebraska cornfield an average of 42,500 pollen grains are produced for each square inch of the field.

The accuracy of some of these estimates may be questioned, but there is no doubt that corn produces pollen in prodigious amounts, as anyone who has covered a tassel with a bag to collect pollen for breeding purposes can attest. The pollen that a single tassel sheds in one day is measurable in terms of volume. The pollen grains themselves are small, barely visible to the naked eye, about 1/250 of an inch in diameter, light in weight, and easily carried by the wind, sometimes for a considerable distance. The late Frederick D. Richey once found some blue kernels in an isolated field grown for seed production in Ohio. Canvassing the county to ascertain from

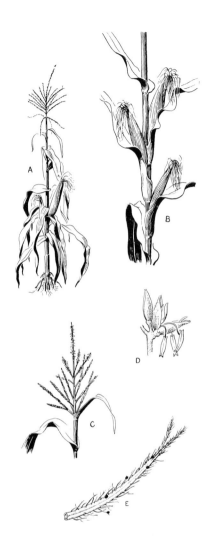

where pollen of a blue-kerneled corn might have come, he could locate only one possible source, and this was more than five miles away.

Since the air in a cornfield is seldom completely still, this means that the pollen from any given plant does not often fertilize the ovules of the same plant. Even the slightest breeze or air current will carry the pollen to silks of other plants. Thus corn, in contrast to the majority of cereals, is a naturally cross-pollinated plant. It is this feature more than any other that makes possible on a large scale the production of hybrid corn, one of the most spectacular developments in applied biolgy of this century.

The lateral inflorescences (Fig. 1.1*A*, *B*), which when mature become the familiar ears of corn, usually in modern corn bear only female spikelets. Like the male spikelets of the tassel, these occur in pairs, both members of the pair being sessile, but one is potentially pedicellate and may become so under certain circumstances.

Also, like the male spikelets, each female spikelet has two flowers, but usually only one of these develops, the other, the lower of the two, remaining rudimentary. In some varieties, however, notably the sweet corn variety Country Gentleman, the lower floret becomes functional. The result is that the ear has twice the normal number of kernels, and these are so crowded on the ear that the usual orderly arrangement of the kernels in distinct rows is completely obscured. Another curious feature of such ears is that about half of the kernels, those borne by the lower flowers, have their germ or embryo side facing, not the tip of the ear as they do in most varieties, but the base. Thus the two kernels that develop from a pair of florets are set back to back on the cob.

Each female flower contains a single ovary terminated by a long style, the pollen-receptive organ commonly known as the "silk." The silks are covered with fine hairs that are admirably designed for capturing wind-blown pollen (Fig. 1.1*E*). Each silk represents a potential kernel and must be pollinated in order for that kernel to develop. Pollen grains falling on the silks germinate, often on the hairs, and thrust out pollen tubes which penetrate the tissues of the silk. The pollen tubes reach the micropyles of the ovaries in fifteen to twenty-five hours after pollination, the time varying with the length

of the silks, the temperature, and other factors. The rate of growth as determined by numerous experiments varies from about two and a half to four hours per inch of growth. Thus the pollen tubes of corn are among the fastest-growing organs in the plant kingdom. As they move downward through the tissues of the silks, the carbohydrates of the pollen grain providing nourishment for their growth, the tubes carry with them in addition to their own nucleus two other nuclei, the male cells or sperms. When the ovary is reached and after the pollen tube enters the opening in the micropyle, the end of the tube ruptures, setting free the two sperms. One of these fuses with a female egg nucleus to produce the *embryo* and the other fuses with a double nucleus to produce the *endosperm*, the food storage organ which will nourish the embryo when the seed germinates. Fertilization of the two female nuclei by two male nuclei is known by the appropriate term of "double fertilization," a process characteristic of virtually all flowering plants, the Angiosperms. The phenomenon was first discovered by Nawaschin in 1898 and described in corn by Guignard (1901), whose observations have been fully substantiated by Miller (1919), Weatherwax (1919), and Kiesselbach (1949).

The sperm nuclei and egg nuclei all carry the haploid number of chromosomes, which in corn is ten. The fusion of the egg and the sperm nuclei to produce the embryo restores the diploid number of 20, which is characteristic of all of the cells of the corn plant except these of the endosperm and the gametes. The endosperm, because it is the product of the fusion of two female nuclei and one male nucleus, is triploid in its chromosomal constitution, and this fact is important in connection with the inheritance of certain endosperm characteristics.

In maize double fertilization is responsible for the phenomenon of *xenia*, a term invented by Focke to describe the immediate effect of foreign pollen. Xenia in corn has numerous manifestations. Thus, if a white corn is pollinated by pollen from a yellow corn the kernels are yellow; if a white or yellow corn is pollinated by a blue corn the kernels are blue; if a sweet corn is pollinated by a field corn the resulting kernels are starchy like those of field corn. It was the phenomenon of xenia which intrigued a number of eighteenth-century

FIG. 1.1 Botanical characteristics of the modern corn plant. *A*, the entire plant (a short-stalked variety) showing the male inflorescence, the tassel terminating the stalk, and the female inflorescences, the ears, in the middle region. *B*, young ears enclosed in husks with the pollen-receptive organs, the silks, protruding from the ends. *C*, typical tassel. *D*, typical male flower with three anthers containing pollen. *E*, a single silk magnified to show hairs and adhering pollen grains. (Drawing by G. W. Dillon.)

corn scientists, including Cotton Mather of witchcraft fame, who in 1716 observed in a neighbour's garden in Massachusetts that when a field of yellow corn had one row planted with red and blue varieties the yellow corn had some blue and red kernels on its ears. He observed also that there were more of the red and blue kernels on the side from which the wind was usually blowing than on the side away from the wind. This was the first reported observation of sexuality in corn (Wallace and Brown, 1956). It is also this phenomenon of double fertilization in corn that has made the corn plant so useful for genetic research. There are many hereditary characters of the kernels which the geneticist can use for his purposes. The kernels on a single ear comprising several hundred individuals can sometimes provide the solution of a problem involving some aspects of the hereditary mechanism. Indeed, two of the rediscoverers of Mendel's law, Correns and DeVries, were both studying the inheritance of endosperm characters in corn when they made their rediscovery (Sturtevant, 1965).

Following fertilization, the fused nuclei begin to divide and redivide. The kernels soon begin to swell as the embryo and endosperm develop side by side. In eighteen to twenty-one days the kernels have reached what is commonly called the "roasting-ear" stage, although nowadays green corn is usually boiled rather than roasted. In another six to eight weeks, depending on the variety, they are full grown and the cells of the endosperm are packed with starch grains, if the corn is an ordinary field corn, and with sugar, starch, and intermediate products if it is a sweet corn.

The kernels are firmly attached to a rigid axis, the "cob," and are not completely covered, as are those of the majority of other cereals, by the floral bracts which botanists call "glumes" and which the layman knows as "chaff." Instead the entire ear is enclosed, often quite tightly, by modified leaf sheaths, the husks or shucks (Fig. 1.1*B*). Thus, while in other cereals the kernels are protected individually, in maize they are covered en masse. The result is that cultivated corn has no mechanism for the dispersal of its seeds and hence is no longer capable of reproducing itself without man's intervention. This is one of its unique characteristics.

The husks are modified leaf sheaths arising from the joints in the lateral branch, which is terminated by the ear. There are as many husks enclosing the ear as there are joints on this branch. It was formerly thought the number of joints on a lateral branch was equal to the number of joints on the main stalk above the point at which the branch arose. This is not strictly true, but it is true that the higher the position of the branch the fewer its joints and also the fewer its husks.

The branch differs from the main stalk in having its internodes, the spaces between the joints, drastically shortened. This in turn results in the leaf sheaths, the husks, being strongly overlapped and tightly wrapped around the ear. In some strains of corn the lateral branches tend to elongate as the ear matures, with the result that the degree of overlapping of the husks is decreased and the ear becomes partly or wholly exposed and is capable of dispersing its seeds. This may have been one of the mechanisms for dispersal in wild corn.

The ear of corn is unique not only in the way it is enclosed by the husks and consequently its lack of means of dispersing its seeds but also in certain other characteristics which have puzzled botanists and which have been the subject of numerous observations, investigations, and conjectures. It is now generally agreed that the ear is the counterpart or homolog of the central spike of the tassel, as was suggested by Mrs. Kellerman (1895) and independently by Montgomery (1906). However, to recognize this homology does not explain its morphological nature. Quite a number of botanists in both the nineteenth and the twentieth centuries have assumed that the ear is the product of fusion of spikes similar to those of the lateral branches of the tassel. There has never been conclusive evidence to prove the fusion theory. The fact that branched ears of corn sometimes occur has been regarded by some students as evidence of "disruption" of a compound structure resulting from fusion, but close examination of various types of branching which occur does not support such an interpretation. Basal branches on an ear, for example, are nothing more than the homologs of the tassel branches. They are not the disrupted members of a fused organ.

Any doubts still remaining about the homology of the ear and central spike of the tassel or of the role of fusion in the development of the ears should have been finally resolved by the beautiful photographs of O. T. Bonnett (1940, 1966) illustrating the early stages in the development of those two organs. These show not only that in their early stages the ear and central spike of the tassel are virtually identical but also that there is no evidence whatever of fusion in the development of either.

Although there is fairly general agreement that the ear of corn is the homolog of the central spike of the tassel there are other possible homologies about which there are differences of opinion. Nickerson (1954), for example, has suggested that the lining of the structure known as the cupule is an adnate prophyll. Galinat (1959) sees the gross structure of the entire plant in maize and its relatives organized upon a basic and repetitive pattern of an organ, the *phytomer*. The general concept of the phytomer is attributed to the eminent nineteenth-century botanist, Asa Gray. Galinat's concept of the phytomer as it applies to maize and its relatives comprises an internode, a leaf, a prophyll, and an axillary bud. Assuming on the basis of this concept that the internodes of the tassels must have homologs of the prophylls, Galinat has identified these as the *pulvini*, which are mussel-like swellings in the axils of the branches. To the extent that the phytomer concept focuses attention on characteristics otherwise generally overlooked, it may be useful.

Weatherwax (1935) has shown convincingly that the ear is essentially the same kind of structure as the spike of other grasses such as the millets, *Setaria*, and *Pennisetum*. From my own studies of certain greatly elongated ears of pod corn (1945) I am certain that Weatherwax's interpretation is correct. The ear of corn differs from the heads of millets and even from the little spikes of foxtail grasses of gardens and lawns primarily in size and in the compaction of its parts. Once this simple fact is recognized, the ear of corn loses a substantial part of its mystery.

One effect of the compaction of the female inflorescence is that the cup-like depressions in the rachis, to which Sturtevant (1899) has given the appropriate name *cupules*, are compressed into somewhat constricted shapes. These vary considerably from one race to another in length, width, and depth; in pubescence and in the prominence of their lateral rims, which the Cutlers (1948) have called *rachis flaps*. Studies of the internal characteristics of cobs have been greatly facilitated by a simple technique developed by Galinat (1963) and also successfully employed by Sehgal (1963), Sehgal and Brown (1965), and Johnston (1966). This consists of sawing the cob longitudinally near the median and filing and sanding the larger part down to a median section. In such sections all of the principal parts of the cob are apparent; the cupules to which each pair of female spikelets is attached; the *rachillae*, the short stems on which the kernels are born; the lower *glumes*, which are often thicker and more highly indurated than the upper; the two remaining floral bracts, the *lemmas* and *paleas*; and a vestigial second floret. Photographs of median cob sections of several kinds of corn are illustrated in Figure 11.6. Such photographs tell much more about the characteristics of cobs than any of the techniques previously employed: grinding down the cobs to a hard surface or dissolving away the softer parts with acid leaving only a skeleton of indurated material.

Although in modern varieties of corn the ear is normally wholly female and the tassel wholly male, there are numerous examples which do not follow this simple rule. The tassels may include female flowers; the ears may be terminated by a spike of male flowers or may have anthers intermixed with kernels. Perfect flowers having both male and female organs also occur, and this is perhaps not surprising, since all of the flowers are potentially hermaphroditic, the male flowers containing vestigial female organs and female flowers containing vestigial male organs.

The mature kernel of corn, technically a *caryopsis*, as are the fruits of all grasses, consists of three principal parts: (1) the thin shell or *pericarp*; (2) the *endosperm*, which represents the food storage organ; and (3) the *embryo* or germ. The pericarp is the thin layer of maternal tissues surrounding the entire seed. This layer is usually colorless, but in some varieties, such as the ornamental "strawberry pop" sold in florist shops in the fall, it is red. In Peru and parts of Mexico and Guatamala there are many other pericarp colors, including brown, orange, cherry, and a variety of variegated patterns. Pericarp colors, because they occur in the maternal tissue, are not subject to xenia. A variety with colorless pericarp is not changed by cross-pollination with a variety with colored peri-

carp; the results of such pollinations are not manifest until the following generation.

The endosperm is the food storage organ. In field corn it consists primarily of starch, which is digested into sugar when germination occurs and growth begins. In flour corns the endosperm is soft and mealy; in flint corns it is hard, and in popcorns harder still. In sweet corns the cells of the endosperm are packed with a mixture of sugar, starch, and various intermediate products. In a peculiar type of corn called "waxy," the starch is composed exclusively of a type called amylopectin, which is used commercially as a substitute for Cassava starch in making tapioca and certain types of mucilages. In color the endosperm may be white, yellow, or orange.

The outer layer of the endosperm, a microscopically thin layer one cell in thickness called the *aleurone*, is of particular interest because of its usefulness in a great variety of genetic studies. In most varieties grown in the United States the aleurone layer is colorless, but many different aleurone colors are known: red, blue, bronze, brown, and various shades of these. The combination of pericarp, aleurone, and endosperm colors can produce a great variety of shades. A blue aleurone, for example, over a yellow endosperm produces a distinct greenish aspect.

The embryo or germ is actually a tiny corn plant in a dormant stage. It consists of three principal parts: the *root*, protected by a root cap; the shoot or *plumule*; and the *scutellum*, which is a digestive organ. The majority of the leaves which the plant will ultimately produce on the main stalk are already formed in miniature in the embryo before the seed is planted. For a seed to germinate it must have a living embryo. Seed of corn usually remains alive for three to five years, and the maximum period of viability under ordinary storage conditions seldom exceeds ten years. However, when seed is kept in cold storage viability may be maintained for twenty-five years or more. There is little, if any, possibility that kernels of corn several thousand years old turned up in archaeological excavations will grow; their embryos have long since lost their capacity to germinate. If they could germinate, however, the endosperm would be quite capable of nourishing them. I once received from an archaeologist in Peru a number of ears of prehistoric corn estimated to be about 2,000 years old. Somewhere along the line the

ears had become infested with larvae of the Angoumois grain moth, and when the corn arrived in Cambridge the moths were beginning to emerge. They were slightly smaller than usual grain moths of this species but otherwise quite normal. Perhaps these prehistoric grains had lost some of their vitamins over the centuries, but in other respects they were still capable of providing an adequate insect diet. The fact that their mothers had laid their eggs on these kernels is proof that they recognized them as perfectly good cereals for the nourishment of their prospective young.

Under the proper conditions of air, moisture, and temperature the seed absorbs water, swells, and increases in weight. In thirty-six to sixty hours the primary root bursts through the pericarp; soon the plumule, the young stalk surrounded by a sheath, the *coleoptile*, also bursts through the pericarp and begins to grow in the opposite direction. The next step is the appearance of several secondary rootlets at the base of the first root. These seedling roots comprise a temporary root system supplying water and some nutrients until the permanent system, which grows from the lowest joint, has developed. At this stage most of the food utilized in the growth of the young seedlings is obtained by the digestion of the starch and other food materials in the endosperm.

That part of the young seedling between the seed and the lowermost joint of the stalk is known as the *mesocotyl*, and practically all of the growth which occurs in bringing the seedling to the surface of the soil occurs in this region. The mesocotyl is a very delicate organ whose elongation is suddenly stopped when the tip of the coleoptile penetrates the surface of the soil and is exposed to light. For this reason the first joint of the stalk from which the permanent root system later develops is always just below the surface of the soil regardless of the depth at which the seed was planted. For this reason, too, the depth to which corn can be planted depends entirely upon the length to which the mesocotyl is capable of growing. In some of the Indian varieties of the Southwest the mesocotyl can grow to a length of twelve to fourteen inches, and seed of such varieties may be planted deeply in order to reach moist soil (Collins, 1914). In ordinary varieties, however, the mesocotyl is incapable of growing more than three or four inches and the seed cannot safely be planted deeper than this. The length of the mesocotyl of prehistoric corn can also be used to determine how deeply ancient people planted their corn; an example of this is described in Chapter 15.

Soon after the pointed coleoptile has emerged from the surface of the soil the leaves begin to unfold and enlarge. Each leaf consists of two main parts: the *blade* and the *sheath* which surrounds the stalk. It is in the leaves and sheaths that the manufacture of food occurs. Much of the water taken up by the roots and transported to the leaves is lost from them by transpiration. A single corn plant may transpire as much as a barrel of water during its lifetime.

The permanent root system, which develops at the lowest joint of the stalk is a fibrous, freely branching system which spreads out in all directions to a distance of three to four feet and may penetrate the soil to a depth of four to seven feet. While the root system is de-veloping, the leaves of the stalk, of which the lower ones are already well developed in the embryo, unfold one by one, alternating between opposite sides of the stalk. The leaf sheaths which surround the stalk also alter-nate in the way in which their margins over-lap. In half of them the left overlaps the right as in a man's coat or shirt; in the other half the right margin overlaps the left as in a wo-man's coat or blouse (Weatherwax, 1948). With the unfolding of the uppermost leaf, the tassel makes its appearance, and soon after it has completely emerged its anthers are ex-serted and begin to shed their pollen. A few days later, in the majority of varieties, the first silks emerge from the lateral ear shoot about midway up the stalk. Pollination occurs, the kernels swell, the ear shoot enlarges to be-come the mature ear, and the cycle is completed.

It is this grain-bearing ear, tightly enclosed in husks, that comprises a substantial part of the mystery of corn. The ear of corn enclosed in its husks has no close counterpart else-where in the plant kingdom either in nature or among other cultivated plants. It is super-bly constructed for producing grain under man's protection, but it has a low survival value in nature, for it lacks a mechanism for dispersal of its seeds. When an ear of corn drops to the ground and finds conditions favorable for germination, scores of seedlings emerge, creating such fierce competition among themselves for moisture and soil nutrients that all usually die and none reaches the reproductive stage.

The mystery has not yet been completely solved in all of its details, but the broad out-lines of the solution are at least in sight. Most of the remaining chapters of this book are concerned with describing the progress that has been made, in its solution but first we should become acquainted with the relatives of corn, for it is research on these that has contributed much to that progress.

2 Theories on the Origin of Maize

I will listen to any hypothesis but on one condition—that you show me a method by which it can be tested. Von Hoffman (in Gregory, 1916)

There have been four principal and several minor theories* regarding the origin of maize: (1) that cultivated maize originated from pod corn, a form in which the individual kernels are enclosed in floral bracts as they are in other cereals and in the majority of grasses; (2) that maize originated from its closest relative, teosinte, by direct selection, by mutations, or by the hybridization of teosinte with an unknown grass now extinct; (3) that maize, teosinte, and *Tripsacum* have descended along independent lines directly from a common ancestor; (4) the tripartite theory of Mangelsdorf and Reeves (1939) that (*a*) cultivated maize originated from pod corn, (*b*) teosinte is a derivative of a hybrid of maize and *Tripsacum*, (*c*) the majority of modern corn varieties are the product of admixture with teosinte or *Tripsacum* or both.

Other hypotheses called "minor" in the sense that they have had little effect upon the thinking and experimentation concerned with the origin of corn are: (1) that cultivated corn originated from papyrescent corn, a type superficially resembling a weak form of pod corn; (2) that corn originated from "corn grass," a mutant form with numerous tillers; (3) that corn is an allopolyploid hybrid originating in southeast Asia by the hy-

*Like other students of maize I use the word "theory" here in its popular or semipopulat meaning. In a strict sense we are not dealing with theories—certainly not with theories involving broad principles—but with hypotheses. Webster's dictionary defines hypothesis as "a tentative theory or supposition, provisionally adopted to explain certain facts and to guide in the investigation of others."

bridization of two ten-chromosome species such as *Coix* and *Sorghum*.

The Pod-Corn Theory

The hypothesis which holds that cultivated maize has been derived from a wild form of pod corn is at once the oldest and among the youngest of the various propositions which have been developed to explain the origin of this unique cereal. More than a century ago the French naturalist, Saint-Hilaire (1829), described as a new variety *Zea maïs* var. *tunicata*, a peculiar type of maize sent to him from Brazil in which the grains were covered by the glumes. He concluded that this was the natural state of maize and that South America (probably Paraguay) was its native home. Virtually all students of maize since Saint-Hilaire have given serious attention to pod corn, have recognized its primitive characteristics, and have either accepted it as the ancestral form or for a variety of reasons have dismissed it from this role. My colleague, Dr. Reeves, and I have reviewed and discussed their viewpoints and conclusions in detail (1939). Here it will suffice to set forth the principal reasons given by various of the earlier students who dismissed pod corn as the ancestral form: (1) it does not breed true; (2) it apparently rises spontaneously in cultures of normal maize; (3) it is frequently monstrous and sterile; (4) it differs from normal maize primarily by a single gene; (5) the hypothesis that teosinte is the ancestral form is the more plausible one.

Still other objections have come from later students of the problem: (1) pod corn is ex-

tremely variable; (2) it is similar to other monstrosities such as teopod and corn grass; (3) it is the product of plant-hormone action; (4) it does not have the characteristics of a wild grass; (5) it could not have existed in the wild (Weatherwax, 1950, 1954, 1955). Despite the earlier objections to the pod-corn theory, Dr. Reeves and I revived it in somewhat modified form in 1939, and despite subsequent objections we later concluded (1959) that the pod-corn theory had greater validity and more evidence in its support than when it was first proposed. One merit of the theory, which some other theories lack, is that it is testable, at least in part, and my associates and I have spent many years and grown thousands of plants in subjecting it to a variety of tests.

The Teosinte Theory

The theory that teosinte has played a major role in the origin of maize may be said to have had its beginnings with the work of Ascherson (1875, 1877, 1880), who showed convincingly for the first time that teosinte is the nearest known relative of maize. Ascherson (1880) considered the ear of corn to be the result of fusion of lateral branches with the central spike. He assumed that the fusion of various spikes of the pistillate inflorescence of teosinte would result in a structure similar to the maize ear. We now know that there is no reliable evidence that the maize ear is the product of fusion.

The theory that teosinte was one of the parents of a hybrid from which maize had arisen had its origin indirectly in the so-called *Zea canina* first described by Watson (1891), which Harshberger (1893) assumed to be the ancestral form but which he later discovered to be a natural hybrid of maize and teosinte. Thereupon (1896) he proposed two hypotheses on the hybrid origin of maize: (1) that it had originated from a cross of teosinte with an extinct closely related grass, and (2) that maize is the product of a cross between wild teosinte and a cultivated race of teosinte, developed, perhaps, under irrigation. Harshberger considered that a hybrid origin of maize would account for the many teratological variations and so-called reversions which appeared under cultivation, and he concluded that even if the view of a hybrid origin of maize were not accepted, at least "the fact that teosinte and maize can be crossed and a fertile progeny results shows that the two plants are united by the close and intimate bonds of kinship." Still later Harshberger (1911) described as the theoretical ancestor of maize a plant whose inflorescence would resemble in many respects that of *Tripsacum*.

Collins (1912) advanced the hypothesis that maize originated from a hybrid of teosinte and an unknown grass belonging to the *Andropogoneae*, one similar to the homozygous form of pod corn which is often earless and bears seeds in the tassel. He listed nine differences between pod corn and teosinte and showed that in most cases modern maize is intermediate. In later papers (1918, 1919a, b, 1931) he presented numerous reasons for considering maize to be of hybrid origin, among them its marked heterozygosity and its pronounced intolerance to inbreeding. The close resemblance between maize and the *Andropogoneae* and the frequent occurrence in maize of characteristics typical of this tribe were the chief reasons for Collins' conclusion that one of corn's parents must have been of the tribe *Andropogoneae*.

One objection to the teosinte theory has been that a species so unpromising for food purposes would never have been domesticated. This objection has been weakened to some extent by the discovery (Beadle, 1939) that seeds of teosinte will "pop" when exposed to heat, shattering the hard, bony shells in which they are enclosed. Unfortunately, there is no evidence, archaeological, historical, or from contemporary sources to show that teosinte was ever employed for food in this manner. When teosinte is used for food,

as it occasionally is in times of food shortage, the fruits are crushed on a metate or with a mortar and pestle and the meal of the crushed kernels is separated from the fragments of the bony fruit cases.*

Recent studies on fossil pollen in Mexico (Barghoorn, et al., 1954; Irwin and Barghoorn, 1965) lend no support to the teosinte theory. Although pollen of both maize and *Tripsacum* was found at great depths, the pollen identified as that of teosinte occurred only near the surface in the upper levels of the drill core, where the abundance of maize pollen suggested that the practice of agriculture had begun.

Recent studies of archaeological maize, like those of fossil pollen, do not support the teosinte theory. On the contrary, they show that the earliest maize was less like teosinte than some modern maize. Archaeological remains do show, however, that maize hybridized with teosinte and that its evolution has been strongly influenced by this hybridization. To this extent teosinte is involved in the ancestry of modern maize.

The Theory of a Common Ancestry

It appears that Montgomery (1906) was the first to propose the theory of a common ancestry, although he did not include *Tripsacum* as one of the entities stemming from the common ancestor. The theory, as we now know it, having maize, teosinte and *Tripsacum* derived from a common ancestor, was formulated by Weatherwax (1918) and further elaborated in subsequent publications (1919, 1950, 1954, 1955). He has described it as "the simple, conservative view that the three plants in question descended from some common ancestor by ordinary divergent evolution of the sort that Darwin wrote about a century ago" (Weatherwax, 1954).

The author goes on to say that a morphological analysis of maize, *Tripsacum*, and teosinte discloses numerous rudimentary structures which are unquestionably vestigial organs lost in the course of evolution. If these rudiments could be replaced by fully developed structures, all three plants would converge toward a common form, and it is evident from this that all three have come somehow from a single ancestral stock.

*Personal communication from the late R. H. Barlow, who spent considerable time in Mexico studying the customs of the Indians.

Reeves and I have pointed out (1959) that restoring the primitive organs of only corn and *Tripsacum*, omitting teosinte, would produce the same common ancestral type as one in which teosinte is included. We have also objected to the theory on the grounds that it does not explain all of the known facts and that it is largely untestable. Weatherwax expresses doubt that the common ancestor or even the immediate ancestor of maize will ever be found or that we shall ever know exactly what sort of a plant it was. Consequently we can, in his opinion, safely speculate about some of its characteristics. The wild maize plant which Weatherwax visualizes was probably a perennial. It had terminal staminate panicles on a few main culms and pistillate panicles in various stages of reduction to spikes on numerous lateral branches. The small ears, with four or eight rows of grain, were probably partly enclosed by enveloping leaf sheaths, and the small grains were partly or wholly covered by the bracts of the spikelets. A plant like this would have had some of the characteristics of the present half-tunicate types, but it would not have the monstrous character of "true" pod corn. We shall see later that the prehistoric wild corn discovered in Mexico had a number of the characteristics of the wild plant visualized by Weatherwax, but it was not a perennial and it did not have tillers. It is possible that other geographic races of wild corn may have had tillers; it is doubtful that any will be found that were perennials.

Minor Theories

The Papyrescent "Semivestidos" Theory

Before discussing in detail our tripartite theory, I should mention briefly two minor theories. The first of these, the papyrescent theory, is little more than a slight modification of the pod-corn theory. Andres (1950) discovered in Argentine maize a type which superficially resembles a weak form of pod corn in which the kernels are partly covered with soft glumes. Apparently unaware that Bonvicini (1932) in Italy had described this character many years earlier and had given it the name "palee sviluppate," Andres called the type "semivestidos" because the elongated glumes partially enclosed the seeds. He suggested that it, rather than pod corn, might be the ancestral form. This type of corn has since been given still a third and more appropriate name, "papyrescent," by Galinat (1957), whose studies showed that the glumes become soft and papery as they mature, resembling the glumes of papyrescent sorghum. Unlike pod corn, which, although sometimes monstrous in its expression, still respresents a combination of normal characteristics found in other grasses, papyrescent is a defect in development which can hardly be regarded as the primitive form of modern corn.

The Corn-Grass Theory

Singleton (1951) has suggested that the ancestral form of modern corn is "corn grass." This anomalous type, the product of a single dominant gene, produces numerous tillers and small "ears" with a high proportion of single spikelets. Many of the kernels are partly enclosed in bracts, but the majority of these are not glumes but spathes. He also suggested that if a plant of corn grass were found in nature it would not be recognized as maize and would almost certainly be regarded as a different species if not a different genus. This may be true, and it illustrates how the maize plant can be drastically changed by a single-gene mutation. If corn grass were the ancestral form, mutation at a single locus could have transformed it from a wild, almost useless, plant to the unique cereal which maize is today.

Although corn grass has some of the characteristics which we might expect to find in an ancestral form—for example, a freely tillering habit—it lacks others, such as the regular development of prominent glumes. At the other extreme it has characters which are not demanded of a hypothetical ancestor. One of these, single spikelets (Galinat, 1954), represents a condition more specialized instead of more primitive than the paired spikelets of modern maize. Another, a well-developed spathe, suggests the ancestral form, not of maize but of *Coix*, whose fruit case has been found by Weatherwax (1926) to comprise a spathe and a segment of the rachis. Corn grass probably is, as Galinat (1954) has suggested, a "false" progenitor of maize, exhibiting certain traits which might have occurred in a remote ancestor of the *Maydeae*.

Finally, the evidence from archaeological maize does not support the corn-grass theory. Most early prehistoric maize has relatively prominent glumes, but it does not have the long spathes of corn grass. The possibility that corn grass is the ancestral form appears to me to be remote indeed.

The Tripartite Theory

As the result of our extensive studies of hybrids of maize with its two American relatives, teosinte and *Tripsacum*, Reeves and I concluded (1939) that teosinte, far from being the progenitor of maize, is instead its progeny, or to be exact, the progeny of a hybrid of maize and *Tripsacum*. This conclusion obviously did not explain the origin of maize itself, but with teosinte eliminated (in our minds) as the ancestor of cultivated corn we were free to consider other possibilities explaining corn's origin, and the idea that pod corn might have been the ancestral form appealed to us. Certain other students of maize had been impressed by the fact that pod corn is often monstrous in the expression of its various characteristics. We thought its monstrosity might be the result of a single relic wild gene being superimposed upon the germ plasm of highly domesticated modern varieties. We were impressed by the fact that pod corn's principal characteristic, enclosing the kernels in floral bracts, is an almost universal feature of wild grasses.

The third part of our tripartite theory was the recognition that teosinte, if it was not the progenitor of maize, had at least played an important role in its evolution under domestication. Since teosinte is common in and around the maize fields in parts of Mexico, where it crosses frequently with maize, and since the maize-teosinte hybrids are highly fertile and cross back easily to either or both parents, it seemed inevitable that there must have been, and still is, a flow of genes from teosinte into maize and hence that many modern varieties of corn must be the product of past hybridization with teosinte.

This tripartite hypothesis has been regarded by some of our critics as unduly complex and "top heavy with assumptions of such character that if any one of them should be rejected the whole structure would fall" (Weatherwax, 1954). We regarded our theory as being no more complex than the situation that it attempted to explain and as less speculative than any other because it was concerned only with existing botanical entities. The theories of Harshberger and of Collins relied in part upon an imaginary grass now extinct.

Weatherwax's common ancestor of the three American *Maydeae*, maize, teosinte, and *Tripsacum*, likewise involves an imaginary grass now extinct. Our tripartite theory is concerned only with corn, pod corn, teosinte, and *Tripsacum*, all of which are still living, and this fact has also given our theory an important advantage over certain earlier ones: it is to some degree testable. It not only explains the facts now available but it has also been useful in guiding our own investigations and those of others. My own research since 1939 has been devoted almost completely to testing experimentally the three postulates of our tripartite theory.

Tripsacum a Hybrid of Maize and *Manisuris*

As a result of our joint studies of the inheritance of characters in maize-*Tripsacum* hybrids, my long-time associate, Walton Galinat (1964), has added what amounts to a fourth postulate to our tripartite theory. He suggests that *Tripsacum* is a hybrid having wild maize as one of its parents and as the other a species of *Manisuris*, a genus of the tribe *Andropogoneae*, which includes the cultivated sorghums and millets, sugar cane, and a number of important pasture and forage grasses. Galinat's bold addition to the tripartite theory, although in some respects highly speculative, relies like the previous three postulates, upon botanical entities which still exist or, in the case of wild corn are known archaeologically; and it is to some degree testable. It promises to be a useful addition to the theory.

If teosinte is indeed a hybrid of maize and *Tripsacum*, then since it hybridizes easily with maize to produce highly fertile hybrids, it has served and can continue to serve as a genetic "bridge" over which there is a gene flow from the hardy *Tripsacum* to cultivated maize. If *Tripsacum* is indeed a hybrid of maize and *Manisuris*, then it in turn is a bridge over which there has been and can continue to be a gene flow from the tribe *Andropogoneae* into the tribe *Maydeae* to which maize, teosinte, and *Tripsacum* belong.

The implications for the improvement of cultivated maize in these postulates are considerable. Numerous attempts have been made by corn breeders to cross corn with the sorghums, the most drought resistant of the major cereals. All of these have so far failed. Perhaps the same ends can be attained by introducing genes from the *Andropogoneae* by employing teosinte and *Tripsacum* as bridges between the tribes. It is possibilities such as these that make corn's American relatives important from both theoretical and practical standpoints and entitles them to chapters in this book.

Geographical Origins

As there has been a wide divergence of opinion about the botanical characteristics of corn's ancestor, there have also been differences of opinion about its geographical origin. Those who have considered teosinte as the ancestor or one of the ancestors of corn have necessarily postulated Mexico and Central America, where teosinte grows, as corn's geographical center of origin. Since maize was almost equally important in Mexico and Peru and there was little evidence of contact between them in prehistoric times, De Candolle (1886) assumed that corn must have originated in an intermediate area, and he thought that New Granada, now Colombia, seemed to fulfill these conditions. Birket-Smith (1943) later reached a similar conclusion, but for somewhat different reasons. Although it was not an essential part of our tripartite hypothesis, Reeves and I postulated that maize had its origin in the lowlands of South America primarily because of the historical references to pod corn in that area. That idea had to be abandoned with the discovery, described in Chapter 15, of fossil corn pollen in Mexico. Data on races and lineages presented in Chapters 9 and 10 will suggest that corn had not one origin but several in both Mexico and South America.

3 Teosinte, the Closest Relative of Maize

In a botanical point of view *Euchlaena* is a
most interesting genus, from its being the
nearest congener of maize, whose American
origin it thus supports. Hooker, 1879

The closest relative of maize is teosinte; about
this there can be no doubt. Teosinte has the
same chromosome number, 20, as maize, and
its chromosomes are similar to those of maize
in their lengths and positions of their centromeres. Teosinte hybridizes readily with maize,
and the first generation hybrids are vigorous
and highly fertile when self-pollinated or
crossed back to either parent. For many years
teosinte was considered by a number of students to be the progenitor of maize. Those
who still hold this view are now a minority,
and recent paleobotanical and archaeological
evidence lends no clear-cut support to their
views. The discovery of fossil maize pollen
about 80,000 years old, discussed in Chapter
15, if it has been correctly identified as that of
maize, would seem to be sufficient evidence
to prove beyond a reasonable doubt that the
ancestor of cultivated corn was corn and not
teosinte. Nevertheless, as corn's closest relative, teosinte remains important in any discussion of corn's evolution or improvement,
and for this reason its history deserves a
chapter in this book and the experimental
work of which it has been the subject deserves
another.

The theory postulated by Robert Reeves
and me that teosinte is not the progenitor of
maize but instead is the progeny of the hybridization between maize and *Tripsacum*
cannot be said to be generally accepted, although no evidence that is clearly in conflict
with it has been adduced in the period of more
than thirty years since it was first proposed
and both experimental and archaeological
evidence is consistent with it. Its great usefulness to Reeves and me has been in eliminating
teosinte as the progenitor of maize in our own
thinking and by so doing to focus our attention on pod corn as the ancestral form and on
the roles of the introgression of teosinte and
Tripsacum in the evolution of maize.

My former student, Dr. H. Garrison
Wilkes, who had chosen teosinte as the subject for his doctoral dissertation, traveled
some 30,000 miles in Mexico, Guatemala, and
Honduras in a Land Rover which had already
attained a degree of distinction by having
served as the principal vehicle employed by
MacNeish in the excavations which turned up
the most primitive prehistoric corn so far
discovered. Wilkes collected teosinte over its
entire range; then spent a summer in comparing his collections when grown under controlled conditions; made an extensive survey
of the literature, comprising more than 800
references; and finally published his findings
and conclusions in a book which represents
the most comprehensive study of teosinte that
has yet been made. In many instances here, I
have used Wilkes' statements almost verbatim. I am also relying heavily on Wilkes for
the review of the early literature on teosinte.

Historical Records

The earliest records of teosinte, according to
Wilkes (1967), occur in the preconquest codices of Mexico. These are not "literature" in
the sense of constituting a written language,
since there was no such language in use by any
of the pre-Columbian civilizations of the New
World. The codices represent a form of picture writing in which the customs of people
and the natural history of their environment
were illustrated, often so realistically that
there is no mistaking their meaning.

There are two preconquest records of teosinte in Mexico: one represented by the
interpretation of picture writing by García-
Icazbalcata in 1882 and the other by the
Vatican codex. In the former, reference is
made to people eating seed like maize called
"cincocopi" and in the latter to eating a wild
maize called "acecintle." The Náhuatl name
for teosinte is "cocopi," and the word used in
the Valley of Mexico is "acecintle" or more
commonly "acece," and in parts of the state
of Guerrero, "atzitzintle."

The most important of the early post-
conquest references to teosinte occur in the
classic, *Historia General de las Cosas de Nueva
España*, of Sahagún (cf. Wilkes). Bernardino
de Sahagún was a Franciscan monk, born in
1499 or 1500, who went to Mexico in 1529 and
remained there until he died at the age of
about 90. Unlike some of his contemporaries
among the conquistadores, he was gentle and
kind to the Indians and sought to understand
their problems. They responded with a will-
ingness to talk and to give him information
about their customs and the "things" which
made up their lives. Book 11 of the *Historia*,
in which teosinte is described, was finished in
1576. It is concerned with Aztec natural history and is the best surviving source on this
subject. It is in this book that "cocopi" is
described (in its translated version) as follows:

There is a plant very similar to maize
called cocopi, the grains of the plant are

parched until they are well carbonized and also some grains of wheat are parched in the same way.

They are milled and made into biscuits and topped with a little ground sage.

This is required for those that pass blood. They drink it three times during the day, once in the morning, the others at noon day and in the afternoon.

This herb grows in the maize field, it is not sown, some grow before planting, and others after planting. It grows in-between maize like rye-grass in the wheat field.

A second important early post-conquest source of Aztec natural history is the thesaurus of Hernández. Francisco Hernández, Physician-in-Ordinary to Phillip II during a seven year's stay in Mexico (1571–1577), wrote sixteen folio volumes on the natural history of that country. The completed work was never published because the Spanish Crown was reluctant to disclose too much information about the New World and the volumes, "superbly bound in blue Morocco, tooled in gold and with locks and clasps and corners—all of high wrought silver," were lost in a fire in the library of the Escorial in 1671. Fortunately various extracts of the work were made from time to time before its destruction, and it is in the Matriti edition, published in 1790, that teosinte is briefly described as follows:

Of Cencocopi, a plant like maize.
   This plant looks like maize but its seed is triangular. Grind the seed and take one ounce dosage to cure dysentery. It is inferior quality maize.

The statement that the plant "looks like maize but its seed is triangular" almost certainly identifies the plant as the Mexican form of teosinte, which has triangular fruits, in contrast to those of the Guatemalan teosintes, in which the shape is more nearly trapezoidal.

Thus the codices and the early chroniclers make it clear that teosinte was well known to the Mexicans, who recognized its similarity to maize and its role as a weed in the maize fields. After Hernández, teosinte does not appear again in the literature until the publication by Schräder of a description of the species in the seed catalogue of the Goettingen horticultural academy (which was republished in *Linnaea* in 1833). The description

was apparently based on plants grown in the Goettingen garden from seed received by D. Muhlenfordt, professor of botany in Hannover, from his brother, a mining engineer who obtained them while working in Mexico. This appears to have been the first introduction of teosinte into Europe. In describing teosinte Schräder gave it the name *Euchlaene mexicana*.

Teosinte was introduced into Europe a second time in the middle of the nineteenth century. In 1849, F. de Brignolia Brunnhoff published a description of it in the seed catalogue of the Botanic Garden of Modina, Italy, and this was republished the same year in the French journal, *Annales scientifiques nature botanique*. The plant was described as a new taxon, *Reana Giovannini*, an annual having its habitat in Mexico. The epithet honors the collector, Dr. Melchior Giovannini.

The third introduction of teosinte in Europe and the first use of the Aztec name, "téozinté," occurred in 1869. In that year J. Rossignon, Director of Public Gardens and Promenades, Guatemala, sent to France seed of téozinté, which he had first seen at the exposition of the Sociedad Económica de Guatemala in January 1869. In the letter that accompanied the seed and which was read before the Imperial Society of Acclimatization of Paris on August 27 by Durieu de Maisonneuve, director of the Botanic Garden at Bordeaux, teosinte was described as a species of reed or rattan indigenous to the warm parts of Guatemala. Inhabitants of Santa Rosa, a department south of Guatemala City, were said to eat the young shoots, feed the leaves to cattle, and use the stems in making huts, hedges, screens, and cane.

The following year seed of teosinte was planted in three places in France. In the botanic garden of Bordeaux it produced a tremendous vegetative growth but failed to flower (cf. Wilkes). Single seeds had also been given to Naudin and to Thuret, directors of small botanic gardens at Collioure and Antibes respectively. The single plant in Naudin's garden was killed by frost before it flowered, but Naudin (cf. Wilkes) was impressed by its potential as a fodder plant. In Thuret's garden (latitude 43° 35′) flowering occurred in October according to Durieu (cf. Wilkes), who identified the plant as belonging to the genus *Reana* and gave it the specific name *luxurians* because of its "marvellous vegetative develop-

ment." That the French were impressed by the possible agricultural value of teosinte as a fodder plant is attested by the forty-one notes on teosinte that appeared in the proceedings of the Society of Acclimatization in the period 1869–1879 and by their efforts to obtain additional seed.

Although teosinte had flowered at Antibes, it had not produced mature seed, and more seed was requested of Rossignon in Guatemala. In his reply to the request, Rossignon had apparently confused teosinte with *Tripsacum*, another relative of maize, because he described it as a perennial forming enormous tufts which was propagated by rooted offshoots of the tufts. Despite this confusion, he sent to France in 1875 seed of teosinte that he had produced by sowing the few which remained in his possession from his original collection. In the letter which accompanied the second lot of seed which he sent to France he commented on the great similarity of "téozinté" to maize and raised the question whether it could be the plant that when cultivated was transformed into maize. This appears to be the first suggestion that teosinte might be the ancestor of cultivated maize.

The second planting, like the first, failed to produce mature seed in France, but another planting in France's tropical island, Réunion, produced mature seed abundantly, some 500 grams from six average stalks. The first mature seed produced in Europe was by Gomez (cf. Wilkes) at Malaga, Spain.

Although the French were the first to recognize the potential value of teosinte as a fodder plant for tropical countries (cf. Wilkes) it was the English who were responsible for its world-wide distribution. In 1877 Schweinfurth obtained seed from the French and grew plants to maturity at Cairo, Egypt. The herbarium specimen which he deposited in the Royal Botanical Garden at Kew (illustrated in plate II of Wilkes) was accompanied by a sketch of a freely tillering plant showing numerous stems terminated by inflorescences and by diagrams of longitudinal sections that show with remarkable accuracy the glumes, lemmas, and paleas of the functional pistillate floret and the floral bracts of a vestigial floret.

A herbarium specimen grown the same year, 1877, was also sent by Schweinfurth to the Muséum National d'Histoire Naturelle. Paris, under the name *Euchlaena luxurians* along with the synonyms *Reana luxurians* and

Tripsacum monostachyum (Wilkes, plate III). The profusely branched tassel and the long trapezoidal fruit cases of these herbarium specimens are typical of Guatemalan teosinte.

From the plants that he grew in Egypt, Schweinfurth supplied seed to Hooker at Kew. Hooker grew a teosinte plant which flowered in December "under stove treatment" in the water lily house of the Royal Gardens. This plant was the basis of the first published illustration of teosinte, which appeared in the Curtis Botanical Magazine in 1879. The following year Hooker, now supplied with a considerable quantity of seed from both Schweinfurth in Cairo and King in Calcutta, distributed teosinte to the "Bahamas and West Indies, Cyprus, South and tropical Africa, Australia, the United States and numerous applicants and correspondents."

The Vilmorin Catalogue offered seed of teosinte commercially in 1877, and the United States Department of Agriculture distributed it in 1886 and 1887. Prior to this distribution, however, seed of teosinte had been produced in southern Florida by J. G. Knapper of Limona, Florida, from seed obtained from Honduras. Consequently, it is not now positively known whether the teosinte grown for fodder in this country and known as Florida teosinte originated from seed distributed by the Department of Agriculture or whether it is descended from an independent introduction from Honduras. In any case the Florida teosinte is quite similar to the teosinte of southern Guatemala, and I like to believe that the Florida teosinte with which my experiments in Texas began in 1928 has a long and illustrious line of descent from Rossignon in Guatemala through Durieu in France, Trouette in Réunion, Schweinfurth in Egypt, Hooker in England, King in Calcutta, the seed firm of Vilmorin in France, famous for breeding the stock beet into the modern sugar beet, and finally our own Department of Agriculture. I am grateful to Wilkes for assembling the facts that make possible such a fascinating genealogy.

To complete the history of teosinte mention must be made of an important article by Kempton and Popenoe (1937), who rediscovered teosinte in Guatemala after doubts had arisen that it existed there or that the seed sent to France by Rossignon was really teosinte. These authors also discovered in their search of the literature that one Sr. Guzman had corrected Rossignon on the source of his teosinte seed, stating that it had come from the town of Los Esclavos and not from Santa Rosa.

Teosinte was widely distributed for agricultural purposes throughout the tropics and the southern United States, but the hopes of the French that it would prove to be "the world's most productive fodder plant" were not realized. Asa Gray, the foremost American botanist of his time, proved to be prophetic when he stated in the *American Agriculturist*, August 1880: "Undoubtedly there is not a more prolific forage plant known; but as it is essentially tropical in its habits, this luxuriant growth is found in tropical or subtropical climates. The chief drawback to its culture with us will be that the ripening of the seed crop will be problematical as early frost will kill the plant. To make the teosinte a most useful plant . . . the one thing needful is to develop early flowering varieties so as to get seed before frost."

Now, more than three-quarters of a century later, my associate, Walton Galinat, is attempting to develop a teosinte that will flower in this latitude, not by breeding for earliness as such, but by breeding out its inherited photoperiodic response to short days which renders it strictly vegetative during its growth in the long days of summer at this latitude. It remains to be seen whether a short-season teosinte will be as productive as other forage grasses such as Sudan grass, although it may well be more palatable to livestock.

Botanical Relationships

The close botanical relationship of teosinte and maize was recognized by botanists and others from their first acquaintance with it. Rossignon, who sent seed of it from Guatemala to France, noted its "great similarity" to maize and wondered whether it could be "the plant which cultivation has transformed to maize." Ascherson, writing in 1875 (cf. Wilkes) had stated that teosinte is the nearest relative of maize and is unquestionably of New World origin. A year later (1876) he reviewed the variation in *Euchlaena* and emphasized its intermediate position between *Zea* and *Tripsacum*.

Like Ascherson, Hooker (cf. Wilkes) recognized clearly the botanical affinities of teosinte when he stated, "It unites the habit of maize (*Zea*) with in many respects the structure of *Tripsacum*" and later made the statement quoted at the start of this chapter. Thus both Ascherson and Hooker anticipated by more than half a century the conclusion that Reeves and I reached about teosinte's intermediate nature after comparing maize, teosinte, and *Tripsacum* in a large number of characteristics. Could we have saved ourselves this time-consuming task had we read Ascherson or Hooker first? The question is academic, for the fact is that we had no strong reason for reading these authors until after our own research on teosinte and its relationship to maize had aroused our interest in its previous history. Ideally, perhaps, a scientist should familiarize himself with all of the previously published literature on a subject before he embarks upon a new research project. In practice this seldom occurs. Much more frequently, scientists, at least those working in biology, are concerned with the solution of a particular problem and when in the course of their research they make an important discovery they often find that the discovery was made, at least in part, many years earlier. The classic example, of course, is the rediscovery of Mendel's laws thirty-five years later by DeVries of Holland, Correns in Germany, and Tschermak in Austria independently of Mendel and of each other.

Although the close relationship of teosinte and maize was early recognized by Europeans and even earlier by the Mexican Indians, the exact nature of that relationship was by no means clear. Rossignon's question whether teosinte could be the plant which cultivation transformed into maize was little more than passing speculation. The first to reach the definite conclusion that teosinte is the ancestor of maize apparently was Ascherson, who postulated that the ear of maize may have been formed by the fusion of the two-ranked spikes of teosinte. Here Ascherson was thinking in terms of morphological phylogeny; he did not suggest that teosinte was the immediate ancestor of cultivated maize—that maize is a domesticated teosinte.

Most responsible for the popular idea that maize had its origin in teosinte and that the ear of corn could be developed from teosinte by a few generations of careful selection was Luther Burbank. His "discoveries" on this subject were published in the widely read

Literary Digest, July 1921, and in the Sunday supplements of various newspapers. From his descriptions and illustrations of his so-called teosinte, it is now clear that he was working, not with pure teosinte but with a maize-teosinte hybrid. To produce maize-like ears from such a hybrid would have required no special plant-breeding ability—indeed it could scarcely have been avoided when the progeny of the hybrid was grown.

Several years later Blaringhem (1924), who had earlier concluded that maize could have originated from teosinte by mutations produced by traumatic influences, found apparent support for this farfetched idea in the experiments of one M. Bento de Toledo at the Instituto Agronomico, São Paulo, Brazil, who in four years or less had succeeded in converting Mexican teosinte into maize. Bento, unlike Burbank, started his experiments with pure teosinte, but his "metamorphosis" of *Euchlaena* into *Zea* began when teosinte became hybridized with maize, as his illustrations clearly show.

The classic case of a maize-teosinte hybrid being erroneously considered, at least briefly, to be the ancestral form of maize is that of the so-called *Zea canina* of Mexico. In 1888 Professor Duges of Guanajuato, Mexico, sent to the Botanical Garden of Cambridge, Massachusetts, an unusual cluster of about a half dozen small "maize-like" ears, each about two inches long, bearing pointed white seeds in distinct rows and enclosed in husks. Dr. Sereno Watson, at that time an assistant to Asa Gray, grew two plants from seed of this cluster and noted that their habit of growth was "wholly unusual." He first thought that these plants might represent the wild form of cultivated maize, but after comparing them with cultivated forms concluded on the basis of "differences in habit of growth, the arrangement of the staminate spikelets and the nervation of their glumes, the form of the glumes of the pistillate flowers and the ready disarticulation of the ripened ear" that these specimens represented a distinct species, which he named *Zea canina* S. Wats. after the local Mexican name. This common name was attributed to the small pointed kernels which have a superficial resemblance to dog's teeth. But Wilkes later found in his extensive experience throughout the Central Plateau of Mexico that the term "maiz de coyote" is applied to both teosinte and to maize-teosinte

hybrids and that its most common meaning is "wild or untamed corn."

Although Watson's creation of a new species of *Zea* on the basis of one specimen and two plants was scarcely justified, he was not the only one to have drawn erroneous conclusions about "maiz de coyote." In 1892 Harshberger, one of the most profound of the late nineteenth-century students of maize, reported on plants of this introduction which he had grown in Philadelphia. The one plant that matured and which bore four fruiting nodes with several small ears per node he considered to be a primitive form of maize and one that supported his conclusion that, from the evidence from archaeology, history, ethnology, and philology, central and southern Mexico was the original home of maize and that, on the basis of botanical and meteorological evidence, wild maize was probably adapted to the highlands of Mexico. This "wild" form of maize seemed to lend proof to his conclusions.

Although Harshberger, like Watson, was in error about the nature of "maiz de coyote," his conclusions about the original habitat of maize—at least of one geographical race—have received substantial support by the discovery (discussed in detail in Chapter 15) of the remains of prehistoric wild corn in once-inhabited caves in the Tehuacán Valley of southern Mexico. Had I been more impressed with these conclusions and the evidence on which they were based when I read Harshberger in 1938, I would have been less surprised than I was at the discovery of prehistoric wild corn in the semi-arid Tehuacán Valley at an altitude of more than 4,000 feet. But since he proved to have been initially so wrong* about "maiz de coyote," I had little confidence in his other conclusions; this proved to be a mistake on my part.

Liberty Hyde Bailey, one of the famous botanical figures of his period, was also misled, although to a lesser degree than Watson, by *Zea canina*. He, like Harshberger, obtained seed from Watson and grew it at the Cornell Agricultural Experiment Station, where he noted variation in the "degree of acclimatization" among the plants. Actually there must have been variation in a number of characteristics, since the seed, being the open pollinated

*He (1896) corrected this error after his trip to Mexico.

production of what we now know to have been a maize-teosinte backcross hybrid, would most likely have been a mixture resulting in part from self-pollination and in part from backcrossing to maize. The second year a cross of *Zea canina* with *Zea* "japonica," an ornamental corn with variegated leaves, showed that *Zea canina* was losing some of its features under domestication. This caused Bailey to question whether this plant was really a distinct species and not instead the "original form of *Zea mays* or at least very near it."

The hybrid origin of *Zea canina* was not suspected by Watson, Harshberger, or Bailey until Duges, who had sent the original specimen to Watson, wrote Harshberger (September 22, 1895) informing him that the specimen he had sent to Watson was a form of teosinte known in Mexico as "Maiz de Coyote." In subsequent correspondence with Professor Segura, director of the Mexican School of Agriculture, Harshberger (1896) learned that "*Zea canina* Watson is not but the result of hybridization of asese with maize." Segura had discovered this for himself several years earlier while harvesting in a field in which teosinte was also growing. He also had hybridized maize with teosinte. Harshberger visited Mexico in 1896, saw the result of Segura's experiments with maize-teosinte hybrids, which by now included a third backcross to maize, and concluded from the study of Segura's specimens that the plant described by Watson was the product of a second backcross to maize of a maize-teosinte hybrid.

Actually the Guanajuato hybrid which created so much botanical confusion was not the first maize-teosinte hybrid reported. Professor Brewer, writing to Sturtevant in 1878, mentions the discovery of Roezl, the German collector, who, in the state of Guerrero, had seen a previously undescribed "*Zea* with small ears, truly distichous and covered by a husk, the grain like some varieties of maize only smaller and harder."

Although published reports on the occurrence of hybrids between maize and teosinte were few, observation of such crossing must have been numerous, because even as early as 1576 Sahagún had noted that teosinte "grows in-between maize like rye-grass in a wheat field," and, since teosinte crosses readily with maize, there must have been numerous hybrids which would not have escaped the keen eyes of the Indians and through them would have become known to botanists. As early as 1887, a year before Professor Duges sent specimens to Watson, the Sociedad Mexicana Historia Natural devoted several sessions to a debate between Ingeniero Segura and Dr. Urbina on the one hand and Drs. Villada and Ramírez on the other over the question of teosinte as the primitive form of maize. It was resolved, with unusual perspicacity, that this question could not be answered except by experimentation, and so began the experiments of the Mexican agronomist Ing. Segura on the artificial hybridization of teosinte and maize, which by 1896 finally revealed the true nature of *Zea canina* and cleared away the confusion in which three eminent American botanists had been involved.

Concluding that *Zea canina* could no longer be regarded as the ancestor of maize, Harshberger (1896) proposed two alternative hypotheses to explain corn's origin. The first postulated that maize originated from teosinte by selection, the second that maize originated from the hybridization of teosinte and an "extirpated closely related grass," which in its botanical characteristics was similar to maize-teosinte hybrids. Later (1900) Harshberger combined these two hypotheses to postulate that a partially ameliorated form of teosinte produced by selection crossed with wild teosinte to produce present-day Indian corn. He found evidence supporting this hypothesis in the observation by Lumholz of hybrids of maize and teosinte in the corn grown by the Tarahumara Indians of northern Mexico and information he had received from Leon that "hybrids were grown by the Mixes and Zapotecs inhabiting the state of Oaxaca and that there the plant is quite common."

Although Harshberger's final hypothesis is a rather hazy one and although it involves an imaginary teosinte "ameliorated" by selection, he was correct in recognizing that teosinte hybridizes readily and frequently with maize and that such hybridization in the past with corn's progenitor, whatever that progenitor might have been, would necessarily have had its effects. The conclusion which Reeves and I subsequently reached, as one part of our tripartite hypothesis, that many modern varieties of maize are the product of hybridization with teosinte, although original with us and not adapted from Harshberger, is

nevertheless related to his hypothesis in recognizing the role of hybridization with teosinte in the evolution of maize.

Few students of maize in recent times have been wholly committed to the idea that teosinte as we know it today is the ancestor of cultivated maize. R. A. Emerson in a personal conversation with me suggested that all that was needed to make teosinte the ancestral form is the existence of a soft-shelled type similar to the soft-shelled varieties of *Coix* which are grown in the Philippines and other parts of Asia. This hypothesis, which, so far as I know, Emerson never published, suffers from the same fault as Harshberger's in postulating a progenitor which is no longer in existence and for which there is neither archaeological* nor historical evidence.

Although not completely committed to the idea that teosinte was the progenitor of cultivated maize, Randolph (1955) has been reluctant to dismiss the hypothesis entirely, primarily because of the similarity of maize and teosinte chromosomes. However, he has also been reluctant to accept the change in nomenclature which Reeves and I had proposed (1942) making teosinte a species of *Zea* instead of a separate genus. Actually the two views seem to me to be somewhat inconsistent; there is no other instance of which I am aware in the history of the evolution of cultivated plants of domestication transforming a wild plant into a wholly different genus.

Nomenclature and Taxonomy

The first Latin name assigned to teosinte is that of Schräder (1832, see Wilkes), who called it *Euchlaene mexicana* and placed it in *Olyearum*, the tribe for dioecious grasses. The final "e" in the genus name was evidently a misspelling or a typographical or other error; later authorities have consistently replaced it with an "a." Schräder's description was based on plants grown in the Goettingen Botanic Garden from seed introduced from Mexico. Teosinte introduced into France from Mexico was described by Brignolia Brunnhoff (1849, cf. Wilkes) as a new taxon, *Reana Giovannini*, which he placed in the tribe *Zeineae*.

---

*We did find in one of the Tamaulipas caves (Mangelsdorf et al., 1967) a teosinte-like spike with kernels enclosed in soft glumes, which for reasons set forth in Chapter 14 we regarded not as an ancestral form but a hybrid of teosinte with one of the alleles at the pod-corn locus.

The third introduction of teosinte into Europe, that from Guatemala to France, was identified by Durieu (1872, cf. Wilkes) as belonging to the genus *Reana* of Brignolia but because of its "marvelous vegetative development" it was considered by him to be a different species and was given the provisional name *Reana luxurians*. Pointing out that the genus *Reana* had earlier been described as *Euchlaena* by Schräder, Ascherson in 1876 transferred *Reana luxurians* Durieu to *Euchlaena luxurians* Durieu et Ascherson. It was Ascherson (1875) who first clearly recognized the relationship of *Euchlaena* to *Zea* and the further fact that teosinte is a connecting link, *Bindegliedzwischen*, between *Zea* and *Tripsacum*, thus revealing the relationship between these two genera. Prior to this the Asiatic genus, *Coix*, had been considered the closest botanical relative of maize, and this in turn had lent support to a then still-prevalent idea that maize was a plant of Old World origin. Since Ascherson recognized maize as being related to both *Euchlaena* and *Tripsacum* and since both were unquestionably New World plants, he concluded that the evidence strongly favored the New World origin of maize.

A monograph was written on the genus *Euchlaena* by Fournier (1876, cf. Wilkes), who recognized three species: *E. mexicana* of Schräder, *E. Giovannini* of Brignolia, and a new species, *E. Bourgaei*, based on a collection from the Valley of Mexico in which the staminate inflorescence consisted of a single terminal spike. As Wilkes has pointed out, this is not a valid criterion for establishing a new species, since such inflorescences are common on depauperate plants less than one meter in height. The plant on which Fournier's new species was based was only 60 centimeters high. His monograph contributed little to developing a valid classification and is mentioned here primarily to show the considerable interest in teosinte that prevailed in the third quarter of the nineteenth century.

A more realistic treatment of teosinte is that of Hackel (1890), who in his monograph of the grass family recognized only one species of *Euchlaena*, *E. mexicana* Schräd. This, with two other New World genera, *Zea* and *Tripsacum*, and four Old World genera *Coix*, *Polytoca*, *Schlerachne*, and *Chionacne*, he placed in the tribe *Maydeae*, which was distinguished from the *Andropogoneae* by the separation of male and female inflorescences on the same plant, as in *Zea* and *Euchlaena*, or by the separation of male and female spikelets on the same inflorescence, as in *Tripsacum*. With only minor changes mentioned below Hackel's classification has been generally accepted, especially with respect to the tribe *Maydeae*, by almost all students of maize until fairly recent times. The *Genera Plantarum* of Bentham and Hooker followed Hackel in placing *Euchlaena* in the tribe *Maydeae* but continued to recognize two species, *E. mexicana* and *E. luxurians*. Hitchcock (1922) placed all forms of annual teosinte under one species, *E. mexicana*. Haines (1924) like Hitchcock recognized only one species of annual teosinte but distinguished the fodder teosinte of Central American origin as var. *luxurians*. Randolph (1959), apparently confusing the common name of Florida teosinte with its specific epithet, referred to it as *Euchlaena floridanum*.

A new species of teosinte entered the picture when Hitchcock in 1910 collected a form with rhizomes near Zapotlan, now Ciudad Guzman, Jalisco, Mexico, which he described (1922) as a new species, *Euchlaena perennis* Hitchcock. Collins, revisiting the type locality, found the new form exactly where Hitchcock had previously collected it—a commentary on the accuracy of Hitchcock's description of the locality—and confirmed the supposition that because of its rhizomatous nature it was a true perennial. A complete search of the general area convinced him (Collins, 1921) that all colonies of this form were limited to an area of not more than one square mile. When Cutler visited the locality some years later he found that the perennial teosinte no longer grew there. Subsequently both Randolph and Wilkes have verified Cutler's observation on its disappearance. Consequently, except for plants maintained by students of maize in experimental gardens and greenhouses, the perennial species is now apparently extinct.

A major change in the classification of teosinte was made when Reeves and I (1942), after extensive experimentation with teosinte and its hybrids with maize, proposed that both annual and perennial teosinte be regarded as congeneric with maize, i.e., species of *Zea*, *Z. mexicana*, and *Z. perennis*. There was, as was to be expected, some objection to this change, especially on the part of those

who through years of use had become accustomed to the word *Euchlaena*. Weatherwax (1955) reflected the thinking of this group when he stated, "Fortunately, we believe, this proposal has met with little favor." This statement, however, is somewhat less than accurate, since among modern taxonomists, who are more concerned with natural relationships than with merely putting a plant in a convenient pigeonhole, there has been general approval of the change and of the reasoning for including teosinte in the same genus with maize. Among those who have adopted the new names or otherwise specifically approved the change in nomenclature are Rollins, 1953; Stebbins, 1950; Celarier, 1957; Sinnott, Dunn, and Dobzhansky, 1958; and Shaver, 1962.

Much to the point in this connection are the reasons which Rollins gave for accepting the changes in nomenclature: (1) the two species are fully interfertile; (2) they have similar chromosome morphology; (3) they exhibit normal genetic crossing over, incontrovertible cytogenetic evidence that the two species belong to the same genus. Rollins saw "No compelling evidence on the morphological side for keeping *Euchlaena* and *Zea* as separate genera."

Darlington (1956) would have gone even farther than did Reeves and I in suggesting that *Euchlaena* should be considered as a subspecies of *Zea mays*. As early as 1913, East, later my mentor at Harvard, had expressed essentially this opinion when he stated that maize and teosinte "were no more than diverse types of the same polymorphic aggregation." This is essentially the conclusion that Reeves and I reached as the result of our cytogenetic studies.

When Reeves and I proposed the change in nomenclature in 1942 we were unaware that the transfer of *Euchlaena* to *Zea* had been made almost half a century earlier by Otto Kuntze (von Post and Kuntze, 1904) for reasons essentially the same as ours. It was Wilkes in his extensive review of the early literature on teosinte who discovered this fact. Since perennial teosinte had not yet been discovered when Kuntze made the transfer, Reeves and I are still responsible for changing the name of that species to *Zea perennis*. The correct and only valid Latin names of the two species of teosinte therefore are: annual: *Zea mexicana* (Schräder) Kuntze, and perennial:

Zea perennis (Hitchcock) Reeves and Mangelsdorf.

I can sympathize with those agronomists and botanists who, having become accustomed through years of use to the generic name *Euchlaena*, are reluctant to accept the change. I have had a similar antipathy to the name *Triticum aestivum*, which, according to the rules of nomenclature, must now be applied to common wheat, after having become accustomed since my college years to the name, *T. vulgare*, which is much more accurate than *T. aestivum*, the valid name, and one which would appear by definition to exclude the winter wheats that are as widely grown as the spring-sown or summer wheats. However, if the international rules of botanical nomenclature, which were adopted after many years of study and debate in a worldwide effort to bring order out of what was rapidly approaching chaos in the Babel of naming species, are to serve their purpose, the valid Latin names which they produce are generally to be preferred and used instead of those found for one reason or another to be in conflict with the rules. So much for the nomenclature; much more important than the name is the fact that teosinte is by all odds the closest relative of maize and so lends itself to some interesting and significant genetic experiments.

Geographical Distribution

The teosinte stocks known to maize breeders in 1920 were of the varieties "Florida," which according to the literature had presumably originated in Guatemala, and "Durango," sent from Mexico by H. V. Jackson of Durango. But earlier collections or observations on teosinte's occurrence in Mexico had been made for a number of localities in Oaxaca, Durango, Chihuahua, Chiapas, Chalco, Guanajuato, San Luis Potosi, and Jalisco. Lopez (1908, cf. Wilkes), recognized the range of teosinte as covering the complete length of western Mexico. Collins and Kempton, in an expedition to Mexico in 1921, verified the occurrence of annual teosinte in the vicinity of Durango and Chalco and perennial teosinte in one locality in Jalisco (Collins, 1921).

The existence of teosinte in Guatemala, which had been accepted in the nineteenth century, began to be doubted in the twentieth until Popenoe discovered the Florida type of teosinte growing near Jutiapa. Visiting Guatemala in 1932, Weatherwax confirmed Popenoe's discovery and also obtained seed of teosinte from Huehuetenango in western Guatemala (Weatherwax, 1935). This was the first indication that teosinte grew in western Guatemala as well as in the central part.

In 1935 Kempton and Popenoe, exploring Guatemala, El Salvador, and Chiapas, Mexico, to determine the range of teosinte in that general area found it "a common sight" in the vicinity of Jutiapa and occurring also at Lake Retana, El Progreso, and Moyuta in eastern Guatemala. Proceeding to Huehuetenango in western Guatemala they found in the valley of the Rio Huista teosinte representing the dominant vegetation on thousands of acres at altitudes between 900 and 1,350 meters. On the trail from San Antonio Huista to Jacaltenango they reported "several thousand acres of little but teosinte are seen" (Kempton and Popenoe, 1937). The following year Collins, Kempton, and Stadelman (1937) visited the Sierra Madre Occidental of Chihuahua, where they discovered teosinte in the two localities, Cerro Prieto and Nobogame, which had been mentioned by Lumholtz.

In 1943 Jenkins discovered teosinte near the newly surfaced road from Mexico to Acapulco at 329.5 kilometers, approximately opposite the village of Palo Blanco. Wilkes has recently found an earlier reference to this population in the Mexican literature. In his monograph on the Rio de las Balsas, Hendrichs (1946, cf. Wilkes) stated that teosinte, known locally as "atzitzintle," is common throughout the Sierra but always near houses or around fields. Seed of teosinte from the Rio Balsas was collected by Hernandez-Xolocatzi in connection with the maize-collecting program of the Rockefeller Foundation (Chapter 9) and was supplied by him to a number of research workers in the United States.

The known range of teosinte in Guatemala was extended by Melhus and Chamberlain (1953) when they discovered teosinte at the headwaters of the Rio de las Esclavos. They also obtained reports of the occurrence of teosinte growing wild in the valley of the Rio Cuilco thirty miles southwest of San Antonio Huista, where it was said to occur under the name "madre de maiz." Wilkes, searching the Cuilco area in 1964, found no teosinte and learned that the same name is applied to a broom-straw variety of sorghum. Another

instance of the name being applied to a plant having no relationship to corn is mentioned in Kempton and Popenoe (1937).

The first report of teosinte south of Guatemala is that of Standley (1950), who found it in Honduras growing in cultivated and abandoned fields at San Antonio de Padua, a small village about eight miles from Pespire, where it was known as "maiz cafe" or "maiz silvestre." The small seeds were said to be often collected, roasted, and mixed with coffee beans as a coffee substitute, hence the name "maiz cafe."

In his extensive travels through Mexico, Guatemala, and Honduras, Wilkes found teosinte in the majority of localities in which it had previously been reported or collected. Adding his own collections, he prepared a map which shows teosinte to occur from southern Honduras to northern Mexico. It may be significant that virtually all of the sites are on the Pacific watershed. A conspicuous exception is the archaeological specimen found in a once-inhabited cave in Tamaulipas and described in Chapter 14. This archaeological teosinte may seem to support the suggestion of Weatherwax (1935) that teosinte was once more widespread in its range than it is now. The teosinte of Honduras, however, is consistent with the conclusion which Reeves and I reached that teosinte is a comparatively new species which is still spreading.*

The extensive collections of teosinte which Wilkes was able to make in his travels with the Land Rover over parts of Mexico, Guatemala, and Honduras combined with his observations in the field and the data he assembled from plants grown at the Waltham Field Station enabled him to recognize six more or less distinct races of teosinte, which are described later in this chapter.

Morphology

In general appearance teosinte is similar to maize, and indeed when the two species are growing together in the maize fields in the

*Wilkes, who has studied teosinte over its entire range, is of the opinion that it may have been spreading up to or slightly after the time of the conquest but that there is now evidence of it having become extinct in some localities since 1850 and even more recently since 1960. Recently Wilkes (1972) has warned that some populations of teosinte in Mexico are threatened by rapid extinction.

Valley of Mexico even the keen eyes of the Mexican Indians cannot distinguish one from the other before they flower (Fig. 3.1). Even after flowering there is a marked similarity in the staminate inflorescences, the tassels, although those of Guatemalan teosinte lack the central spike characteristic of maize and in the Mexican teosinte this structure is not conspicuous. Given abundant space and adequate soil fertility, the plants of most races of teosinte produce tillers more profusely than those of maize and the tillers are often as tall as the main stalk. Teosinte also has a tendency to produce branches at the upper nodes of the plant, in some cases at all the nodes. Under less favorable conditions, however, the strong branching tendency may not appear.

The most striking differences between maize and teosinte are found in the lateral pistillate inflorescences. In maize these are usually borne singly, although the potential for developing secondary spikes is present in many races. Individual spikes of maize are many ranked, of teosinte two ranked; the maize ear is solid, the teosinte spike is brittle and disarticulates at maturity; the caryopses of maize—except those of pod corn—are not enclosed, those of teosinte are enclosed in fruit cases that comprise segments of the rachis and indurated lower glumes.

Perhaps the most fundamental morphological difference between teosinte and maize is in the pistillate spikelets, which are solitary in teosinte and in maize occur in pairs (Fig. 3.2). The difference is the result of the abortion in teosinte of one member of the pair, and in this respect teosinte is morphologically not more primitive than maize but more highly specialized. It is significant in this connection that the only contrasting character distinguishing teosinte and maize which behaved as a Mendelian unit character in the maize-teosinte crosses studied by Collins and Kempton (1920) were the paired and single spikelets; the paired spikelets were dominant in the F_1 and in the F_2 segregated in a ratio of approximately 3:1. Langham (1938) obtained similar results. Teosinte and maize are virtually identical in the diagnostic features of the leaves. The short cells of the epidermis form long rows over the veins, and the stomata have triangular subsidiary cells. The presence of cross-shaped silica bodies and balanoform microhairs indicate an affinity to the *Andropogoneae* (Metcalfe, 1960).

FIG. 3.1   A plant of teosinte, the closest relative of maize, flowering in October in Texas from an August sowing. Teosinte resembles corn in its chromosome number, n = 10, and in having its staminate and pistillate spikelets borne in separate inflorescences. It differs from corn in its tendency to produce numerous tillers, in having a strong photoperiodic response, and in the nature of its "ears," which are illustrated in Figure 3.2. This plant is of the "Florida" variety, which is similar to the race "Guatemala" described by Wilkes, 1967.

Physiological Characteristics

A striking difference between teosinte and maize in a physiological character is in their response to the photoperiod. Teosinte is a short day plant. Under continued long days it becomes profusely tillered and assumes an almost perennial form of growth. Maize is generally regarded as day neutral, although short days are effective in speeding floral initiation, especially in tropical varieties, some of which do not flower before frost in Massachusetts without short-day treatment. Before the phenomenon of photoperiodism was well understood the flowering in teosinte during the growing season of northern latitudes in time to mature seed before frost was unusual. At our experimental field at Forest Hills in Boston it flowered about the middle of October, which is also the average time of the first killing frost. Soon after the publication of the classic studies of Garner and Allard on response to photoperiod, Emerson (1924) successfully induced flowering in teosinte by exposing it to only ten hours of light daily. Durango and Chalco teosintes showed silks almost exactly thirty days after treatment began. Florida teosinte required about forty-five days of treatment to bring it into silking. Short-day treatment is now standard procedure among geneticists conducting experiments with teosinte and its hybrids with maize. In my experiments in Texas, teosinte plants growing in metal pails were placed on a platform truck which was wheeled into a dark house at five each evening and out at seven in the morning. Our procedure in Massachusetts is even more simple. The plants are grown in the field and are covered each evening with ordinary metal trash barrels which are removed the next morning. Following an explosion in the Harvard-MIT electronic accelerator in 1965 an aerial photograph which the *New York Times* published of the scene showed, as one of the conspicuous features, rows of trash barrels in a garden adjoining the accelerator. Here one of my students, Ramana Tantravahi, was subjecting several varieties of teosinte to short-day treatment as the first step in attempting to cross teosinte and *Tripsacum*. The "trash-can garden" aroused the curiosity of one television news broadcaster, who had Tantravahi briefly on his program that evening explaining the purpose of the trash barrels. Another minor incident associated with the trash barrels occurred when the chief of Harvard's police department, aware of the nature of the experiment, forbade the Cambridge Fire Department to move its trucks into the area occupied by the trash barrels and so saved the graduate student's summer experiment from disaster. Seldom in recent academic history has a student been more grateful to a police officer.

Another category of physiological characteristics involves susceptibility to infection by pathogenic fungi. Teosinte and maize are usually similar in this respect, while *Tripsacum* often differs from the two. With respect to attack by insects as with infection by fungi teosinte is similar to maize. Of more than seventy-five species of insects known to feed on maize in Guatemala, most but not all have been found feeding also on teosinte. Only a small percentage of these also feed on *Tripsacum* (Painter, 1955).

Chemical Analyses

The earliest nineteenth-century analyses of teosinte were concerned with sugar content of the plant. Sugar is produced in the vegetative culms, but compared to sugar cane the yields are poor. Later analyses of the seed and whole plant indicate that as a fodder plant teosinte is rich in fats and carbohydrates and relatively low in protein when compared to other fodder plants.

The protein content of teosinte seeds is about twice the average of corn, and Sosa (1935, cf. Wilkes) has suggested their use as a human food. However, since it requires about six teosinte seeds to weigh as much as an average corn kernel the suggestion seems impractical. Maize-teosinte hybrids have popcorn-like kernels with an intermediate protein content. The F_2 generation is highly variable with respect to protein content, and there is no significant correlation between segregation for protein content and morphological characters distinguishing the two parents.

Teosinte has the same amino acids as does maize but in levels per unit weight about twice the average values for maize. Methionine, which is often deficient in the predominantly vegetarian diet of underdeveloped countries, is more abundant in teosinte than in maize,

FIG. 3.2 The pistillate spikes of teosinte differ from ears of corn in being distichous instead of polystichous and in bearing solitary instead of paired spikelets. The spikes usually bear 6 to 12 kernels, each enclosed in a shell consisting of a rachis segment and a lower glume. Both structures are highly indurated and are about as hard as the shells of acorns. The segments disarticulate readily at maturity, furnishing an effective means of dispersal. The inheritance of the characters in which teosinte differs from corn involves the majority of their chromosomes.

FIG. 3.3 The author's wife examining a plant of teosinte in an abandoned cornfield near the village of Chalco, Mexico, 1943. Of the six races of teosinte described by Wilkes (1967) this is the most maize-like.

23 Teosinte, the Closest Relative of Maize

and Melhus et al. (1953) have suggested breeding teosinte-modified corn inbreds for a higher methionine content. Attempts in Hungary to accomplish this have met with little success; protein content dropped sharply in the backcrosses to maize (see Wilkes). Other experiments in improving protein content through the introduction of teosinte germ plasm have been made in the U.S.S.R. (see Wilkes). There is little doubt that varieties and hybrids of corn with higher protein content could be developed through the use of teosinte germ plasm, but whether this could be accomplished without some sacrifice in total yield of grain is another question.

Resemblances in Teosinte and Maize Chromosomes

The resemblances between teosinte and maize are not confined to external morphological characteristics; one of the most remarkable similarities between the two species is in the number, morphology, and behaviour of their chromosomes.

Chromosome Numbers

The first cytological studies of teosinte are those of Kuwada (1915), who found the somatic chromosome number of Florida teosinte to be 20. The haploid chromosome numbers in annual and perennial teosintes were found by Longley (1924) to be 10 and 20 respectively. He concluded that the basic chromosome number in *Zea* and its relatives is 10 instead of 12, as had been suggested by Kuwada (1911). These numbers have been confirmed by later investigators for the same or additional varieties of teosinte (Beadle, 1932a, b; Reeves and Mangelsdorf, 1935; Arnason, 1936; O'Mara, 1939; Ting, 1964, 1967; Wilkes, 1967). There is no doubt that the basic chromosome number of teosinte is the same as that of maize.

Chromosome Lengths

In their lengths the chromosomes of teosinte are quite similar to those of maize except that in F_1 hybrids certain teosinte chromosomes appear to be slightly longer than their maize homologs. Unequal length in a pair of chromosomes was first noted by Kuwada (1915) in a hybrid of Florida teosinte and maize. In hybrids of maize with several Mexican and Guatemala teosintes, Longley (1937) found numerous configurations in which the teo-

sinte chromosome appeared to be slightly longer than its maize homolog, perhaps because of the presence of heterochromatic knobs. Later (1941) Longley showed that the presence of a knob did increase the length of the chromosome arm. The significance of the knob is discussed below.

Knobs

Among American *Maydeae* some races or species of maize, teosinte, and *Tripsacum* bear deeply staining bodies thought to consist of heterochromatic material and appropriately called "knobs" on some or all of their chromosomes. These vary in number, size, and position on the chromosomes, some being terminal and others internal. They do not, however, occur at random in the chromosomes but only in certain positions and never in close proximity to the centrometers. The number of positions is fairly large but is limited. It is significant that some of the knob positions in maize chromosomes are the same or approximately the same as those in teosinte chromosomes (Longley and Kato, 1965). This was first noted by Longley (1937), who found that the internal knobs on teosinte chromosomes were in about the same positions as those of the maize homologs. Longley also showed that the Mexican teosintes in general were more maize-like in their chromosome knobs than the Guatemalan teosintes, which have more terminal knobs than either maize or Mexican teosinte.

Because chromosome knobs are never found in close proximity to the centromeres, Longley (1939) postulated a gradient control over knob formation. He concluded that the knob-forming region is farther removed from the centromere in teosinte than in maize. The fact that chromosomes of southern Guatemalan teosintes have terminal knobs like those of *Tripsacum* Longley (1941) regarded as evidence that *Tripsacum* and teosinte had diverged from a common ancestor. Teosinte from northern Guatemala, although possessing terminal knobs, also has internal knobs on the four shortest chromosomes. Mexican annual teosintes have knobs in the same internal positions as maize. These facts Longley accounted for by postulating mutations in the knob-forming gradients, first to the internal positions of the four shortest chromosomes, later to all of the internal positions occurring in the Mexican teosintes.

In connection with the study of variation in chromosome knobs in Guatemalan maize varieties, we, James Cameron and I, took issue with Longley's hypothesis of gradient control, pointing out that recombination of the genetic factors that he postulated would create a great variety of new positions; there is no evidence that this occurs. Rhoades (1955) recognized the fact that knobs are not distributed at random over the ten chromosomes, but he also emphasized the fact that there is no experimental evidence in support of Longley's gradient hypothesis.

Reeves and I in our 1939 monograph pointed out that teosinte is intermediate between maize and *Tripsacum* in the number and position of its chromosome knobs and regarded this as part of the evidence supporting our hypothesis of the hybrid nature of teosinte. Randolph (1955) criticized this hypothesis on the grounds that there is no way in which the terminal knobs of *Tripsacum* and Guatemalan teosinte could have become internal—there is no known mechanism by which the terminal part of a chromosome can assume an internal position. To this we replied (1959) that on any theory of the origin of maize, the change in terminal positions of knobs in *Tripsacum* to the internal positions characteristic of maize must somehow have occurred, and thus all theories, not only ours, are faced with the same problem of accounting for the change.

Randolph (1955) also contended that the chromosome knobs of maize could not have been derived from *Tripsacum* because in hybrids between these two genera, there is little pairing between their chromosomes. There is now direct evidence of the transfer of knobs of *Tripsacum* to maize chromosomes, and it was one of Randolph's students, Marjorie Maguire, who first provided this evidence. In a derivative of a maize-*Tripsacum* hybrid a segment of a chromosome from *Tripsacum* paired regularly with a segment of chromosome 2 of maize, and this was accompanied by crossing over between the segments. One of the products of such crossing over was the transfer of a knob from the *Tripsacum* chromosome to the homologous segment of maize (Maguire, 1957). Galinat and his associates have reported other instances of transfers of knobs from *Tripsacum* to maize chromosomes.

In summary, there are three highly significant facts about the chromosome knobs of teosinte which should be kept in mind in considering theories on the relationship of teosinte to its relatives, maize and *Tripsacum*: (1) Guatemalan teosintes have many knob positions similar to those which occur in *Tripsacum*; (2) Mexican teosintes have some knob positions similar to those occurring in maize; (3) there is evidence that in maize-*Tripsacum* hybrids the knobs of *Tripsacum* can, through crossing over or some form of chromosome exchange, be inserted into maize chromosomes.

Chromosome Affinities

The similarity between teosinte and maize chromosomes in their lengths and positions of their chromosome knobs is also expressed in their pairing affinities in maize-teosinte hybrids. In such hybrids the pairing is usually regular and complete except when chromosome inversions or terminal loop configurations are involved or when the cross is with perennial teosinte, which has twice the normal haploid number of chromosomes. The early observations of Kuwada (1919) and Longley (1924) have been repeatedly confirmed by numerous investigators.

Arnason's (1936) studies were unusual in being based in part on hybrids of teosinte with maize homozygous for known reciprocal translocations. In such crosses structural differences between maize and teosinte chromosomes might modify the cross-shaped configurations expected in translocation heterozygotes. His crosses of translocation stocks with Florida teosinte indicated perfect pairing of chromosomes 1, 2, 6, and 7 and nearly perfect pairing of chromosome 5. Structural differences appeared to prevent chiasmata formation in chromosomes 8 and 9. Hybrids of Durango teosinte with translocation stocks indicated normal pairing in chromosomes 1, 2, and 6 but, as was the case in the Florida teosinte crosses, aberrant behavior in chromosomes 8 and 9. The aberrant behavior in chromosomes 8 and 9 was probably due to chromosome inversions discussed below.

The behavior of the chromosomes in hybrids of maize with perennial teosinte was studied by Longley (1924), who observed trivalents, bivalents, and univalents during meiosis and who later (1941) found partial synapsis between homologous chromosomes in the F_1 hybrids of perennial teosinte and maize, while in hybrids of perennial teosinte with teosinte from southern Guatemala there was little or no synapsis. Both Longley (1924) and Emerson (1929) found the triploid hybrid of maize and perennial teosinte to be partially sterile—which is not surprising since maize triploids are themselves partially sterile—while tetraploid hybrids of perennial teosinte by tetraploid maize were highly fertile (Emerson and Beadle, 1939; Longley, 1934).

Nonpaired Regions and Nonhomologous Associations

Although the pairing between teosinte and maize chromosomes is usually regular and complete except for inversions, there are other minor irregularities. Longley (1941) found unpaired terminal regions in several chromosone pairs. Univalents at diakinesis are an indication of failure to pair at pachytene. Ting (1964) determined the frequency of univalents in crosses of six Mexican teosintes with the same inbred strain of maize. The results were as follows:

Teosintes crossed with Wilbur's Flint	Percentage of microsporocytes having no univalents
Chilpancingo	83.8
Arcelia	98.7
Xochimilco	89.4
Chalco	91.3
Durango	99.2
Nobogame	87.7
Average	91.7

The percentage of microsporocytes having no univalents is an indirect measure of the completeness of the previous pairing. On this basis the chromosomes of Durango teosinte, with 99.2 percent of the microsporocytes of its hybrids with maize having no univalents, would appear to be almost completely homologous to those of maize, while those of Chilpancingo, with 83.8 percent univalents in the microsporocytes of the hybrids, would have the least homology of the Mexican teosintes.

Wilkes (1967) in similar hybrids of various teosintes with Wilbur's Flint found pairing and synapsis between homologous maize and teosinte chromosomes at pacytene to be nearly complete, except around knobs and in the terminal arms of chromosomes 8 and 9.

Inversions

An apparent exception to the remarkable similarity between teosinte and maize chromosomes is in the occurrence of inverted segments in certain of the teosinte chromosomes. Both Arnason (1936) and Longley (1937) observed a lack of homology between chromosomes 8 and 9 of Florida teosinte and their maize homologs. This was probably due to the inversions which have subsequently been identified on these chromosomes. Beadle (1932b) found crossing over between yg_2 and wx on chromosome 9 of Durango and Florida teosintes to occur infrequently and suggested that an inverted segment might be responsible, although no direct cytological evidence of such an inversion was then available. O'Mara (1942) observed an inversion on the short arm of chromosome 9 of Florida teosinte and another possible inversion on chromosome 8. Ting, who has made an intensive study of pachytene chromosomes in hybrids of various teosintes with an inbred strain of Wilbur's Flint as the common maize parent, found inversions in the majority of teosinte varieties studied involving most frequently chromosomes 8 and 9 and less frequently chromosomes 3 and 5. Ting also assembled data on the frequency of dicentric chromatid bridges and acentric fragments at anaphase. These are generally regarded as evidence of previous crossing over within an inversion loop. It is significant that the majority of these inversions are homozygous in teosinte varieties.

Wilkes (1967) has observed terminal loop configurations on chromosomes 8 and 9 in hybrids of Wilbur's Flint with teosinte from Chalco and the Central Plateau. He has also observed bridges and fragments in lines in which no inversion configurations were apparent, perhaps the result of inversions too short to be otherwise detected. An inversion on the short arm of chromosome 8 of approximately the same length as that of the chromosome 8 inversion in teosinte varieties, and probably the same inversion, has been found by McClintock (1933, 1960) in North American maize varieties, in four Mexican varieties, and in one from Bolivia. This inversion has probably spread from teosinte to maize rather than in the reverse direction because it is common in teosinte and comparatively rare in maize and because it has not been reported in the ancient indigenous Mexican varieties, Chapalote and Nal-Tel.

Both Ting and Wilkes have speculated on the evolutionary significance of these loop configurations in teosinte, without, however, reaching any firm conclusions. Ting (1964) suggested that they confer a selective advantage by preserving well-adapted gene combinations within the inverted segments, and Wilkes emphasized the fact that they are most frequent in populations where hybridization with maize is most extensive. Since I still regard teosinte as essentially maize in which the characteristics that distinguish it from maize are determined not by single genes but by blocks of genes, possibly originally from *Tripsacum*, I see these inversions as a mechanism for preserving some of these blocks of genes.*

Sterility of F₁ Hybrids

The close relationship of teosinte and maize and the similarities in their chromosomes are reflected in the high degree of fertility of their F_1 hybrids. The degree of fertility is usually estimated by determining the percentage of empty pollen grains produced by the hybrid. Data on fertility of hybrids of annual teosintes with maize were first reported by Beadle (1932b), who found fertility to be approximately normal in hybrids of Durango and Chalco teosintes with maize but only 68 percent in hybrids of Florida teosinte with maize, a hybrid which we now know to involve an inversion. Reeves and I (1939) confirmed Beadle's observation on hybrids of Durango and Florida teosintes with maize. Fertility determinations have subsequently been made by O'Mara (1942), Rogers (1950), Lambert (1964), and Wilkes (1967).

*Dr. Marcus M. Rhoades has suggested in a letter that if these inversions do indeed contain blocks of genes differentiating teosinte from maize they should show strong linkage with characteristics distinguishing the two species. We have no direct evidence on this point. However, we have found linkage of teosinte characters with marker genes involving chromosomes on which inversions have been identified: *Wx* on chromosome 9 in a Florida teosinte—maize cross (Mangelsdorf and Reeves, 1939) and with marker genes on chromosomes 8 and 9 in Nobogame-maize and Durango-maize crosses (Mangelsdorf, 1947). With respect to these two chromosomes, Rogers (1950) found strong linkage with only one teosinte characteristic, photoperiodic response, and this only with the marker gene on chromosome 8. More research on this problem is obviously needed.

To the extent that fertility is an indication of closeness of relationship it would appear that the Mexican teosintes are quite closely related to maize, the Guatemalan teosintes slightly less so, and certain teosinte varieties still less so in relation to other teosinte varieties. The data of Rogers (1950) illustrate this fact especially well since the crosses with maize of four teosinte varieties which he studied had a higher fertility, with one exception, than the hybrids of the same four teosinte varieties with other teosinte varieties. In the case of Durango, Nobogame, and Chalco teosinte, their crosses with other teosintes had substantially lower fertility than their crosses with maize. In the case of Florida teosinte, there was little difference in the crosses studied by Rogers, but Beadle had earlier (1932) found the cross between Durango and Florida teosinte to have a noticeably higher degree of sterility than the hybrid of Florida teosinte and maize.

How are these results to be explained? They are probably the product, at least in part, of the inversions carried by the teosinte varieties, not only the longer ones which have been identified cytologically by Ting and others but also the cryptic ones postulated by Wilkes, which are too short to be detected in pachytene configurations but which occasionally make their presence known by the occurrence of bridges and fragments at anaphase. There are probably more heterozygous inversions in teosinte × teosinte crosses than in teosinte × maize crosses. Teosinte varieties may also differ from one another in certain compound loci, which in hybrids can give rise to "unequal" crossing over, disturbances in function as one of the products. The important point and to me a highly significant one is that varieties and races of teosinte are diverging from each other while continuing to maintain the closest possible relationship with maize.

Races of Teosinte

As a result of his intensive and extensive studies of teosinte both in the field and in experimental cultures, Wilkes recognized and described six more or less distinct races: Nobogame, Central Plateau, Chalco, Balsas, Huehuetenango, and Guatemala. Each of these races possesses a number of characteristics which although not unique to that particular race have a higher frequency of occurrence in that race than in others. In this

respect the races of teosinte are the counterparts of the races of maize discussed in Chapter 9.

The six races can be distinguished by the shapes of their fruit cases, which consist of rachis segments and indurated lower glumes, and by the sizes of their mature caryopses. Teosinte from southern Guatemala has trapezoidal fruit cases; races from northern Guatemala and from Mexico have fruit cases which are distincly triangular. These races may differ in the shape of the triangle, which varies from slightly rounded to sharply pointed or blunt. The "maizoid" (most maize-like) races have much larger caryopses than the "tripsacoid" (least maize-like). There are also differences with respect to the presence or absence of a central spike in the tassel, the degree of tillering, and in other vegetative features of the plants, but these are more subject to environmental influence than are the fruit cases.

The general growth habit and the pistillate spikes and mature fruits of the six races are illustrated in Figures 3.4 to 3.10.

The most maize-like race of teosinte is, as might be expected, the one which grows as a weed in the maize fields in the general vicinity of Chalco in Mexico and survives as a "mimic" of the maize. Its plants are tall, thick culmed, wide leaved and sparsely tillered. The tillers are shorter than the main stalk. The least maize-like teosinte is that of southern Guatemala, which is easily distinguished from other races by its trapezoidal fruit cases. Plants of this race tiller profusely, and there is no distinct central stalk. The tassel lacks a central spike. This race resembles *Tripsacum* more than any of the others. Almost equally primitive, however, is the race from the Balsas region. This race has the smallest fruit cases, the most slender culms, and most narrow leaves of any of the races. In its general aspect it looks more like a "wild" plant than any of the other races. Yet it is more maize-like than Guatemalan teosinte in having a recognizable main culm and a distinct central spike in its tassel.

Teosinte's Role in the Origin of Maize

The theories mentioned in Chapter 2 and in this chapter that teosinte is the primitive ancestor of maize are in general rather crude attempts to explain the origin of maize and

FIG. 3.4 Chalco is the most mazoid race of teosinte. The plants are tall, thick-culmed, wide-leaved, sparsely tillered, and easily cultivated in the temperate zone. The tillers are shorter than the well-branched main culm. The central spike of the moderately branched tassel is many ranked and as long as the branching space. The fruitcase is large and triangular, the apex appearing pointed or crimped because the rachis is so condensed. This unique maize-mimetic race of teosinte is found growing in the maize fields of the Valley of Mexico. Plant scale: 1 cm. = 12.4 cm. (Wilkes, 1967.)

CM   1   2

FIG. 3.5 The weedy form of the race Central Plateau is more luxuriant than the wild form of this race. The plants are tall, sparsely tillered, and possess both large seeds and wide leaves. In the weedy form the fruitcase is triangular, twice the size and weight of the wild form, and often appears crimped at the apex. This form of the race is adapted to cultivated fields and is found at several stations on the Central Plateau. Plant scale: 1 cm. = 11.3 cm. (Wilkes, 1967.)

FIG. 3.6 The wild form of the race Central Plateau is less maizoid and not as luxuriant as the weedy member of this race. The plants are of medium height, moderately tillered, and well branched. The tassel is moderately branched, and a central spike is present, but not prominent. The fruitcase is triangular, and the kernel is half the weight, about 0.025 gm., of the form adapted to cultivated fields. This form of the race is found on limestone outcroppings and other sites protected from grazing animals on the Central Plateau. Plant scale: 1 cm. = 9.6 cm. (Wilkes, 1967.)

CM   1   2

FIG. 3.7 Nobogame is the most northern of the naturally occurring races and is the earliest flowering when cultivated in the temperate zone. The plants are short, moderately tillered, with sparsely branched tassels. The triangular fruit-cases are small and more pointed at the apex than those from the Rio Balsas. The native habitat of this race is along the margin of maize fields, in willow thickets and other sites protected from grazing animals in the Tarahumare Valley of the Sierra Madre Occidental. Plant scale: 1 cm = 8.6 cm. (Wilkes, 1967.)

CM   1   2

FIG. 3.8 Balsas is the most tripsacoid of the Mexican races of teosinte. The plants are short, thin-sulmed, narrow-leaved, profusely tillered, and of medium to late maturity. The tillers are often taller than the main culm. The tassel branches are lax and droop, while the central spike is only slightly more condensed than the branches. The fruitcases are triangular with a blunt apex and are the smallest of any race. The race is native to the foothills of a large region in Michoacan and Guerrero, Mexico drained by the Rio Balsas. Plant scale: 1 cm. = 10 cm. (Wilkes, 1967.)

CM 1 2

FIG. 3.9 Huehuetenango is the latest flowering and most profusely tillered of all the teosintes with a triangular rachis-segment. The plants are tall, erect, and vegetatively prolific. Some of the tillers are often taller than the main culm. A central spike is borne in the tassel but is shorter than the branching space. The race is native to the hillsides and deserted milpas along the Rio Huista and Rio Dolores in the Department of Huehuetenango, Guatemala. Plant scale: 1 cm. = 13.1 cm. (Wilkes, 1967.)

CM   1   2

FIG. 3.10 Guatemala teosinte is easily distinguished from all other races by a trapezoidal rachis-segment and a perennial-like growth habit. There is no central culm in this race, and several fruiting canes appear to be equally well developed. The basal nodes do not elongate, increasing the *Tripsacum*-like appearance of this race. There is no central spike in the tassel, which is borne erect and does not droop, as in all other teosintes. The race is found on the broad valleys and hills of the Departments of Jutiapa, Jalapa, and Chiquimula in southeastern Guatemala and a single population in Honduras. Plant scale: 1 cm. = 11.5 cm. (Wilkes, 1967.)

CM 1 2

are based primarily on a single fact, that teosinte is obviously the closest relative of maize. Because of this close relationship any theory on corn's origin must also account for teosinte.

Although not the first to recognize the close relationship of teosinte and maize—credit for that must go to the Mexican Indians—Ascherson apparently was the first to postulate that maize is a descendant of teosinte. He (1880) interpreted the maize ear as the result of the fusion of four-rowed spikes and in teosinte saw the starting point for such a fusion phenomenon.

The phenomenon of fusion was also relied upon by Hackel, who regarded the ear of maize as a "monstrous or teratological development" resulting from a cluster of teosinte spikes growing together to form a spongy unbranched cob on which somewhat indistinct double rows separated by shallow furrows run lengthwise.

Quite similar to Hackel's was Harshberger's (1893) conception of the maize ear, and in *Zea canina* he saw a series of ears that might represent steps in the fasciation from a four-rowed to a many-rowed cob. "The arrangement of the grains," he stated, "corresponds to the separate spikes of the consolidated cob. These structures and teratological arrangements point to the probable union of several spikes into a thick, fleshy axis, with grains on the circumference, each paired row limited at the side by a long shallow furrow, a row corresponding to a single spike of *Euchlaena* or *Tripsacum*." When *Zea canina* turned out to be a hybrid of maize and teosinte, Harshberger modified his theory on the origin of maize but continued to regard the four-rowed ear as a starting point of modern maize. He proposed two alternative theories on the origin of maize: (1) that it originated from a cross of teosinte with an extinct closely related grass, and (2) that maize is the product of a cross between teosinte and a cultivated race of teosinte developed, perhaps, under irrigation.

Schuman (1904), a former student of Ascherson, noted that cultivated plants often differ from their wild relatives by teratological development of those organs most useful to the cultivator. Since maize and teosinte are alike in having terminal staminate and lateral pistillate inflorescences but since maize differs greatly from teosinte in its grain-bearing ear

—the structure for which it is grown—Schuman concluded that maize originated from teosinte as a teratological form which had become stabilized during a long period of selection. Like Ascherson, Schuman attributed the teratological structures, the maize ears, to the fusion of lateral branches of teosinte with the central spike.

The eminent German morphologist Goebel (1910) postulated a sequence of morphological forms starting with an ancestor with distichous spikes similar to that visualized by Harshberger except that it was perfect flowered and not strictly pistillate. Once the inflorescence became pistillate and polystichous and assumed a lateral position, it also became thickened because of its more favorable position on the plant for obtaining nutrients. Goebel's theory, like the previous one and the later one of Worsdell (1916) involving an ancestral form of distichous spikes, relied on fasciation to explain the ear of maize.

We now know that all of the theories involving the fusion of distichous spikes—those of Ascherson, Hackel, Harshberger, Schumann, Goebel, and Worsdell suffer from one fatal defect: the ear of maize is completely lacking anatomically at maturity or at any previous stage in its ontogeny of any evidence of fusion. Weatherwax pointed this out as early as 1918, and later investigators have confirmed his earlier observations (Weatherwax, 1935; Mangelsdorf, 1945; Kiesselbach, 1949; and Bonnett, 1953, 1966; Reeves, 1950, 1953b; and Laubengayer, 1948). Especially convincing are the splendid photographs of Bonnett showing successive stages in the ontogeny of the maize ear which are completely devoid of any evidence of fusion. Also significant is the fact that Laubengayer, working as a graduate student under Eames, who at one time was convinced that the ear of maize is the product of fusion, was unable to find any evidence to support his mentor's conviction.

It may be appropriate here to point out that the homology of the ear and central spike of the tassel is now generally recognized. If there were previous doubts about this homology they should have been resolved by Bonnett's (1953) studies, which show the two structures to be indistinguishable in the early stages of their development. And since they are homologous, if the ear were the product of fusion it

would seem that the central spike must also be a fusion product. Even the most ardent proponents of the fusion theory have never gone so far as to postulate this. The fact is that there is no convincing evidence that either the ear or the central spike is a product of fusion, and since all morphological theories involving teosinte spikes as the starting point also involve fusion, the whole idea of teosinte as the ancestral form is not supported by the evidence from morphology. It may also be appropriate to repeat here that morphologically teosinte is in some respects less primitive and more highly evolved than maize. The strongly indurated tissues of the rachis segments and the lower glumes of teosinte fruits represent a form of specialization. Even more significant, however, is the complete supression in teosinte of one member of a pair of pistillate spikelets along with its vestigial lower floret.

Collins recognized the importance of this specialization in teosinte when (1912) he combined elements of previous theories in formulating the hypothesis that "maize originated as a hybrid between teosinte and an unknown grass belonging to the tribe *Andropogoneae*." This hypothesis is based on two assumptions: (1) "that the prototype of maize was some relative or ancestor of teosinte" and (2) "that the prototype should be sought in a grass possessing the characteristics of tunicate maize." He considered maize to be the only member of the *Maydeae* which bridges the gap between the specialized floral characteristics of the staminate and pistillate spikelets of teosinte and the flowers, resembling those of certain *Andropogoneae*, of homozygous pod corn. Collins concluded that a hybrid origin accounted best for the natural variability of maize and the "pronounced evidence of teosinte blood" in the maize of Mexico and Central America.

In advancing this hypothesis Collins came close to anticipating two parts of our tripartite theory (1939): (1) that pod corn has had a role in the evolution of maize and (2) that there is teosinte "blood" in the maize of Mexico and Central America. As already mentioned, we regard pod corn as the ancestral form and "teosinte blood" as occurring in many modern varieties including those of the United States.

Six years later Collins (1918) presented what he considered to be supporting evidence to his theory of the hybrid origin of maize:

(1) maize is intolerant of inbreeding, a common characteristic of hybrids, and is generally dependent on cross-fertilization for normal and vigorous development, and teosinte shows no such intolerance;* (2) the characters that distinguish maize from teosinte do not show Mendelian ratios following hybridization with maize. This makes the hypothesis that maize originated from teosinte by mutation highly improbable. Collins and Kempton later (1920) published data on the segregating generation of a cross between Florida teosinte and Tom Thumb popcorn which showed that of the 33 characters which they studied distinguishing the two species, only one, the paired pistillate spikelets of maize, seemed to show clearcut Mendelian segregation.

Earlier, Collins had studied the pistillate inflorescences of this cross with particular attention to fasciation and failed to find any evidence of it in the formation of the ear. This led to his suggestion that the polystichous spike might be formed by the "yoking" of alicoles—now generally called cupules—accompanied by a shortening and twisting of the spike. Both Weatherwax (1920) and I (1945) presented reasons for regarding the yoking hypothesis of Collins as untenable.

In her book on the origin of cultivated plants, Schiemann (1932) drew heavily upon the several papers of Collins in her own interpretation of the origin of maize, which was even more bold than his. The pedigree chart she prepared shows *Euchlaena perennis* crossing with a sorghum species to form the

*Wilkes has since found that the race Chalco, the most maize-like of the teosinte races, is somewhat intolerant to inbreeding.

ancestor *Zea antigua* of modern maize. *Euchlaena perennis* gave rise to *Euchlaena mexicana*, which in turn hybridized with *Zea antigua* to form the present-day popcorn, an idea which she gained from Kuleshov (1929). The popcorns then crossed with an ancient flour corn to produce the flint corns, and these in turn hybridized with modern flour corn to produce the youngest race, dent corn. Schiemann's personal experience with maize was obviously less than her intimate familiarity with the European small grains on which she is an eminent authority. The genealogy she proposed for modern maize is of interest primarily because it is the most elaborate of all of the theories involving a hybrid origin and it is the first to suggest clearly different levels of teosinte introgression in the evolution of maize.

In 1931 Collins reviewed the various theories concerned with the origin of maize and indicated where on the basis of available evidence each fell short. He objected to Blaringhem's idea of gross mutation because crosses of maize and teosinte showed blending inheritance and not the unit characters expected from mutations. Direct selection from teosinte he regarded as improbable because teosinte with its bony fruit case is not a promising material from which to develop a cultivated food plant. Beadle (1939) has since shown that, when heated, teosinte grains explode like grains of popcorn, shattering their enclosing fruit cases, from which they are then easily separated. Weatherwax's theory of an independent evolution of maize from a perfect flowered form did not account for the high degree of maize-teosinte hybrids. Collins continued to favor his own theory of a hybrid

origin, conceding that the exact nature of that origin remained to be determined. One fact, however, he considered to have become forcibly clear: "The many resemblances between maize and teosinte together with the fact that the two forms interbreed with perfect freedom makes it certain that whatever the origin of maize it must be intimately associated with teosinte or some near relative of that plant."

In the light of recent evidence from archaeological remains described in Chapters 14 and 15, Collins. statement is completely true but not in the sense in which he meant it. The "near relative" of teosinte involved in the origin of cultivated maize is probably wild maize.

In reviewing the theories that have assigned to teosinte a role of one kind or another in the origin of maize, it becomes clear that these have comprised several categories: (1) the origin of the maize ear through the fusion of teosinte spikes first proposed by Ascherson and adopted with various modifications by Hackel, Harshberger, Goebel, and others; (2) the hybridization of teosinte with various unknown grasses (Harshberger, Collins, Schiemann); (3) teosinte and maize descending along independent lines from a common ancestor (Montgomery, Weatherwax). To these we may now add a fourth category involving teosinte: the hypothesis by Reeves and me (1939) that teosinte is a hybrid of maize and *Tripsacum* but that it has subsequently crossed with maize in the development of modern varieties. This theory, discussed at length in the following chapter, has teosinte serving as a genetic "bridge" over which genes from *Tripsacum* have become incorporated into maize.

4 The Genetic Nature of Teosinte

> Any hypothesis which has so much plausibility as to explain a considerable number of facts helps us to digest those facts in proper order, to bring new ones to light, and to make crucial experiments for the sake of future inquiries. Westaway, 1937

Of the three parts of our tripartite theory on the origin of maize the postulate that teosinte is a hybrid of maize and *Tripsacum* is by far the most difficult to test experimentally. The most direct test would be to cross maize and *Tripsacum* and, by growing large populations for a number of generations, produce a new true-breeding hybrid that might be classed as a synthesized teosinte. However, because we knew so little about the species of *Tripsacum* or races of maize that might have served as parents of the postulated hybrid, we regarded the possibility of successfully creating teosinte by hybridization with its putative parents as too remote to justify this line of attack. Instead, we sought by various kinds of experiments involving the hybridization of maize and teosinte to determine the genetic nature of teosinte. Ours were not, however, the first maize-teosinte hybrids to be studied by genetic techniques.

Early Experiments on Maize-Teosinte Hybrids

Because teosinte crosses readily with maize to produce highly fertile hybrids in which the pairing of the chromosomes is remarkably regular and because the two parents of such hybrids differ in a number of conspicuous morphological characteristics, it was almost inevitable that maize-teosinte hybrids would sooner or later be recognized as providing unusually favorable materials for studying the inheritance of differences between genera or species. The first such experiments are those of Collins and Kempton (1920), who studied the inheritance of 33 characters differentiating the two parents in a cross of Florida teosinte and Tom Thumb popcorn. In only one of these, paired versus single spikelets, did the segregation appear to follow a regular Mendelian pattern, the single spikelets of teosinte behaving as a simple recessive. In all other characteristics in which the two parents differed, the F_1 tended to be intermediate and the F_2 results were complicated and irregular, suggesting multifactorial inheritance. Thus even these early experiments provide convincing evidence against the often-repeated suggestion that maize might have been derived from teosinte through a few large-scale mutations.

Another significant result of these early experiments is the finding of numerous correlations between characteristics. No character studied was found to be completely independent of all other characters. Every character measured showed significant correlation with one or more others, which in turn were correlated with still others. A third significant fact not especially emphasized at the time is that in a relatively small F_2 population one plant similar to teosinte in its essential botanical characteristics occurred. This indicates that the number of hereditary units distinguishing the two species is small. The appearance of one of the parental types through recombination in a small population of 127 plants suggests that not more than four principal hereditary units were segregating in the hybrid.

The first experiments involving the inheritance of specific Mendelian characters of maize in hybrids of maize and teosinte are those of Kempton (1924), who crossed maize possessing the recessive characters, crinkly, ramosa, and brachytic with both annual and perennial teosintes. All three of these characters proved to be recessive in the hybrids as they are in maize. Among the F_2 progeny, two of the characters, crinkly and ramosa, segregated in normal monohybrid ratios, whereas the third, brachytic, appeared in only 12 percent of the segregates. Of the many marker genes subsequently employed in maize-teosinte hybrids and studied for various purposes, all have had alleles in teosinte, and the majority have segregated normally in F_2 or back-cross populations. An exception to this general rule are the crosses with perennial teosinte. The reason for the exception here is that this teosinte has twice the normal chromosome number of maize and annual teosinte. Collins and Longley (1935) made ingenious use of this fact when, among seven F_1 hybrids of maize × perennial teosinte, they found one with forty instead of the usual thirty chromosomes of such hybrids, indicating that twenty instead of ten chromosomes had been introduced by the maize parent. By employing the waxy gene as a marker in the maize parent, they estimated the degree of autosyndesis with respect to chromosome 9 in the hybrid by determining the percentage of *wx* pollen—identified by staining red with iodine—a characteristic first discovered by Weatherwax. In a formula devised to calculate the coefficient "t" of autosyndesis, complete autosyndesis was represented by 1 and complete allosyndesis by −1. In the F_1

plants "t" equaled 0.80; F_2, 0.77, and F_3 ranged from 0.54 to 0.82.

Crossing over between Maize and Teosinte Chromosomes

Because teosinte contains the dominant alleles of recessive characters in maize and the recessive alleles of certain dominant characters, it is possible to determine the amount of crossing over between teosinte and maize chromosomes in hybrids. Emerson and Beadle (1932) crossed Chalco, Durango, Florida, and perennial teosintes with stocks of maize carrying two marker genes on each of several chromosomes. Except for the cWx region of chromosome 9 of Florida and Durango teosintes, which is now known to involve an inversion, the percentage of crossing over was of the same order as it is in maize. Actually the average percentages of crossing over are slightly lower in maize-teosinte hybrids than they are in maize. The differences, although not statistically significant individually, are frequent enough to suggest the need of additional experiments in which the environment and residual heredity are as nearly alike as possible in the crosses to be compared.

It was these experiments of Emerson and Beadle, which were known to me before the results were published, that suggested the possibility of determining whether the characteristics distinguishing the two genera are controlled by genes borne on the chromosomes. I assumed at the time, as did most botanists familiar with teosinte, that it represented a good genus, *Euchleana*, distinct from that of maize, *Zea*. Here then was a unique opportunity to study the inheritance and linkage relations of characters distinguishing two genera. The maize parent would contribute the genes marking the chromosomes, teosinte would contribute their alleles, chromosome pairing in the hybrid would be expected to be approximately normal, and crossing over within the chromosomes would be of about the same order as it is in the maize parent.

Linkages of Generic Differences with Marker Genes

Thus it was that in the early thirties as part of a joint project with Reeves, I crossed Florida teosinte with maize stocks having two marker genes on chromosomes 2, 4, and 6 and one marker gene, wx, on chromosome 9, which by that time we knew to contain a region, suspected of being an inversion, in which crossing over was remarkably reduced. F_1 hybrids involving chromosomes 2, 4, and 6 were backcrossed to double recessive stocks; the chromosome 9 hybrid to recessive wx. Crossover genotypes were compared with the noncrossover parental type with respect to a number of characteristics distinguishing the two parental taxa.

The data from these experiments are reported in detail in our 1939 monograph. In brief they show that the genes controlling the characteristics distinguishing the two parents are not distributed at random over the chromosomes. There was little evidence of linkage of such characters with the marker genes on chromosomes 2 and 6 but strong evidence of linkage on chromosomes 4 and 9. We concluded from these results that the differences between teosinte and maize are controlled by a limited number of hereditary units, of which the major ones appeared to be not more than four. Because they were not always inherited intact and because they affected a number of different characteristics, we concluded that these units were not single genes but blocks of genes. Later experiments tended to confirm the conclusion. The data then available indicated that two of these were located on chromosome 4, one on chromosome 9. Because the phenotypic effects of these blocks of genes were similar to the effects of the *Tripsacum* chromosomes in our maize-*Tripsacum* hybrids, we regarded them as blocks of genes translocated to corn from *Tripsacum*, and we considered the inheritance of such blocks to provide strong support for our hypothesis that teosinte is the product of the hybridization of maize and *Tripsacum*—that it is essentially maize into which a small number of blocks of *Tripsacum* genes have been incorporated.

Additional Linkage Experiments

Although the data from these early experiments were consistent with our hypothesis of the hybrid origin of teosinte, they were far from conclusive, so in 1938 in collaboration with John S. Rogers, then my graduate student at Texas A & M, I began a series of experiments to obtain more evidence on the genetic nature of teosinte. These experiments comprised three principal categories: (1) obtaining additional data on the linkage relations of the postulated segments; (2) determining whether the segments we had studied in Florida teosinte also occurred in other teosinte varieties; (3) extracting the segments from teosinte and introducing them into an inbred strain of corn.

The linkage tests were made with Florida teosinte as one parent and involved twelve marker genes on chromosomes 1, 3, 4, 5, 6, 8, 9, and 10. Since our earlier experiments had involved chromosomes 2, 4, 6, and 9 we had now tested linkage on nine of corn's ten chromosomes; only chromosome 7 was not represented in the test crosses. The results showed that one of the segments, but apparently only one—not two as we had previously concluded—is located on chromosome 4 and shows linkage with the alleles of three different marker genes, su, Tu, and gl_3, on that chromosome. A second segment showed strong linkage with the allele of the gene, P, for pericarp color on chromosome 1 and a slight indication of linkage with Bm_2 at the opposite end of this chromosome. A third segment showed a clear-cut linkage with Wx on chromosome 9. The fourth segment was not definitely located by these tests, but there were strong indications of linkage with chromosome 3 and weaker indications of linkages with chromosomes 6 and 10.

Genetic Differences between Teosinte Varieties

To determine whether other varieties of teosinte were similar to Florida teosinte in containing blocks of genes which could be recognized by their effects and mode of inheritance we designed what we considered to be a rather elegant experiment. We crossed Florida, Durango, and Nobogame teosintes with a very uniform inbred strain, Texas 4R-3. We then crossed the F_1 hybrids to another uniform inbred strain, Texas 47R-2. By employing inbred strains we reduced the variation resulting from residual heredity to a minimum, since all of the plants in the backcross generation had half of their genes from 47R-2 and all of their remaining maize genes from 4R-3. Thus the backcross segregates differed only in the genes received from teosinte. Use of a second inbred strain in the backcross created hybrid vigor in all of the segregates and eliminated the deleterious effects of inbreeding, which might have been quite marked had the growing season been unfavorable as it sometimes is in Texas. However, the summer of 1939, when the backcross progenies were grown, was unusually favorable, and there was no selective elimination of genotypes because of drought.

The results of this experiment, which have never previously been published, are set forth in Table 4.1. They show that the backcross progenies involving Florida and Durango teosintes could be classified into the five classes expected if the population were segregating for four segments, and the data fit a theoretical backcross ratio of 1:4:6:4:1 involving genotypes containing 0, 1, 2, 3, or 4 segments, respectively, quite well.

The results in the cross involving Nobogame teosinte were, however, quite different. In the first place, the F_1 hybrids of this teosinte with maize were much more maize-like than the F_1 ears of the other two hybrids and more nearly resembled genotypes classified as containing three segments in the backcrosses involving Florida and Durango teosinte. Also, the segregation in the backcross comprised a more narrow range than in the Florida and Durango backcrosses. It appeared that the Nobogame teosinte contained only three of the segments occurring in Florida and Durango teosinte and the segregation conformed closely to the 1:3:3:1 ratio expected from a three-factor backcross. Repeating these experiments later, Rogers (1950) obtained similar results. There is no doubt that Nobogame is a less "potent" teosinte than other varieties in the genes distinguishing maize and teosinte which it carries. Photographs illustrating the most maize-like and the most teosinte-like ears obtained in these three backcross populations are illustrated in Figure 4.1.

Extracting Blocks of Genes from Teosinte

The third group of experiments which Rogers and I initiated in 1938, some of which thirty years later are still in progress, involved the transfer by repeated backcrossing to the inbred 4R–3 of the individual blocks of genes from Florida, Durango, Nobogame, and a fourth variety of teosinte which had come to us under the name of "New." When the different stocks became relatively isogenic (after three backcrosses following the first cross)

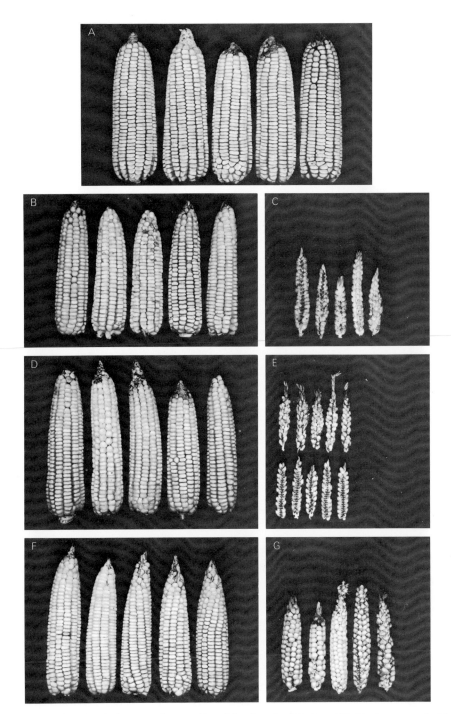

FIG. 4.1   Ears of a single cross of two uniform inbred strains (4R–3 × 47R–2), A, compared with the most maize-like and the most teosinte-like segregates resulting from crossing one of the inbreds with teosinte and backcrossing the F_1 hybrid to the other inbred. B and C (4R–3 × Florida teosinte) × 47R–2; D and E, the same cross involving Durango teosinte (in E five F_1 ears—lower—are included for comparison); F and G, the same cross involving Nobogame teosinte. Note that the most teosinte-like ears from this cross were much more maize-like than the corresponding ears from the two preceding crosses. It was this experiment that led to the conclusion that Florida and Durango teosinte differ from maize by four principal blocks of genes and Nobogame teosinte by only three blocks.

39   The Genetic Nature of Teosinte

TABLE 4.1 Frequency of Classes of Segregates in Backcrosses of Maize Inbred 47R–2 × (Maize Inbred 4R–3 × Teosinte) Compared with Theoretical Frequencies*

No. of segments	Description of ears	Variety of Teosinte in F_1					
		Florida		Durango		Nobogame	
		No.	Theoretical	No.	Theoretical	No.	Theoretical
0	Resembling maize	5	7	9	9	13	18
1	Approaching maize	25	27	30	35	52	54
2	Intermediate	39	40	55	53	61	54
3	Approaching F_1	31	27	38	35	19	18
4	Resembling F_1	8	7	9	9		

*Assuming four segments in Florida and Durango teosintes and three in Nobogame.

TABLE 4.2 Classification with Respect to Presence, Absence, or Intermediate Nature of Teosinte Segments in Backcrosses Involving Segregation for One Segment

Lot no.	Chromosome involved	No. of plants with segment			
		Present	Intermediate	Absent	Total
737	3	56		49	105
742	4	41	21	45	107
743	4	11	12	18	41
744	4	17	8	18	43
745	3	26		27	53
746	3	30		23	53
749	—	28		25	53
750	—	23		29	52

they were selfed to produce stocks homozygous for the blocks of genes involved.

We were able to extract from Florida teosinte four more or less distinct types corresponding, so we supposed, to the four segments which we had postulated as result of our previous experiments. These experiments showed clearly that the principal differences between maize and teosinte involve hereditary units which usually behave in inheritance almost as though they were single Mendelian factors. These units do not always behave as single genes, however, for they are capable of breaking up into smaller units with less conspicuous effects. This was demonstrated by an experiment in which we crossed various strains of 4R–3 which were homozygous for single segments with the original 4R–3. The hybrid—essentially 4R–3 heterozygous for a single segment—was then crossed by a second uniform inbred strain, usually A158.

A population resulting from such a cross would be expected to have two types of ears —and only two types—if there was no crossing over within the segment. If crossing over occurred, however, we should expect three types of ears: (1) those containing no segment; (2) those containing the intact segment; (3) intermediates exhibiting the effect of part of the segment.

Both classes of segregating populations were found, as shown by the data in Table 4.2. Three populations segregating for a segment, later found to be located on chromosome 3, comprised two types in approximately equal numbers. Two other populations, not identified with respect to the chromosome on which they were borne, behaved in a similar manner. In contrast, three populations, involving a segment later found to be located on chromosome 4, comprised three classes with the intermediates—the presumed crossovers—representing 21 percent of the total.

In our classification of the chromosome 4

population, we were ably assisted by flocks of hungry sparrows that pecked away the kernels from the tips of the ears which projected beyond the end of the husks. The mutilated kernels, furnishing a perfect medium for it, then became infected by a black mold. The result was that the one-segment ears were readily recognized by their barren, blackened tips.

The different segments are similar but not completely identical in their effects. All of them, regardless of the variety of teosinte from which they were derived or the chromosome on which they are borne, increase the prominence and induration of the lower glumes. All of them have a tendency to reduce the number of kernel rows and the size of the kernels. But the segments are not completely identical in their effects even in these characteristics, and they differ considerably in others.

A highly significant fact is that the segment from Florida teosinte, which produces the most conspicuous effect and which is located on chromosome 4, has close counterparts in all three of the other teosinte varieties, Durango, Nobogame, and New, from which segments have been extracted by repeated backcrossing to a common inbred strain. Ears of this type compared with the parental 4R-3 are illustrated in Figure 4.2. The ears of modified strains of 4R-3, which are homozygous for this segment, are slender and as long or longer than the ears of the parental 4R-3. The lower glumes are prominent and strongly indurated, causing the cobs, when the kernels have been removed, to resemble a wood rasp with rigid, slightly curved teeth.

Linkage Relations of Extracted Blocks of Genes

The next step, already anticipated in part by the discussion above, was to determine the linkage relations of these blocks of genes which had been extracted from four varieties of teosinte and incorporated in the inbred strain 4R-3. This work was greatly facilitated by the use of two multiple-gene linkage-tester stocks, one with yellow, the other with white endosperm color, which I had spent some fifteen years in developing, first in Texas and later in Massachusetts and which carried marker genes on nine of its ten chromosomes. Teosinte derivatives were crossed with the yellow multiple tester, and the F_1 hybrids were backcrossed to the white multiple tester. The ears of the backcross populations were classified into two or three classes if the derivatives appeared to carry only one segment but into four or more classes if the derivatives carried two or more segments. I tested a total of 24,930 chromosomes in such backcrosses with the results shown in Table 4.3.

The results are generally quite consistent with the conclusions we had previously drawn about the number of segments in the different varieties of teosinte. For example, in the crosses involving segments from Florida teosinte, clear-cut evidence or strong indications of linkage were found with four and only four chromosomes, 1, 3, 4, and 9. With one exception the situation was similar in the test crosses involving segments extracted from Durango and New teosinte. In contrast, evidence or strong indications of linkage occurred with only three chromosomes in the test crosses involving segments derived from Nobogame teosinte. These results are in agreement with those obtained in our 1939 experiments, which had indicated that Nobogame teosinte contained only three major segments instead of the four found in Florida and Durango teosintes. And the frequency with which chromosome 4 shows linkage is also consistent with our previous conclusion that all four teosinte varieties carry on this chromosome a segment with especially conspicuous effects.

The data in Table 4.3 were collected in experiments conducted in 1946. These experiments were repeated in 1947 with a com-

TABLE 4.3 Summary of Linkage Relations of the Multiple-Factor Segments Derived from Four Varieties of Teosinte and Introduced into the Inbred Strain 4R3

Variety of teosinte	No. of segments	Linkage with chromosome number									Total no. of chromosomes tested
		1	2	3	4	6	7	8	9	10	
Florida	1	−	−	+	−	−	−	−	−	−	1,134
	1	−	−	+	−	−	−	−	−	−	1,530
	1	−	−	−	+	−	−	−	−	−	1,575
	1	−	−	−	+	−	−	−	−	−	1,512
	2	−	−	+	−	−	−	−	+	−	1,512
	2	−	−	+	+	−	−	−	−	−	828
	2	−	−	I	+	−	−	−	−	−	1,386
	2	+	−	−	+	−	−	−	−	−	675
Summary	12	+	−	+	+	−	−	−	+	−	10,152
Durango	1+	−	−	I	+	−	−	−	−	−	567
	1+	I	−	−	+	−	−	−	I	−	756
	2	+	I	+	−	−	−	−	−	−	1,305
	3	−	−	−	+	−	−	−	+	−	1,494
Summary	7	+	−	+	+	−	−	−	+	−	4,122
New	1	−	−	−	−	−	I	−	I	−	1,539
	1+	I	−	−	+	−	−	−	−	−	855
	2	I	−	−	+	−	I	−	−	−	1,575
	2	−	−	−	+	−	−	−	I	−	1,440
Summary	6	I	−	−	+	−	I	−	I	−	5,409
Nobogame	1	−	−	−	+	−	−	−	−	−	1,359
	1	−	−	−	+	−	−	−	−	−	765
	2	−	−	−	+	−	−	−	I	−	1,521
	2	−	−	+	+	−	−	−	−	−	1,602
Summary	6	−	−	+	+	−	−	−	I	−	5,247
Grand summary	31	+	−	+	+	−	I	−	+	−	24,930

+ = Linkage.
I = Indication of linkage.
− = Independent inheritance.

FIG. 4.2 Left to right: an ear of 4R–3 compared with ears of strains of 4R–3 that have been modified by the substitution of chromosome 4 from Florida, Durango, Nobogame, and "New" teosintes. Note the marked similarity of the last three ears.

TABLE 4.4 Effects of Different Numbers of Teosinte Segments on Characters which Differentiate Maize and Teosinte

Character	No. of teosinte segments				
	0	1	2	3	4
Ear characters					
Length of ears, cm.	14.5	13.1	11.2	9.7	8.3
No. of kernel rows	11.5	9.7	8.6	7.5	4.7
No. of ovules per row	30.8	26.5	22.0	17.6	15.3
Total ovules per ear	357	256	190	132	71
Plant character					
No. of tillers	1.0	1.7	2.4	2.7	3.2
Height of tallest tiller, cm.	108	137	148	155	179
No. of leaves	22.2	30.2	38.1	42.7	50.1
Percent secondary tassel branches	28.4	32.5	36.2	38.8	40.7

Source: Mangelsdorf and Reeves, 1939.

TABLE 4.5 Segregation of Characters of Maize-Teosinte Crosses

Characters segregating	No. of individuals	
	Maize-Durango F$_2$	Maize-Nobogame F$_2$
Many-ranked central spike	166	291
Two-ranked central spike	13	35
Total	179	326
Many-ranked pistillate spike	112	213
Two-ranked pistillate spike	65	86
Total	177	299
Pistillate spikelets mainly paired	96	192
Pistillate spikelets mainly single	81	107
Total	177	299
Weak response to short day*	160	301
Strong response to short day†	21	17
Total	181	318

*Flowered before September 15.
†Flowered after September 15.

pletely different set of teosinte derivatives in test crosses involving 14,022 chromosomes. The results were essentially the same as those of the previous year. There was evidence of linkage with chromosomes 4 in eleven of the thirteen derivatives tested, with chromosome 3 in three of the derivatives, and with chromosome 9 in one.

A Second Cycle of Transferring Blocks of Genes

Finding the Texas inbred 4R-3 to be somewhat late in maturity in Massachusetts, I transferred a number of the segments extracted from teosinte to an earlier-maturing, well-adapted inbred, Minnesota A158. After the transfer had been completed, I repeated several of the linkage tests which had previously been made with the 4R-3 derivatives, and in backcross populations that were large enough to reveal clear-cut evidence of linkage the same chromosomes proved to be involved. What we have done then is to: (1) extract chromosomes or segments of chromosomes from four varieties of teosinte; (2) introduce these into a uniform inbred strain by repeated backcrosses; (3) determine their linkage relations; (4) extract the segments from the first inbred strain; (5) introduce them into a second inbred strain by repeated backcrossing; (6) determine their linkage relations and find them still to be the same.

Polygene Segments as Supergenes

These blocks of genes which distinguish teosinte and maize are the counterparts of the "supergenes" (closely linked clusters of loci acting, except when crossing over occurs, as one gene) discussed by Ford (1964), Turner (1967), and others. Turner, employing rigorous mathematical analysis, has concluded that in a system of balanced polymorphism, selection for epistatic effects tends to produce supergenes and that these are a major feature of evolution. If teosinte originated as a hybrid of maize and *Tripsacum*, the polygene segments of teosinte are actually blocks of genes originally from *Tripsacum*. They do, however, behave as supergenes, and in the system of polymorphism represented by cornfields in which teosinte survives and flourishes as a weed, they are subject to the same selective forces as the supergenes discussed by Ford or Turner. In such an environment, natural selection would certainly tend to preserve them; it might even tend to make them more potent.

It has recently been suggested (Beadle, in conversation) that the hereditary units are not blocks of genes but single genes with numerous nucleotides. The fact that all of them affect characteristics of both ears and plants (Table 4.4) might be considered as evidence against his hypothesis. The fact that one of the units, that on chromosome 4, involves at least 20 percent of crossing over (Table 4.2) and produces mutagenic effects over a substantial part of the short arm of that chromosome (Chapter 12) would seem to rule out the possibility that it represents a single gene.

Failure to Reconstitute Teosinte

Since this record is concerned with failures as well as successes, I should at this point describe an experiment that failed. When we concluded that teosinte differs from maize primarily by four segments or blocks of genes, we thought that it might be possible to reconstitute teosinte by recombining these four segments after they had been introduced into the uniform inbred 4R-3. We had additional reasons for believing that this might be done, because Collins and Kempton, in a population of 127 F₂ plants, had obtained one teosinte-like plant combining what they considered to be the four essential botanical characteristics of teosinte. Later Kempton (1924) illustrated three teosinte-like spikes occurring in a population of 409 plants. Still later I obtained similar results in my F₂ populations of crosses of maize with Durango and Nobogame teosintes. Genotypes combining all of the essential botanical characteristics of teosinte in F₂ populations could not occur in such frequencies if a large number of independently inherited genes differentiating the two parents were involved.

Despite the extensive data indicating that only a relatively small number of blocks of genes control the differences between teosinte and maize, our attempt to synthesize teosinte by recombining the extracted blocks of genes was not a success. There are probably two reasons for this: (1) we did not have all of the essential blocks of genes in our substitution strains; (2) we had not taken into consideration the importance of the modifier complex.

Data presented later show that, although there may be four principal segments differentiating maize and teosinte, there are other

FIG. 4.3 *Left*, an ear of the inbred strain 4R-3. *Right*, an ear of 4R-3 modified by the introduction of chromosome 3 of Florida teosinte. *Center*, an ear heterozygous for the introduced chromosome. In this particular combination the maize genes are almost completely dominant over the teosinte genes.

smaller segments or genes involved in the differences. By introducing the blocks of genes of teosinte into an inbred strain not only had we failed to introduce the necessary modifying factors but, since the inbred strain was homozygous for the greater part of its residual heredity, there was no opportunity for selection to create a new complex of modifying factors. By combining what we thought to be the four segments, we did succeed in producing a true-breeding type similar to the F₁ hybrid, but we failed to resynthesize the parental teosinte.

In the generations of backcrossing to a uniform inbred strain and in classifying the genotypes in the linkage tests, the segments are recognized by their dominant effects in the heterozygous condition. The segments also have recessive effects, and these are illustrated by comparing the modified strains of 4R-3 in which the segments are heterozygous with modified strains in which they are homozygous. One such comparison is illustrated in Figure 4.3. Since the strains containing the homozygous segments differ from the original 4R-3 much more conspicuously than those in which the segments are heterozygous, it would appear that the segments are more strongly recessive than dominant in their effects. Other experiments described later in this chapter tend to support this conclusion.

Mendelian Segregation in F₂ Generations

Shortly after the publication of our monograph, D. G. Langham, who had conducted his thesis research on maize-teosinte hybrids under the direction of R. A. Emerson at Cornell, reported (1940) that in crosses of maize and Durango teosinte, three maize characters, weak response to length of day, paired pistillate spikelets, and many ranked ears, all acted as dominants to the contrasting teosinte characters and that all segregated in the F₂ generation as simple Mendelian characters. Since these results were quite different from those obtained earlier by Collins and Kempton, I felt that further studies of this kind were needed, and so in 1944 I grew an F₂ generation of a maize-Durango teosinte and a maize-Nobogame teosinte cross. The maize parent in both crosses was our nine-gene tester stock described previously in this chapter.

This was one of those wartime years when, because of a shortage of labor, my wife and I became true "dirt farmers," plowing the experimental field and performing all of the subsequent tillage and weeding operations. This was also the year when a hurricane blowing first in one direction and later in another left our field a tangle of criss-crossed, prostrate stalks, which had to be propped up one by one and accurately identified with respect to their genotypes before their ears could be harvested. In spite of these difficulties we were able to classify all of the plants, about 500 in number. The results of this classification did not agree with those of Langham, as the data in Table 4.5 show. In no case did the segregation fit closely a simple Mendelian ratio, although it approached such a ratio in many-ranked versus two-ranked pistillate spikes. Also, the two crosses differed significantly in their segregation; teosinte-like plants with respect to three of the four characters studied occurred with greater frequency in the maize-Durango cross than in the maize-Nobogame cross. This furnished additional support for the conclusion reached earlier that Nobogame is less "potent" than Durango teosinte in the genes which distinguish it from maize. We also had evidence from other crosses that the results in an F₂ population may vary with the varieties of maize as well as with the varieties of teosinte serving as the parents of the cross. It must have been an unusual sample that produced the clear-cut Mendelian

FIG. 4.4 The most teosinte-like segregate (*A*) and the most maize-like (*E*) in an F$_2$ population of 318 plants of the cross Nobogame teosinte × maize. Specimens *A* and *B* illustrate the transition from solitary to paired spikelets; *A* and *C* the change from a two-ranked to a many-ranked spike. Specimens *A*, *B*, *C*, *D* represent fragile disarticulating rachises and *E* a solid one. In this cross and a similar one involving Durango teosinte as one parent, the transition from the spike of teosinte to something resembling an ear of corn involved genes on almost all of the marked chromosomes.

44   The Genetic Nature of Teosinte

segregation in Langham's crosses which led him to conclude that simple Mendelian inheritance is involved. This obviously was not the case in our crosses, or in those of Rogers described later. The most teosinte-like segregate, the most maize-like segregate, and various intermediates from the Nobogame teosinte × maize cross are illustrated in Figure 4.4. The inheritance of the differences that distinguish maize and teosinte was not always clear cut. In several instances one ear had had two ranks and paired spikelets and a second ear on the same plant had four ranks and solitary spikelets. One such pair is illustrated in Figure 4.6.

Additional evidence that the characters which distinguish teosinte from maize are not simple Mendelian ones is provided by the results of the linkages with marker genes that the F_2 data revealed. These include, in addition to the characters which Langham studied, a "glume score," which attempts to estimate in a wholly empirical fashion the degree of development of the indurated lower glume, a characteristic of teosinte, and a "disarticulation score," also an empirical value that attempts to estimate the readiness with which the rachis segments separate from each other. Both of these characters are important in differentiating maize and teosinte.

The data from these test crosses are set forth in Table 4.6. They show that there are genes controlling the two-ranked condition (of either the ear or the central spike of the tassel or both) on chromosomes 1, 2, 6, 8, and 9 of Nobogame teosinte and chromosomes 2, 3, 4, 8, and 9 of Durango teosinte. Genes responsible for teosinte's strong response to short days definitely occur on chromosome 10 of Durango and possibly also on chromosome 10 of Nobogame and chromosome 8 of Durango. Genes controlling the differences between paired and single spikelets are found on chromosomes 4 and 8 of Durango and on chromosome 4 and possibly also 8 of Nobogame.

Apparently there are numerous genes acting upon the development of the lower glumes, for the glume score shows linkage with the marker genes on chromosomes 4, 6, 7, 8, and possibly also 3 and 10 in the maize-Nobogame cross and with the marker genes on chromosomes 3, 4, 8, 9, and 10 in the maize-Durango cross. A similar situation prevails with respect to the disarticulation

FIG. 4.5 Four F_1 maize-teosinte hybrids, showing how the hybrid varies with both the maize and teosinte parent. A, Nobogame teosinte × multiple-gene tester; B, Guarany × Nobogame; C, Durango teosinte × multiple-gene tester; D, Durango × Guarany. The Durango hybrids are more teosinte-like than the Nobogame hybrids, the Guarany hybrids more so than the multiple-gene tester ones. We have since isolated from the Guarany race a chromosome carrying cryptic genes for single spikelets and other teosinte characters (see Chapter 11.)

score, which is correlated with the marker genes on chromosomes 3, 4, 6, and possibly 8 in the maize-Durango cross and with 4, 8, and possibly 1, 3, and 7 in the maize-Nobogame cross.

In summary there are clear-cut linkages of one or more teosinte characters with marker genes on chromosomes 1, 2, 4, 6, 7, 8, and 9 in the Nobogame cross and with markers on 2, 3, 4, 6, 8, 9, and 10 in the maize-Durango cross. There are more linkages with the chromosome 4 marker gene than with any other. This is in agreement with the data presented in Table 4.3, which shows that all four varieties of teosinte tested have a segment on chromosome 4 with prominent effects. Finally it is significant that the various teosinte characters are strongly correlated with each other, a fact which Collins and Kempton (1920), treating the problem in a much different way, noted many years ago. These strong correlations of teosinte characters with each other are consistent with the concept of teosinte differing from maize by blocks of genes affecting several of the characters which differentiate the two species.

Five Races of Teosinte Analyzed

John S. Rogers, who had been my graduate student at Texas A & M in 1938–40, became my student again at Harvard in 1946–47 after completing several years of wartime service. As a thesis problem he chose to carry still further these studies of the inheritance and linkage relations of characters differentiating maize and teosinte which I had made with only two varieties, Nobogame and Durango. He repeated these two crosses and added three more: crosses of maize with Chalco, El Valle, and Huixta teosintes. The Nobogame teosinte employed in Rogers' experiment is essentially the same race as the Nobogame described by Wilkes; Durango probably belongs to Wilkes' race "Central Plateau"; Chalco teosinte is approximately the same as Wilkes' race of the same name. The seed of El Valle teosinte came to us from Venezuela, where it had been grown as a forage crop. It actually is a strain of Florida teosinte, which probably came originally, as discussed in Chapter 3, from the region of Jutiapa in Guatemala. Huixta teosinte is from the vicinity of San Antonio Huixta in northwestern Guatemala. These two teosintes correspond to the Guatemalan races described by Wilkes as Guatemala and Huehuetenango respectively. Thus five of the six races which Wilkes recognized were employed in Rogers' experiments. Only the Balsas race, which at that time had not yet been recognized as a distinct race, was omitted.

Rogers' results (1950a, b, c) agree in general with those of our earlier experiments. He found the F_1 hybrids of the maize-Nobogame cross to be more maize-like than the F_1 hybrids of the maize-Durango cross. In the backcrosses the most teosinte-like segregates of the Nobogame cross were more maize-like than the most teosinte-like segregates of the Durango cross. Rogers concluded as he and I had in our 1939 experiments that Nobogame teosinte is less "potent" than Durango in its "teosinte" characters.

The data on linkage relations reported by Rogers are set forth in Table 4.7, which represents the summary of four different tables published by him. The character "ear grade" employed by Rogers is based on classifying the pistillate spikes into five classes with respect to their resemblance to the maize parent at one end of the range of variation

TABLE 4.6 Linkage Relations of Teosinte Characters with Each Other and with Marker Genes on Nine Chromosomes of Maize

	Characters	Bm₂	Lg₁	A₁	Su₁	Y	Gl₁	J₁	Wx	G₁	T.R.	S.D.	P.S.	G.S.	D.S.
						Nobogame Teosinte × Maize F₂									
Durango Teosinte × Maize F₂	Bm₂ —Chromosome 1		−	−	−	−	−	−	−	−	+	−	−	−	I
	Lg₁ —Chromosome 2	−		−	−	−	−	−	−	−	+	−	−	−	−
	A₁ —Chromosome 3	−	−		−	−	−	−	−	−	I	−	−	I	I
	Su₁ —Chromosome 4	−	−	−		−	−	−	−	−	−	−	+	+	+
	Y —Chromosome 6	−	−	−	−		−	−	−	−	+	−	−	+	−
	Gl₁ —Chromosome 7	−	−	−	−	−		−	−	−	I	−	−	+	I
	J₁ —Chromosome 8	−	−	−	+	−	−		−	−	+	−	I	+	+
	Wx —Chromosome 9	−	−	I	−	−	S	−		−	+	−	−	−	−
	G₁ —Chromosome 10	−	−	−	−	−	−	−	−		−	I	−	I	−
	T.R. —Two-Ranked Spikes	−	+	+	+	−	−	+	+	−			+	+	+
	S.D. —Response to Short Day	−	−	−	−	−	−	I	−	+	+		+	+	I
	P.S. —Paired Spikelets	S	−	−	+	−	S	+	S	−	+	+		+	+
	G.S. —Glume Score	−	−	+	+	−	−	+	+	+	+	+	+		+
	D.S. —Disarticulation Score	−	−	+	+	+	S	I	−	−	+	I	+	+	

+ = Linkage. − = Independent inheritance.
I = Indication of linkage. S = Significant deviation not due to linkage.

and to the F₁ hybrids at the other. It corresponds essentially to the classification with respect to number of segments that we had made in our earlier studies. Three other characters that Rogers employed, two-ranked spikes, paired spikelets, and photoperiodic response, are the same as those which I had used in the crosses involved in Table 4.6. Consequently Rogers' data on these characters in Table 4.7 are comparable with mine in Table 4.6. Rogers did not use my disarticulation score but added a wholly different character, tiller number. The symbols employed in the two tables are the same except that in Rogers' table a single plus sign indicates a deviation from independent inheritance significant at the .05 level of probability and a double plus sign represents a deviation significant at the .01 or lower level.

With respect to ear grade, Rogers' data show linkage with the marker gene Su on chromosome 4 in all teosintes except El Valle. The absence of linkage in this cross is explained by the fact that there was a marked deficiency of sugary seeds so that there was no meaningful comparison between the Su and the su plants. In other respects the data are in reasonably close agreement with the data set forth in Table 4.3, which shows the linkage relations of extracted segments. There

is no doubt that chromosome 4 in all varieties of teosinte so far tested carries more genes or more potent genes affecting the characters of the pistillate spike than do any other of the nine chromosomes tested. Essentially the same can be said about glume score. In all teosintes tested (except El Valle, for the reason mentioned above) chromosome 4 carries more genes or more potent genes than any other.

The data on the inheritance of two-ranked spikes are less consistent. Here chromosome 3 appears to be the most potent, but linkages with 1, 2, 4, 6, 8, and 9 also occurred. The genes for paired spikelets show linkage with gl₁ on chromosome 7 in Nobogame teosinte and also linkage with a on chromosome 3 of Durango teosinte. Photoperiodic response shows more and stronger linkages with chro-

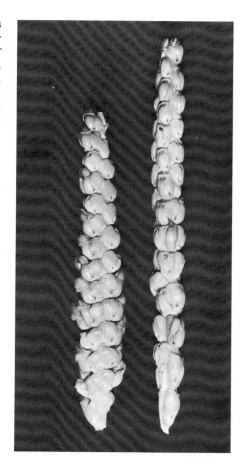

FIG. 4.6 Two spikes from the same plant of the F₂ generation of a maize-teosinte cross. The spike at left is distichous, with paired spikelets; the other is four ranked, with single spikelets. The total number of spikelets is approximately the same in the two specimens. The difference in the morphology of the two spikes having the same genotype shows that what is inherited is not specifically paired spikelets or many ranks but a tendency to proliferation which may take either one of two paths. Actual size.

mosome 10 than with any other chromosome, but 8 ranks a close second. Weaker linkages occur on chromosomes 1, 2, and 6. Tiller number shows numerous strong linkages, but some of these may be fictitious, since several of the maize marker genes have a direct effect in reducing tiller number. This is especially true of *lg* on chromosome 2, but it is also true to a lesser extent of *su*, *j*, and *g* on chromosomes 4, 8, and 10 respectively. So far as I know *bm₂* on chromosome 1 and *gl* on chromosome 7 have no such direct effect in reducing tiller number, and since these marker genes show numerous and strong linkages with tiller number we may conclude that chromosomes 1 and 7 carry numerous or potent genes for this character.

In summary, linkages between the characters distinguishing maize and teosinte and marker genes occurred on all nine chromosomes tested. The strongest and most numerous linkages were with the marker gene on chromosome 4, and this is undoubtedly the most potent chromosome distinguishing teosinte from maize. Other chromosomes showing strong linkages are 1, 3, 7, 8, and 10. The remaining chromosomes tested show weaker or less numerous linkages. Chromosome 5 was not represented in these tests, but in the earlier tests which Rogers and I conducted in 1939 one of the markers that we employed was Bm_1 on chromosome 5. We found no strong indications of linkages of teosinte characters with this marker.

In analyzing these linkage data two facts should be considered: (1) Whether a deviation from independent inheritance is statistically significant depends in part upon the size of the population; the larger the population the more likely is a weak linkage to be revealed as a significant deviation from independent inheritance. (2) There are probably three more or less distinct categories of hereditary units involved in these crosses: (*a*) relatively long segments such as those on chromosome 4; (*b*) inversions such as those which occur on the short arm of chromosome 8 and which probably are responsible for the strong linkages with the marker gene *J*, which is located on the long arm of that chromosome; (*c*) modifying factors which may well be distributed over all ten chromosomes; linkages with these would not be detected in small populations but some of them would be revealed in larger populations.

TABLE 4.7 Linkage Relation of Characters Differentiating Maize and Teosinte with Marker Genes on Nine Chromosomes

Differentiating characters in five races of teosinte	Marker genes on nine chromosomes									
	1	1	2	3	4	6	7	8	9	10
	Bm_2	P	Lg_1	A_1	Su_1	Y	Gl_1	J_1	Wx	G
Ear grade										
Nobogame	I		—	+	+ +	—	+ +	—	—	—
Durango	+	I	+	—	+ +	—	—	—	—	—
Chalco	+	—	—	—	+ +	—	—	—	I	I^s
El Valle	—		+	+	—	I	I	—	+	—
Huixta	—		—	I	+ +	—	—	—	—	I
Glume score										
Nobogame	—		—	—	+ +	—	—	—	—	—
Durango	—	—	—	—	+ +	—	—	—	—	—
Chalco	I	—	—	—	+ +	—	—	—	—	—
El Valle	—		—	I	—	+	+	—	I	I
Huixta	—		—	I	+ +	I^s	—	—	—	—
Two-ranked spikes										
Nobogame	—	I^s	—	+ +	—	+	—	—	—	—
Durango	+ +	—	+ +	—	—	—	—	—	—	—
El Valle	I		I	+ +	—	I	—	I	—	—
Paired spikelets										
Nobogame	—	—	—	—	—	—	+	—	—	—
Durango	—	—	—	+	—	—	+	—	—	—
El Valle	I		I	I	—	—	—	I	I	I
Photoperiodic response										
Nobogame B.C.	—		—	S	S	—	—	—	I^s	—
Nobogame F₂	S	—	—	I^s	S	—	—	—	—	—
Durango B.C.	—	—	—	—	—	—	—	—	—	—
Durango F₂	—	—	—	—	—	—	—	—	—	+
Chalco B.C.	—	—	—	—	—	+	—	+	—	+ +
Chalco F₂	+	+	—	S	—	+	—	+ +	—	+ +
El Valle B.C.	—		+ +	—	—	I	+	+	—	+ +
El Valle F₂	S		—	—	—	+	—	+ +	—	+ +
Huixta B.C.	I^s		+	—	—	—	—	+ +	—	I
Huixta F₂	—		+	—	—	—	—	+	—	+ +
Tiller number										
Nobogame B.C.	+		+	—	—	—	+ +	—	+	—
Nobogame F₂	+	—	+ +	—	+ +	—	+ +	+ +	—	+ +
Durango B.C.	+ +	—	+ +	—	—	+	—	—	—	—
Durango F₂	+	—	+ +	—	—	+	+ +	I^s	—	—
Chalco B.C.	+	—	+ +	—	—	—	+ +	—	—	—
Chalco F₂	+	—	—	+	I^s	—	—	—	—	—
El Valle B.C.	+ +	—	+ +	—	—	—	+ +	—	—	—
El Valle F₂	+ +	—	+	—	—	—	—	—	—	—
Huixta B.C.	+ +	—	+ +	—	+ +	—	+ +	—	—	—
Huixta F₂	+ +	—	+ +	—	+ +	—	+ +	—	—	—

Source: Rogers, 1950.
+ + = Strong linkage, P < 0.01.
 + = Linkage, P 0.05–0.01.

— = Independent inheritance.
S = Strong deviation not due to linkage
I = Indication of linkage.
I^s = Small deviation not due to linkage.

What is Teosinte?

Summing up the facts about teosinte and the evidence gained from extensive experimentaton presented in this and the preceding chapter, we can see that the relationship between teosinte and maize is unique and is one which has no close counterpart in other plants. The two species have the same chromosome number; they hybridize readily; the fertility of the hybrids is high; the pairing of their chromosomes is regular and virtually complete; except for inversions, crossing over between their chromosomes is of the same order as it is in maize itself. In parts of Mexico, teosinte exists as a weed in the maize fields, where it survives by behaving as a mimic, imitating the maize so closely in coloration and other features that it cannot be distinguished from its model before flowering. As a mimic, teosinte is unique in hybridizing freely with its model, thus producing a gene flow in both directions between the two.

Teosinte is intermediate between its closest relative, maize, and its more distant relative, *Tripsacum*, or similar to one or the other of these two species in a long list of cytological, morphological, and physiological characteristics. It was this fact more than any other that led us to conclude (Mangelsdorf and Reeves, 1939; Reeves and Mangelsdorf, 1959) that teosinte is the product of hybridization between maize and *Tripsacum*. We were not the first, however, to recognize teosinte's intermediate nature. Edgar Anderson, with remarkable perspicacity, recognized it almost at once when he saw corn, teosinte, and *Tripsacum* growing together in our experimental cultures in Texas. Much earlier Ascherson (1875), one of the early students of teosinte, regarded it as a connecting link between *Zea* and *Tripsacum* and in so doing recognized that *Tripsacum*, an American plant, is more closely related to maize than is *Coix*, an Asiatic genus whose relationship had previously influenced thinking about maize as a plant of Old World origin. Hooker (1878) likewise recognized teosinte's intermediate nature when he stated, "It unites the habit of maize (*Zea*) with in many respects the structure of *Tripsacum*." This is especially true of the principal botanical characteristics which distinguish teosinte from maize: (1) two-ranked pistillate spikes; (2) solitary pistillate spikelets; (3) brittle rachises; (4) strong induration of the tissues of the rachis and lower glumes; (5) profuse tillering. All of these are characteristics that teosinte has in common with *Tripsacum*.

Criticisms of the Theory of the Hybrid Origin of Teosinte

The evidence from our extensive experiments over a period of almost forty years is consistent with the hypothesis of the hybrid origin of teosinte, but it is still short of proof. Also the theory has been strongly criticized by several long-time students of the problem, especially Weatherwax (1950, 1954, 1955) and Randolph (1952, 1955). In our papers commemorating the Darwin centennial, published in 1959, Reeves and I gave serious consideration to these criticisms and found all of them to be either answerable or, in our opinion, not valid. I do not wish here to repeat in detail the debate in which we engaged at that time but will mention several of the more pertinent criticisms and our replies to them.

(1) The cross of maize and *Tripsacum* that we succeeded in making by shortening the silks of maize to about the same length as the styles of *Tripsacum* would not have occurred in nature. In reply we suggested five conditions under which natural crossing might have occurred: (*a*) ears that protrude beyond the husks and so expose their uppermost silks for their entire lengths; (*b*) silks exposed to pollination at their bases through the mutilation of husks by insects or by holes bored by larvae; (*c*) depauperate plants with no ears but with terminal inflorescences that are predominantly pistillate; (*d*) short tillers that are often terminated by inflorescences which are partly or wholly pistillate; (*e*) homozygous pod corn in which the terminal inflorescences of the main stalk and tillers may be partly pistillate.

(2) A cross that succeeds in such low frequency as it did in our experiments could not have occurred in nature. Our reply: Farquaharson (1957), a student of Weatherwax, found that some plants of *Tripsacum* when used as female parents crossed readily with corn without the use of special techniques, and she concluded: "It seems highly probable that this cross has occurred occasionally in nature." Later Galinat (see Chapter 5) found that crosses of corn with *T. floridanum* were much more easily made than those with *T.*

dactyloides, the species used in our experiments.

(3) The frequency of interchange or crossing over between corn and *Tripsacum* chromosomes in their hybrids is not sufficient to give rise to new intermediate species by introgression. Our reply: Our own experiments had shown some evidence of interchange. Maguire, a student of Randolph, showed clearly (1952, 1957) that some form of exchange had occurred between chromosomes and that the terminal knob of a *Tripsacum* chromosome was occasionally transferred to a corn chromosome. The latter observation tends to support our suggestion that the chromosome knobs of teosinte may have come from *Tripsacum*. Maguire's observation on homology between maize and *Tripsacum* chromosomes was later confirmed by Galinat, who also found much evidence of it on additional chromosomes. Some of this will be illustrated in Table 5.1.

It is gratifying to note here that two of the items of evidence supporting our theory were produced by students of Weatherwax and Randolph, our principal critics, and that neither was deterred from expressing an opinion contrary to that of her mentor. Throughout this long controversy there has always been a wholesome exchange of information, opinions, and criticisms in the tradition founded by the late R. A. Emerson and effectively maintained for forty years by his former student, Marcus Rhoades.

I have sometimes been asked, "If teosinte is a hybrid of maize and *Tripsacum* why do you not synthesize it by crossing the putative parents?" This question was first put to me in 1939 by Theodosius Dobzhansky on board the Queen Mary when we, along with a number of other American geneticists and our wives, were enroute to Europe to attend the Seventh International Congress of Genetics in Edinburgh, where I was scheduled to present a paper explaining our tripartite theory on the origin of maize. My reply to Dobzhansky was essentially that set forth at the beginning of this chapter: We knew so little about the species of *Tripsacum* or the kinds of maize that might have been the parents of such a hybrid that it would scarcely, at this point have been worthwhile to undertake such a direct attack. We were not dealing here with a suspected allopolyploid hybrid which might be easily synthesized by crossing its putative parents, since we regarded teosinte as essentially maize with a relatively few blocks of genes from *Tripsacum*. I also pointed out that in the segregating offspring of a species hybrid, as in games of chance, the odds against the fortuitous repetition of a particular complex combination are almost overwhelming.

The most critical evidence against the theory of the hybrid origin of teosinte may eventually come from the laboratory at Harvard University of my long-time colleague, Elso Barghoorn. There one of his graduate students, Umesh C. Banerjee, studying the surface sculpture of grass pollen grains under the great magnification of the electron microscope, has found the patterns in maize to be quite different from those of *Tripsacum*. The exines of the pollen grains of both species are beset with short protuberances called "spinules," but in maize these are regularly distributed over the surface while in *Tripsacum* they tend to occur in clusters. If teosinte is a hybrid of maize and *Tripsacum* we might expect its spinule pattern to be intermediate between those of its putative parents. However, in those races of teosinte that Banerjee has so far studied, the pattern is virtually identical with that of maize. If this proves to be true of all races of teosinte, I shall begin to have serious doubts that teosinte is a hybrid of maize and *Tripsacum* and I shall be compelled to consider alternative theories.*

Alternative Possibilities

Charles Darwin once wrote: "I have steadily endeavored to keep my mind free so as to give up any hypothesis however much beloved." I, too, have tried—I hope with some degree of success—to keep an open mind with respect to hypotheses however long maintained. Our hypothesis of the hybrid origin of teosinte has proved to be quite useful in stimulating us to investigate the role of pod corn in the evolution of maize and the introgression from teosinte in the formation of modern races and varieties. The hypothesis is not so "beloved,"

*Since this was written the studies of Barghoorn and Banerjee, working with Galinat have demonstrated beyond a reasonable doubt, that teosinte is not a hybrid of maize and *Tripsacum*. Perhaps I may be permitted to enjoy some degree of satisfaction in the fact that it is my colleagues and not my critics, who have shown that this part of our tripartite hypothesis is no longer tenable.

however, that I am averse to giving it up if and when the evidence so demands. What are the alternatives?

Teosinte the Ancestral Form?

The most obvious alternative is the nineteenth-century theory, recently revived by Miranda, Beadle, Galinat, and others, that teosinte is the ancestral form—that cultivated corn is essentially a domesticated teosinte. The reasons for favoring the theory are obvious. As mentioned above, teosinte is the closest relative of maize; it hybridizes readily with maize, and the hybrids usually are quite fertile; teosinte has the same chromosome number as maize and the pairing of chromosomes in the hybrid is virtually complete; crossing over between maize and teosinte chromosomes is of the same order as it is in maize.

In view of these striking similarities it may seem strange that a number of careful students of the problem—those who have relied on all of the evidence instead of only the most conspicuous part of it—eventually dismissed teosinte as the ancestral form. The list includes Harshberger, Collins, Weatherwax, and Mangelsdorf and Reeves. One of the attractive and wholesome attributes of science is that it recognizes no infallible authorities on any subject, and so it is quite possible that all of these authorities are wrong in the conclusions they have reached. It may be a mistake, however, to overlook or ignore the evidence and reasons on which these conclusions were based.

One of the most common reasons for skepticism regarding the teosinte theory is that teosinte is so unpromising as a source of food that it seems unlikely that it would ever have been domesticated. In addition to its tendency to shatter its ripe fruits and render large-scale harvesting difficult if not impossible, teosinte has its seeds enclosed in hard bony shells from which they cannot be removed by ordinary threshing operations. Beadle (1939) found that when exposed to heat the seeds of teosinte pop and burst from their indurated shells, and he suggested that this may have been a primitive use that led to teosinte's domestication. The idea is plausible but, strangely, there is no evidence, either archaeological or contemporary, that teosinte was or is now used in this way; in contrast there is evidence both archaeological (Chapter 14) and historical

FIG. 4.7 Kernels of *Tripsacum*, like those of teosinte, (see Beadle, 1939) "pop" when exposed to heat and burst from their hard, bony shells converting a most unpromising food into a palatable and nutritious one. There is, however, no archaeological evidence that either teosinte or *Tripsacum* was utilized this way.

(Chapter 7) that corn was processed by popping.

If teosinte, because of its ability to pop, was domesticated as a food plant, why was corn's other American relative, *Tripsacum*, not also utilized? It has the same type of fruit and caryopsis, and it is much more widespread than teosinte in its geographical distribution, occurring in abundance in South and Central America, Mexico, and extending into the United States as far north as Massachusetts. Yet the only evidence that *Tripsacum* may sometimes have been used by the Indians is the discovery by Gilmore (1931) of *Tripsacum* seeds among the plant remains in the Bluff Dweller caves of the Ozarks. When Beadle's paper came to hand in 1939 I suspected at once that the collection and storage of *Tripsacum* fruits might be accounted for by the fact that they were occasionally utilized by popping. Putting the matter to a test I quickly found that *Tripsacum* seeds pop as readily as those of teosinte and the popped grains are easily separated from the bony shells (see Figure 4.7).

Actually the ability to pop is probably common in all starchy seeds that are small and corneous. The people of primitive cultures undoubtedly recognized this fact, but they put into cultivation only those species which possessed other promising characteristics as well. In the Old World, seeds of rice, sorghum, and the millets are commonly popped. In some parts of India there is a village "popper" who processes rice kernels by stirring them in hot sand and takes his toll of the finished product. There is no tradition of any kind in the New World with respect to the popping of teosinte.

Beadle has also suggested (personal communication) that teosinte fruits may have been ground and the product—meal mixed with shell fragments—baked into "teotortillas" and consumed in this form. One objection to this idea is that food-preparing artifacts apparently were not yet in existence at the time, 7000 to 10,000 B.C., when the domestication of teosinte is assumed to have occurred. A more serious objection is that the indigestible shells represent a substantial part—about half by weight according to the figure kindly supplied me by Beadle—of the fruits and a considerable amount of the mixture would have to be consumed daily in order to obtain an adequate food supply. Consequently, if teosinte had ever been extensively used in this way the feces of the consumers would be filled with the remains of the shell fragments. Studies of the coprolites—prehistoric feces—found in archaeological sites show no evidence of this. One fragment of teosinte shell was found in feces dated at 1850–1200 B.C. in Romeros Cave in Tamaulipas, Mexico, and several entire fruits—presumably consumed for medicinal purposes rather than for food—were found in later specimens. But prehistoric stools packed with shell fragments, the product of a diet consisting in substantial part of crushed teosinte fruits, have never, so far as I know, been discovered. Until they are there is good reason to doubt that teosinte was ever used extensively in the manner that Beadle has suggested.

Demonstrating the courage of his convictions, Beadle, using himself as an experimental subject, once consumed 150 grams of ground teosinte fruits per day for two days without discernible ill effects but also without contributing substantially to his dietary requirements. Only about half of the mixture, 75 grams, could be counted as digestible. Assuming the digestible portion to have about the same energy value as whole ground cornmeal, 355 calories per 100 grams, Beadle's daily ration of teosinte contributed about 266 calories. This is less than one-sixth of the minimum daily requirement for adults. Would an amount of this order stimulate people in primitive societies to domesticate teosinte? I doubt it. Collins (1919) put the matter succinctly when he stated, "There are hosts of wild grasses which have never been domesticated, any of which would seem more promising material for the primitive plant breeder than teosinte."

Morphologists, especially, object to the teosinte theory because teosinte is more specialized than maize in a number of its morphological characteristics including single spikelets, distichous spikes, and the indurated tissues of its pistillate rachises and glumes. Consequently, to have cultivated maize evolving from teosinte is to assume a series of reverse mutations. Furthermore, maize is usually dominant to teosinte in at least two of these morphological characteristics: polystichous spikes and paired spikelets, suggesting that the reverse mutations were dominant in nature. To have a cultivated plant originating from its wild progenitor by a series of dominant reverse mutations is, I think, unique in the evolution of crop plants.

The cytoplasm of teosinte differs from that of maize according to the experiments of Mazoti (1958). Could maize have become differentiated from teosinte in its cytoplasm as well as its morphological characteristics in the few millennia that have passed since domestication is assumed to have begun?

Despite the renewed popularity of the teosinte hypothesis, little new evidence in its support has been adduced. The specimens employed by Miranda (1966) to illustrate a supposed evolutionary series from teosinte to maize can all be duplicated in almost any F_2 population of a maize-teosinte cross. My former colleague, Walton Galinat, in a bulletin (1970) well illustrated with photographs and excellent drawings, has presented some new observations which he regards as supporting the teosinte hypothesis. These are concerned with the cupule, a depression in the rachis, that is characteristic of the three American Maydeae: maize, teosinte, and Tripsacum. A number of earlier students, notably Sturtevant* (1899), Collins (1919), and Cutler (1946), have been intrigued by the cupules and have sought to find some significance in the variation in cupule shape in different races of maize. They have not made much progress in this direction, partly, I suspect, for the reason that cupule shape is somewhat "equivocal," varying greatly with the genetic background, to some extent with the environment, and in corn with the position on the cob. An extreme example of such variation occurred in an ear of pod corn that I described and illustrated many years ago (Mangelsdorf, 1945). This ear, the product of

*Sturtevant, who first used the term, defined the cupule as a "corneous alveolus" in the cob.

crossing pod corn with a long-eared corn from Paraguay, protruded beyond the husks and became greatly elongated. At the tip of the ear paired spikelets were arranged in whorls on a very slender rachis. The elongation of the internodes of the rachis had so flattened the cupules that, defined as depressions in the rachis, they no longer existed. This specimen, which is illustrated in Figure 7.3, shows that although the cupule may not be the product of compaction its shape is definitely influenced by the degree of compaction.

Another objection to Galinat's conclusions is that some of the most primitive corn, the earliest prehistoric corn from Tehuacán, does not have the long, narrow cupule shape of teosinte. On the other hand, this shape can be found in some of the most highly evolved races. The evolutionary sequence depicted in his article is a contrived one based on selected specimens and does not represent any known archaeological sequence. In this respect it is somewhat misleading, especially to those who do not read the legends carefully. A third objection to Galinat's article involves the assumptions that he has made with respect to the change from the distichous spike of teosinte to the polystichous ear of maize. In teosinte the spikelets facing in opposite directions alternate on the axis. Galinat has assumed that this also occurs in the ear of maize, that the spaces sometimes occurring between cupules of a maize cob represent the dorsal surfaces of spikelets on the opposite side of the cob. So far as I can determine there is nothing in the anatomical studies of the maize ear by Bonnett (1940, 1966), Laubengayer (1949), Reeves (1950), or Galinat (1959) to support this assumption, and he presented no new anatomical evidence showing how the distichous spike of teosinte could have become the polystichous ear of corn with kernel-row numbers ranging from 8 to 40.

Beadle, working with Galinat, has classified F_2 populations of maize-teosinte hybrids with respect to several characteristics differentiating the two genera. Their data tend to confirm the earlier results of Collins and Kempton (1920) and Mangelsdorf and Reeves (1939) in showing that four—possibly five—principal hereditary units are involved in the segregation, and their results differ from those of Langham, who found the segregation in similar crosses to be more simple. Collins and Kempton found every characteristic that they studied to be correlated with at least one other characteristic, and Mangelsdorf and Reeves showed the characters of the ear to be correlated with various plant characters such as height of the stalk, number of tillers and leaves, percentage of secondary tassel branches, and several others. I think it unfortunate that Beadle and Galinat confined their classification to only a few of the numerous characteristics in which maize and teosinte differ and so gained a greatly oversimplified impression of the true nature of segregation in maize-teosinte crosses.

Recently Beadle, supported by a grant from the National Science Foundation (1970), organized a teosinte-mutation hunt in Mexico, the principal object of which was to find variations in teosinte of a kind that might have been selected by man in the domestication of teosinte and its transformation into cultivated maize. The original plan was for a party of about twenty hunters to examine approximately a million teosinte plants to determine the frequency of "mutation" from single to paired spikelets, distichous to polystichous spikes, brittle to solid rachises, and indurated to soft shells. Only about 75,000 plants were actually examined; among these no clear-cut "mutations" from teosinte to maize characteristics were found. This is, perhaps, fortunate because the enterprise had a serious defect—no provision was made for distinguishing mutations from segregates of maize-teosinte hybrids, and many hybrids were found. From the standpoint of showing how teosinte might have been transformed into cultivated maize, the hunt was disappointing. As a safari, however, it was a notable success, the counterpart, on a modest scale, of Heyerdahl's voyage on the "Kon-Tiki," proving little or nothing scientifically but providing its participants with a memorable adventure.

I could cite a number of additional reasons for being skeptical about the teosinte theory but will mention only two, which if accepted —as I accept them—virtually rule it out. These are the fossil corn pollen and the prehistoric wild corn described in Chapter 15. Realizing that their theory is not compatible with this evidence, its advocates argue that "there must be something wrong" with it or that "it is not to be taken seriously." I shall discuss these curious convictions further in Chapter 15. Here I shall only say that in al-

FIG. 4.8 Plants of the teosinte race Balsas growing at the base of a rock cliff at kilometer 55 on Mexican highway 51 between Iguala and Teloloapan in Guerrero, Mexico. It was in sites such as this in which the teosinte mutation hunt organized by George W. Beadle occurred. Numerous hybrids of teosinte and maize were found but no soft-shelled forms or "macromutations" converting teosinte into maize. (Photograph by H. Garrison Wilkes.)

most fifty years of reading and research I have found no *factual* evidence of any kind, archaeological, ethnographic, linguistic, ideographic, pictorial, or historical on the domestication of teosinte. The only evidence that teosinte might be the progenitor of cultivated maize is that it is corn's closest relative. An alternative interpretation of this close relationship is proposed in the following section.

Maize the Progenitor of Teosinte?

As the result of reexamining various ideas—my own and those of others—about the role of teosinte, I have come to recognize an alternative possibility, namely, that teosinte is essentially a mutant form of maize. Once we consider seriously the possibility that maize evolved from teosinte as the result of a few large-scale mutations, then it becomes apparent that the change might equally well have been in the opposite direction. Indeed, of the two possibilities, the latter is the more plausible. As previously mentioned, morphologists have long recognized that in a number of morphological characteristics maize is the more primitive and teosinte the more highly evolved of the two. Furthermore, the mutations required to transform teosinte to maize are still largely in the realm of speculation, while some of those required to convert maize to teosinte are well known. Mutants in maize involving a change from the polystichous to distichous condition have been described by East and Hayes (1911), Tavcar (1935), and Burdick (1951). Hepperly (1949) has described a line in which some of the ears have odd numbers of kernel rows and has demonstrated that these have single spikelets, a teosinte character. A mutant showing a strong photoperiodic response to short days has been described by Singleton (1946). I once regarded these teosinte characters in maize

as the product of the previous introgression of teosinte into maize through repeated hybridization. However, they can as well be considered as evidence that maize is capable of mutating in the direction of teosinte. Galinat (1971) is now engaged in combining a number of these mutants in an effort to synthesize teosinte. If this succeeds, it will show, he argues, that cultivated maize might have been derived from teosinte by mutations in the reverse direction. What it will show, even more clearly and directly, is that teosinte may be the product of combining mutations in maize.

Another fact favoring the theory that teosinte originated from maize is that there is virtually no limitation of time in this postulated evolution; it may have occurred over a period of millions of years. Grasses and grazing animals began to become abundant in the Miocene epoch some 20 to 30 million years ago. If Galinat's most recent theory (1970) that *Tripsacum* is an allopolyploid hybrid of a species similar to teosinte in its characteristics and one similar to *Manisuris* is valid—and I regard it as quite plausible—time must be allowed for this allopolyploid to have become differentiated into at least five diploid species and later through the hybridization of these into at least four tetraploid species (see Chapter 5). And even more time must be allowed for the still earlier divergence of teosinte from maize.

In contrast the concept of maize as a domesticated form of teosinte supposes that the evolutionary changes that now differentiate the two occurred after agriculture was invented. In both the Old World and the New the archaeological evidence indicates that domestication is a relatively recent development in man's cultural history, beginning sometime after about 10,000 B.C. in the Old World (Renfrew, 1969; Murray, 1970) and perhaps somewhat later in the New World

(MacNeish, 1964). Apologists for the teosinte hypothesis may argue—and they do—that domestication probably began much earlier than the archaeological evidence now indicates. This is sheer speculation with no basis in fact.

I now propose that serious consideration be given to the hypothesis that the remote ancestor of maize was an Andropogonaceous plant quite similar in its principal characteristics to our genetically reconstructed ancestral form of pod-popcorn except that it was perfect flowered. The change from the perfect-flowered to the monoecious condition is not difficult to visualize. It has occurred in many genera of grasses. It was probably after maize had become monoecious that there occurred a series of mutations—from the polystichous spike to a distichous one, from paired spikelets to solitary ones, from the herbaceous glumes to indurated ones—all of which are in the direction of evolutionary specialization—that combined to differentiate teosinte from maize.

Once maize and teosinte had diverged they remained spatially isolated until the domestication of maize and the subsequent spread of its culture brought them into sympatric relations and hybridization, accompanied by a flow of genes in both directions, became common (see Chapter 11). In the meantime the maize line of descent had become differentiated into a number of geographical races of which several were domesticated (see Chapter 10). Hybridization between these races, with teosinte and perhaps also with *Tripsacum*, produced explosive evolution and the wealth of variation that exists today.

Perhaps the most important conclusion that Reeves and I reached in 1939 is that the ancestor of cultivated corn was corn. I am now pointing out the distinct possibility that corn was the ancestor not only of cultivated corn but also of teosinte.

5 *Tripsacum*, a More Distant Relative of Corn

Tripsacum is unfortunately one of those genera which present special difficulties to the collectors and has consequently been rather neglected by them. Making an accurate and complete record of a *Tripsacum* plant on an ordinary herbarium sheet is like attempting to stable a camel in a dog kennel.

Cutler and Anderson, 1941

Our success in hybridizing maize and *Tripsacum* (1931) followed by our hypothesis that teosinte is a hybrid of maize and *Tripsacum* has served to focus attention—our own and that of others—on this distant relative of corn which previously had received but little study. There had, for example, never been a thorough taxonomic treatment of this genus, although one might have expected it to appeal to monographers as a subject worthy of attention because of its diversity and widespread distribution throughout this hemisphere, not to mention its importance as a relative of America's principal crop plant. It is difficult to believe that taxonomists were deterred, as Cutler and Anderson suggest, from studying this interesting genus because of the limitations imposed by the dimensions of the standard herbarium sheet. Yet it is true that except for a brief synopsis of the genus by Hitchcock in 1906 and an outline of the species by Nash in 1909, there has been no complete treatment of the genus as a whole until that of Cutler and Anderson (1941). Although these authors modestly called their survey "preliminary" it is quite comprehensive in reviewing the previous nomenclature, examining and listing the specimens in the herbaria of the United States, and in describing the species then known, including a new one not previously recognized.

Cutler and Anderson recognized seven species ranging in geographical distribution from Connecticut in the northern United States to Paraguay in South America. Two additional species have since been described by Hernandez and Randolph (1950). The genus appears to have its center of diversity

and probably of origin in Mexico and Central America—of the nine recognized species six, possibly seven, occur in Mexico. All of the species of *Tripsacum* are perennial herbs with well developed rhizomes which may be either underground or partially exposed. The plants have numerous tillers or shoots, some of which are short, sterile, and leafy, and others long, fertile, and branching. In favorable sites the fertile stems tend to branch at nearly all of their nodes, and each branch may terminate in an inflorescence which varies from a single spike to a freely branched panicle. In most species the inflorescences borne on the axillary stems have fewer branches than those terminating the main stalk.

All species of *Tripsacum* are monoecious, having their male and female spikelets in separate positions on the same spike. The male or staminate spikelets are borne on the upper part of the spike and are paired, as in maize and teosinte. The female or pistillate spikelets are borne below and are solitary as the result of the abortion of one member of the pair whose vestigial remains, scarcely visible to the naked eye, are easily identified by microscopic examination.

The staminate spikelets are paired and the solitary pistillate spikelets are borne alternately along the axis of the rachis. This disarticulates readily when the fruits are mature and even sooner if the styles are not pollinated. The styles are relatively short compared to those of maize or teosinte and are bifurcated almost throughout their entire length, the two branches being fused only at the point of attachment to the ovary wall. Both branches are profusely covered with

hairs. The caryopses are completely enclosed in fruit cases consisting of segments of the rachis and lower glumes. Both structures become highly indurated at maturity, in some species being about as hard as the shells of acorns. One of the characteristics of *Tripsacum* is that the induration occurs only in those segments in which caryopses are developing. Rachis segments in which fertilization of the ovule has not occurred or in which the caryopsis has failed to develop to maturity remain less indurated and lack the "polished" appearance of the segments containing normally developed caryopses. Perhaps akin to the induration of the tissues of the rachis segments and lower glumes is a tough, relatively thick rind of the flowering stems of *Tripsacum*. Even the slender stems of *T. zopilotense* and *T. floridanum* are wiry, and the thick stems of the robust tropical species, although not hollow, approach the canes of some of the reeds in toughness.

Although there is great variation in *Tripsacum* species in their robustness, ranging from the somewhat delicate growth habit of *T. zopilotense* with its slender stems and narrow leaves to the tall, thick-stalked, broad-leaved *T. laxum*, all species have in common a characteristic that I can only describe as "hardiness." Some of the Mexican species, for example, are quite resistant to drought; individual plants are found growing on steep canyon walls in gaps in the limestone outcrop where such soil as there is is likely to be on the alkaline side. The tetraploid *T. dactyloides*, on the other hand, is tolerant of excessive moisture and is often found at the edge of drainage ditches or marshes where the soil

FIG. 5.1 A plant of *Tripsacum zopilotense* grown at the Fairchild Tropical Garden, Miami, Florida. This species, which is characterized by narrow, drooping leaves, is the most delicate of the *Tripsacum* species. (Photographed by R. Tantravahi.)

FIG. 5.2 Mr. Raymond Baker, a famous hybrid corn breeder, is here examining the terminal inflorescences of *Tripsacum dactyloides*, the most common species in the United States. (Courtesy Pioneer Hi-Bred Corn Co.)

at times becomes completely saturated and is usually acid. Transplanted to an upland site, this *Tripsacum* is tolerant of drought. Mexican species are quite tolerant of heat; the *Tripsacum* from Connecticut and Massachusetts can survive even the coldest winters. Most species are relatively resistant to injury by insects and diseases.

Although not aggressive when compared with weedy perennial grasses such as Johnson grass, *Sorghum halepense*, in subtropical regions, or couch grass, *Agropyron repens*, in the temperate zone, some species of *Tripsacum* have a remarkable capacity to take advantage of an improved environment and are quite capable of moving rapidly into sites which are especially suitable for their growth. Among these are the well-drained shoulders of modern highways. I do not recall that, as a student at Kansas State College in 1917 to 1921, I ever saw *Tripsacum* along the railroad or the old highway between Topeka and Manhattan, although I made the trip many times. Now it has become the dominant vegetation for long stretches on the highway. In North Carolina there are many places on the highways where *Tripsacum* grows in almost solid stands. When *Tripsacum* is planted in an experimental garden where it is fertilized and watered and free of competition with weeds, it grows luxuriantly. The individual plants may become massive clumps several feet in diameter at the base with numerous tall stems terminated by graceful fascicles of spikes. Under these favorable conditions *Tripsacum* becomes a truly noble grass altogether worthy of its relationship to maize.

Descriptions of the Species

Except for the addition of two new species, mentioned above, discovered by Hernandez and Randolph, the classification of *Tripsacum* is still essentially that of Cutler and Anderson (1941). Randolph's (1971) extensive field studies of *Tripsacum* in Mexico and Guatemala have extended the known range of several species and shown that hybridization among them may have been much more extensive than had previously been described.

Of the nine recognized species of *Tripsacum* five are now known to be predominantly diploid with a haploid chromosome number of 18 and four are tetraploid with a "haploid" number of 36. The distinction between diploid and tetraploid species is, however, not hard

and fast, since most of the so-called diploids have some tetraploid forms. Brief descriptions of the species approximately in order of their robustness within these two categories follow. For more detailed botanical descriptions of the Mexican and Guatemalan species I refer the reader to Randolph's article (1971).

The Diploid Species

Tripsacum zopilotense Hernandez X. & Randolph. This is the most delicate of all of the *Tripsacum* species. Its stems are slender and seldom more than a meter tall. The leaves are narrow and usually pendant, the lower ones markedly petiolate, the upper less so or not at all. Auricles are well defined. Sheaths are completely glabrous; spikes are usually unbranched, erect and slender, the pistillate portion about one-third of the total length. Staminate spikelets, 5–6 mm. long, usually sessile, their glumes normally lanceolate, relatively soft and pointed.

At first glance plants of *T. zopilotense*, like those of *T. floridanum*, are easily mistaken for those of *Manisuris cylindrica*, which has the same narrow leaves, slender stems, and slender unbranched terminal spikes. *Manisuris*, however, is not monoecious and its functional spikelets are all perfect flowered. Because of its general resemblance to *Manisuris cylindrica*, this species may be regarded as the most primitive one of the genus. It is approximately the kind of plant that might be expected as an allopolyploid hybrid of *Manisuris* and wild corn or teosinte. Originally found in the Cañada del Zopilote, Guerrero, Mexico, and once thought to be confined to this area, it has since been collected by Randolph (1971) in other localities in southwestern Mexico.

T. floridanum Porter ex Vasey. This species, like *T. zopilotense*, has a delicate habit resembling that of *Manisuris cylindrica*. Its flowering stems are 50 to 100 cm. in height; the leaves are narrow and not petiolate. The sheaths are glabrous; the auricles indistinct. The terminal spikes are usually unbranched; the staminate spikelets stiff and blunt; both members of a pair are usually sessile. Growing together in a garden, *T. floridanum* and *T. zopilotense* are usually easily distinguished by the stiff, erect leaves of the former. Some plants of this species, however, have pendant leaves similar to those of *T. zopilotense*. In the herbarium specimens it is distinguished by its nonpetiolate leaves, absence of an auricle, and by its stiff, blunt, staminate spikelets. In other respects the two species are quite similar.

T. floridanum is confined to several counties in southern Florida and one of the adjoining Florida Keys. One collection has been reported from east Texas. Cutler and Anderson considered this to have been probably introduced from Florida, since no connections between the two points have been made, but one wonders who could have introduced it and for what reasons. I have examined this specimen and doubt that it is *floridanum*, since it cannot be positively distinguished from the narrow-leaved form of *dactyloides* described below.

T. floridanum is found most commonly among pine trees on soils derived from coral rock. Plants very often occur at the base of pine trees, a fact that Galinat attributes to birds perched on the trees' lower branches dropping seeds while cracking the indurated shells with their beaks.

Tripsacum dactyloides L. 2n. The most common species in the central and western United States is a tall grass, 2 to 3 meters in height, with thick, knotty rhizomes. The leaf blades are usually 1 to 2 cm. wide. The terminal inflorescences are usually branched, with as many as six branches. This species extends from Texas to the plains and prairies of the cen)ral United States and as far north and east as Michigan.

T. dactyloides var. *occidentale* Cutler & Anderson. Cutler and Anderson made a new variety of *T. dactyloides* of southwestern Texas west of the Pecos River which they distinguished from the species by its staminate glumes, which are more than 9 mm. long, soft, and tapering to an acute tip. Although no Mexican localities are included in the collections listed by them for this variety, its range undoubtedly extends into northeastern Mexico. Even more distinct from the species in some respects than the variety *occidentale*, and therefore in my opinion entitled to separate varietal status, is the *Tripsacum* of the Gulf Coast region of Texas. In its narrow leaves and sparsely branched terminal inflorescences this form is quite similar to *T. floridanum*. Indeed when I first encountered it in 1927 I assumed that it was *T. floridanum* until Dr.

A. S. Hitchcock, to whom I had sent a specimen, identified it as *T. dactyloides*.

T. maizar Hernandez X. & Randolph. This is by all odds the most robust of the so-called diploid species. Its stalks are tall and thick; the leaves are wide and not petiolate or glabrous. The lower sheaths are strongly pubescent, the upper less so or not at all. The auricle is well defined. The terminal inflorescences are profusely branched with secondary and tertiary branching, 18 to 50 branches, some lacking pistillate spikelets. The staminate spikelets are soft and pointed, one member being pedicellate. *T. maizar*, like *T. zopilotense*, was first found in Guerrero but occurs also in other parts of southwestern Mexico, Randolph (1971).

In its general aspects *T. maizar* is the most maize-like of the *Tripsacum* species, hence its name. It is the only *Tripsacum* species in which the terminal inflorescence contains a central spike and some of the branches are completely staminate. I have a strong suspicion, but no really tangible evidence and certainly no proof, that *T. maizar* is the product of past introgression of maize into *Tripsacum*.

T. australe Cutler & Anderson. This *Tripsacum*, which Cutler and Anderson (1941) recognized as a new species, is described by them as follows: "Plant slender to robust, nodes usually enlarged; leaves 1–4 cm. wide, somewhat petiolate, blades smooth, usually glabrous, sheath with distinct to semi-distinct auricles, outer surface glabrous below, lanulose-tomentose above, at maturity barely clasping the culm, culm lightly to heavily lanulose-tomentose; inflorescence of 1–4, rarely more, spikes, staminate spikelets sessile." *T. australe* is distinguished from *T. dactyloides* by the dense pilosity covering parts of the culm and sheath and by its somewhat petiolate leaves. Its range is restricted to the countries of South America, including Bolivia, Brazil, British Guiana, Colombia, Ecuador, Paraguay, and Venezuela. Once thought not to occur in Peru, it has recently been collected there (Grobman, 1967).

The Tetraploid Species

T. dactyloides 4n. The tetraploid *T. dactyloides* differs from the diploid both in being generally somewhat more robust, having

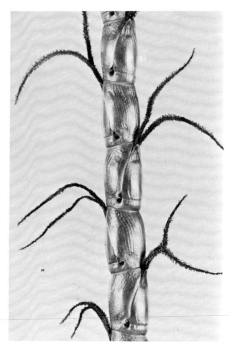

FIG. 5.3 Part of the pistillate portion of a lateral inflorescence of *Tripsacum dactyloides*, illustrating the short, bifurcated styles beset with hairs and the indurated rachis segments and lower glumes in which the seeds when mature will be enclosed. (Mangelsdorf and Reeves, 1939.)

Fig. 5.4 A plant of the Texas-Gulf-Coast form of *Tripsacum dactyloides*. In its stiff narrow leaves and unbranched terminal spikes this form is quite similar to *T. floridanum*.

larger pollen grains, and in its geographic range of distribution. Employing pollen grain size of herbarium specimens as a criterion, Tantravahi (1968) found the tetraploid *T. dactyloides* to be the only *Tripsacum*, except for *T. floridanum* in Florida, in the southeastern and eastern part of the United States. Along the eastern seaboard it extends from Florida to Massachusetts. Along the Gulf Coast it is the only species from Mississippi to Florida. In eastern Texas and the central United States both the diploid and tetraploid forms occur.

T. lanceolatum Rupr. ex Fourn. In its general characteristics *T. lanceolatum* is somewhat more robust than *T. dactyloides*, and its leaves, as its specific name suggests, are more slender and pointed. The flowering stems are 0.5 to 2 m. in height. The leaves are 10 to 30 mm. wide and are not petiolate. The sheaths are sparsely hirsute and the auricle very indistinct. The terminal inflorescence is unbranched or bears 2 to 5 branches. *T. lanceolatum* is one of the most variable of the species. In northeastern Mexico it resembles *T. dactyloides*; in southern Arizona it includes a form, *T. Lemmoni*, which was previously recognized as a distinct species and which may again, after further study, be assigned specific rank. Part of the variability of this species may be the result of hybridization with other species with which its range overlaps. *T. lanceolatum* has been collected in Arizona, in more than half of the states of Mexico, and in Guatemala.

T. laxum Nash ex Ascherson. This species has a rather wide range of variation, but it can usually be distinguished from other species by its smooth sheaths and wide, nonpetiolate leaves. The flowering stalks are 1 to 3 m. tall. The leaves are 30 to 80 mm. wide. The auricle is indistinct. The terminal inflorescences usually have about six branches, but their number may be as high as twelve. In the wild, *T. laxum* occurs in Mexico, Guatemala, and San Salvador, but under cultivation as a forage crop in the tropics it has a much wider distribution. One of the cultivated forms is sterile. Ting (1960) has shown that this form is not a tetraploid but a triploid, presumably a hybrid between a tetraploid and a diploid species. Randolph (1971) recognizes only the

sterile form and regards this as chiefly or exclusively a cultivated plant. He states that *T. laxum* is often confused with *T. latifolium*.

T. latifolium Hitchc. This species is distinguished from the other tetraploid species less by its characteristics than by its geographic distribution. This is the principal species of the West Indies, but it occurs also in Central America and southern South America. The West Indian plants are slender with unbranched terminal inflorescences but the Central American forms are quite robust with flowering stalks 1 to 4 m. tall, leaves 20 to 80 mm. wide, the lower leaves petiolate, sheaths glabrous, auricle indistinct. In northern South America this species is difficult to distinguish from *T. australe*.

T. pilosum Scrib. & Merr. In some respects this is the most robust of the *Tripsacum* species. Flowering stalks 2 to 6 m. in height; leaves 20 to 100 mm. wide, not petiolate; sheaths tuberculate-hispid; auricle distinct; terminal inflorescence profusely branched, averaging about 15 branches but sometimes as many as 22. The stiff hairs of the lower sheaths, which are very irritating to the hands of the collector, are the most conspicuous feature of *T. pilosum*. In Mexico *T. pilosum* has approximately the same range as *T. lanceolatum*, but it is more selective in its habitat and grows in more protected sites. Its range extends into northwestern Guatemala.

The Origin of Tetraploid Species of Tripsacum
Randolph (1955) has suggested that the Mexican and Central American tetraploids of *Tripsacum* originated as the result of the hybridization of two diploid species, *T. maizar* and *T. zopilotense*. Tantravahi, employing a computer and a technique of linear discrimination made a multiple-character analysis, using eight morphological characters, of the *T. lanceolatum pilosum* complex and its postulated diploid parents. This showed that the parental species are morphologically distinct and clearly distinguishable from each other. The graphs for the tetraploids were found to lie between those of the diploids showing that in their morphological characters these tetraploids are intermediate between their postulated diploid parents and indicating that at least these two tetraploids, *T. lanceolatum* and *T. pilosum* are of hybrid origin.

Cytology of *Tripsacum* Species
In both chromosome numbers and in the morphology of its chromosomes, *Tripsacum* differs from its American relatives, teosinte and maize, in which the haploid number is 10. In the diploid species of *Tripsacum* the basic number is 18 and in the tetraploid species, 36 (Longley, 1924; Reeves and Mangelsdorf, 1935; Graner and Addison, 1944; Dodds and Simmonds, 1946; Hernandez and Randolph, 1950). Supernumerary accessory chromosomes known as B type chromosomes and thought to consist primarily of hererochromatin have been found by Tantravahi (1968) in *T. floridanum*, *T. zapolotense*, and *T. maizar*.

The 18 *Tripsacum* chromosomes vary considerably in length, the longest being about five times as long as the shortest. The longer chromosomes are within the range of maize chromosomes with respect to their lengths. Ting (1960) has pointed out that the chromosomes of *T. australe* could be considered as representing two groups of nine: group A comprising the nine longer and group B the nine shorter chromosomes. The shortest chromosome in the A group is about the same length as the shortest maize chromosome.

Some or all of the chromosomes of *Tripsacum* species bear deeply staining bodies known as "knobs." The number and position of the knobs varies with the species and with the geographic race within the species. Thus in *T. floridanum*, Longley (1937) identified 21 knobs and both Chaganti (1965) and Tantravahi (1968) identified 12 but not in identical positions. Prywer (1963) found the chromosome-knob number in *T. zopilotense* to vary from virtually none to high numbers. She found (1954) *T. maizar* to have knobless chromosomes, and Tantravahi (1968) identified three knobs. Graner and Addison (1944) described the chromosomes of *T. australe* as knobless, and Ting, studying a collection from a different part of South America, identified five knobs. An important difference between the knobs of *Tripsacum* and those of maize and teosinte is in their positions. Those of maize and the Mexican teosintes are predominantly internal, and those of *Tripsacum* are, with a few exceptions, terminal. The difference has an important bearing, as we shall see later in this chapter, on our hypothesis of the origin of teosinte as a hybrid of maize and *Tripsacum*.

FIG. 5.5   Dr. U. J. Grant, a Rockefeller Foundation corn breeder, stands beside a plant of *Tripsacum laxum* grown at Medellin, Colombia. This is one of the most robust of the *Tripsacum* species.

Diploid Tripsacum an Amphidiploid?

Anderson (1944), noting that the basic chromosome number, 18, of *Tripsacum* is twice that which Reeves and I (1935) had found in *Manisuris* (n = 9), suggested that the "diploid" *Tripsacums*, those with 18 pairs of chromosomes, may themselves be tetraploid and that if these *Tripsacums* have the cytological formula XXYY, where X and Y stand for two sets of 9 chromosomes, *Manisuris* might represent the XX genome. There is some support for this suggestion in the resemblance which has long been recognized of *Manisuris* to *Tripsacum*; in fact *M. cylindrica* was assigned to the genus *Tripsacum* and called *T. cylindrica* by Michaux in 1803 (cf. Hitchcock, 1935).

The older literature shows many other instances of the assignment of *Manisuris* and *Tripsacum* to the same genus. Galinat and I, going through the sheets of *Manisuris* in the Gray Herbarium, found numerous sheets on which the specimen had been initially identified as *Tripsacum*. The most striking difference between the two genera is that *Manisuris* has perfect flowers and *Tripsacum* is monoecious, bearing staminate and pistillate spikelets in separate parts of its inflorescences. We pointed out (Mangelsdorf and Reeves, 1939) that these differences may not be sufficient justification for considering the relationship of the two genera to be remote.

There was some support for Anderson's suggestion in the observation by Randolph (1955) that diploid *T. dactyloides* has two pairs of the nucleolus-organizing chromosomes, suggesting that it is an allopolyploid. This particular condition has not been observed in other species, but perhaps equally significant is the fact that in *T. floridanum* Longley (1937) and later Tantravahi (1968) identified chromosome 16 as the one carrying the nucleolus organizer, and Chaganti (1965) found the organizer on chromosome 10. Similarly the organizer occurs on chromosome 16 in *T. maizar* (Tantravahi) and *T. australe* (Ting, 1960) but on chromosome 10 in *T. laxum* (Tantravahi). Anderson (1944) is also responsible for the suggestion that the tetraploid *T. dactyloides* of the United States is an allopolyploid between related "diploids" XXYY and XXZZ. This would account, Anderson suggested, for the 30 configurations he found quite commonly in the 36-chromosome *Tripsacums* from Virginia and southern Indiana: 6 quadrivalents and 24 bivalents. Already mentioned is the suggestion of Galinat *et al.* (1964) that *Tripsacum* is an ancient allopolyploid hybrid of *Manisuris* and wild maize which has lost one of corn's 10 chromosomes.

Crossing Relationships

The first attempts to hybridize *Tripsacum* with its relatives were those of Collins and Kempton (1914, 1916). Their cross of *Tripsacum* × maize produced only matroclinous plants, presumably the result of apomictic development of *Tripsacum* ovules. The cross *Tripsacum* × teosinte was even more surprising, for it gave rise to a single plant which proved to be pure teosinte like the pollen parent, a condition for which the authors suggested the term "patrogenesis."

The first true, living hybrids of *Tripsacum* with one of its relatives were the products of the crosses which I made in Texas. In 1929, as part of my joint project with Reeves, I had applied pollen from Texas *Tripsacum*, now known to be a diploid, to the silks of maize. Within 24 hours the silks had wilted as they normally do when corn pollen is applied, and microscopic examination showed that the pollen grains of *Tripsacum* had germinated and many of the pollen tubes had entered the styles. No seed development occurred, however, and fertilization presumably had not been accomplished. Since the silks of the Texas *Tripsacum* are usually less than an inch long and those of maize are generally several inches to more than 12 inches long, the distance to be traversed by the *Tripsacum* pollen tubes in reaching the micropyles in maize is many times as great as that normally required of the tubes in fertilizing their own species. The obvious solution to this difficulty was to shorten the styles of maize, since these are receptive to pollen over their entire lengths. This was first tried in a group of ten plants. In five of these the pollinations were made on silks of normal lengths. In the other five, an incision was made through the husks on one side of the ear and the silks exposed by the incision were pruned back to less than an inch in length before the *Tripsacum* pollen was applied. The results were decisive. In the five ears with silks of normal length no development of any kind occurred, but on the five ears with part of their silks shortened by pruning, every one produced some seeds and these were confined to the narrow region of the ear in which the silks had been shortened.

Spurred on by these encouraging results we carried on extensive crossing experiments the following year (1930). Seven commercial varieties and a number of genetic-tester stocks were used as female parents. We made the crosses in an isolated garden well removed from the main experimental corn field and on its windward side. Tassels were removed from all of the corn plants before they began to shed pollen. Before pollinating with *Tripsacum* pollen, we removed the entire shuck by a circumcision at the base of the ear; all of the silks were then pruned back with shears. After pollination the ear was covered with an artificial shuck to prevent contamination by wind-blown corn pollen from other fields and to forestall insect damage and excessive drying. Ordinary crepe paper wrapped over a strip of cotton wound around the base of the ear furnished a substitute shuck that expanded readily to accommodate the growth of the ear. Because of these artificial techniques employed in the first successful hybridization of maize and *Tripsacum*, both Weatherwax and Randolph have concluded that this hybridization could never have occurred in nature. I shall show later that this conclusion does not necessarily follow.

Altogether 382 ears with an estimated 184,925 silks were pollinated by *Tripsacum* in 1930. The total number of mature seeds obtained was 84, an average of 4.54 per 10,000 pollinated silks. The varieties of maize differed widely, however, in their "crossability" with *Tripsacum*. From the Horton variety we did not obtain a single mature hybrid seed; from the Surcropper variety we obtained an average of 6.6 mature hybrid seeds for every 10,000 silks pollinated.

Even the best of the mature seeds were small and shrivelled, weighing on the average only about one-seventh as much as normal maize seeds of the same variety. Planted in soil in the field probably few of them would have grown. Even between moist blotters in a germinator only a few would have been capable of producing living seedlings. This was in the days before embryo culture employing well-balanced nutrient solutions had become an established practice. What we did instead was to soak the seeds in water long enough to soften the pericarp, remove the pericarp, dip

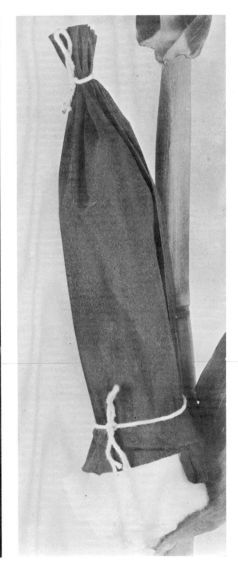

FIG. 5.6 The techniques we employed in 1930 in producing the first hybrids of maize and *Tripsacum*. The husks were removed (*left*) and the long silks were pruned to an inch or less in length (*center*). After pollen of *Tripsacum* was applied an artificial husk of crepe paper was provided (*right*) to prevent contamination, desiccation, and insect damage. Because artificial techniques were used in effecting this hybridization it has been argued that natural crossing between these two genera could never have occurred in nature. (Mangelsdorf and Reeves, 1939.)

the seeds briefly in a fungicidal solution, and plant them on sterile agar in petri dishes. These were kept in a glass-walled chamber at an optimum temperature. Since the seedlings were exposed to light as soon as germination began, development of chlorophyll ensued immediately and with it the manufacture of food. This no doubt enabled the young seedlings to overcome, partially at least, the handicap of a poorly developed endosperm and a weak root system. By employing this technique we were able to obtain, from the 84 seeds, 45 seedlings, of which 29 survived transplanting to the field—29 living seedlings from 185,000 silks pollinated. No wonder that some botanists have contended that this hybridization could never have occurred in nature. In addition to the 29 seedlings of maize × diploid *Tripsacum*, we had one seedling of maize × tetraploid *Tripsacum*. Because this hybrid had been planted several months earlier than the others it was the first to bloom.

After almost forty years I can still recapture part of the thrill with which I watched our first maize-*Tripsacum* hybrid come into flower. This must have been the counterpart on a modest scale of the thrill that the New World's early explorers experienced when encountering for the first time a new river, a new lake, or a new mountain which no white man had ever seen before. Each day I made a trip to the greenhouse to make sure that no new pests—of which we, in Texas, had more than our share—had attacked the precious plant and to see how much the flowering stem had grown. One night I dreamed that the inflorescence had emerged from the surrounding leaf sheaths and in my dream I had a vivid picture (although not in color) of its characteristics. When the first inflorescence finally did appear it was almost exactly as I had dreamed it. This, perhaps, is no cause for wonder, since I was by this time thoroughly familiar with the characteristics of both maize and *Tripsacum* and would have expected the inflorescence to be intermediate between its two parents in its characteristics, as it indeed proved to be.

Although intermediate in the sense that it showed some of the characteristics of its maize parent the hybrid resembled its *Tripsacum* parent much more closely. This was not surprising, since it had received 36 chromosomes from *Tripsacum* and only ten

from maize. Like those of *Tripsacum* the staminate and pistillate spikelets were borne on the branches of the inflorescence, the former above and the latter below. The pistillate spikelets were enclosed in an indurated shell consisting of a segment of the rachis and the lower glume. The styles were almost twice the length of *Tripsacum* styles and were divided only part of their length; the fusion of the remaining part was weak. Reeves' cytological studies showed that at diakinesis there usually were 18 bivalents and ten univalents, the former presumably representing the chromosomes of *Tripsacum* and the latter those of maize. Trivalents, the product of maize chromosomes associating with a pair of *Tripsacum* chromosomes, occurred rarely. This hybrid proved to be sterile and set no seeds when pollinated by either of its parents, although we made hundreds of pollinations on it.

Although our first hybrid of corn and *Tripsacum* turned out to have approximately the characteristics that we might have expected from our knowledge of its parents, the hybridization had one wholly unexpected result. Our first article on the hybridization of maize and *Tripsacum* was published in the *Journal of Heredity*. A short news item based on this article and reporting the hybridization was released at about this same time by Science Service. Francis Dahl, a cartoonist for the *Boston Herald* with a talent for noting unusual news items and employing them as subjects for cartoons, picked up the Science Service item and used it in an imaginative way in a cartoon that appeared in the Herald January 12, 1932 (Fig. 5.7). I have always found the last picture in the series "Expert Crosses Grass with Impunity" especially amusing.

Another person who took more than passing interest in our successful hybridization of corn and *Tripsacum* was Henry Wallace, who in 1933 became Secretary of Agriculture under Franklin D. Roosevelt. I think it was during his first year in office that I was invited with Frederick D. Richey, recently appointed Chief of the Bureau of Plant Industry but previously Principal Agronomist in charge of corn investigations in the U.S.D.A., to visit Mr. Wallace in his office. He expressed great interest in our maize-*Tripsacum* hybrids, asked some penetrating questions about what contributions, if any, *Tripsacum* might make

to the improvement of corn, and suggested to Mr. Richey that the role of corn's relatives, teosinte and *Tripsacum*, be discussed in the section on corn improvement in a forthcoming yearbook of the Department of Agriculture, which was to be devoted to the subject of "Better Plants and Animals." This is how it happened that in the 1936 Yearbook the chapter on corn improvement contained a section on the botanical relatives of corn that was well illustrated with our photographs of teosinte, *Tripsacum*, and their hybrids with corn.

Zea × *T. dactyloides* (2n)

Since Reeves and I studied this hybrid more extensively than any other and since it contributed to our hypothesis on the origin of maize, I shall describe our studies in some detail.

Development of the Hybrid Seeds

Reeves' anatomical studies of the hybrid seeds showed that development was approximately normal in the early stages. By the eleventh day after pollination, however, retarded or abnormal development of the hybrid seeds was apparent. The most advanced of these showed differentiation in the embryo, but both embryo and endosperm were smaller than those of normal maize seeds of the same age. Other embryos remained completely undifferentiated. Fourteen days after pollination virtually all kernels were abnormal in their embryos or endosperms or both; some had begun to atrophy. It is not surprising that so few viable hybrid seeds were finally obtained from this cross.

The Hybrid Plants

The majority of the hybrid plants finally obtained were quite vigorous. In general appearance they resembled their *Tripsacum* parent more closely than the maize parent. The plants had many tillers and numerous flowering stalks. The terminal inflorescences had several branches but no central spike. The lateral inflorescences were unbranched. The spikes, like those of *Tripsacum*, were staminate above and pistillate below. The pistillate spikelets were embedded in hollow segments of the rachis. The most conspicuous influence of the maize parent was in the long styles, which were about six times as long as those of the *Tripsacum* parent.

FIG. 5.7   A cartoon by Francis W. Dahl that appeared in the Boston Herald, January 12, 1932, based on a release by Science Service reporting our successful crossing of corn with the wild grass *Tripsacum*. (Permission of the artist.)

FIG. 5.8 Plant of the hybrid *Zea × Tripsacum dactyloides*, 2n, (*right*) compared with its maize parent (*left*). Grown in Texas in 1931, this is the first known hybrid of these two forms. Mangelsdorf and Reeves, 1939.)

Cytology

The chromosome number of the hybrid plants was, as expected, 28, representing 18 chromosomes from *Tripsacum* and 10 from maize. During the first meiotic division the degree of chromosome pairing in late prophase was variable. It was rarely possible to find a cell in which there was no pairing, but when pairing did occur it was often feeble except for two or three pairs which had a tendency to unite more closely and regularly. Reeves tabulated 58 cells with respect to the number of pairs observed, with the following results: no pairs, 0; one pair, 2; two pairs, 5; three pairs, 21; four pairs, 21; five pairs, 8; six pairs, 1. Whether this pairing was between different maize chromosomes, between different *Tripsacum* chromosomes, or between maize and *Tripsacum* chromosomes could not be definitely determined.

Studies of the anaphase stage showed a great variety in types of chromosome behavior, of which the most significant was one which resulted in the formation of unreduced gametes. In some cells the chromosomes failed to move toward the poles, and all of them became enclosed in a nuclear membrane. The partition wall that formed following this failure of nuclear division was sometimes located entirely to the side of the restitution nucleus. This resulted in one daughter cell receiving no chromosomes, the other receiving all of the chromosomes of the two parents. It was the gametes formed from these unreduced nuclei—and only these —that functioned and made it possible to backcross the hybrid by both of its parents and to cross it as well by teosinte.

In spite of the formation of unreduced gametes the hybrid was completely pollen sterile, probably because the proportion of potentially functional pollen in the anthers was so small that the anthers failed to develop normally. Female gametes carrying the unreduced chromosome number were functional when pollinated by *Tripsacum*, maize, and teosinte. The fertilities which occurred were 1.3, 3.4, and 5.8 percent respectively.

The Triploid Hybrid (*Zea × Tripsacum 2n*) × *Zea*

The triploid hybrid plants, as might be expected, were more maize-like than those of the diploid hybrids in virtually all of their

62   *Tripsacum*, a More Distant Relative of Corn

characteristics. The leaves were broader, the number of tillers fewer, the terminal inflorescences had a central spike, the lateral inflorescences were covered with husks. The ears bore both staminate and pistillate spikelets, but the proportion of the latter was much higher than in the diploid hybrid and the spikelets were not solitary but paired. The kernel-row number was most commonly four but six- and eight-rowed ears also occurred.

This hybrid had 38 chromosomes, the number expected from combining two maize genomes of 10 chromosomes each with one genome of 18 chromosomes from *Tripsacum*. Reeves' studies of diakinesis figures showed univalents up to 18 in number. We assumed these to represent the *Tripsacum* chromosomes. Bivalents up to 10 in number we assumed to represent 10 pairs of maize chromosomes. In other cells trivalents up to 4 in number were observed; we assumed these to represent an association between a *Tripsacum* chromosome and a pair of maize chromosomes. Many irregularities, including the production of supernumerary nuclei, occurred in the formation of the gametes. This hybrid, like the preceding one, was completely pollen sterile. It was also female sterile when pollinated with either diploid or tetraploid *Tripsacum* but when pollinated with maize it was fertile to the extent of 21.3 percent.

A Comparison of Diploid and Triploid Hybrids of *Zea* and *Tripsacum* with the Parental Genera

We had by now produced, grown, and obtained data on three *Zea-Tripsacum* hybrids: (1) *Zea* × 4n *Tripsacum*; (2) *Zea* × 2n *Tripsacum*; and (3) (*Zea* × *Tripsacum*) × *Zea*. With respect to the genomes received from *Zea* and *Tripsacum*, these hybrids were of the constitutions ZTT, ZT, and ZZT respectively. Combined with the two parental genera, they comprised a series of five forms in which the percentage of *Zea* germ plasm varied from 0 to 100. A comparison of these genotypes in various morphological characteristics in which the two parents differed by wide extremes seemed to offer an opportunity to answer two questions: (1) Are the two genera differentiated primarily by dominant genes? (2) Is a genome of 10 *Zea* chromosomes equivalent to a genome of 18 *Tripsacum* chromosomes or is it the individual chromo-

FIG. 5.9 A plant of the triploid hybrid (*Zea* × *Tripsacum dactyloides*) × *Zea*. This plant, the first of its kind to be produced, had 38 chromosomes comprising two genomes of 10 chromosomes each from *Zea* and one genome of 18 from *Tripsacum*. Grown in Texas *ca.* 1932. (Mangelsdorf and Reeves, 1939.)

63   *Tripsacum*, a More Distant Relative of Corn

somes which are equivalent to each other?

In terms of genomes the five genotypes were 0, 33.3, 50.0, 66.7, and 100 percent *Zea*, with an average of 50 percent. In terms of chromosome numbers they were 0, 21.7, 35.7, 52.6, and 100 percent *Zea*, with an average of 36.7 percent. The average relative values for the quantitative maize characteristics scored were 0, 33.2, 42.0, 73.6, and 100, and the average of these was 49.6 percent. We concluded that it is the genomes and not the individual chromosomes which are equivalent and that there is not marked dominance of the characteristics of either of the parental genera.

The Progeny of the Triploid Hybrid
(*Zea* × 2n *Tripsacum*) × *Zea* × *Zea*

Pollinating the silks on 1,758 pistillate spikelets of the triploid hybrid produced 374 seeds, from which 178 mature plants were grown. Since there had been fairly regular pairing of the 20 maize chromosomes in the hybrids and since the 18 unpaired *Tripsacum* chromosomes appeared to have segregated more or less at random, we expected this population of plants to be an extremely variable one. All plants were expected to have 20 *Zea* chromosomes and to differ primarily in the number of extra *Tripsacum* chromosomes which had been added to the normal *Zea* chromosome complex. The number of extra *Tripsacum* chromosomes might theoretically vary between 0 and 18. If the population were large enough we should expect to find the variation ranging from normal corn plants on the one hand to plants duplicating the parental triploid hybrid on the other.

Unfortunately it was impossible for various reasons to make chromosome determinations of all of the plants in this population. I did collect pollen from the majority of them and later attempted to estimate the number of extra *Tripsacum* chromosomes in the individual plants by determining the amount of empty pollen. I am certain now that this was at best a crude and unreliable criterion, since *Tripsacum* chromosomes vary considerably in their effect on producing empty pollen. By a strange coincidence, however, it now turns out that the conclusions that I reached by this unreliable method were not far wrong. The pollen fertility determination indicated that the most common number of extra chromosomes in these backcrossed plants was one,

and the number of plants having no extras or having two extra chromosomes was about equal. Chaganti (1965) in a later study of a similar population found essentially the same kind of distribution.

Perhaps the most important result of studying this backcross population was the finding by Reeves, in one of the 2n + 1 segregates (now known as an addition monosomic) which carried the dominant allele of the marker gene *su* on chromosome 4, that there was synapsis at the pachytene stage between this *Tripsacum* chromosome and certain maize chromosomes. The synapsis was most common with chromosome 4, but chromosomes 1, 2, and 9 were sometimes involved.

More recent studies of maize-*Tripsacum* hybrids by Randolph and his student, Marjorie Maguire, and by my student, Raju Chaganti, not only have confirmed the observations which Reeves and I made before 1939 but have also extended them. Randolph (1955) found, as we had, that the hybrid of maize with diploid *Tripsacum* produces viable seed exclusively from unreduced female gametes and that the backcrosses to the triploid hybrid resulted in some individual plants which have 20 maize chromosomes plus one *Tripsacum* chromosome.

Particularly significant have been the studies of Maguire on later generation 2n + 1 backcross derivatives of the maize-*Tripsacum* hybrid which Randolph had produced. Although much of her work on the monosomic addition stocks was concerned primarily with theoretical cytology—how chromosomes behave under unusual circumstances—her observations are also important in shedding light on the relationship of *Tripsacum* and maize.

The *Tripsacum* chromosome in one of the monosomic addition stocks which Maguire studied was about 42 percent as long as corn's longest chromosome. Its long arm, terminated by a conspicuous knob, paired regularly with the short arm of corn's chromosome 2, and the *Tripsacum* chromosome carried the dominant alleles of two marker genes, lg_1 and gl_2, located on the short arm of corn's chromosome 2. Through some mechanism, which Maguire was apparently reluctant to call either crossing over or translocation, the segment of the *Tripsacum* chromosome, equivalent to about 24 to 25 percent of the length of chromosome 2, was trans-

ferred to that chromosome along with its terminal knob. The transfer, whatever its nature, was a reciprocal one, and the *Tripsacum* chromosome now carried an equivalent segment of corn's chromosome 2 without a terminal knob. In plants heterozygous for these transferred segments designated as Z and T, a strong homology was indicated by the fact that synapsis was complete in 93 percent of the microsporocytes examined (in a later study the proportion was even higher, 249 of 250 cells exhibiting normal synapsis). Yet crossing over between the marker genes on the segments was rare. Introducing a third marker gene, ws_3, into the cultures, Maguire found that there was no crossing over between this gene and lg_1—which in corn is 11 crossover units away—and only 0.8 percent crossing over between lg_1 and gl_2, which in corn are about 19 crossover units apart. The map distance between ws_3 and gl_2 inidcated by Maguire's data is 1.5 crossover units compared to 30 units in corn.

On the question of homology Maguire (1962) concluded it is conceivable that "the two segments have maintained sufficient homology through their evolutionary divergence to allow synapsis which appears normal under microscopic examination, but that minute differences have accumulated which prevent an exact juxtoposition which might be necessary for crossing-over to occur." Plants homozygous for the *Tripsacum* segment were vigorous and fertile. The TT plants were short and stocky with stiff leaves and few tassel branches. The ears had a smaller number of rows of ovules and a smaller number of ovules per row. The silks were usually split for an appreciable distance, a characteristic in an attenuated form of *Tripsacum* in which the silks are split throughout their entire length.

On the question of substituting *Tripsacum* chromatin for corn chromatin Maguire (1961) concluded "it appears that a *Tripsacum* chromosome segment can be substituted for a corn chromosome segment equal to about 3% of the total length of the corn genome." Although Maguire is of the opinion that the genomic differences between the two genera, *Zea* and *Tripsacum*, and the difficulty in crossing them support the view of Randolph (1955) and Weatherwax (1955) and although she stated that "recent natural exchange of chromosome material between corn and *Trip*-

sacum has probably happened rarely if at all" her findings are quite consistent with and indeed lend substantial support to our theory that teosinte may be the product of the hybridization of maize and *Tripsacum*.

While Randolph and Maguire were engaged in studies of maize-*Tripsacum* hybrids, Galinat and I were also making progress. An important contribution was Galinat's discovery that *T. floridanum*, the *Tripsacum* species endemic to southern Florida, could be crossed with corn much more readily than any species that had previously been tried. When Galinat brought back from Florida plants of this species which he had collected from several localities in the Everglades National Park and proposed to cross these with corn I gave him no encouragement. I had collected plants of this species near Homestead, Florida, in 1941 and had maintained them in the greenhouse for several years. I made no attempt to cross this species with corn because, of all of the *Tripsacum* species then known, *T. floridanum* resembled maize the least. Assuming that botanical resemblances are to some extent, at least, a measure of relationship, I also assumed that *T. floridanum* was less closely related to corn than the species which we previously employed in our crossing experiments and as a consequence would be less crossable.

My assumption proved to be completely wrong. Galinat found that *T. floridanum* had a higher degree of crossability with corn than any *Tripsacum* species so far tested. Of the 35 ears pollinated with *T. floridanum* pollen all yielded at least a few hybrid kernels and some of these were so well developed that they germinated and produced viable seedlings without the benefit of embryo culture. Not only did *T. floridanum* cross readily with corn but the F_1 hybrids were highly fertile, the first generation about 85 percent female fertile and the first backcross generation only slightly less so. Galinat produced so many backcrossed hybrid seeds that he was able to distribute them to other experimenters by the hundreds.

Why *T. floridanum*, one of the most primitive of the *Tripsacum* species, should cross so readily with corn is not yet determined. One explanation, suggested by Galinat and a quite plausible one, is that it is a peripheral species which, far removed in its range from the center of origin of corn, had lost some of

FIG. 5.10 Ear of maize (*A*) and pistillate spikes of *Tripsacum* (*B*), Florida teosinte (*C*), *Zea* × *Tripsacum* (*D*), and the trigenomic hybrid (*E*) combining the chromosome complement of the three parental species. In most of its characteristics the hybrid showed the character that was common to two of its three parents. Thus its pistillate spike is similar to that of *Tripsacum* and teosinte.

65 *Tripsacum*, a More Distant Relative of Corn

the genetic factors which, in earlier stages of their evolution, had served as barriers between the species.

The hybrids of corn and *T. floridanum* furnished the material for extensive cytogenetic studies of Raju S. K. Chaganti, a graduate student from India, and became the basis for his doctoral dissertation, which was later published in full. He found a high frequency of chromosome associations at meiosis in the F_1 hybrid. By comparing these with frequencies in a haploid maize plant and in the haploid genome of *Tripsacum* in a triploid ZZT hybrid, in both of which pairing frequencies were low, he reached the conclusion that pairing in the F_1 hybrid is allosyndetic, i.e., between maize and *Tripsacum* chromosomes. He found the transmission of extra *Tripsacum* chromosomes in the progeny of the first backcross to be nonrandom. Plants with one to four *Tripsacum* chromosomes occurred in much higher frequencies than would be expected on the basis of the random assortment and transmission of the 18 *Tripsacum* chromosomes. The frequency polygons which Chaganti plotted from his data on the transmission of *Tripsacum* chromosomes shows a remarkable similarity to the polygon plotted on the estimates of extra *Tripsacum* chromosomes based on pollen fertility which I had made earlier (Mangelsdorf and Reeves, 1939). We now know that pollen fertility is not a reliable criterion for estimating the presence of extra *Tripsacum* chromosomes, since Maguire (1957) has shown that certain single *Tripsacum* chromosomes can cause complete pollen sterility. My estimates apprently represent one of those rare cases of reaching the right conclusion for the wrong reason.

Chaganti found various *Tripsacum* chromosomes to produce a variety of effects upon the meiotic cycle of maize. These included asynapsis and a consequent reduction in chiasma frequency; breakdown of metaphase, post metaphase, and cytokinesis; chromosome breakage, multinucleolation, supernumerary division of microspores, plasmodial mycrosporocytes, timing imbalance, and others. Since these abnormalities did not occur in the F_1 hybrid where all 18 *Tripsacum* chromosomes were present, he explained their occurrence as the result of "the breaking up of a coadapted genome and the superimposition of its segments on a foreign genome that bring forth an array of interactions in the host

genome" (Chaganti, 1965). An obvious corollary of this explanation is that inherited defects in the meiotic cycle such as the asynapsis and supernumerary cell divisions described by Beadle (1930, 1931), may likewise be the product of superimposing chromosome segments on a foreign genome.

Chaganti, who before coming to the United States from India had made a cytogenetic study of the genus *Coix*, considered by some nineteenth-century botanists to be the closest relative of maize, concluded that the Old World members of the tribe *Maydeae* comprise two distinct groups, neither of which is closely related to the American *Maydeae*. He considered it logical to assume that maize originated from a New World Andropogonaceous ancestral stock with n = 10 chromosomes as the basic number. Like other students of the tribe, Weatherwax, 1954, Randolph, 1955, and Stebbins, 1956, he concluded that the *Maydeae* represent "an artificial assemblage so far as natural relationships go." Cytogenetically the *Maydeae* are monoecious *Andropogoneae*.

It was during this period, too, that the successful cross between my nine-gene multiple tester and *Tripsacum* was made. After perfecting this stock and using it to detect linkages in maize-teosinte crosses, I attempted repeatedly to cross it with *Tripsacum* but without success, partly for the reason, I suspect, that the several lines of the tester were all too highly inbred and there was little or no variation in the gametes which they produced. It remained for Galinat to succeed where I had failed. He did so by employing an F_1 hybrid of two different lines as the female and by enlisting the invaluable assistance of Dr. Floyd P. Hager, a retired bacteriologist, who, in connection with his own experiments in making wide crosses among lilies, had become quite expert in the techniques of embryo culture.

Perhaps spurred on in part by my offer to award each of them a gold medal if they would successfully cross the multiple-gene tester with *Tripsacum*, Galinat and Hager finally succeeded in producing one living plant of this hybrid. In recognition of their success I presented each of them not with a gold medal but with a Peruvian gold tie clasp depicting one of the ancient Incan fertility Gods.

Unfortunately, this hybrid, unlike earlier maize-*Tripsacum* hybrids which produce

some functional unreduced female gametes, was completely sterile. Its failure to form unreduced gametes was probably due to the extensive pairing of the maize chromosomes with those of *Tripsacum* which Ting (unpublished report) observed at meiosis. This pairing may have been due to blocks of genes originally from teosinte or *Tripsacum* which had become incorporated in the tester stock by bringing together marker genes from so many different genetic stocks. In our linkage tests discussed in Chapter 4 both Rogers and I found evidence of teosinte characters being linked in the coupling phase with the recessive marker genes introduced by the tester. We also now know that teosinte introgression has mutagenic effects, a phenomenon discussed in more detail in Chapter 12, and that the mutants that maize geneticists employ in their experiments are often associated with the introgressed blocks of genes in which they occurred. In any case there was extensive pairing of maize chromosomes with those of *Tripsacum* in this hybrid, and it is probably for this reason that it was completely sterile.

By treating a tiller of this plant with colchicine, Galinat succeeded in doubling its chromosome number, thus producing an amphidiploid hybrid with 10 pairs of corn and 18 pairs of *Tripsacum* chromosomes. This plant, although not producing rhizomes, has a strong perennial tendency, and Galinat has maintained it as a clone since 1961 through the propagation of basal tillers.

The amphidiploid proved to be highly fertile when pollinated by corn. Its functional gametes carried 10 maize 18 *Tripsacum* chromosomes. Pollinated with the multiple-gene tester the next generation plants are triploid with a genomic constitution ZZT. All should be homozygous for the marker genes carried by the recurrent tester parent, but the effects of the recessive markers should be concealed if the *Tripsacum* genome carried their dominant alleles. This proved to be the case. None of the triploid ZZT plants exhibited the effect of any of the recessive marker genes throughout the entire plant. Some plants, however, have shown sectors of recessive tissue, most commonly japonica striping involving the gene *j* on chromosome 8, less commonly absence of plant color involving the gene *a* on chromosome 3. The exact mechanism causing the appearance of these sectors has not been determined, but it probably involves some form of somatic segregation in which a *Tripsacum* chromosome or a part of it fails to be transmitted to a daughter cell in one of the mitotic divisions.

The triploid plants were again pollinated by the multiple-gene tester to produce a population of plants all presumably possessing 20 maize chromosomes and various numbers of additional *Tripsacum* chromosomes. Plants receiving no *Tripsacum* chromosomes would be expected to show all seven of the recessive characters of the recurrent multiple-gene tester stock. Plants receiving extra *Tripsacum* chromosomes carrying one or more of the dominant alleles of the marker genes would be expected not to show the recessive character covered by the alleles on the extra chromosomes. Ninety-two plants of the segregating backcross population were grown in 1962, thus finally executing an experiment that I had first planned in the thirties in Texas when I began to bring together genes for recessive characters on as many chromosomes as possible.* That the experiment finally succeeded, however, was due at least as much to Galinat's skill as to my planning.

The experiment was designed to serve several purposes: (1) to determine whether *Tripsacum* with its 18 chromosomes—almost twice the number in corn—carried one or two dominant alleles of each recessive marker in corn; (2) to identify the individual *Tripsacum* chromosomes both genetically and cytologically; (3) to determine whether a *Tripsacum* chromosome carrying the allele of one of the recessive marker genes on a corn chromosome also carried the alleles of other recessive genes on that same chromosome. By this time, Chaganti, who was making a cytogenetic study of the maize–*T. floridanum* hybrid, joined us in our study of the multiple-tester–*T. dactyloides* crosses, and we were able to make progress toward these three objectives.

If the *Tripsacum* chromosomes segregate at random at meiosis and if there is no selective gametic or zygotic elimination, then there

*It is not unusual in genetic experiments for the development of the appropriate stocks to require more time than the experiment itself. There are several classical examples of this in research employing Drosophila as the experimental organism. The long lapse between the planning of my experiment and its final execution was due in part to having to develop first a tester adapted to Texas conditions and subsequently to convert it to one adapted to New England conditions.

should be, in a backcross population of the triploid hybrid by the multiple-gene tester, 50 percent of dominants for those loci for which *Tripsacum* carries one dominant allele and 75 percent dominants for loci for which *Tripsacum* carries two alleles. The data do not fit closely either of these theoretical expectations, probably because there is a high degree of selective elimination of the gametes carrying extra *Tripsacum* chromosomes. We considered it significant, however, that the frequencies of dominants were similar for six of the seven marked loci, average 32.2 percent dominants, while the seventh, J_1, had almost twice this frequency, 60.5. We concluded tentatively that *Tripsacum* may carry two alleles for this marker but only one each for the other six, which represented chromosomes 1, 2, 3, 4, 7, and 9. It turned out later that Tripsacum carries only one allele for the marker gene *J* on chromosome 8, the apparent higher percentage of dominants being due to difficulty in classifying japonica plants, a difficulty not confined to maize-*Tripsacum* hybrids.

By this time we had adopted the term "homeolog"* to designate these extra *Tripsacum* chromosomes which carried the dominant alleles of the maize markers. Chaganti was able by cytological examination of about one-third of the plants to confirm our tentative conclusion that *Tripsacum* had only one homeolog for each of the marked maize chromosomes. The evidence was as follows. First, one of the genetic addition monosomics, carrying Wx^T, the allele of *wx* on maize chromosome 9, bore an ear with 68 percent waxy and 32 percent nonwaxy kernels. This result we attributed to single homeolog transmission with the elimination of some homeolog-carrying gametes. Chaganti found this plant to have only one extra *Tripsacum* chromosome. Secondly, chromosome counts in many of the other plants revealed the presence of various numbers of unmarked *Tripsacum* chromosomes in addition to those which could be identified genetically as homeologs of the seven recessively marked chromosomes of the WMT maize parent. Since 3 of the 10 maize chromosomes were unmarked by reces-

*A modified spelling of "homoeologue," a term first used by Huskins (1941) to designate the chromosomes in the different genomes of hexaploid wheat which are counterparts of each other but not homologs.

sive genes, some of these additional chromosomes might represent the homeologs of these 3 unmarked maize chromosomes, but if a plant carried more than 3 unmarked *Tripsacum* chromosomes the presence of these must be accounted for in other ways. For example, one plant carried 13 extra *Tripsacum* chromosomes, of which 7 were marked by dominant alleles. Of the remaining 6, not more than 3 could be assumed to be homeologs of the 3 unmarked maize chromosomes. Thus at least 3 of the remaining 6 must be nonhomeologs. Another plant had 12 addition chromosomes, of which only 3 were marked by dominant alleles, leaving not less than 6 or more than 9 as unmarkable nonhomeologs. A third plant carried five addition chromosomes, none with dominant markers, of which not less than 2 nor more than 5 must be considered unmarkable.

Of the 163 addition chromosomes identified by Chaganti, 85 were found by Galinat to carry dominant alleles of marker genes. If we assume that *Tripsacum* has two distinct genomes, one comprising nine markable chromosomes and the other nine unmarkable ones; and if we assume further that markers are lacking in the WMT stock the expected number of dominant alleles exhibited by the 163 chromosomes is 100. This differs significantly from the 85 identified. However, as has been mentioned above, the classification for japonica striping was not satisfactory and was believed to be inaccurate in some cases. Excluding *j* as one of the markers, the theoretical ratio of marked and unmarked chromosomes becomes 109:54. The observed ratio was 100:63, which represents a good fit.

The results of these observations were published by Galinat et al. (1964). The authors regarded this close fit between theoretical and observed ratios of marked and unmarked *Tripsacum* chromosomes as "good, indeed almost conclusive" evidence that *Tripsacum* contains two distinct genomes. One of these was assumed to comprise homeologs to nine maize chromosomes, the other to comprise nine chromosomes so remotely related to maize that they were not homologous to maize chromosomes and carried no dominant alleles of the maize markers. These cytogenetic data, combined with the fact that *Tripsacum* is intermediate between maize and *Manisuris* in a number of characters, were regarded as evidence supporting the hypothesis of Galinat (1964) that *Tripsacum* is an ancient allopolyploid hybrid of wild maize and *Manisuris*. This hypothesis, although bold and new, is not without precursors. The resemblance of *Manisuris* and *Tripsacum* has long been recognized. Indeed, *Manisuris* has sometimes been regarded as a species of *Tripsacum*. The older literature shows many instances of *Manisuris* and *Tripsacum* being assigned to the same tribe, and both Weatherwax (1935), and Randolph (1955) have suggested that this should be done again. Finding that *Manisuris cylindrica* has a haploid chromosome number of 9, half that of *Tripsacum*, Reeves in our 1939 monograph stated "This number suggests a relationship of *Manisuris* and *Tripsacum* since in *Tripsacum* the reduced numbers known at present are 18 and 36. The addition of *Manisuris* to the series gives us 9, 18, and 36." Anderson (1944) was more specific, suggesting that the 18-chromosome *Tripsacum* may be an allopolyploid having a genomic formula of XXYY in which the XX genome represents nine pairs of *Manisuris* chromosomes. He suggested further that the 36-chromosome *Tripsacums* arose as octoploids through the hybridization of an XXYY tetraploid and an XXZZ tetraploid. Anderson made no suggestion about the origin of the YY and ZZ genomes, and there is no reason to believe that the "Z" symbol which he employed to designate one of the genomes was intended to represent *Zea*. In Galinat's hypothesis, however, it does represent exactly that.

Randolph (1955), like Anderson, suggested that species of *Tripsacum* with n = 18 chromosomes "probably should be considered as natural tetraploids, and both *Tripsacum* and *Manisuris* should be placed in the same tribe." His finding (1955) of two nucleolus-organizing chromosomes in *T. dactyloides* lends some support to the concept of *Tripsacum* as an allopolyploid as does also the fact that in *T. floridanum*, although only one such chromosome has been found in any given plant, it is chromosome 10 in some collections (Chaganti, 1965) and chromosome 16 in others (Longley, 1937; Tantravahi, 1967).

After observing the apparent homology between a segment of *Tripsacum* chromosome and the short arm of chromosome 2 of maize and the ability of the *Tripsacum* chromatin to substitute for the corresponding maize chromatin, Maguire (1961) came close to suggesting that *Tripsacum* might contain a genome of maize chromosomes when she stated: "But if similarities of the sort demonstrated here are widespread, then chromosomes of *Tripsacum* must be basically more like those of corn than a comparison of their idiograms would suggest. There is evidence to support the view that 36 chromosome diploid *Tripsacum* actually arose as an allotetraploid (Randolph, 1955). In this case the basic genome of *Tripsacum* would be 9 chromosomes compared to the 10 of corn." Galinat's hypothesis on the origin of *Tripsacum* is more specific than that of previous writers, partly because he had earlier (1956) made a morphological study of the inflorescences of maize and its relatives, including *Manisuris* and several genera of the *Andropogoneae*, and so was well acquainted with their characteristics. And in our joint studies of the prehistoric maize of the Tehuacán Valley described in Chapter 15, he had become familiar with the characteristics of prehistoric wild corn. He was thus able to see that *Tripsacum* possesses a combination of characteristics of *Manisuris* and wild maize.

The Galinat hypothesis of *Tripsacum's* origin has important implications for maize improvement. If there has been introgression of *Tripsacum* into maize in the past either directly or through teosinte as a genetic "bridge," then such introgression has involved on the one hand germ plasm from the tribe *Andropogoneae*, to which the drought-resistant, productive cultivated sorghums belong, and on the other hand genes of wild maize, which is now extinct. And if there is indeed a genome of wild maize in *Tripsacum* it has been subjected to some very rigorous natural selection in diverse environments for long periods of time during *Tripsacum's* evolutionary divergence and spread throughout the hemisphere. Corn breeders might do better to seek genes for corn's further improvement in *Tripsacum* than in sorghum, which despite numerous attempts has never been hybridized with maize.*

*Since this was written Galinat (1970) has modified his hypothesis on the origin of *Tripsacum*. He now postulates that one of the ancestors of *Tripsacum* was teosinte-like rather than maize-like. The new hypothesis greatly diminishes the importance of *Tripsacum* in the improvement of corn. Under the original hypothesis *Tripsacum* could be considered a source of genes from a wild corn now extinct. Under the more recent version it is a source of genes from teosinte. There is no reason for corn breeders to seek such genes in *Tripsacum* when they can obtain them much more easily directly from teosinte itself.

In connection with *Tripsacum's* role in the improvement of maize, I am reminded of a conversation that I had in 1939 with H. K. Hayes, of the University of Minnesota, a pioneer in the development of hybrid corn and a profound student of corn improvement. We were both attending a meeting at Columbia, Missouri, of the corn breeders of the U.S.D.A. and the Middle Western states. Hayes raised the question whether a part of our program should be devoted to discussing the possible role of teosinte and *Tripsacum* in corn improvement. I said that we still knew too little about those relatives of corn to make such a discussion meaningful and suggested that it might become appropriate in another 25 years or so. The 25 years have passed and such a discussion is now, indeed, relevant.

It may be well to remember at this point that the Galinat hypothesis is still only that. It is useful in stimulating new interest in *Manisuris* and in other members of the tribe *Andropogoneae*; it is useful in being testable in various ways. Strong evidence in its support would be the successful hybridization of *Manisuris* with either maize or *Tripsacum* or with both. Reeves and I attempted such hybridization on a small scale in the thirties, and although we found *Manisuris* pollen germinating on the styles of *Tripsacum* and both maize and *Tripsacum* pollen germinating on the styles of *Manisuris*, there was no development of hybrid seeds. Although no hybrids involving these genera have since been produced, the possibility of crossing them should not be ruled out until crosses have been attempted between several different species of *Manisuris* and all of the known species of *Tripsacum*.

More recent data from the maize-*Tripsacum* addition monosomics are consistent with the hypothesis but also show that the genomes of *Tripsacum* are not as distinct as was first supposed. Galinat (1971) found that one *Tripsacum* chromosome, designated as 9, carries the alleles of six marker genes on the short arm of maize chromosome 2. Another *Tripsacum* chromosome, tentatively designated as 4, carries the alleles of five marker genes on maize chromosome 7. A third *Tripsacum* chromosome, designated as 5, carries the alleles of eight marker genes on maize chromosome 9. On the other hand the *Tripsacum* chromosome that is designated as 9, which carries alleles of six marker genes on maize chromosome 2, does not carry the

TABLE 5.1 Cytogenetic Correspondence of Some *Zea* and *Tripsacum* Chromosomes

Chromosome no. in Maize	Identified loci common to maize and *Tripsacum*
1	Bm_2^* (but not $Sr Br An_1$)
2 S	$Ws_3 Lg_1 Gl_2 b Sk Fl_1$
2 L	V_4
3	A_1
4 S	Su_1 (but not $La: Gl_3 Bm_3 Ra_3 J_2$)
2 L	Gl_3 (but not $La Su_1: Bm_3 Ra_3 J_2$)
5	Pr
6	$Py Sm?^*$
7 S & L	$O_2 V_5: Ra_1 Gl_1 Ii$
8	J_1
9 S & L	$Yg_2 C Sh_1 Bz_1 Wx: Gl_{15} Bk_2 Bm_4$
10	G

Source: Galinat, 1971.
S = Short arm; L = Long arm;
* = Preliminary data based on monosomic;
: = Centromere position.

allele of the seventh, v_4. And *Tripsacum* chromosome 7, which carries the allele of the marker gene *su* on the short arm of maize chromosome 4, does not carry the alleles of any other marker genes for that chromosome. Likewise, *Tripsacum* chromosome 13, which carries the allele of gl_3 a marker on the long arm of chromosome 4, does not carry the alleles of any other chromosome 4 marker genes. Thus the alleles of the marker genes of maize chromosome 4 must be distributed over at least three *Tripsacum* chromosomes and perhaps more.

The evidence so far as it now goes does not disprove the hypothesis of *Tripsacum* comprising two distinct genomes, one of which is closely related to maize—or according to Galinat's most recent hypothesis—to teosinte. I regard it as especially significant that three of *Tripsacum*'s chromosomes carry alleles of 6, 5, and 8 maize marker genes respectively. In any case the intergenomic mapping in which Galinat and his associates are engaged is, I think, the most elegant cytogenetic-cytotaxonomic research in progress today.

Crossing Over between Maize and *Tripsacum* Chromosomes

Maguire (1962) found crossing over between the short arm of corn chromosome 2 and the homologous segment of a *Tripsacum* chromosome to be greatly reduced—only about one-twentieth as much as in maize. Galinat's unpublished data indicate a similar but somewhat less drastic reduction. He has found crossing over to occur in all of the chromosomes so far tested: 2, 4, 7, and 10 at roughly one-tenth the frequency of that in maize. To me the significant fact is that the occurrence of crossing over between maize and *Tripsacum* chromosomes has now been conclusively demonstrated. This fact has important implications with respect to both the origin of teosinte and the future improvement of maize, subjects which will be discussed in detail in other chapters.

The Trigenomic Hybrid of *Zea*, *Tripsacum*, and Teosinte

Knowing that the functional gametes produced by the hybrids of *Zea* × *Tripsacum* were all unreduced gametes having one set of 10 *Zea* chromosomes and one set of 18 *Tripsacum* chromosomes, it was at once apparent to us that if the hybrids could be crossed with teosinte the resulting progeny would all be trigenomic, having one genome from each of the three parents, which at that time were still regarded as distinct genera.

Pollen of Florida teosinte applied to the styles of 208 pistillate spikelets of the hybrid produced 12 well-developed seeds. This represents a fertility of 5.8 percent, which is slightly higher than that resulting from pollinations made on that same hybrid with maize pollen. The difference is probably of no significance, since the teosinte pollinations were made in the greenhouse under somewhat more favorable conditions than the maize pollinations, which were made in the field. The evidence does suggest, however, that the hybrid is at least as fertile crossed with teosinte as it is with maize.

The trigenomic hybrid plants were quite uniform, indicating that the gametes received from the hybrid parent were all unreduced. The plants had 38 chromosomes as was expected, 10 from *Zea*, 10 from teosinte, and 18 from *Tripsacum*. The behavior of the chromosomes at meiosis was exactly what might have been predicted from our previous knowledge of meiosis of maize-teosinte and *Zea*-*Tripsacum* hybrids. There were usually 10 bivalents at diakenesis and metaphase of the first division, and these undoubtedly represent

the 10 chromosomes each from maize and teosinte. There were usually 18 univalents representing the *Tripsacum* chromosomes. Trivalents and a smaller number of univalents sometimes occurred. To the extent that chromosome pairing is a measure of relationship, these results showed clearly that maize and teosinte are much more closely related to each other than either one is related to *Tripsacum*.

In its morphological characteristics the hybrid showed in almost every case the character that was common to two of its three parents. Thus it resembled maize and teosinte in its annual habit of growth, absence of rhizomes, pedicels on one staminate spikelet, pistillate spike enclosed in husks, and relatively wide leaves. It resembled teosinte and *Tripsacum* in the absence of central spikes in the tassel, two-ranked pistillate spikes, solitary pistillate spikelets, brittle rachis, caryopses enclosed in rachis segments, and profuse tillering. In certain "quantitative" characters, such as length of styles, the hybrid struck almost a perfect average. Maize, teosinte, and *Tripsacum* in this study had average style lengths of 233, 76, and 14 mm. respectively. The average of the three is 108 mm. The styles of the trigenomic hybrid averaged 95 mm.

The close and regular pairing of the maize and teosinte chromosomes on the one hand and on the other the "dominance" in the hybrid of those inflorescence characters which distinguish both teosinte and *Tripsacum* from maize remind one again of the conclusion reached in the preceding chapter: teosinte has essentially the chromosomes of maize but genes producing some of the characteristics of *Tripsacum*.

Unfortunately no studies of pachytene configurations were made of this hybrid. Such analyses might have shown whether the *Tripsacum* chromosomes showed any affinity with the teosinte chromosomes, which might have been distinguished from the maize chromosomes by their knobs. When pollinated by its three parents no seeds were set with teosinte or *Tripsacum* pollen and only one with maize pollen. This would suggest greater irregularities in the formation of the gametes than the studies of meiosis had indicated.

In retrospect we can see that we did not give this hybrid as much study as it deserved. More extensive pollinations with teosinte or maize might have resulted in progenies in which certain teosinte chromosomes would have been interacting with certain *Tripsacum* chromosomes. Since we have not succeeded in producing living hybrid plants of teosinte × *Tripsacum*, the progeny of the trigenomic hybrid might reveal affinities between teosinte and *Tripsacum* chromosomes.

Failure to Hybridize Teosinte × Tripsacum

Among the many crosses that Reeves and I attempted in our studies of maize and its relatives were those of the two species of teosinte, annual and perennial, with several species and forms of *Tripsacum*. Our numerous attempts to cross perennial teosinte and *Tripsacum* consistently resulted in failure, but crosses of annual teosinte with *T. dactyloides*, both the 2n and the 4n forms, produced seeds. Reeves' cytological study of the developing seeds of the cross teosinte × *T. dactyloides* 2n showed 28 chromosomes in the embryo and 38 in the endosperm, indicating that the seeds were true hybrids. Although we obtained no living plants from any of the hybrid seeds we were reasonably sure that hybrid plants could be produced by crossing on a more extensive scale than we had so far practiced. The fact that teosinte and *Tripsacum* chromosomes could be combined to produce the vigorous trigenomic hybrid described above seemed to lend some support to our opinion. For one reason or another, however, our attempts to produce living hybrid plants of teosinte and *Tripsacum* were pursued no further.

My last graduate student, Ramana Tantravahi, after reading the extensive literature that had been amassed on maize and its relatives pointed out that failure to produce hybrid plants of teosinte × *Tripsacum* represented a conspicuous gap in the research that had been conducted over a long period of years, and he proposed a study of teosinte × *Tripsacum* hybrids as the basis for his doctoral thesis problem. It seemed reasonably certain that employing the techniques of embryo culture, in which he had become proficient, it would be possible to obtain hybrid plants from seeds resulting from teosinte × *Tripsacum* crosses.

In a preliminary experiment in 1964 and in a much more extensive one in 1965, in a well-isolated garden in Cambridge surrounded by university buildings, Tantravahi crossed several races of teosinte by both *T. dactyloides* and *T. floridanum*. In 1965 he pollinated 10,398 silks of teosinte by *Tripsacum*. About three-fourths of these, 7,531, were crosses by *T. dactyloides*. From these he cultured 264 embryos; three of these grew into normal plants, all of which proved to be teosinte, presumably the product of apomictic development. Forty-five of the embryos "germinated" and produced weak roots but failed to differentiate further. No living hybrid plants of teosinte × *Tripsacum* were obtained from this cross. The results of the crosses with *T. floridanum* were similar except that they produced a larger number of apomictic teosinte plants.

Tantravahi suggested (1968) that the crosses failed because teosinte has developed strong genetic barriers to hybridization with *Tripsacum*. The ranges of teosinte and *Tripsacum* in Mexico and Guatemala are quite similar (see Wilkes, 1967) and in many localities both species are exposed to cross-pollination with the maize growing in cultivated fields, with the result that teosinte hybridizes frequently with maize. Tantravahi concluded that "If teosinte hybridized indiscriminately with maize on the one hand and *Tripsacum* on the other, it would cease to be teosinte and there would result a complex of polymorphic forms completely obliterating the specific boundaries."

More recently Galinat has conducted some ingenious experiments to determine the relationship between teosinte and *Tripsacum* chromosomes. He has crossed disomic addition maize stocks carrying a pair of *Tripsacum* chromosomes with maize stocks into which chromosomes or parts of chromosomes from teosinte had been incorporated. The preliminary results so far obtained indicate a rather high degree of affinity between the teosinte and *Tripsacum* chromosomes. These results are somewhat different from those that Reeves observed in our trigenomic hybrid in which there was little association between the teosinte and *Tripsacum* chromosomes. Despite the extensive cytogenetic research that has been conducted in the past half century the relationships of the American *Maydeae*, maize, teosinte, and *Tripsacum* to each other are not yet completely clear.

6 Corn's Old World Relatives

The lack of direct information as to the past
history of maize places a greater emphasis
on the botanical evidences. In the plant
itself there are numerous lingering reminders
of what it used to be, and the significance of
these is more pointed when we find many
of them in functional form in some of the
relatives of maize. Weatherwax, 1954

In the three preceding chapters I have dealt at length with corn's closest relatives, teosinte and *Tripsacum*. These, like corn, are New World species which before 1492 were confined to America. In this chapter I shall deal briefly with those Old World species which botanists in the past have regarded as having some degree of relationship to maize. The disparity in the space devoted to the New and Old World relatives is intentional. I became reasonably certain from our early researches —Reeves' and mine—that the problem of the origin and evolution of maize is primarily, indeed almost wholly, a New World problem; that the Old World genera have played virtually no part in corn's evolution *under domestication*. Now after more than 40 years of research on the problem I am more than ever persuaded that this is true.

We should, however, make a distinction between corn's origin and evolution under domestication and its previous phylogeny. Students concerned with the longer-range problem, the evolution of the genus *Zea* through geological eras and periods, must necessarily give some consideration to its Old World relatives. Consequently a brief consideration of these seems in order.

The family *Gramineae*, to which corn belongs, is one of the largest in the plant kingdom, comprising, according to some estimates, as many as 8,000 species. The family is as complex as it is large, and despite the work of many systematists a wholly satisfactory classification of its genera and species has not yet been constructed. The words of Sir Joseph Hooker, one of the eminent nineteenth-century botanists, are still appropriate.

He found the grasses "dreadfully difficult and systematically a chaos of imperfect descriptions, erroneous identifications, confused synonymy and imbecile attempts" (see Bor, 1965). One fact, however, most authorities agree upon: the division of the family into two subfamilies, proposed by Robert Brown in 1814, is valid. These groups, now known as the *Panicoideae* and *Poacoideae*, are distinguished by a combination of differences in the articulation of the pedicels and the number of florets in the spikelets. The subfamilies are further divided into categories called "tribes." The subfamily *Panicoideae*, to which maize belongs, is generally regarded as having three tribes: the *Maydeae*, which includes maize and its relatives; the *Andropogoneae*, a much larger tribe, which includes the cultivated sorghums, sugar cane, and many important range, pasture, and forage grasses; and the *Panicae*, a large tribe including the cultivated millets and many other valuable grasses.

The principal distinguishing feature of the *Maydeae* is monoecism, the separation of the male and female spikelets in different inflorescences or in different parts of the same inflorescence. Monoecism is now known to be far from rare among the world's grasses, and hence, to borrow the words of Paul Weatherwax (1954), a life-long student of the characteristics of this tribe of grasses, "is not as good a criterion of classification as it was once thought to be. It has appeared too many times and in too widely divergent groups of the *Gramineae* for it to have any weight in marking major categories of the family. It is futile, therefore to base any serious con-

sideration of the ancestry of maize on this outmoded concept of the phylogenetic unity of the *Maydeae*."

Weatherwax has also shown that the oriental *Maydeae*, although having monoecism in common with maize and its New World relatives, are quite different in certain morphological features: the development of the florets and the nature of the fruit cases. On the latter characteristic Weatherwax (1926) has this to say:

In all the Maydeae the fruit is wholly or partly covered by an indurated shell, which is an especially attractive superficial indication of relationship. Its relative absence in Zea may be explained by the unusually complicated covering of husks, or as a result of conscious selection by man. But this general occurrence of a hard shell is a deceptive analogy, rather than a homology. The indurated structure is a combination of a glume and an alveolus of the rachis in Tripsacum and Euchlaena, a spathe in Coix, and a glume in Polytoca, Schlerachne and Chionachne. A tendency toward induration of something connected with the fruit seems, therefore, to be all that the genera have in common, and this is possessed by so many other genera of grasses as to be of little significance in determining tribal relationships.

In his classification of the Old World *Maydeae* which are native to southeastern Asia: China, Burma, India, Java, and the Philippines, Henrard (1931), like Weatherwax (1926), recognized differences in the nature of their fruit cases and employed these as distinguishing features. The following key adapted from Henrard shows the principal characteristics of the fruit cases of the five genera:

Key to the Old World *Maydeae* after Henrard

1. Female spikelets totally enclosed at the base by an ovoid or globose osseus part of the bract.　　*Coix*
1. The fruit case chiefly formed by an indurated lower glume.
   2. Lower glume deeply 3-lobulate at the tip with the middle lobe longest.　　*Trilobachne*
   2. Lower glume not cleft or only shortly 2–3 dentate at the apex.
      3. Lower margins of the grain enclosing a cavity at the bottom of which is found the hilum.　　*Polytoca*
      3. No cavity at the base of the grain, hilum not hidden by margins of the caryopsis.
         4. Margins of the lower glumes overlapping enclosing the rachis segment.　　*Schleracne*
         4. Margins free not overlapping, the rachis segments visible.　　*Chionacne*

It is evident from these brief descriptions that with respect to their fruit cases the Old World *Maydeae* comprise two distinct groups: (1) *Coix*, which has its caryopsis enclosed in a modified spathe, (2) *Trilobachne*, *Polytoca*, *Schleracne*, and *Chionacne*, which have their caryopses enclosed in indurated lower glumes. Considering only the nature of the fruit cases, *Coix* would seem to be less closely related to the American *Maydeae* than are the remaining four genera. Before the recognition of teosinte's relationship to maize, *Coix* was regarded by some botanists as corn's closest relative and this in turn was in part responsible for the idea held by some students of plants that maize, like *Coix* must be a plant of Old World origin.

The genus *Coix* includes three species, of which *C. lachryma jobi*, known as "Job's tears" because of the supposed resemblance of the shape of its fruits to tear drops, is the most widely distributed. The fruits of the hard-shelled varieties are often used as ornaments most commonly strung as beads. The soft-shelled races are cultivated as cereals in many parts of the tropics, especially among hill tribes of Asian countries, who make a porridge and brew a beer from them. *Coix*—the hard-shelled form—has escaped from cultivation and persists as a weed or feral plant in parts of the American tropics. So far as the

lower glumes are concerned the genus *Trilobachne* seems to have more in common with tunicate maize, which I regard as one of corn's ancestral forms (Chapter 7), than any of the other genera. A short article by Bhagwat and Deodikar (1961) on this genus shows other similarities to maize which may or may not be significant. Another feature which the American *Maydeae* have in common, at least in part, with the Old World relatives and which can lead to erroneous conclusions are chromosome numbers. In maize and annual teosinte the haploid number is 10; in perennial teosinte, 20. In the Old World *Maydeae* the numbers, so far as they have been determined, are, 5, 10, and 20. This suggests that the basic chromosome number in the *Maydeae* may be 5, the higher numbers representing several degrees of polyploidy. So far as I know the chromosome number of *Trilobachne* has never been determined.

The suggestion by Anderson (1945) that maize might have originated in Asia as an amphidiploid of a 5-chromosome *Sorghum* and a 5-chromosome relative of *Coix* has never been taken very seriously even by its author, who emphasized that it was only one possible hypothesis. As such it cannot be completely dismissed by the evidence that several authorities have cited against it (Randolph, 1955; Venkateswarlu, 1962). However, the facts that the 10 chromosomes of maize show no evidence of comprising two sets and that the chromosomes of the 5-chromosome *Sorghum* and *Coix*, so far studied, are quite different from those of maize certainly lend no support to the hypothesis. Also, there is certainly no possibility that maize originated in Asia in recent times from such a cross. The fossil maize pollen discovered in Mexico and described in Chapter 15 seems to rule out this possibility quite conclusively. In rejecting the hypothesis that corn is a hybrid of *Coix* and *Sorghum* we should not overlook the fact that *Zea* and *Sorghum* may be more closely related than their present taxonomic positions—assigned to different tribes—would indicate. Recent students of *Zea Mays*, Weatherwax, Randolph, and Reeves and I, have regarded corn as being more closely related to the *Andropogoneae* than to the Old World *Maydeae*. Crossing relationships tend to support this opinion.

Crossing Relationships

When two species can be crossed regardless of the techniques that may have been employed in effecting the cross this is evidence that there is some degree of relationship between them. Failure to cross may not, however, be proof of a lack of relationship. Crossing is sometimes prevented by rather simple devices of minor phylogenetic significance. Some races of maize, for example, when serving as female parents do not cross with other races because of a single cross-sterility gene, *Ga*, on chromosome 4. The reciprocal cross, however, is easily made. The fact that corn can be crossed with both of its New World relatives, teosinte and *Tripsacum*, shows that these three taxa are related. The fact that it has never been successfully crossed with any of the Old World *Maydeae* strongly suggests that its relationship to them is more remote.

Early in our investigations of maize and its relatives we attempted to cross *Coix* with corn, using the latter as the female parent after pruning its silks as we had done in our successful hybridization of corn with *Tripsacum*. Although there was some evidence that the pollen tubes of *Coix* entered the shortened silks of corn, they produced no effect whatever, not even nucellar swelling. Venkateswarlu (1963) made about 1,200 pollinations on corn with pollen of *Coix aquatica*; no hybrids were produced. Harada et al. (1954, 1955), employing *Coix* as a female parent, obtained some seeds. These proved, however, not to be viable and Venkateswarlu (1963) has suggested that they may have been the product of apomixis, which is fairly common in the grasses and is known to occur in *Coix* (Goodman, 1965). I know, from conversations that I have had with them, that a number of other investigators have attempted to cross corn with *Coix*, in some cases on an extensive scale, always without success.

Apomixis may also be the explanation of one instance of apparent successful crossing of teosinte by *Trilobachne*. We know from Tantravahi's attempt to cross teosinte with *Tripsacum* described in Chapter 5 that apomixis occurs in teosinte. In any case I have seen no further reports confirming the success of this cross. My own attempts to cross corn and *Tripsacum* with *Polytoca mycrophylla* produced neither seeds nor nucellar swelling. While attempted crosses of corn with the Old World *Maydeae* have consistently met with failure, corn has been crossed with sugar cane, *Saccharum officinale*, a genus of the tribe *Andropogoneae*. This cross, first reported by Janaki-Ammal (1938), has since been repeated by Li and his associates (1954). Sugar cane has also been crossed with *Sorghum*, first by Venkatraman and Thomas (1932) and subsequently by a number of other investigators.

Since both *Zea* and *Sorghum* can be crossed with a third genus there must be some degree of relationship between them. Agronomists familiar with the characteristics of these two grasses have long hoped to hybridize them in order to combine the outstanding qualities of both in the hybrids. So far no actual hybrids have been produced, but agronomists continue to hope. Several years ago Gerrish (1967) made a survey of attempts to hybridize maize with cultivated sorghums. Of 40 of the larger American crop research groups to which he addressed his inquiry, 18 indicated an effort of some kind in this direction and 16 furnished descriptions of the scope, techniques, and success—or lack of it —of their work. Three principal types of crosses were made, employing as females: (1) diploid maize with shortened silks, (2) tetraploid maize with shortened silks, and (3) cytoplasmic male-sterile sorghum. Altogether several thousand plants were involved in the reported experiments. In most cases no seeds were produced, but in several instances the pollinations resulted in poorly developed seeds that either failed to germinate or produced weak seedlings that failed to survive.

The results, though scanty, lend some encouragement to continued hope. With the development of improved media for embryo culture of the grasses any hybrid seeds that are produced, however poorly developed, now stand some chance of yielding viable seedlings. Also, the sorghum varieties employed in these experiments are highly evolved cultivated varieties of *Sorghum vulgare*. Some of the more primitive varieties of this species and some of the other species of *Sorghum* might profitably be included in future crossing experiments.

Other Andropogoneae Possibly Related to Maize

Virtually all students who have worked intensively with maize and its relatives have concluded that the *Maydeae* are actually no more than a branch of the *Andropogoneae* (Weatherwax, Randolph, Mangelsdorf and Reeves), but so far none has undertaken the challenging task of determining where, within this tribe, they should be placed. Speaking of the other two tribes of the subfamily *Panicoidea* and the need for further research in this group, Goodman (1965) has expressed the opinion that "Until a more accurate classification is developed for these tribes of grasses, any search for relatives of maize is almost like searching for a needle in a haystack when one is not certain he is looking in the right haystack." This statement is obviously an intentional exaggeration designed to emphasize the present unsatisfactory state of our knowledge about the tribes of grasses related to maize. And it is true that much still needs to be done in constructing a reliable

classification of these tribes of grasses, but much good work has been and is being done. On the basis of what is already known I, for one, would concentrate on the *Andropogoneae* rather than the *Paniceae*. In view of the hypothesis (Anderson, 1944; Galinat et al., 1964) that corn's relative *Tripsacum* is a hybrid having *Manisuris* as one of its parents, I should certainly explore the species of this genus as well as some other members of the subtribe, *Rottboelinae*, to which it belongs. One of the species of this subtribe to be considered is *Elyonurus tripsacoides*, whose spikelets are quite similar in structure to those of tunicate teosinte (Galinat, 1956).

In this connection I should emphasize the importance of recognizing pod corn as an ancestral form in determining the relationship of *Zea Mays* to other genera. Collins (1912) noted many years ago that homozygous pod corn has characteristics similar to those of the *Andropogoneae*. The similarity is especially striking in our genetically reconstructed form, the fertile pod-popcorn described in Chapter 7. Except for its monoecism the reconstructed ancestral form is quite similar to some of the species of *Sorghum*, especially to the papyrescent form *S. membranaceum*, in a number of its characteristics, including the nature of its lower glumes. Because of its interesting and significant botanical characteristics, because of its role in the origin and evolution of corn, and because of the extensive experiments that my colleagues and I have conducted on it, pod corn should have a thorough treatment in this book and will be the subject of the following two chapters.

7   Pod Corn, the Ancestral Form

From a botanical standpoint I am disposed
to accept *Zea tunicata* as a primitive form. . . .
We hence are inclined to believe that when
a truly wild *Zea* is discovered it will be of
podded form, the kernels small and very
flinty. Sturtevant, 1894

Having dismissed teosinte as the progenitor of maize, Reeves and I concluded that the ancestral form of corn was probably pod corn. Pod corn is a peculiar type in which the individual kernels are enclosed in bracts, which botanists call "glumes" and which the layman knows as "chaff." This covering of the grains by floral bracts is a primitive characteristic quite common and indeed almost universal in wild grasses, but it is not confined to them. The world's principal cereals other than corn: rice, wheat, barley, rye, the sorghums, and millets, with the exception of certain little-grown varieties, all have their grains enclosed. Modern cultivated corn is unique in bearing naked kernels. Its floral bracts have been reduced to scale-like glumes too short to enclose the grains.

Genetically pod corn represents the expression of a locus on chromosome 4 designated by the symbol for tunicate, *Tu*, which is partly but not completely dominant. In the heterozygous genotype, *Tutu*, the grains are parly or completely covered by the glumes and the staminate inflorescences are fertile. The homozygous genotype, *TuTu*, is usually earless, but when an ear is produced its floral bracts are often so monstrous that the florets are sterile. Its terminal inflorescences are also often monstrous and sterile but sometimes bear grains on the branches. These facts, not generally known until the present century, account for some of the conclusions reached by various nineteenth-century botanists.

Virtually all students of corn of this century as well as the previous one have assumed that wild corn must have had its grains enclosed and protected by glumes; consequently when

a kind of corn—now known as pod corn—was discovered in which the grains were so enclosed, it was only natural that some of the students of corn would regard it as the primitive or ancestral form. Others recognized its primitive characteristics but for one reason or another dismissed it as the ancestor of cultivated corn. The statement by Weatherwax (1954) "The idea that pod corn is essentially the ancestral form and that it once grew as an undomesticated plant has always been easy for the popular mind to accept" does not tell the whole story. It was not only the popular mind that pod corn appealed to but also the professional. Virtually all serious students of corn's origin have seen in pod corn at least one characteristic, glume-covered grains, which wild corn must have had, and virtually all have given serious consideration to the possibility that pod corn could have been the ancestral form.

History of Pod Corn

There are several early descriptions of corn which, with the benefit of hindsight, could be considered to be those of pod corn. However, the earliest clear-cut and unmistakable reference to pod corn and the first suggestion that it might be the ancestral form is that of Saint-Hilaire (1829), who, in a letter to the French Academy of Sciences, described a peculiar type of maize which he had received from Brazil. This maize, which he named *Zea Maïs* var. *tunicata*, was distinctive in having its kernels enclosed in "envelopes," a condition which Saint-Hilaire regarded as primitive and one which he assumed to have been lost during domestication. Saint-Hilaire showed

this ear to a young Guarany Indian from Paraguay, then living in France, who acknowledged the corn as belonging to his country and added that it grew there in the humid forest. As De Candolle later commented "This is very insufficient proof that it is indigenous." Nevertheless Saint-Hilaire considered the evidence important and concluded that maize had its origin in Paraguay. He thought that in the natural state its grains were covered as are those of other grasses but that it soon loses these through culture.

In the light of Saint-Hilaire's communication it is now possible to conclude that there is a still earlier reference to pod corn in Paraguay. This is found in the statements of Azara (1809), who was the Spanish commissioner to Paraguay from 1771 to 1801 and who upon his return to Spain completed a manuscript describing the country and various features of its natural history. In a chapter on cultivated plants he described four kinds of corn which he saw in Paraguay. One of these, known as "abaty-guaicurú," is almost certainly pod corn, for he states "chaque grain est enveloppé à part par de petites feuilles qui ressemblent entièrement aux grandes qui enveloppent l'épi entier."

The fourth kind of corn Azara described as follows:

I do not recall the name which they give to the fourth kind whose stem, much more slender, is terminated, not by an ear, but, as in millet, by a kind of whip of many cords of which each one is covered with grains absolutely resembling those of maize but much smaller. Neither do I know the particular uses to which it is applied. I know

only that when this sort of whip is boiled in fat or oil the grains burst without becoming detached and there results a superb bouquet capable of adorning at night the head of a lady. (My translation, original quoted in Mangelsdorf and Reeves, 1939.)

An Experimental Verification of an Historical Reference

When I first read the passage from Azara in 1938 I had already worked extensively with pod corn and was familiar with its characteristics in both heterozygous and homozygous genotypes. Thus it struck me immediately that the fourth kind of corn described by Azara may well have been the homozygous form, which is often earless and which sometimes bears grains in its terminal inflorescence. I subsequently tested this possibility experimentally. I crossed pod corn with popcorn and repeatedly backcrossed the hybrid to popcorn until the grains became small, hard, and capable of popping when exposed to heat. When the terminal grain-bearing inflorescences of this corn were immersed in hot oil the kernels exploded but remained attached to the tassel to produce the "superb bouquet" which Azara had so vividly described. A photograph of one of these specimens is reproduced in Figure 7.1. We once used this photograph as a Christmas card; it was the only one of its kind.

Although this may not be proof that Azara's fourth kind of corn was pod corn, it comes close to being that. I have called it "an experimental verification of an historical reference," and it has generally been accepted as such. Weatherwax (1954), for example, who had once thought that Azara's fourth kind of corn might be a species of *Amaranthus*, later concluded "There can be no reasonable doubt that these two varieties described by Azara were the heterozygous and the homozygous forms of pod corn."

Other Early References

Because these two writers, Azara and Saint-Hilaire, are clearly concerned with pod corn and because they both place it in Paraguay, a still earlier reference may be regarded as describing pod corn. This occurs in the account of Dobrizhoffer (1822, first published in 1784), a Jesuit missionary to Paraguay from 1749 to 1767. This author, like Azara, described four kinds of corn in Paraguay, one called "bisingallo the most famous of all, the

FIG. 7.1 An experimental verification of an historical reference. By combining the characteristics of pod corn with those of popcorn it is possible to produce plants with tassels bearing small, hard kernels enclosed in glumes and capable of popping. When such tassels are dipped in hot oil the kernels explode to produce the "superb bouquet" that Azara (1809) described in connection with one of the four kinds of corn of Paraguay.

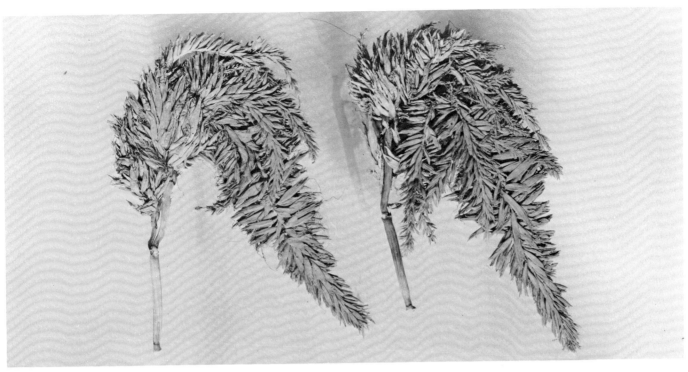

FIG. 7.2   Tassels of homozygous tunicate plants in which the residual heredity is that of inbred strains. Under these conditions the expression of the pod-corn locus is relatively constant within the strain but differs from strain to strain. *Upper:* tassels of *TuTu* on A158; *lower:* on P39.

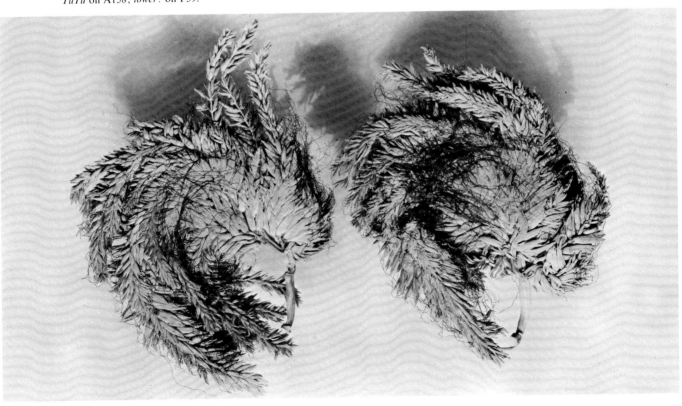

FIG. 7.3 One of the causes of monstrosity in modern homozygous pod corn is vegetative vigor. This may be a manifestation of hybrid vigor or it may be the product of good growing conditions. This photograph illustrates the effect of the environment on the expression of the tunicate character. *Left*, a tassel from a homozygous tunicate plant grown in a well-fertilized experimental field. *Right*, a tassel from a plant of the same inbred strain grown in a simulated wild habitat without fertilizer and in competition with an agressive, weedy grass.

grains of which are angular and pointed." The words "most famous of all" suggest an unusual kind of corn, and the word "bisingallo" is significant because it is only a slight variation of the word "pinsingallo" the name under which Bonafous (1836) later received a specimen of pod corn from South America. Similar words: "pisingallo," "passankalla," "pisincho," "pichinga," and "piksenkalla" are applied today in Peru, Bolivia, and other countries of South America to varieties of popcorn with pointed kernels (Brieger et al., 1958; Parodi, 1935). Consequently, it is not clear whether Dobrizhoffer was describing pod corn or popcorn. I shall present evidence later that the race of pointed popcorn was originally also a pod corn, and if this is true it would account for the fact that words related to "pinsingallo," which was definitely applied to pod corn in the nineteenth century, are now used for the pointed popcorns.

It was these early references to pod corn in Paraguay, those of Dobrizhoffer somewhat vague, but of Azara and Saint-Hilaire quite definite, that led me to initiate the search mentioned in the Preface for wild corn in Paraguay and other parts of the South American lowlands. Another early reference to pod corn and the first clearly defined illustration of this unusual corn appears in the elegantly printed and lavishly illustrated monograph of Bonafous published in 1836. The pod corn which Bonafous described and which he called *Zea cryptosperma* was sent to him from Buenos Aires by a French planter in South America who evidently supplied very little information about the corn except to state that it was known as "pinsingallo" and, because of the difficulty in separating the grains from the glumes, it was grown but little. Bonafous, who considered corn to be of Old World origin, evidently did not regard this as wild corn.

By the second half of the nineteenth century it was well known that pod corn does not breed true, and this fact influenced the conclusions of both Darwin and De Candolle. Darwin (1868) thought it almost certain that the original form of corn had its grains protected as they are in pod corn but doubted that pod corn is the original form because it produces either common or husked maize and it is not credible that a wild species when first cultivated should vary so quickly and in so great a degree. De Candolle (1882), like

Darwin, was aware of pod corn's failure to breed true; he stated that it has been cultivated in Europe and that "it often passes into the ordinary state of maize" and concluded that "this form, which might be believed a true species, . . . is hardly even a race." By this time teosinte was also well known, and its close relationship to maize was generally recognized. Pod corn was no longer the only entity to be considered as the ancestral form. Among the late nineteenth century botanists only Sturtevant appeared to have been not unduly impressed by the close relationship of corn and teosinte or by pod corn's inability to breed true, for in 1894 he made the statement that I have used as the epigraph for this chapter.

Collins (1912) realized that pod corn mus be fitted into the picture, "that the prototype of maize should be sought in a grass possessing the peculiar characteristics of *Zea tunicata*," and again, "The definite tendency toward a brittle rachis in podded plants and the demonstration that this character may serve as a natural means of dissemination, are additional evidence that the podded varieties of maize are not meaningless monstrosities, but definite reversions to a perfect-flowered ancestor." And still again (1917), "in every particular by which *Z. tunicata* departs from normal maize, it does so by replacing the specialized characters of maize with characters common to practically all other grasses." But Collins could not overlook corn's close relationship to teosinte and so arrived at a compromise in which he had maize originating from a hybrid of teosinte and an unknown pod-corn-like plant. Weatherwax (1935) assumed that primitive corn had the seeds partially enclosed in bracts, but he did not call this pod corn.

It should by now be obvious to the reader that the hypothesis that cultivated maize originated from a form of pod corn is not original with us. The original idea we owe to Saint-Hilaire. We adopted the pod-corn theory only after our extensive studies of maize-teosinte hybrids had suggested to us that teosinte is itself a hybrid of maize and *Tripsacum* and could therefore be dismissed as the ancestor of corn. Pod corn then appeared to be the only plausible alternative to teosinte as the progenitor unless we resorted to a strictly hypothetical ancestral form which is now extinct. We have, however, made two contributions to the pod-corn theory by (1)

showing that the ancestral form could not have been a heterozygous genotype which does not breed true; (2) postulating that modern pod corn is not the original wild corn. On this point we stated: "The pod corn found occasionally as a mixture in modern cultivated varieties is certainly not the wild pod corn which the wandering Indians discovered millennia ago. Modern pod corn is the result of superimposing a single 'wild' gene with perhaps a few closely linked modifiers upon a genetic complex which has been tremendously altered by centuries of domestication." Since 1939 a substantial part of the research in which my colleagues and I have been engaged has been concerned with subjecting the pod corn hypothesis to a variety of experimental tests. These are described in the remainder of this chapter and in the following one.

Pod Corn on Uniform Genetic Backgrounds

One of the objections to the pod-corn hypothesis is that pod corn is "equivocal" (Weatherwax, 1954). In this context the word equivocal is more or less synonymous with "extremely variable." It is true that pod corn is often variable, but it is scarcely more so than other characteristics of the corn plant including monoecism and the nature of the ear, two features that distinguish corn from other major cereals. We assumed that pod corn's variability is the product of variation in the genetic background on which the pod-corn gene is superimposed as well as variation in the environment to which it is subjected. To test these assumptions we incorporated the pod-corn gene into two uniform inbred strains by repeated backcrossing. The expression of this gene is quite different in the two inbred strains, but within either one of them it is quite uniform. This is illustrated in Figure 7.2. This simple experiment shows that pod corn is not at all equivocal when the genetic background and the environment are held constant.

In another experiment we were able to show to what extent phenotypic variation in a pod-corn genotype is the product of differences in environment. The results are illustrated in Figure 7.3 in which two tassels of an inbred strain homozygous for the *Tu* gene are compared, one grown in a well-fertilized experimental field, the other in a simulated wild habitat. This comparison should make it obvious that in considering pod corn as the

ancestral form we must try to visualize it as it grew in the wild and not see it only as it develops in a modern agricultural field, well fertilized and free of competition with other species.

Various Expressions of the Pod-Corn Locus

The pod-corn locus is pleiotropic; it has numerous phenotypic expressions, some of which are constant, others quite variable depending in part upon the residual heredity and in part upon the environmental milieu. The locus always causes the development of glumes partially or completely enclosing the kernels. Since this is a primitive characteristic it is not surprising that it has attracted the attention of so many students of maize. Indeed no serious student of the origin and evolution of maize can well afford to ignore this principal characteristic of pod corn.

Another common effect of the locus, although a somewhat variable one, is to release the compaction that is characteristic of the ears of many modern varieties of corn. This in turn has various secondary effects, such as branching of the ear, the development of the lower florets in the pistillate spikelets, the elongation of the branch commonly known as the "shank" on which the ear is borne, resulting in the emergence of the ear from its husks. Related to this is the elongation of the rachis of the ear which results in the pistillate spikelets becoming so widely separated that the ear is clearly revealed as a typical grass spike (Figure 7.4) with no indication whatever of the fusion of parts. Sometimes the rachillae, the kernel-bearing stalks, also are elongated and thus provide a biologically magnified view showing, more dramatically than any drawing that I have ever seen, the botanical nature of the pistillate spikelet and the arrangement of its parts (Figure 7.5).

Producing a Fertile, True-breeding Pod Corn

Another objection to the pod-corn hypothesis has been that it is monstrous and sterile, especially in the homozygous condition. We assumed that this monstrousness and sterility is the result of superimposing a wild relict gene upon the genetic background of highly heterozygous and vigorous modern varieties. Today's pod corn is comparable to a 1900 chassis, a horseless carriage, powered by the sophisticated engine of a late model car. We assumed that pod corn would be less

← FIG. 7.4 By enlarging certain parts, the pod-corn gene, *Tu*, is sometimes useful in providing a biologically magnified view of certain of corn's principal botanical characteristics. This specimen, the terminal portion of an elongated ear of Guarany pod corn, shows, after removal of a row of paired spikelets, that the ear of corn is basically a typical grass spike with the spikelets arranged in an orderly fashion on a central stem, the rachis. The ear of corn is not, as some morphologists have postulated, the product of fusion. This specimen also shows that the cupules, depressions in the rachis, regarded by some students as evidence that corn is a domesticated form of teosinte, can virtually disappear when the rachis is elongated (× 2).

monstrous and would exhibit normal grass characteristics when combined with other "wild" genes still extant in modern corn varieties. We hoped to find these in varieties of popcorn. This hope has been realized.

The pod-popcorn described above which served to test the historical reference of Azara was developed in Texas from a cross of pod corn and a popcorn of the "white rice" type. After moving to Massachusetts I continued to carry along this stock in the hope of producing a fully fertile, true-breeding non-monstrous pod corn. I did succeed in producing such a pod corn but subsequently discovered that there had been a mutation at the tunicate locus resulting in a less extreme form of pod corn, which we called "half tunicate" and to which we assigned the symbol tu^h because its expression in the homozygous state was quite similar to the expression of the full tunicate locus, *Tu*, in the heterozygous state. Plants homozygous for the full-tunicate locus continued to be pollen sterile.

In 1952, while I was visiting Dr. Lorenzo R. Parodi in Buenos Aires, he presented me with an ear of pod corn which was also a popcorn of the race known as "Argentine." This seemed to offer promising material for developing a true-breeding pod-popcorn. This ear proved to be heterozygous, and when some of its progeny plants were self-pollinated they produced homozygous genotypes in the next generation which were less monstrous than those commonly encountered. I hoped by selection to produce a fertile true-breeding pod corn superimposed upon the genetic background of Argentine pop. Again I succeeded and again the homozygous genotype proved to be the product of a mutation to the half-tunicate form. Although the half-

tunicate genotype was not what we had set out to develop it proved to be useful for another purpose, that of comparing six different tunicate genotypes with respect to a number of characteristics. This experiment is described in the next chapter.

By this time Galinat and I had made a study of the early Bat Cave corn described in Chapter 14 and I was more than ever convinced that the ancestral form of cultivated corn was a form of pod corn. Obtaining a collection of popcorn varieties from Dr. Oliver Nelson of Purdue University, who had been studying the occurrence of the cross-sterility gene *Ga* and its alleles, I crossed these with a uniform stock of pod corn. A comparison of the F_1 hybrids showed that several popcorn varieties, notably Lady Finger and Baby Golden, carried genes which we called "minus modifiers" that tended to suppress the monstrous expression of pod corn. One of these genes from the Baby Golden variety was so conspicuous in its effects that we termed it a "tunicate inhibitor" and gave it the symbol *Ti*. Linkage tests showed this gene to be located on chromosome 6 with about 40 percent of crossing over with *Y*, the gene for yellow endosperm color. The effects of this gene in reducing the glumes of homozygous half tunicate are illustrated in Figure 7.6.

Crossing the Argentine pod-pop with both Lady Finger and Baby Golden and subsequently intercrossing these two hybrids, we finally succeeded in 1958 in developing a number of lines of popcorn which were homozygous for the *Tu* gene on chromosome 4, the *Ti* gene on chromosome 6, and the assemblages of minus modifiers contributed by the Argentine pop and Lady Finger varieties. These lines were fully male fertile, and most of them shed pollen profusely, but some were temperamental about their shedding, doing so only when the temperature and humidity were at an optimum. I suspect that wild corn may also have had this characteristic.

The Genetically Reconstructed Ancestral Form, the First Model

Our purpose in combining the genes of pod corn and those of several varieties of popcorn was twofold: (1) to test our assumption that pod corn is not monstrous and sterile when the *Tu* gene is superimposed on appropriate genetic backgrounds; (2) to reconstruct the

ancestor of corn. Our purpose in reconstructing the ancestor of corn was to retrace so far as possible some of the principal steps that had been involved in its evolution under domestication. We did this in the hope of gaining a better understanding of the corn plant as one of those unique biological systems which man employs on a grand scale to convert the energy of the sun, the carbon dioxide of the air, and the water and minerals of the soil into food. The first objective was attained when we developed a fertile true-breeding pod-popcorn. We felt that we were on the right track, with respect to the second objective, when we saw that the majority of our fertile, true-breeding pod-popcorns had additional characteristics which we regarded as primitive. The plants, when grown on fertile soil, instead of having only a single stalk, as do most modern corns, had several and in this respect resembled the majority of wild grasses including all of the known relatives of corn both American and Old World. The plants were shorter than those of ordinary corn because one of the effects of the pod corn gene is to shorten and thicken the upper internodes of the stalk. This shortening causes, or at any rate is accompanied by, the development of a terminal inflorescence which bears both male and female flowers, the female flowers at the base and the male flowers above on the same tassel branches (Figure 7.7). These branches are quite fragile when mature and break apart easily when disturbed by the wind and by birds. They thus provide one of the most important primitive characters which cultivated corn lacks, a mechanism for dispersing its seeds.

Plants of homozygous pod corn frequently do not have ears—most of their energy is apparently concentrated in the terminal inflorescences—but when they do have ears these tend to be borne in high positions on the stalk, in some plants at the joint of the stalk immediately below the tassel. This elevation of the position of the ear has profound effects, which are illustrated by the diagram in Figure 7.8. This diagram which is based on data from several many-eared plants of Argentine pop, shows how a number of the characteristics of the ears are determined by their positions on the stalk: (1) the higher the position the smaller the ear, partly for the simple mechanical reason that the stalk at this position is slender and is incapable of

FIG. 7.5 As the pod-corn gene *Tu* sometimes causes an elongation of the rachis, the central stem of the ear (Fig. 7.4), so in other ears it may cause elongation of the rachilla, the central stem of the individual spikelet, thus providing a biologically magnified view showing the arrangement of the floral bracts beginning basally: lower glume, upper glume, lemma, palea. Both teosinte and *Tripsacum* differ from corn in having highly condensed rachillae.

FIG. 7.6 *Left*, an ear of pod corn of the homozygous genotype *tu^{md}tu^{md}*. *Center*, a similar ear which is homozygous for both *tu^{md}* and *Ti*, the tunicate-inhibiting gene. *Right*, an ear similar to the center ear with some kernels removed to show the relatively long soft glumes that partially surround the kernels but do not completely enclose them.

bearing a heavy load; it would be mechanically impossible for the large modern ear of corn to be borne near the slender tip of the stalk; (2) the higher the position the more likely is the ear to have both male and female flowers; (3) the higher the position the shorter the lateral branch or "shank" upon which the ear is borne; (4) the shorter the branch the fewer the joints from which the husks arise; (5) the fewer the husks the less completely is the ear enclosed. Thus an ear borne immediately below the tassel has only a few husks; it is enclosed while the young seeds are developing but as these mature the husks flare open, allowing the ear to disperse its seeds. In short, a simple change in position, determined by a single gene change, can provide two mechanisms for seed dispersal: one for seeds borne on the fragile branches of the tassels and another for high ears not completely enclosed at maturity by husks.

The ears borne in high positions on the stalks of our genetically reconstructed ancestral form had female spikelets on their lower parts and male spikelets on the remainder. If our reconstruction is valid, should not the prehistoric ears of primitive corn also bear male spikelets? A reexamination under the microscope of the cobs of the prehistoric corn from Bat Cave and La Perra Cave described in Chapter 14 showed that at least some of them once did and that these have been lost in handling. Some of the ancient cobs have stumps, previously unnoticed, of slender stems on which male spikelets were undoubtedly borne. Thus our genetically reconstructed ancestral form taught us to look for an important botanical characteristic in prehistoric cobs that we had previously overlooked. It also showed us the significance of the ears bearing terminal spikelets which are still found in certain primitive races of corn in the countries of Latin America: Nal-Tel and Chapalote of Mexico, Pollo of Colombia, and Confite Morocho of Peru. Finally, it may explain some curious ears which had previously puzzled us which were molded in bas-relief on a prehistoric Zapotec funerary urn from Mexico. An even more convincing specimen which Wilkes found in Mexico and which subsequently came into my possession is illustrated in Figure 16.10.

On the basis of the data available by 1958, Galinat drew, for a paper I presented at the annual meeting of the American Philosophical Society, a series of six corn plants illustrating some of the principal environmentally induced and genetically controlled changes which we supposed to have occurred during domestication. His drawing is reproduced in Figure 7.8. The first three plants illustrate the genetically reconstructed ancestral form as it would be expected to develop in three different environments. The first plant, a short single-stalked plant with a slender unbranched tassel bearing both male and female flowers but no ear, was intended to represent a wild corn plant growing in nature in a site of low fertility and in severe competition with other natural vegetation. Such a plant would barely reproduce itself. The second plant represents the same genotype grown under primitive agricultural conditions. Here it is still single stalked but is capable of producing a branched tassel and a single small ear borne high upon the stalk. The third plant represents the genetically reconstructed ancestral form grown under modern agricultural conditions in a well-fertilized soil and free of competition with weeds. In such an environment it has several stalks, each bearing small ears. Plants like these might also have occurred sporadically in the wild under unusually favorable natural conditions.

The ability of the wild corn plant to respond in a spectacular fashion to freedom from competition with other vegetation and to high levels of fertility is undoubtedly one factor that led to its domestication. The ability to take full advantage of the improved environment usually afforded by an agricultural system is one of the characteristics found in almost all highly successful domesticated species. Many wild species do not have this trait; they cannot stand prosperity. I am reminded here of the thousands of dollars spent during World War II in a futile effort to domesticate goldenrod as a source of natural rubber.

Since the corn plant is genetically plastic as well as responsive to an improved environment, domestication may soon have brought other changes, which are illustrated in the last four plants in Figure 7.8. One of the most important of these was a mutation at the pod-corn locus on chromosome 4. This single genetic change had numerous effects: it reduced the glumes which in wild corn partly or completely surrounded the kernels; the energy released from the production of these vegetative floral bracts now went into the development of a larger cob, which in turn bore more and larger kernels; the mutation also lowered the position of the lateral inflorescences, and this had profound effects of several kinds, which can be understood by referring again to Figure 7.9. This diagram shows (1) the lower the ear the stronger the stalk at ear-bearing position and the greater its capacity to support larger ears; (2) the lower the ear the more likely it is to bear only female flowers which develop kernels when pollinated; (3) the lower the ear the longer the shank, the branch on which it is borne, and this in turn has a number of secondary effects; (4) the longer the shank the more numerous its nodes or joints and the husks which arise from them; (5) the greater the number of husks the more completely the ear is enclosed and the less capable it is of dispersing its seeds.

In short, a rather simple change but a very important one, the lowering of the position of the ear (comparable perhaps to moving the engine of a primitive airplane from a position behind the wings to one in front of them) has separated the sexes and made for a larger, strictly grain-bearing ear which is completely protected by the husks and is no longer capable of dispersing its seeds. Thus, a mutation at a single locus on chromosome 4 has made the corn plant less able to survive in nature but much more useful to man.

The last two plants illustrated in Figure 7.8 show some of the changes which human selection has subsequently effected. Selection for large ears has tended to eliminate secondary stalks and to reduce the number of ears per stalk. The fifth plant in Figure 7.8, represents a typical New England flint corn in which the secondary stalks have been reduced to low tillers known to the farmer as "suckers," which in the days of cheaper labor

FIG. 7.7   A tassel and ear of a true-breeding homozygous pod corn. The shortening of the internodes of the upper part of the stalk causes or is accompanied by a tassel that bears seeds at the bases of the tassel branches. The branches are quite fragile and are easily broken by the wind or by birds attempting to eat the seeds. Thus, introducing the pod corn gene has given modern corn a means of dispersal that it must once have had and lost during domestication. This plant had a second means of dispersal in a sub-tassel ear—its silks shown in the photograph—which, borne in a high position on the stalk, had only a few husks. These would open at maturity allowing the small ear to disperse its seeds.

ENVIRONMENTALLY INDUCED CHANGES

GENETIC CHANGES OCCURRING DURING EVOLUTION UNDER DOMESTICATION

FIG. 7.8 Environmentally induced and genetically controlled changes in the corn plant. *Left*, a plant of pod-popcorn as it might have grown in nature in a poor site in competition with other natural vegetation (compare this with Figure 7.10). *Second*, a similar plant grown under primitive agricultural conditions. *Third*, a similar plant in a fertile site free of competition with weeds. Compare this with the plant illustrated in Figure 8.5. *Fourth*, a popcorn plant that has lost the pod corn gene and hence bears its ears in lower positions on the stalk. *Fifth*, human selection for larger ears has tended to eliminate the secondary stalks, reducing them to "suckers," as in this typical plant of a New England flint corn. *Right*, this trend has been carried still further in Corn Belt dent corn, which is usually single stalked, bearing a single large ear in the middle region of the stalk.

were often removed under the erroneous impression that their removal was a kind of beneficial pruning operation. The last plant in Figure 7.8 represents a typical Corn Belt dent corn, which, before machine harvesting became common, was predominantly single stalked, usually bearing one large ear at approximately the middle region of the stalk.

We realized when the 1958 article was written that we had not yet completely reconstructed wild corn or duplicated exactly the most primitive specimens from either Bat Cave or La Perra Cave—the glumes of the pod-popcorn were still too prominent to match those of the prehistoric specimens. One thing we had accomplished, however, was the development of what was probably the world's most unproductive corn. This seemed to us useful in suggesting that we were on the right track in attempting to retrace corn's evolutionary paths. In further experiments, discussed later, we succeeded in developing other genetically reconstructed ancestral forms more similar to the prehistoric specimens.

Also, when Galinat made the drawings illustrating our evolutionary series we had not yet produced the two short single-stalked plants illustrated on the left of the series. However, knowing from long experience how corn plants behave under conditions of low fertility, inadequate moisture, or severe competition with weeds, I was reasonably certain we could produce such depauperate plants by subjecting our reconstructed ancestral forms to an unfavorable environment. The results of experiments in this direction are described below.

The Reconstructed Ancestral Form in a Simulated Wild Habitat

We first attempted to produce depauperate plants of our pod-popcorn by planting the kernels 3 inches apart, instead of the usual 12 inches, in a row in the experimental nursery. This planting was not completely successful because one of the adjoining rows failed to grow, thus giving the thickly planted row more space than we had intended for it to have. We did, however, succeed in this planting in producing a plant similar to the second plant in the series illustrated in Figure 7.8. However, in another experiment in the same season we unexpectedly produced a depauperate plant similar to the first plant illus-

FIG. 7.9 Diagrammatic longitudinal section based on data from plants with many-eared stalks showing how the position of the ear on the stalk affects its characteristics. The higher the ear, the smaller its size, the fewer the joints on its stem, and hence the fewer its husks and the more likely it is to be terminated by a staminate spike. Drawing by W. C. Galinat.

trated in Figure 7.8. This occurred in a large planting of a population consisting of heterozygous tunicate and nontunicate plants of about equal numbers made to determine the rate of mutation of the gene *Tu* to its allele *tu^h*. Members of the planting crew had been instructed to plant as many full rows as possible from the seed of each individual cross and to bring back to the laboratory any seed remaining. I discovered later that in several instances the planters, having only a dozen or so seeds left over, did not bother to return them to the laboratory but instead threw them into the grassy sod at the edge of the field. Because we had an abundance of rainfall following this planting the seeds which had been scattered in the sod germinated. The plants they produced, however, were stunted and no more than one to two feet in height. The nontunicate plants bore sparsely branched tassels and no ears; they were completely female sterile. The heterozygous tunicate plants were earless, but they bore unbranched tassels with female spikelets containing kernels at their base with male spikelets above. These plants were the exact counterparts of our reconstructed ancestral form growing in nature in a poor site in severe competition with other vegetation. One of these plants was short enough to be mounted intact and without being folded on a standard

herbarium sheet. This may have been the first entire corn plant to be so treated.

The following year we conducted a somewhat more sophisticated experiment involving a simulated wild habitat. Seed of an inbred strain of homozygous pod-popcorn as well as a population comprising nontunicate and heterozygous tunicate genotypes were started in small pots—actually paper drinking cups —eight seeds per pot. After the seedlings had emerged they were planted in a weedy fence row, in which one of the components was the aggressive perennial couch grass, *Agropyron repens*. With no fertilizer, no cultivation, no irrigation, or other cultural attention, the plants flowered and produced seed. Most of the plants bore unbranched tassels with female flowers below and male flowers above. Several of these are illustrated in Figure 7.10.

A comparison of the nontunicate and heterozygous tunicate plants illustrated in Figure 7.10 proved to be of particular interest. Both types were earless, but the heterozygous plants bore an average of 77.8 pistillate spikelets per plant, while the nontunicate plants had almost none, an average of 3.6 per plant. We considered this convincing evidence that the pod-corn character, which is a handicap in an agricultural environment, may have a substantial selective advantage in a wild habitat.

This experiment also demonstrated clearly a fact of which we had been only vaguely aware before: that the monstrousness of modern pod corn is in part the product of environment. The glumes of pod corn are floral bracts, which morphologically are modified leaves. Like other vegetative parts they respond to favorable environments. This is illustrated by comparing the tassels of several pod-popcorn strains grown in the simulated wild habitat of the weedy fence row with tassels of the same strains grown in well-fertilized and irrigated soil of the experimental nursery. This comparison leaves no doubt in my mind that much of the monstrosity of modern pod corn is the product of vegetative luxuriance resulting in part from the favorable conditions of an agricultural environment and probably also in part from the hybrid vigor that is characteristic of most modern corn.

We kept a close watch the following spring on the simulated wild habitat to see if self-sown seeds had survived the winter to pro-

FIG. 7.10 A colony of corn plants comprising *Tutu* and *tutu* genotypes grown in a simulated wild habitat without fertilization or cultivation and in competition with an agressive, weedy perennial grass, *Agropyron repens*. Under these conditions none of the plants produced ears but the heterozygous tunicate plants (those with the heavier tassels) had an average of 77.8 functional pistillate spikelets per plant and the nontunicate an average of 3.6 per plant.

duce a new generation of pod-corn plants. Apparently none had, or, if so, they had been discovered by field mice. In our experimental nursery, however, many seedlings emerged, some of which undoubtedly would have grown into mature plants had it not been necessary to remove them in preparation for the current year's planting. In the rigorous winter climate of New England, subjected to repeated snows, besieged by hungry birds, some of our homozygous pod-popcorn lines had managed to protect and to disperse part of their seeds, thus perpetuating their kind. By combining pod corn and popcorn we had succeeded in developing a corn capable, in suitable environments, of surviving in the wild. No wonder that I felt that our earlier conclusion, Reeves' and mine, that wild corn must have been both a pod corn and a popcorn had now been experimentally confirmed. Only later were we to prove that the pod-corn locus is considerably more complex than we had first supposed and that our initial hypothesis about pod corn as the ancestral form was somewhat too simple. On this subject more in the following chapter.

8 The Nature of the Pod-Corn Locus

The fact that our pod corn had, in the course of our experiments, mutated several times to a weak form of pod corn, which we called "half-tunicate," led us to suspect that its locus may be a compound one and that the mutations which occurred were, in fact, the product of crossing over, which separated the components of the locus.

The Locus Dissected and Reconstructed

To test this possibility we crossed a uniform inbred strain, A158, into which the Tu locus had been incorporated by nine generations of crossing and backcrossing, with another inbred strain carrying two recessive marker genes on chromosome 4: su on one side and gl_3 on the other side of the Tu locus. The F_1 plants heterozygous for the three loci represented by the genotype, $SuTuGl_3/sutugl_3$, were then backcrossed to a second inbred strain homozygous for the three recessive loci. A total population of 10,248 plants of this backcross was grown over a period of three seasons, 1958–60. Approximately half of these plants were expected to be heterozygous tunicate; 5,273 such plants were found. There were, in addition, four plants that appeared to be heterozygous half tunicate. All of these proved by progeny tests to be crossovers, two being of the genotype $Sugl_3/sugl_3$, and two of the genotype $suGl_3/sugl_3$. This represents a "mutation" or crossover rate of one in 1,319.

Since the average percentage of crossing over between the loci Su and Gl_3 is 34 percent, the chances of a mutation at the Tu locus being accompanied by a crossing over are approximately one-third if the two events are independent. The chances of all four muta-

tions being accompanied by crossing over are 1/81. If this were the first case on record of a locus being separated by crossing over, we should not regard these odds as proving a relationship between mutation and crossing over. However, since the components of other loci in both maize and *Drosophila* have been separated through crossing over, the probability that the four mutations occurring in this experiment were indeed the product of crossing over seemed to us great enough to justify our proceeding to further steps.

The next step was to determine whether the two components were identical or different in their phenotypic effects. If identical, it could be concluded that the Tu locus is one which, like the classical case of the *Bar* locus in *Drosophila* described by Sturtevant, had arisen through a duplication without subsequent differentiation of function of the ancestral wild locus. Such a duplication could have occurred at any time during the domestication of maize. However, if the two components proved to be different in their phenotypic effects, as are the components of the pseudoallelic loci discussed by Lewis, it would suggest that divergence in function had occurred, and it seems unlikely that this degree of gene evolution could have taken place in the few thousand generations during which maize has been cultivated. Consequently, if the components proved to be identical, we would assume that the Tu locus is the product of unequal crossing over which occurred during domestication and that the wild locus is the one producing the half-tunicate effect. If, however, the two components proved to be different, we would

assume either that the wild locus was Tu or that there had been two kinds of wild corn, one represented by the left-hand component tentatively designated tu^l and the other by the right-hand component designated tu^d.*

From the outset the two components appeared to differ slightly in their effects, but whether this was actually the case or the product of differences in their residual heredity could only be determined with certainty by comparing them on the same genetic background. This was accomplished by incorporating both into the uniform inbred strain A158 through repeated backcrossing. After the fourth backcross the difference between the components in their phenotypic effects was clear. The heterozygous genotype tu^dtu consistently had longer, more prominent glumes, both staminate and pistillate, than the genotype tu^ltu, and its kernels were noticeably more difficult to remove from the cobs. Even more pronounced were the differences between homozygous genotypes produced by selfing heterozygotes after three generations of backcrossing. The genotype tu^ltu^l proved to be similar to one involving our earlier mutations to half tunicate. The genotype tu^dtu^d however, produced monstrous inflorescences, both staminate and pistillate, of which the majority of the former and all of the latter were completely sterile.

Before the final proof of the differences between the two components of the Tu locus

*We discovered later that the right-hand component tu^d is itself compound, comprising two components tu^m and tu^d. Since this was not known when these experiments were conducted, I shall use the symbol tu^d here for the double component.

had been established, we had proceeded to an additional experiment—one designed to determine whether the Tu locus could be reconstituted by restoring its components to their original positions on the same chromosome. In 1961 the heterozygous genotypes $tu^l tu$ and $tu^d tu$ were crossed. The progeny of one such cross was grown in 1962. It was expected that approximately one-fourth of the progeny plants would be double heterozygotes in the $trans$ configuration, $tu^l tu\ tutu^d$. In a population of 133 plants, 24 proved to be of this type. Pollen from several of these plants was applied to plants of two inbred strains, A158 and NY16. It was assumed that the progeny of these test crosses would consist of the heterozygous genotypes $tu^l tu$ and $tutu^d$ in approximately equal numbers and that the great majority of the plants would fall into these two categories. It was assumed further, however, that there would be rare crossovers between the two components and that these would be of two complementary types, $tu^l tu^d$ and $tutu$. The occurrence of both types in approximately equal numbers would furnish virtually conclusive proof that mutations at the Tu locus are the product of crossing over between its components.

Because most modern corn, including the two inbreds used in this experiment, is of the genotype $tutu$, extraordinary precautions were required to eliminate any possibility whatever of contamination. Consequently, stocks to be crossed were grown, not in our regular experimental plots at Forest Hills, but in a small garden in Cambridge completely isolated from all other maize and surrounded by university buildings. In the segregating progeny grown to provide the pollen parent genotype, $tu^l tu/tutu^d$, the plants of the remaining three genotypes, $tu^l tu/tutu$, $tutu^d/tutu$, and $tutu$, were removed as soon as they were identified and well before they had reached the pollen-shedding stage. Likewise, all tassels were removed from the two inbred strains before their pollen had matured, and in addition their ear shoots were covered before their silks had emerged. It seemed certain that the only pollen to which the silks of the inbreds were exposed at any time was that of the selected double heterozygotes.

In the winter of 1962–63, a population of hybrid plants resulting from the pollinations on the inbred NY16 was grown in a planting near Homestead, Florida. When classified in April, three plants in a total of 2,333 proved to be similar to heterozygous tunicate $Tutu$, showing that the tunicate locus had been reconstituted. However, the other crossover class, $tutu$, did not occur. Its absence in a population of this size is not statistically significant but made it necessary to grow additional populations. This was done in the summer of 1963, a population of 956 additional plants involving NY16 as a parent and 6,801 plants involving A158 as a parent being grown. In the total population of 10,090 plants of these two crosses, 8 were identified as $Tutu$ and 7 as $tutu$.

The rate of "mutation" involved in reconstituting the locus, 1 in 1,261, is of the same order as that, 1 in 1,319, which occurred in the experiment involving the dissection of the locus. This is further evidence that both types of "mutation" are the product of crossing over, since reverse mutations not involving crossing over are seldom as frequent as direct "point" mutations.

The experiment on reconstructing the tunicate locus shows why pod corn, which Weatherwax and others have assumed to be a mutant form, has never been reported in pedigreed cultures, although millions of ears of inbred strains and their first-generation hybrids have been studied by corn breeders. Pod corn, of the type represented by the Tu locus, can appear as a mutant only in stocks of half-tunicate maize. If our genetic analysis of its locus is valid, it cannot occur as a mutant in modern commercial nontunicate maize.

We—Galinat and I—concluded at this point that there may have been two kinds of wild corn, one of the genotype $tu^l tu^l$, the other $tu^d tu^d$. When these were brought together under domestication by the American Indians, hybridization would have produced—as it did in our experimental cultures—two new types: (1) an extreme form of pod corn which the Indians in parts of both South and Middle America preserved (and still do) for its supposed magical properties; (2) a nonpodded corn, similar to modern corn in lacking conspicuous glumes, which is more productive and in other ways more useful than pod corn as a cultivated food plant (Mangelsdorf and Galinat, 1964).

The prehistoric wild corn, uncovered by archaeological excavations in the Tehuacán Valley of Southern Mexico, described in Chapter 15, appears to be a weak form of pod corn similar to that of the genotype $tu^l tu^l$ combined with an inhibiting factor in our cultures. This wild corn is the progenitor of two still-existing but somewhat primitive races of corn in Mexico, $Chapalote$ and Nal-Tel. But the Tehuacán wild corn is quite distinct in a number of characteristics from a third primitive race, a Mexican popcorn known as $Palomero\ Toluqueno$. The Tehuacán wild corn lacks tillers and has glabrous leaf sheaths and round kernels, brown or orange. $Palomero\ Toluqueno$ has tillers, pilose leaf sheaths, and pointed white kernels; it probably stems from a different race of wild corn. I shall suggest in a later chapter that there may have been at least six different races of wild corn.

Postscript to the Dissection Experiment

After the above description of this experiment was published we discovered that the two tu^d mutants which had occurred were not identical in their expressions. The first one which we had identified produced monstrous inflorescences both male and female in the homozygous state. The second mutation, which occurred a year later, proved to be a much weaker form of pod corn similar to but not identical with the tu^l mutant. From these facts we concluded that the tu^d component is itself a complex locus, to which we have assigned the symbols tu^m and tu^d, and that the Tu locus must therefore comprise at least three components, tu^l, tu^m, and tu^d.

In 1966 and 1967 we attempted to separate the components of the $tu^m tu^d$ locus through crossing over with the markers Su and Gl_3 which we had employed in the earlier dissection experiment on both sides of the Tu locus to identify the crossovers. In a population of about 1,500 plants heterozygous for this locus, we found several ears in which the development of the glumes was so weak that the ears were suspected of being heterozygous for the tu^m component. Progeny tests, however, proved these to have been phenocopies. At this writing we have not yet succeeded in isolating the tu^m component. If and when the third component of the locus is isolated it will be possible to produce eight different homozygous genotypes with respect to the components of the tunicate locus. These are *lmdlmd, lmlm, ldld, mdmd, ll, mm, dd, tutu*.

A Comparison of *Tu-tu* Genotypes in Isogenic Stocks

Our discovery of an intermediate allele at the *Tu-tu* locus, which we initially called half tunicate and to which we assigned the symbol tu^h, made it possible to produce and to compare in a uniform genetic background six distinct genotypes: $TuTu$, $Tutu^h$, $Tutu$, tu^htu^h, tu^htu, and $tutu$. The alleles Tu and tu^h were incorporated into the uniform inbred strain A158 of the genotype $tutu$ by repeated backcrossing. The Tu allele was crossed six times to this strain and the tu^h allele seven times. This should have made the several stocks virtually isogenic except for the alleles and for genes so closely linked to them that they were not separated by the several generations of backcrossing.

Four of the desired genotypes, $Tutu^h$, $Tutu$, tu^htu, and $tutu$, were produced by intercrossing the two stocks which were heterozygous for the alleles Tu and tu^h respectively, and these four genotypes occurred at random in two adjoining rows in the experimental nursery. The two remaining genotypes, $TuTu$ and tu^htu^h, were produced by selfing the two heterozygotes; these were grown in adjacent rows. Thus the genotypes were compared in a uniform environment as well as on a uniform genetic background. An average of six plants were scored in each genotype, except that for characteristics of the cob only one ear of the genotype $tutu$ was dissected. The ear selected for this purpose was close to the average in its other characteristics.

Ears and tassels of the six genotypes are illustrated in Figures 8.1 and 8.2, and the data describing their characteristics, which previously have appeared in complete form only in the Maize Genetics Cooperation News Letter (Mangelsdorf and Mangelsdorf, 1957), are set forth in Table 8.1. The data show that the genes Tu and tu^h are stongly pleiotropic, affecting many characteristics of both the terminal and the lateral inflorescences. They show also that the allele tu^h has exactly the same kinds of effects as Tu, only in a lesser degree. This is not surprising, since we now know from our experiment in dissecting the Tu locus that this is a compound locus apparently comprising three components, of

FIG. 8.1 Ears of six genotypes involving the *Tu-tu* locus on an isogenic background of inbred A158. From left to right: *TuTu, Tutu^h, Tutu, tu^{h}tu^h, tu^{h}tu, tutu*.

FIG. 8.2 Tassels of six genotypes involving the *Tu-tu* locus on an isogenic background of inbred A158. *A, TuTu; B, Tutu^h; C, Tutu; D, tu^{h}tu^h; E, tu^{h}tu; F, tutu*.

89   The Nature of the Pod-Corn Locus

TABLE 8.1   A Comparison in Isogenic Stocks of Six Genotypes Involving the *Tu-tu* Locus

Characteristics	Genotypes					
	TuTu	*Tutu^h*	*Tutu*	*tu^h tu^h*	*tu^h tu*	*tutu*
Tassels						
Weight tassels, gm.	28.9	18.3	12.7	9.2	6.7	4.9
Weight central spikes, gm.	6.6	4.8	3.5	2.9	2.1	1.5
Weight central spikes/tassels	22.8	26.2	27.6	31.5	31.3	30.6
Weight penduncles, gm.	2.64	2.20	1.30	1.11	1.09	1.06
Weight rachises, gm.	4.6	2.9	2.1	1.4	1.3	1.1
Percent rachises	15.9	15.8	16.5	15.2	19.4	22.5
Weight spikelets, gm.	24.3	15.4	10.6	7.8	5.4	3.8
Average length glumes, cm.	2.7	1.9	1.7	1.4	1.2	1.0
Percent pistillate spikelets	79.9	0.9	0.0	0.0	0.0	0.0
Ears						
Weight ears, gm.		21.6	59.5	89.7	126.6	125.2
Weight cobs, gm.		6.4	20.6	24.5	24.0	22.7
Weight glumes, gm.		5.3	16.5	17.7	14.2	8.7
Weight rachises, gm.		1.1	4.1	6.8	10.5	10.9
Percent glumes		82.8	80.1	72.2	57.5	44.4
Percent rachises		17.2	19.9	27.8	42.5	55.6
Weight kernels, gm.		15.2	39.0	65.2	102.6	102.5
Percent kernels		70.3	65.6	72.7	81.0	81.8
Number of kernels		76	282	371	532	477
Average weight kernels, gm.		20	13	17	19	21

TABLE 8.2   A Comparison in Isogenic Stocks in Characteristics of the Tassels of Various Genotypes Involving the *Tu-tu* Locus

Genotype	Components	No. of Components	Weight, gm.				Length 10 spikelets, cm.	% pistillate spikelets	% proliferated spikelets
			Tassels	Central spikes	All spikelets	Ten basal spikelets			
1. *TuTu*	*lmd/lmd*	6	37.3	4.1	28.7	0.45	31.5	7.1	62.0
2. *TuTu^md*	*lmd/--md*	5	21.4	5.0	16.1	0.38	27.0	7.2	50.0
3. *Tutu^h*	*lmd/--h**	4	13.5	2.9	10.4	0.22	22.1	3.1	2.0
4. *Tutu^d*	*lmd/--d*	4	14.4	3.1	11.0	0.21	21.6	2.9	1.3
5. *tu^md tu^md*	*-md/-md*	4	27.0	4.2	20.9	0.26	24.7	16.9	36.0
6. *Tutu*	*lmd/---*	3	10.8	2.2	8.0	0.12	17.0	0.6	0
7. *tu^md tu^h*	*-md/--h*	3	9.8	2.9	7.0	0.16	19.7	2.3	0
8. *tu^md tu^d*	*-md/--d*	3	14.7	3.0	11.3	0.13	17.4	0.3	0.3
9. *tu^md tu*	*-md/---*	2	11.3	2.2	8.3	0.09	13.8	0.2	0
10. *tu^l tu^l*	*l--/l--*	2	8.9	2.1	6.6	0.12	16.5	0	0
11. *tu^h tu^h*	*--h/--h*	2	10.5	2.1	7.9	0.11	15.3	0	0
12. *tu^h tu^d*	*--h/--d*	2	10.6	2.4	7.7	0.10	15.1	1.0	0
13. *tu^h tu*	*--h/---*	1	7.6	1.9	5.5	0.10	13.1	0	0
14. *tu^d tu*	*--d/---*	1	6.2	1.7	4.5	0.08	12.7	0	0
15. *tutu*	*---/---*	0	5.8	1.3	3.9	0.07	11.4	0	0

*Data discussed in the text indicate that the *h* and *d* components are identical.

which *tu^h* (later designated as *tu^d*) is one. The six genotypes compared in this experiment therefore involve 6, 4, 3, 2, 1, and 0 components respectively.

In proceeding through the series from *TuTu* to *tutu* several profound changes are apparent. First, the tassels become progressively smaller, declining in weight from 28.9 gm. to 4.9 gm. The central spikes and branches of the first two genotypes, *TuTu* and *Tutu^h*, are distinctly drooping because of the sheer weight of the spikelets which they bear. Accompanying the decrease in the weight of the terminal inflorescences, the tassels, is an increase in the weight of the lateral inflorescences, the ears. The relationship between the decrease of the one and the increase of the other, however, is not equivalent. For each gram of weight lost in the tassel, there is a gain of more than 5 gm. in the ear. This suggests that the lateral position of the ear has an advantage over the terminal position of the tassel in laying down dry matter. I once assumed this advantage to be the result of the ear's location in the center of photosynthetic activity instead of at the periphery, and I supposed that in this respect corn had an advantage over and was potentially more productive than other cereals whose grain-bearing inflorescences are borne terminally. Some rather simple experiments that we later conducted showed that the lateral position of the ear does indeed confer an advantage, but the advantage results from the food-manufacturing capacity of the leaves *above* the ear.

A second important change in the series is that the central spike of the tassel, like the tassel itself, becomes less massive through the series, its weight declining from 6.6 gm. to 1.5 gm. In relation to total weight of the tassel, however, the central spike becomes more prominent, its percentage of the total weight increasing progressively from 22.8 to 30.6 percent. Another change is from a partially pistillate terminal inflorescence to a wholly staminate one, accompanied by a shortening and thickening of the upper internodes of the stalk. The tassel of the *TuTu* genotype has 79.9 percent of pistillate spikelets, the next genotype only 0.9 percent. The remaining four genotypes are wholly staminate. In this characteristic the change from one end of the series to the other is not progressive but more in the nature of the crossing of a threshold below which the tassels are wholly staminate.

FIG. 8.3 An ear of corn comprises three principal zones. The outer zone (*A*) is represented by the kernels. When these are removed the zone of floral bracts (*B*) is exposed. Removal of these exposes the rachis (*C*), which provides the grain-bearing surface and encloses the system of supply.

We found virtually no evidence of pistillate spikelets in the tassels of prehistoric corn from any of the sites discussed in Chapters 14 and 15. If there were races of wild corn with mixed terminal inflorescences they have not yet been discovered in archaeological remains.

The tendency to produce pistillate spikelets in the tassels is associated, perhaps causally, with a shortening and thickening of the upper internodes of the stalk. This is shown in part by the progressive decrease through the series in the weight of the peduncle, the internode on which the tassel is borne. As in the case of the percentage of pistillate spikelets, a threshold effect seems to be involved. Beginning with the genotype *tutu*, there is little difference in the first three genotypes, in fact no difference in these would be shown if we reduced the figures to one digit beyond the decimal point. The genotypes *TuTu* and *tutu^h*, however, have much heavier peduncles. The shortening and thickening is not confined to the peduncle; it affects also the second and

third internodes below the tassel. The relationship between the shortening and thickening of the internode and the development of the pistillate spikelets is probably similar to that observed by Werth (1922), who found that short tillers were more likely to bear pistillate spikelets than long ones.

A decrease in the average length of the staminate glumes was also found in the series. This is readily apparent to the observer and is clearly shown by the data, the average length decreasing progressively from 2.7 cm. to 1.0 cm. There is a corresponding decrease in the length of the pistillate glumes, which is reflected in the weight of the floral bracts of the ear, discussed below.

There was also an increase in the weight of the rachis in relation to the pistillate glumes and other floral bracts. This is particularly significant because the rachis not only contains the system of supply, the vascular bundles, but also provides the grain-bearing surface. This may be better understood perhaps if we dissect an ear of corn into its three principal regions illustrated in Figure 8.3: the region of (1) the grains; (2) the glumes and other floral bracts; (3) the rachis. In the extreme form of pod corn the rachis is quite slender and the region of the glumes is quite extensive, completely overlapping the region of the grains. In the nontunicate type, *tutu*, the region of the glumes is greatly reduced and the rachis is much thicker, allowing it to contain a larger system of vascular bundles and to provide a substantially larger surface on which the kernels are borne. The larger grain-bearing surface opens up new evolutionary paths: it makes it possible for the rachis to carry either larger kernels or more rows of kernels, as shown in Figure 8.4. The corn plant has made use of both these paths. The Corn Belt dent in the United States is an example of a large rachis bearing many rows of grain; the giant-seeded Cuzco race of Peru, which has a rachis of approximately the same size, is an example of a large rachis bearing large grains.

A further change through the series was an increase in the number and average weight of the grain, which accompanied the increase in size and weight of the rachis. The correlation between the two is fairly high, the weight of the kernels in four of the five genotypes being consistently between 9 and 10 times the weight of the rachis. Perhaps it should be

FIG. 8.4 An increase in the diameter of the rachis during domestication would permit an increase in either the number of kernel rows or the size of the kernels. Corn has followed both of these evolutionary paths, as is shown by the cross-sections of a 12-rowed flint corn and an 8-rowed flour corn, the latter the famous Cuzco of Peru.

stated that we do not regard the monstrous pod corn represented by the genotype *TuTu* in this particular series as a model of the ancestral form. It probably does have some of the essential characteristics of wild corn—tassel seeds, long glumes, slender rachises—but in highly exaggerated forms. Consequently, a comparison of the six genotypes on this particular genetic background provides a kind of biologically magnified view of some of the changes which have occurred in evolution under domestication.

Characteristics of Additional Genotypes at the *Tu-tu* Locus

Following the discovery that the tunicate locus is a compound one apparently comprising three components, *tu^l*, *tu^m*, and *tu^d*, we produced additional genotypes involving this locus and compared these in several characteristics of the tassels. The purpose of these

TABLE 8.3 Three Comparisons in Characteristics of the Tassels of Genotypes Involving the tu^h and tu^d Components

Genotype	No. of Components	Weight of tassel, gm.	Weight central spike, gm.	Weight spikelets, gm.	Length 10 Glumes, cm.	Weight 10 Glumes, mg.
			Characteristics			
$Tutu^h$	4	13.5	2.9	10.4	22.1	0.22
$Tutu^d$	4	14.4	3.1	11.0	21.6	0.21
$tu^h tu^{md}$	3	9.8	2.9	7.0	19.7	0.16
$tu^d tu^{md}$	3	14.7	3.0	11.3	17.4	0.13
$tu^h tu$	1	6.2	1.7	4.5	12.7	0.08
$tu^d tu$	1	7.6	1.9	5.5	13.1	0.10
Av. tu^h		9.8	2.5	7.3	18.2	0.15
Av. tu^d		12.2	2.7	9.3	17.4	0.15

comparisons was twofold: (1) to determine whether the components of various combinations of them are identical or different in their effects; (2) to determine to what extent the number of components is correlated with the degree of expression of various phenotypic characteristics. The data showing the comparisons are set forth in Table 8.2.

One of the first questions to be asked is whether the earlier mutation which we had called half tunicate and to which we had assigned the symbol tu^h is identical in its effects with either of the tu^l or tu^d components which we had identified by their linkage relationships. If this locus is not a continuum, capable of being separated into an indefinite number of fractions, but comprises, as our experiments have indicated, three distinct components then the earlier tu^h mutation should be similar in its effects to the tu^l or tu^d mutants produced in our experiment on the dissection of the locus. It is quite unlikely to be the tu^m component, because this could have been separated from the other two only by double crossing over.

A comparison of the data on genotypes 11 and 12 involving the components tu^h and tu^d in Table 8.3 shows that they are virtually identical in their characteristics, while genotype 10 involving the tu^l component is somewhat different. Other differences not shown by the data are that the ear of genotype $tu^h tu^h$ is more nearly cylindrical in shape and has sorter glumes on basal kernels than the ear of $tu^l tu^l$. Also ears of the genotype $tu^l tu^l$ often "cast off" approximately the upper half

of the ear before maturity, while those of the genotype $tu^h tu^h$ and $tu^d tu^d$ reach maturity still intact. There is little doubt that tu^h and tu^d are identical and both are different from tu^l. The difference between the genotypes involving the components tu^l and tu^d is well illustrated in a comparison of three ears of each with respect to several characteristics. The data showing their difference are set forth in Table 8.4. Because of its characteristic mentioned above of casting off part of the ear before maturity, the genotype $tu^l tu^l$ has mature ears somewhat lighter in weight and bearing fewer kernels than those of the genotype $tu^d tu^d$. There is also substantial difference in the rachis/cob ratio. In at least our 1965 planting the $tu^l tu^l$ genotype consistently showed stronger tunicate characteristics than the $tu^d tu^d$ genotype.

Since the tu^m component of the Tu locus has not yet been positively identified or introduced into an isogenic stock, we cannot be completely certain about the expression of its phenotypic characteristics. However, the characteristics of genotypes involving both the tu^m and tu^d components indicate that tu^m differs from both tu^l and tu^d in its expression. For example, genotype 9, which is heterozygous for components tu^m and tu^d, although having heavier tassels than genotype 13 and 14 (which are heterozygous only for component tu^d), has basal glumes of the same weight. Also there is little difference in the ears; the glumes of the genotype heterozygous for both tu^m and tu^d are slightly longer than those of the genotype heterozygous for tu^d,

but if populations of the two were thrown together it should be difficult to separate them with complete accuracy. These several facts show that tu^m has little effect in the heterozygous state.

A comparison of homozygous genotypes, however, tells an altogether different story. Genotype 5, which is homozygous for tu^{md} and has four components, has a much stronger expression of both tassel and ear characteristics than genotypes 3 and 4, which also involve four components and are homozygous for tu^d. The difference is that genotype 5 is homozygous for tu^m but lacks tu^l. This shows that homozygous tu^{md} is much stronger in its expression than heterozygous tu^{lm}. The general conclusion is that the tu^m component has less effect than either tu^l or tu^d when heterozygous but a greater effect when homozygous. This in turn suggests that the effects of tu^m are largely recessive rather than dominant.

The data in Table 8.2 show that there is a strong correlation, with several notable exceptions, between the number of components involved in a genotype and the degree of expression of the characteristics of its phenotype. The most conspicuous exception is the genotype 5 discussed above, which is homozygous for tu^{md}. This genotype, like genotype 1, $TuTu$, has monstrous tassels and ears, both of which are usually sterile. Part of the tassel sterility in this genotype is the result of proliferation of the spikelets, which instead of producing fertile flowers, either pistillate or staminate, put forth secondary inflorescences bearing a number of sterile spikelets. Both the tassels and ears of this genotype are more extreme in their characteristics than those of genotype 2, which is also homozygous for tu^{md} and in addition is heterozygous for tu^l, thus having five components instead of four. Yet both the tassels and ears of genotype 2 are fertile. The only explanation which occurs to me for this curious situation is that the tu^l component interacts in some way with the tu^m component to modify it when it is homozygous, and this may suggest that tu^l and tu^m have evolved together, a possibility discussed below.

The Components of Wild Corn

The only wild corn so far discovered is the prehistoric corn from caves in the Tehuacán Valley in Mexico discussed in Chapter 15.

TABLE 8.4 A Comparison in Characteristics of the Ears of Genotypes Involving the *l* and *d* Components of the Tunicate Locus

Genotype	Weight ears, gm.	No. of grains	Percent rachis
$tu^l tu^l$	92.6	270	30.2
$tu^l tu^l$	94.4	240	29.0
$tu^l tu^l$	83.0	454	26.3
Average	90.0	321	28.5
$tu^d tu^d$	105.2	477	41.1
$tu^d tu^d$	121.4	418	33.8
$tu^d tu^d$	126.1	500	34.4
Average	117.6	465	36.4

This is a weak form of pod corn, one that might involve either $tu^l tu^l$ or $tu^d tu^d$ combined with the tunicate-inhibiting factor Ti on chromosome 6. When, in our first experiments in dissecting the Tu locus, we found that the tu^d component (which later turned out to be compound and which we designated) tu^{md} was quite different and much stronger in the expression of its characteristics than the tu^l component (Mangelsdorf and Galinat, 1964) we concluded that there must once have been two kinds of wild corn. We concluded further that the intercrossing of these races of corn was responsible for the creation of the present Tu locus.

Now that we recognize three components at this locus, what conclusion is to be drawn with respect to races of wild corn? Are we to conclude that there were once three distinct races differing in the nature of their tunicate alleles and that their intercrossing has ultimately produced the present Tu locus comprising three more or less distinct components? This obviously is one possibility. Another possibility—perhaps one somewhat less plausible—is that the remote ancestor of maize, the progenitor of the several postulated wild races, carried all three components at the Tu locus along with various major and minor modifying factors and that in the course of its evolution it gave rise to two or more races differing in the components they had inherited. I am not aware of any counterpart of such a situation in other organisms. The evolutionary trend appears to be in the direction of more rather than less complex genetic loci. A third possibility—the most simple and plausible with the facts now at

hand—is that the remote ancestor carried only one of the Tu components, that the progenitor of the Mexican races, Chapalote and Nal-Tel, inherited this component, and that the progenitor of the Mexican race, Palomero Toluqueño, also inherited this component, which sometime in the course of evolution became double, after which the two components became somewhat differentiated in their functions. These changes are assumed to have occurred before domestication began.

If we accept this model at least tentatively, the next question is which are the two components of the postulated tunicate locus of the wild progenitor of Palomero Toluqueño? In our 1964 paper we concluded that this race had the component tu^d, which we now know to be tu^{md}, and we have subsequently developed a genetically reconstructed ancestral form employing tu^{md} as the tunicate locus. But the double locus might equally well have been tu^{lm}. So far as crossing over data are concerned, the possibilities are equal. We have had altogether six mutations of the Tu locus. Three of these have separated tu^l from tu^{md}, and three have separated tu^d from tu^{lm}. The only data that would suggest one model as more plausible than the other are those involved in the comparison of genotypes 2 and 5. Both are homozygous for tu^m and tu^d, but genotype 2 is in addition heterozygous for tu^l and on that account might be expected to be more monstrous and sterile than genotype 5. Instead it is less so. Apparently, component tu^l interacts in some way with component tu^m. This could be the consequence of the two components having evolved together. This evidence is far from conclusive, but it may be suggestive. Perhaps we shall know more about the problem when we have isolated both tu^m and tu^{lm} and have incorporated them into our isogenic stock in a homozygous state.

Wild Corn a Pod Corn

Whatever the final conclusions regarding the components of the Tu locus involved in the wild ancestors of cultivated corn, there can no longer be much doubt that some or all of them were forms of pod corn. Reeves and I originally regarded the Tu locus itself as the wild gene and in testing our hypothesis we have succeeded in developing fertile, true-breeding strains of this form of pod corn. Both Randolph (1952) and Weatherwax (1954, 1955) have been skeptical not only of

our conclusions but also of the validity of our genetically reconstructed ancestral form. Weatherwax considers it conceivable that pod corn may have originated as a mutant and that the progress which my associates and I have made in "bestowing fertility by repeated crossing with certain varieties of corn which are thought to have primitive characteristics" may be interpreted as "imparting fertility to a plant which never before had it rather than restoring fertility which had been lost" (Weatherwax, 1955). Reeves and I (1959) once rejected this suggestion that we had through selection imparted fertility to a type of maize which never before possessed it. Now that we know, however, that the Tu locus is a compound one, probably originating during domestication, we obviously cannot be certain whether this type of pod corn was originally sterile or fertile. If it originated early when cultivated corn was still predominantly popcorn the tunicate mutation may well have been fertile; if it originated late when cultivated corn had lost many of its popcorn characteristics and its modifiers for the Tu locus, it may have been sterile and capable of being perpetuated only in a heterozygous state. In 1959 we concluded that "There is no type of pod corn so monstrous or so sterile that it cannot be changed in a few generations to a normal, fertile type by introducing modifying and inhibiting genes from varieties of popcorn."

This conclusion is still valid. We also now recognize as valid, however, Weatherwax's conclusion (1954) that pod corn—the extreme form that is usually monstrous when homozygous—may have originated as a mutant. But we also now know that this extreme form of pod corn could originate as a mutant only in stocks carrying the genes of the weaker forms of pod corn that involve the components of the Tu locus. Unless we assume that all three of these components of the Tu locus originated during domestication—an assumption not consistent with the archaeological evidence—we can only conclude that some of the wild ancestors of corn must have been forms of pod corn.

The Genetically Reconstructed Ancestral Form: Later Models

When our experiment in dissecting the Tu locus showed that it contained two components which were different in their effects

93   The Nature of the Pod-Corn Locus

we concluded, as I have mentioned earlier, that there must once have been two distinct kinds of wild corn. Whether the existence of a third component in the *Tu* locus means that there were once three kinds of wild corn with respect to this locus is not yet clear, but before the existence of a third component was discovered we had already proceeded to develop two distinct genetically reconstructed ancestral forms, one homozygous for the tu^d component and the other for the tu^{md} component. Both of these were combined with the tunicate inhibiting gene, *Ti*, on chromosome 6 and with the other tunicate modifying factors which we had introduced into our earlier model of a reconstructed ancestral form.

The most interesting of the various lines of genetically reconstructed wild corn which we have developed is one in which the major part of the germ plasm had been contributed by the Mexican popcorn race, Palomero Toluqueño. This race, one of the four Ancient Indigenous races of Wellhausen et al., is characterized by highly colored, strongly pilose leaf sheaths, conical ears, small, slender, pointed kernels with white endosperm, and a peculiar pericarp color, which we have sometimes described as "dirty" but for which the word "dingy" would probably be more appropriate. Some ears of this race have an additional curious feature: the members of the pair of spikelets tend to spread apart, with the result that the right-hand members of a row of paired spikelets tend to "interlock" with the left-handed members of the adjoining row of paired spikelets. This characteristic also occurs in some lines of pod corn which are not popcorn and is one of the facts which suggest that this race of popcorn was also once a pod corn. Another fact pointing in this direction is that the kernels of this race have exactly the right shape to fit snugly into pods of tunicate corn and their arrangement on the ears is one which has been described as "imbricated," i.e., the tips of the kernels tend to overlap the bases of the kernels immediately above in a manner somewhat resembling shingles or tiles on a roof. This characteristic also is found in pod corn and is especially well illustrated in the first published picture of pod corn in the monograph of Bonafous.

Kernels of modern pod corn are often pointed; they are squeezed into this shape during their development by the pressure of the glumes. Indeed, when occurring in the

FIG. 8.5   A freely tillering plant of the genetically reconstructed ancestral form. Selection for minus modifiers of the expression of alleles at the *Tu–tu* locus resulted also in an increase in the tendency to produce tillers. This plant had sixteen stalks, the largest number arising from a single seed encountered in our experiments.

tassels if they fail to become pointed they spread the glumes apart, exposing the kernels and thus inviting the depredations of insects and birds. Such a characteristic would have a low survival value in nature. If there were a race of wild corn in which the kernels were completely enclosed in glumes then it is highly probable that its kernels were slender, pointed, and slightly curved to fit snugly within the glumes and that this shape was determined genetically and not alone by the pressure of the glumes. If such a race of wild corn lost the enveloping glumes through a mutation at the *Tu* locus, we would expect it, for a time at least, to retain the characteristic slender, pointed, slightly curved shape of the kernels, their inbricated arrangement, and the spreading of the members of a pair of spikelets. All of these characteristics are found in the Mexican race, Palomero Toluqueño.

Combining the genes for the recognized characteristics of this race with the tu^{md} component of the tunicate locus has accentuated some of these features and has revealed others which might also have been characteristic of wild corn. The capacity for producing tillers, for example, has been greatly enhanced. In 1967 we spaced plants of one of the lines of

our reconstructed wild corn three feet apart in a border row having no competition on one side. Under these conditions the plants tillered profusely and bore numerous ears. One plant, illustrated in Figure 8.5, had 15 tillers in addition to the primary stalk, which was virtually indistinguishable from the tillers. Each stalk bore at least one ear, and the majority bore two. The total number of ears on this plant was 26. Since the average number of kernels per ear was about 250, the total number of kernels produced by this single plant was over 6,000. I know of no modern corn variety in which the individual plants normally produce this many seeds. This conclusion may seem to be in conflict with our earlier statement that in developing a genetically reconstructed ancestral form we had succeeded in breeding the world's most unproductive corn. The earlier conclusion is still valid, however, with respect to closely spaced plants; it obviously is not valid for plants widely spaced. It is not inconceivable that the difference in behaviour between wild corn plants grown in unfavorable sites in nature and those growing from seeds accidentally dropped near man's habitation, provided not only with space but also with the fertility contributed by garbage and excrement, was one factor in the invention of agriculture. Even the most primitive people could scarcely fail to see that freeing the corn plant of competition with other vegetation made it much more productive of food and much more useful to man.

The capacity for tillering of our reconstructed ancestral form may also be a characteristic of wild corn which enabled it to survive under unfavorable conditions in nature and to compete successfully with other species in the wild. We have some evidence from other experiments (Galinat, 1967) that when tillering and nontillering strains are grown together in the same hill, the tillering strain has an advantage and the nontillering strain suffers. Also we have found that our freely tillering reconstructed ancestral form thrives on soils which have not been fertilized. New England soils in general are low in fertility; even the Indians recognized this when at planting time they put a fish in every hill of corn. In New England modern commercial varieties of corn grow poorly on land that has not been well fertilized at least once, usually before planting and often a second time during the growing season. But the freely tillering reconstructed ancestral form does quite well without any fertilizer. Apparently the tillers serve to develop a large and vigorous root system.

Tillers may also be a factor in adapting the reconstructed wild corn to low temperatures in the early stages of development when the plants assume a rosette type of growth and expose considerably more leaf area to sunlight than nontillering plants. It is a fact that in the winter planting in Florida made in November where the young plants often are subjected to several weeks of cold weather in December, the reconstructed wild plants fare much better than the nontillering plants of modern corn varieties. There is probably a similar advantage to tillering in plants grown at high altitudes. The Mexican race, Palomero Toluqueño, the principal contributor to the germ plasm of our reconstructed wild corn, is adapted to altitudes of 2,200 to 2,800 meters, the highest elevation at which corn is grown in Mexico.

Another feature which has been accentuated by bringing in the germ plasm of Palomero Toluqueño is the terminal staminate spike of the lateral inflorescences. The modern ears of this race do not usually bear such staminate spikes, but evidently the race still carries genes for this characteristic, and these are expressed when the residual genetic background is appropriate. In any case many of the ears of the reconstructed wild corn do have prominent terminal male spikes.

Still other characters which have been brought to light by combining the components of the pod-corn locus with genes from popcorn varieties are various mechanisms for seed dispersal. I have already mentioned two mechanisms which appeared in our first models of the reconstructed ancestral form: seeds borne in the tassel (Fig. 8.6) and ears only partially enclosed in husks borne in high positions on the stalk. To these we may now add several other means of dispersal. (1) Tillers terminated by mixed inflorescences when the tassel of the main stalk is strictly staminate (Fig. 8.7). (2) Ears in the middle position in which the husks, although numerous, flare open at maturity, exposing the ear. Selection for this character was practiced by Corn-Belt breeders in earlier times when hand harvesting was the rule; it greatly facilitated the husking operation. Some plants of the ancient indigenous race of Mexico, Nal-Tel,

FIG. 8.6 In these strains of pod-popcorn, produced by combining intermediate alleles at the *Tu–tu* locus with *Ti* the tunicate-inhibiting gene, the tassels are not at all monstrous and some of them bear kernels in profusion. The branches of such tassels are quite fragile and are easily broken by the wind or the activity of birds, thus providing a means of dispersal that modern corn lacks. Other means of dispersal that have been observed in the new genetically reconstructed "wild" corn are illustrated in Figures 8.7 and 8.8.

95 The Nature of the Pod-Corn Locus

Fig. 8.7   In some strains of pod-popcorn, especially those carrying germ plasm from the Mexican popcorn race Palomero Toluqueño, the unbranched tassels of the tillers are staminate above and pistillate below. At flowering time the pistillate spikelets are protected by the enclosing leaf sheaths but after the silks have been pollinated and the kernels develop, the peduncles elongate and the kernels become exposed. Being borne on a somewhat fragile rachis they are provided with a means of dispersal. Other means of dispersal have been observed in the genetically reconstructed "wild" corn and are illustrated in Figures 8.6 and 8.8.

FIG. 8.8   The stems of the lateral branches in some strains of pod-popcorn tend to elongate, causing the inflorescences which are staminate above and pistillate below to become exposed, thus providing a means of dispersal of the seeds. This is one of four such means observed in the genetically reconstructed "wild" corn. Others are shown in Figures 8.6 and 8.7.

still exhibit this trait. (3) Ears, which elongate after their ovules have been fertilized so that at maturity the terminal part of the ear protrudes beyond the ends of the husks, are capable of dispersing part of their seeds. (4) Elongation of the shank or peduncle on which the ear is borne, causing the ear at maturity to be partly or completely exposed and capable of dispersing its seeds. Several of these mechanisms contributing to the capability of seed dispersal are illustrated in Figures 8.6 to 8.8.

Some of the plants of our later-model reconstructed ancestral form are admirably designed for both cross- and self-fertilization. If the tassel of the main stalk is partly pistillate, the silks of the pistillate spikelets are exposed long before staminate spikelets exsert their anthers and shed pollen. There is no possibility of self-fertilization of these spikelets, but the silks of the pistillate spikelets of the tillers as well as those of the ears in the middle position become exposed at about the same time that the staminate spikelets of the tassel of the primary stalk shed pollen. These will be largely self-pollinated if the plant happens to be isolated from other plants, but some of them will be cross-pollinated if there are other corn plants in the vicinity.

In its adaptation to both cross- and self-pollination the reconstructed wild corn is quite similar to *Tripsacum*, in which the first styles to appear on terminal inflorescences must necessarily be cross-pollinated while the later appearing styles are amenable to either cross- or self-fertilization. These plants are also similar to *Tripsacum* in the elongation of the lateral branches after pollination occurs. In *Tripsacum* as in some lines of the reconstructed wild corn, the lateral inflorescences are well enclosed in leaf sheaths when the styles first emerge but are completely exposed and capable of dispersing their seeds at maturity.

I once suggested (1958) that the reconstructed wild corn, because of the characteristics which it has in common with *Tripsacum*, might well be classified as an annual form of *Tripsacum*, or conversely since corn was the first of the two to be given a Latin name, *Tripsacum* could be classified as a perennial form of the genus *Zea*. So far as morphological characteristics are concerned this suggestion still has merit. But if Galinat's hypothesis,

discussed in a previous chapter, of the origin of *Tripsacum* as a hybrid of *Manisuris*, and wild corn proves to be valid, classifying *Tripsacum* as a perennial form of corn would not tell the whole story of their natural relationships. In this connection, however, I should perhaps point out that an amphidiploid hybrid of our reconstructed wild corn with certain species of *Manisuris*, if it were intermediate between the two parents, would in the majority of its characteristics resemble very closely the two most primitive species of *Tripsacum*, *T. zopilotense* and *T. floridanum*. An exception would be the rhizomes of *Tripsacum*, but perhaps these may have been somehow induced by polyploidy. Perennial teosinte, which is a polyploid form of annual teosinte, has rhizomes, although its ancestor, an annual, does not. Finally our reconstructed wild corn has a characteristic which so far as I know has never before been observed: its glumes are bicolored, the main part of the glumes white and the margins red. The effect is similar to the agouti pattern in rodents. I do not suggest that it is a counterpart in plants of such a pattern in animals, although it is conceivable that it served a similar purpose. The agouti pattern provides excellent camouflage for animals in the wild. The bicolor pattern in the glumes of wild corn might well tend to make seeds falling on the ground less conspicuous to birds than seeds enclosed in white glumes or glumes of solid colors.

The bicolor pattern apparently represents one expression of the gene responsible for dingy pericarp color, an expression not previously observed because the long glumes of pod corn have not previously been combined with the pericarp color of the Mexican race Palomero Toluqueño. This is the counterpart of the situation of certain inherited hair colors in man; the genes for greying of the hair, for example, are not expressed in men who become totally bald at an early age.

In concluding this chapter, I wish to call attention to the rather complex pedigree of our genetically reconstructed wild corn. More than half of its germ plasm was derived from the Mexican popcorn race Palomero Toluqueño, which has contributed colored and pilose leaf sheaths, sparsely branched tassels, a tendency to tiller, and small, slender, pointed kernels with white endosperm, spreading spikelets and a dingy pericarp color,

FIG. 8.9 Ears of the genetically reconstructed "wild" corn showing the variegated pattern of the glumes, a characteristic associated with "dingy" pericarp color. Both of these ears show spreading of the members of pairs of spikelets, a characteristic of the popcorn race Palomero Toluqueño, one of the races included in this strain of the genetically reconstructed ancestral form. One ear has lost its staminate tip. (About $\frac{3}{4}$ actual size.)

FIG. 8.10 By combining one of the alleles, tu^d at the tunicate locus with Ti, the tunicate-inhibiting gene, and crossing repeatedly with the popcorn variety Argentine pop, we have produced a reconstructed wild corn that has a number of the principal characteristics of the Tehuacán wild corn described in Chapter 15: short ears terminated by staminate spikes, round kernels with pericarp color partly surrounded but not completely enclosed by long, soft glumes, which are visible in the specimen on the right from which some of the kernels have been removed. Compare these ears with that of the artist's reconstruction illustrated in Figure 15.24 and with the specimens of prehistoric ceramics illustrated in Figures 16.8 and 16.9 (actual size.)

which also has an expression in the glumes of pod corn. Baby Golden has contributed a major minus modifying factor, Ti, located on chromosome 6. Both Argentine pop and Lady Finger have contributed other minus modifiers. It is possible that we could develop an even more primitive reconstructed wild corn by bringing in genes from additional races of popcorn, and indeed we are attempting this by introducing the slender fragile cobs of the Peruvian race, Confite Morocho. But the reconstructed wild form which we have already succeeded in developing has enough primitive characteristics to enable it to survive without man's ministrations in a suitable environment. Indeed even in our bleak New England climate many self-sown seeds survive the cold winter and produce seedlings the following spring.

Perhaps all of these facts still do not prove that wild corn was both a pod corn and a popcorn, as Sturtevant concluded in the late nineteenth century and as Reeves and I concluded independently several generations later. I will venture to predict, however, that no really plausible reconstructed wild form can be developed by any other means than by combining the components of the tunicate locus with genes from popcorn varieties.

We have developed still another reconstructed wild ancestral form, one designed to resemble the prehistoric wild corn of the Tehucán caves, described in Chapter 15. We produced this by introducing one of the half-tunicate components of the Tu locus and the tunicate inhibiting gene Ti into a single-stalked, multi-eared popcorn with red pericarp color. By selection we were able to develop lines with ears quite similar to those of the Tehuacán wild corn, terminated by staminate spikes and their kernels partly enclosed in glumes. One of these is illustrated in Figure 8.10.

I have now concluded that there were even more than two distinct races of wild corn, and in a later chapter I shall suggest that there may have been at least six. To develop reconstructed ancestral forms of all of these would be an interesting exercise in plant breeding but one that would scarcely justify the time, effort, and expense involved. In developing two such reconstructed ancestral forms we have already demonstrated—at least to our satisfaction—that although wild corn is now extinct, its genes are still extant and can be

recombined to produce primitive types that, in some of their principal characteristics, resemble their remote ancestors.

A Tunicate Teosinte

With the revival of the theory that teosinte is the ancestral form of cultivated maize, indeed *is* wild maize, the advocates of the theory are faced with a number of facts that require explanation. One of these—one of the most critical—is that teosinte, its seeds enclosed in hard, bony shells, is not promising as a food plant and even less so as a subject for domestication. In answer to this objection Beadle has suggested that teosinte might have been used by popping or by grinding the entire fruits and consuming the mixture of meal and shell (see Chapter 4). The ideas of the late R. A. Emerson on teosinte as the ancestral form of cultivated maize are somewhat more realistic than some of the more recent ones. Emerson supposed that the domestication of teosinte might have begun with the discovery of soft-shelled forms similar to those of *Coix*, an Old World cereal grown in the Philippines and some other parts of Asia. To the extent that the theory of parallel variation in cultivated plants, proposed by Vavilov, Russia's famous student of the subject, has validity this supposition is quite sound.

Many years ago in studying the expression of the tunicate locus on different genetic backgrounds I introduced the Tu gene into Florida teosinte through crossing followed by repeated backcrossing to the teosinte parent. The product was a soft-shelled teosinte whose fruits are illustrated in Figure 8.11 and described in its accompanying legend. Later Galinat introduced one of the half-tunicate alleles (the one now designated as tu^d) into Nobogame teosinte in the hope of developing an annual forage grass that would also be to some extent grain bearing. He succeeded in producing a true-breeding, soft-shelled teosinte, but it proved to be so attractive to birds that it could scarcely be propagated as a crop. I suspect that a soft-shelled teosinte occurring in nature would likewise invite the depredations of birds and have little chance of survival in competition with the hard-shelled forms.

It is of some interest in this connection—and it may be highly significant—that the pod-corn genes, individually as tu^d or collectively as Tu, are dominant or epistatic to the

FIG. 8.11   At least two alleles at the pod-corn locus, *Tu* and *tu^h*, are dominant or epistatic to the cupulate fruit cases of teosinte. The immature spike (*A*) is the product of introducing the *Tu* gene into teosinte; it is heterozygous for this allele. As the spike matures, the rachis becomes fragile, breaking into segments (*B*), in which the caryopses are enclosed in the glumes and the rachis segments are attached as appendages. These fruit cases differ strikingly from those of teosinte (*C*), in which the caryopses are enclosed in indurated rachis segments and lower glumes. If teosinte is a mutant form of maize, a possibility suggested in Chapter 4, one of the major differences between the two species may be controlled by the *Tu–tu* locus on chromosome 4.

FIG. 8.12   A plant of the cross (maize × *Tripsacum floridanum*) × tunicate maize. In crosses with *Tripsacum*, as in those with teosinte, the *Tu* gene and its allele *tu^h* are dominant or epistatic to the condition in corn's relatives. This results in massive, profusely branched tassels illustrated in the photograph and in replacing with herbaceous glumes the indurated rachis segments and lower glumes characteristic of *Tripsacum* and teosinte.

genes that determine the characteristic fruit cases of teosinte and even in the heterozygous condition cause teosinte's indurated shells to be replaced with the herbaceous bracts of pod corn. In this respect, as in several others, pod corn is distinctly more primitive than teosinte, and it is not surprising that genes responsible for the primitive condition should be dominant or epistatic to those determining the more specialized characteristics. Considering these facts one may postulate, as I have in Chapter 4, that the ancestral form of teosinte, like that of cultivated corn, was a form of pod corn. It is also interesting, and may be significant, that the pod-corn locus is dominant or epistatic to the genes that determine the fruit cases of corn's more distant relative, *Tripsacum* (see Figure 8.12). This is not surprising if one of the parents of *Tripsacum*, an allopolyploid hybrid, was, as Galinat (1970) has suggested, a teosinte-like plant.

It is encouraging to learn that in the teosinte-mutation hunt that Beadle organized in 1971, one of the mutants for which the hunters were instructed to look was a soft-shelled form. On this hunt none was found, but if ever one is, I predict, with some degree of confidence, that it will, if and when genetically analyzed, prove to involve one of the alleles at the *Tu* locus. If so, even the supporters of the teosinte theory will finally find it necessary to take the pod-corn theory into consideration, and they may well find the data presented in this chapter to be of critical importance.*

*Since this was written, Beadle (1972) has suggested that a mutation in teosinte from the usual hard-shelled condition to a soft-shelled tunicate form may have been involved "as the first step in a teosinte-to-corn transformation". And he stated further: "To elaborate on the tunicate trait, I believe it to have had a far more important bearing on the origin of corn than has previously been recognized." In view of the extensive research, described in this chapter and the preceding one, in which my colleagues and I have been engaged for more than thirty years in testing our 1939 hypothesis that the ancestor of cultivated corn was a form of pod corn, I regard this statement as remarkable.

9 Races of Maize

The problem of races and their recognition
is indeed almost the same in *Zea Mays* as in
mankind. Anderson and Cutler, 1942

One of the most important and far-reaching developments in shedding light on the origin and evolution of corn occurred within the past two decades: the collection, classification, and description of the races of maize of the countries of this hemisphere and the publication of these descriptions in eleven well-illustrated volumes. This project had its beginnings, as I mentioned in the preface, with the initiation by the Rockefeller Foundation of an agricultural program in Mexico. When, in 1943, the foundation, in cooperation with the Mexican Ministry of Agriculture, began a program of practical maize improvement it became evident almost at once that a survey of the native maize varieties was needed to serve as an inventory of the material available to the plant breeders. A systematic program of collection, originally wholly utilitarian in purpose, was begun. Varieties were assembled from all parts of Mexico and, in controlled experiments, were compared for productiveness, disease resistance, and other characteristics of agricultural importance.*

As the collection grew and the extraordinary diversity of corn in Mexico began to be revealed, the need for a taxonomic classification that would bring some semblance of order out of the bewildering multiplicity of varieties became apparent. Accordingly botanical, genetic, and cytological studies, to supplement the agronomic investigations, were undertaken and, to make the collections as nearly complete as possible, special efforts

*My remarks in the first several pages of this chapter are part of what, as one of its authors, I wrote in *Races of Maize in Mexico*, Wellhausen et al., 1952.

were made to obtain little-known varieties of doubtful agronomic importance from remote localities. Gradually it became possible to discern relationships between varieties and to group these into more or less well-defined natural races. Since relationships are implicit in any natural system of classification, a definite attempt was made to determine the origins and relationships of the recognized races.

What had begun as a strictly utilitarian venture of limited scope thus evolved into the study of the evolution in a single geographical region of America's most important cultivated plant. One result was that the corn breeders in Mexico found themselves the possessors of a far more useful inventory than they had originally sought of the breeding material available in that country and were able to approach the breeding problems with some degree of confidence in their choice of stocks.

In late December of 1948 at the invitation of Edwin Wellhausen I went to Mexico to spend six weeks working with him, with my former student, Lewis M. Roberts, who had joined Wellhausen in the corn improvement project, and with Ephraim Hernandez, who had done the major part of the collecting, on a monograph classifying and describing the Mexican races of corn. The collection was laid out on all of the available space in a recently constructed laboratory building at the National School of Agriculture at Chapingo, arranged first by regions of origin and then on the basis of resemblances. Grouping of the ears was correlated with data on characteristics of plants grown in the field. This intensive

study led eventually to the recognition of 25 more or less distinct races divided into four major groups: Ancient Indigenous, Pre-Colombian Exotic, Pre-historic Mestizos, and Modern Incipient.

Previous Classifications of Maize

The classification of cultivated plants in general has not kept pace with classification of natural species, and the reasons are not far to seek. The principal one lies in the kinds of variation that occur in the two categories of plants. In nature variation is usually discontinuous and the majority of natural species have become separated from each other by well-defined morphological gaps. Where this has not occurred the taxonomist is likely to regard the species as "not good." In cultivated plants, on the other hand, discontinuous variation is rare and is more often the exception than the rule. Frequently there are no sharp lines of demarcation between the varieties or races that comprise a cultivated species or genus. This is especially true when the variation is entirely intraspecific. It is almost inevitably true when, in addition, the species is one in which natural cross-fertilization accompanied by a continuous interchange of genes between populations is the rule.

Since all maize not only belongs to a single species but is also largely cross-fertilized, it offers more than the ordinary number of difficulties to the taxonomist. Hence it is not surprising that the classification of maize, in spite of its importance, should have been so long neglected. Taxonomists who shun cultivated plants as not botanically important may actually be avoiding difficult problems not easily solved by traditional taxonomic methods. The variation in cultivated plants is frequently so bewildering that additional techniques, including those of the geneticist, the cytologist, and the agronomist, are needed to bring some degree of order out of apparent chaos. Of no cultivated plant is this more true than of maize.

The earliest attempt at comprehensive treatment of the problem of maize classification is that of E. Lewis Sturtevant (1899), who at the turn of the century published a monograph entitled "Varieties of Corn." Sturtevant cataloged the variability of maize then known to him into six main groups, five of which were based upon the composition of the endosperm of the kernel. This classification had been used almost without modification for a period of fifty years, and for almost the same length of time interest in advancing the classification of maize had remained dormant. One reason for this was the fact that the classification was—and still is—commercially useful. Sturtevant's five categories based on characteristics of the kernels: dent, flint, flour, sweet, and pop, are recognized in commerce by persons having little or no botanical knowledge. His sixth category, pod corn, was primarily a botanical curiosity.

In recent years, partly as the result of an accumulated body of knowledge on the genetics and cytology of maize and partly also—I like to believe—as an outgrowth of our own new hypotheses concerning the origin of maize and its relatives, there came a revival of interest in the classification of maize. Contributing especially to this revival were Edgar Anderson and Hugh C. Cutler of the Missouri Botanical Garden. In a series of papers published jointly and separately, these botanists brought to bear upon the problem new evidence from botany, genetics, and archaeology. In their first paper on maize classification (1942), they pointed out that Sturtevant's classification, while useful, is largely "artificial," since it is based almost entirely upon characteristics of the endosperm, several of which— floury and sweet—are now known to be primarily dependent for their expression on single loci on single chromosomes. A "natural" classification, according to Anderson and Cutler, should be one based upon the entire genetic constitution and one integrating the maximum number of genetic

facts. As a mere cataloging device a natural classification may be no more useful than an artificial one, but as a means of showing relationships and tracing origins it can be infinitely more valuable. Anderson and Cutler sought characteristics which would be more useful in reflecting the entire genotype than did Sturtevant's endosperm characteristics and in this search made an important contribution in showing that the maize tassel, whose central spike has long been recognized as the homolog of the ear, is valuable in studying and classifying the variation in maize.

Indeed these maize taxonomists came to regard the tassel as the most useful organ of the maize plant for purposes of classification, stating (1942) that "the tassel of *Zea Mays* presents us with more easily measured characters than all of the rest of the plant combined." We were not convinced that this was true when we began our classification of Mexican maize in 1948; I am still far from convinced that it is true. More nearly correct, I think, is the statement made by De Candolle (1886), the distinguished nineteenth century student of cultivated plants, "a cultivated species varies chiefly in those parts for which it is cultivated. The others remain unmodified or present trifling alterations."

Since the ear is the part for which corn is cultivated, it is not surprising that it has become the most highly specialized organ of the maize plant and is the structure which, more than any other, distinguishes *Zea Mays* from all other species of grasses. By the same token it is reasonable to suppose that the ear and not the tassel should offer more diagnostic characters useful in classification than any other part of the plant. This has proved to be the case. In all of the volumes that have been published describing the races of maize of this hemisphere, the most numerous data and illustrations are concerned with the ears and their various parts: kernels, cobs, etc. Nevertheless, characters of the tassels are also useful in arriving at a natural classification, and all recent works on the classification of maize have employed them to some extent.

Our intensive studies of the collections of Mexican maize in December 1948 and the early months of 1949 led to the conclusion that the domestication of a native wild corn or corns had produced four different races that had been grown in Mexico from time immemorial and that had maintained their

identities throughout the centuries to the present time. All four of these races: Palomero Toluqueño, Nal-Tel, Chapalote, and Arrocillo Amarillo, comprising the group which we called Ancient Indigenous, were primitive in at least one respect, all were popcorns. Although nowhere widely grown, they were maintained by the Indians, who recognized them as being a traditional part of their agriculture and who, for sentimental or other reasons best known to themselves, continued to grow these corns, although there were more productive races at their command.

We assumed that somewhere in the remote past, perhaps at about the time of Christ or a few centuries earlier or later, four other races of maize were introduced from farther south. These four, called Cacahuacintle, Harinoso de Ocho, Oloton, and Maiz Dulce, comprised the group that we designated as Pre-Colombian Exotic. Hybridization of these introduced races with the Ancient Indigenous races and with corn's closest relative, teosinte, perhaps also with its more distant relative, *Tripsacum*, produced an almost explosive diversification in Mexican corn resulting in the creation of 13 new races to which Wellhausen et al. (1951, 1952) gave the name Prehistoric Mestizos (mestizo being the Mexican word for racial hybrid). Finally we recognized four races that had come into existence within historical times; these made up the group called Modern Incipient. There was still a fifth group entitled "Poorly Defined Races," which was later shown to be involved in the diffusion of maize from Mexico to the southwestern United States.

The Mexican Ministry of Agriculture published the classification and descriptions resulting from these intensive studies in April 1951 in a well-illustrated volume of 237 pages entitled "Razas de Maiz en Mexico." An English edition of the same work published in 1952 by the Bussey Institution of Harvard University contained in the foreword the following significant statement. "Maize is the basic food plant in most of the Americas and its diversity, the product of thousands of years of evolution under domestication, is one of the great natural resources of this hemisphere. To lose any part of that diversity is not only to restrict the opportunities for further improvement but also to increase the difficulties of coping with future climatic changes or with new diseases or insect pests. The modern corn breeder, therefore, has a responsibility not only to improve the maize in the country in which he works but also to recognize, to describe and to preserve for future use, the varieties and races which his own improved productions tend to replace and in some cases to extinguish."

Other Countries Follow Mexico's Example

Both the Spanish and English editions of this monograph were widely distributed and aroused much interest among maize workers throughout the hemisphere. One of these, Dr. Friedrich Brieger of the University of São Paulo Agricultural College, Piracicaba, Brazil, began to explore the possibility of raising funds to finance similar programs of collection and classification in other countries of Latin America, especially in his own. In the course of a trip through the United States, Brieger discussed the problem with Dr. Ralph Cleland of Indiana University, who was then completing a three-year term as chairman of the Division of Biology and Agriculture of the National Academy of Sciences—National Research Council. Convinced by Brieger of the importance of collecting and preserving the races of corn of this hemisphere, Cleland in turn persuaded the Academy—Research Council to name a committee to investigate the problem and make recommendations. The result was the appointment of a Committee on the Collection and Preservation of Indigenous Strains of Maize. This group, composed of leading corn breeders, geneticists, botanists, and administrators, soon sought and obtained a grant of $86,000 from the Point IV Program (then called the Technical Cooperation Administration). Three centers, each with cold-storage facilities, were established in Mexico, Colombia, and Brazil, and a fourth—to store "standby" samples from the others—was set up in the United States. The grand total of collections assembled under this program eventually reached 11,353. These represented 32 different countries, from Canada to Chile.

A second grant of $90,000 was obtained from Point IV (by that time the International Cooperation Administration) to classify and describe the varieties. This resulted in the publication of ten treatises on the races of maize of Cuba, Colombia, Central America, Brazil and other eastern South America countries, Bolivia, the West Indies, Chile, Peru,

Ecuador, and Venezuela. Together with the original work on the races of Mexico, they represent a virtually complete inventory of the corn of this hemisphere, and the collections in storage represent virtually all the germ plasm available to corn breeders. It is possible that a few races grown only in isolated localities remain to be discovered, but there is no doubt that the majority of corn races have now been classified and described. New collections which have come to my attention in recent years have contained no elements not related to these already described in the eleven publications on races.

It turns out that this program of collecting was undertaken none too soon. Already in parts of Mexico and Colombia the improved varieties and hybrids have replaced the native sorts, some of which are now difficult to find. Another five or ten years and some varieties might have become extinct and their particular combination of genes, the product of centuries of evolution under domestication, forever lost. These eleven volumes not only represent a virtually complete inventory of the maize of this hemisphere but are also a monument to effective international cooperation. All of them include among their authors one or more agricultural scientists of the country whose maize the volume describes. All of them acknowledge the invaluable assistance rendered by other native workers, in the aggregate running into the hundreds, in collecting and summarizing field and laboratory data.

It was the first of these eleven publications, *Races of Maize in Mexico*, that employed a unique pattern of recognizing authorship which was later followed by the majority of the remaining volumes. This was the invention of J. George Harrar, then the field director of the Rockefeller Foundation's Agricultural Program. It was obvious to him that my participation in the preparation of this volume should be recognized; it was equally obvious to me that Wellhausen, who had been studying the Mexican collections intensively for more than five years, when I joined his team in 1948, must be the senior author. Where to put my name in the sequence? Harrar, with the precocious wisdom of a young Solomon, suggested setting it apart at the end, preceded by the phrase "in collaboration with." This proved to be a brilliant solution in setting a pattern for effective

collaboration between workers of the countries concerned and botanists from the United States invited to participate in producing a classification of the maize of a given country. To librarians and bibliographers, however, it has proved to be (as I predicted that it would be) an annoyance, if not an abomination. A number have cited the authorship as it appears on the title page; some omit the phrase "in collaboration with," thus making me the junior author in a series of four, five, or six authors. This does not disturb me unduly. What does annoy me, however, is to have my name omitted entirely, apparently on the curious assumption that a collaborator cannot also be a co-author. This minor annoyance, however, is more than overshadowed by the satisfaction that I have had in participating in the preparation of four of the eleven volumes on races of maize, those describing the maize of Mexico, Central America, Colombia, and Peru. It is this first-hand experience plus the experience of describing a number of collections of prehistoric corn, including the prehistoric wild corn of Tehuacán, Mexico, that now gives me the temerity to attempt to summarize for the hemisphere as a whole some of the salient facts contained in the reports on the separate countries; specifically to determine, so far as possible, the relationships between the races of one country and those of others; to trace origins back to primitive progenitors; to recognize and delineate evolutionary paths showing, if possible, how ancient maize has evolved into modern maize. Classifying the races of maize is comparable to classifying the races of man, and the task is no less formidable and no less fraught with hazards; nevertheless it has been undertaken.

A total of 305 names of races are listed in the eleven volumes describing the races of maize of this hemisphere exclusive of those in the United States, which, for bureaucratic reasons, could not be included in the project financed by our foreign-aid programs. Since races of maize are oblivious of national boundaries, some duplication between the races of one country and those of others, especially adjoining ones, is to be expected. What I have done is to attempt to recognize this duplication in some instances through similarities in local names, for example, Chullpi, Chulpi, Chuspillo for a race of sweet corn and Pisankalla, Pisincho, Pichinga for a

FIG. 9.1 Variation in the size, shape, color and endosperm texture of the ears of maize varieties of the hemisphere. Much of this variation has now been classified in a series of monographs published between 1952 and 1963 describing the recognized races of the countries of Latin America. Some of these are illustrated in the photograph. *A*, Chullpi, the Peruvian sweet corn considered to be the progenitor of modern sweet corn varieties. *B*, Piricinco, a Peruvian race that often has odd numbers of kernel rows. *C*, Conico, the most common race of the Mexican plateau. *D*, Tuxpeño, a Mexican race considered to be one of the ancestors of the U.S. Corn Belt dent. *E*, Jala, a Mexican race characterized by having the world's largest ears. *F*, U.S. Corn Belt dent, the world's most productive corn, considered to be a hybrid of *D* and *G*. *G*, Northern Flint descended from a Mexican 8-rowed flour corn and considered to be one of the parents of U.S. Corn Belt dent. *H*, Cuzco Gigante, a Peruvian race with the world's largest kernels.

race of popcorn with pointed seeds. Much beyond these obvious resemblances I have not depended on similarities in names, since I have little confidence in linguistics carried too far.

More commonly the duplication was recognized by the resemblance in the ears as they are depicted in the excellent photographs with which the volumes are illustrated. I did not, however, regard resemblances in appearance alone as establishing relationships; I also examined the data for other similarities, including adaptation to altitude, earliness, tillering, plant color, pilosity, hardness of the kernels, etc.

My procedure was to take the races of one country, Mexico, for example, and look for resemblance between these and the races of all other countries. Of the 32 races described in Races of Maize in Mexico, 8 proved to have close counterparts among the races of maize of Guatemala, 6 among Colombian races, 5 among Peruvian races, 2 among races of Brazil, and strangely, considering Cuba's geographical proximity to Mexico, none among the Cuban races. The same procedure was then followed with respect to the races of Guatemala, and for those of all other countries. Of the 32 Mexican races, 20, almost two-thirds, are "endemics," races not found in other countries. This suggests that Mexico is one of the principal centers of the origin, evolution, and diversification of maize. This is also true of Peru; some 30 of its 48 races occur only in that country. At the other extreme is Guatemala, which though adjoining Mexico has only one race of 13, which is not found in any other country and even this one, because of a close relationship to another race, is not strictly endemic. Our analysis permits us also to list the synonyms, in various countries or at least in the different publications, for any given race. This shows that the number of distinct races is not 300 or more, as the listings in the individual publications might suggest, but nearer one-third that number.

In analyzing the publications on races of maize from different countries or regions, I found it necessary to make allowance for the factor of national pride. The native workers of several of the countries of this hemisphere like to believe that theirs is the country of origin not only of the majority of the races of maize occurring there but also those of adjoining and even more remote areas. I first discovered this phenomenon in Mexico in 1943 when, speaking to a group of students from the National School of Agriculture at Chapingo, I narrowly escaped being lynched (figuratively speaking) when I gave it as my considered opinion that, if there was only one center of origin of maize, that center was in South America. The Mexican's fervent belief that corn's origin must be sought in Mexico has since been vindicated in part by the discovery of prehistoric wild corn in that country.

I encountered the same expression of national pride in Peru when I worked there with Alexander Grobman and his group of Peruvian *Maiceros*. My suggestion that the large-seeded flour corn for which Peru is famous, the race Cuzco Gigante, might be the product of post-Columbian hybridization between a smaller version of this race and an 8-rowed corn, Pardo, introduced by the Spaniards into the Peruvian coast from Mexico, where it is known as Tabloncillo, evoked open scorn. The idea that the most highly evolved Peruvian race might be half Mexican in its origin was not only thoroughly repugnant to my Peruvian friends but also a grievous offense to national pride. Later, when they could find no better hypothesis to explain the origin of Cuzco Gigante and at the same time came to realize that the Mexican Tabloncillo may be partly South American in origin, the idea gradually became more acceptable and was adopted.

The same phenomenon of national pride with respect to origin of races is found in Brieger et al., 1958, who, although taking issue with virtually everything that I have written on the origin and evolution of maize, still cling to the 1939 hypothesis of Reeves and mine that corn originated in the lowlands of South America, an idea which I was compelled to modify many years ago with the discovery of fossil corn pollen in Mexico. The fact is that Brazil, far from being the center of origin of maize, has no indigenous races of its own. All present-day Brazilian corn apparently has had its origin elsewhere. I have little hope, however, that Brazilian botanists will accept this unpalatable conclusion within the foreseeable future.

The Pointed Popcorns

A significant fact, that quickly became apparent when I began to analyze the description of races contained in the eleven publications on the races of maize of the countries of this

FIG. 9.2   An ear of the primitive Mexican popcorn race Palomero Toluqueño. Its most distinctive features are the relatively long, pointed kernels with "dingy" pericarp color. The ears are commonly cone shaped. There is often a tendency for the kernels of a pair of spikelets to spread apart, as shown at the base of the ear illustrated here. This race or its somewhat contaminated derivatives are found in virtually every country of the American mainland. Its lineage is represented by the Mexican races Cónico, Cónico Norteño, Chalqueño, and Pepitilla. In Peru the characteristics of pointed seeds and spreading spikelets are found in the races Confite Puntiagudo, Paro, Huancavelicano, and Rabo de Zorro. (One-half actual size; from Wellhausen et al., 1952.)

hemisphere, is that there is one race which occurs or has left its mark in all of the countries from which collections have been made except those of the West Indies. This is Palomero Toluqueño, one of the popcorn races of Mexico described by Wellhausen et al. (1952) and included in their category of Ancient Indigenous races. This race, or a derivative of it, occurs under various names in Mexico, Guatemala, Colombia, Ecuador, Peru, Bolivia, Chile, Argentina, Venezuela, and Brazil (and also in the United States). The race is one which obviously became widely diffused, and if there is in this instance any correlation between area and age the diffusion must have begun long ago.

As mentioned in the previous chapter Palomero Toluqueño has a number of distinctive characteristics: a high kernel-row number, the kernels pointed and imbricated (like the overlapping tiles of a roof), the kernels relatively long and slender, the members of a pair of kernels arising from a pair of spikelets having a tendency to spread apart so that the right-hand kernels in one row of paired kernels may press against the left-hand members of the adjoining row to produce what appears at first glance to be a distinct row of paired kernels but which actually represents the neighboring rows of two ranks of paired kernels. This characteristic is shown in the basal part of one of the ears illustrated in Wellhausen et al. The basal kernel rows are often irregular, and inverted kernels (those developing from the lower, usually nonfunctional floret and having their embryos facing the base instead of the tip of the ear) are sometimes present. The endosperm of the kernel is often "glassy" in texture; the endosperm is white; genes for partial inhibition of yellow endosperm are present. The aleurone is usually colorless, partly because of the aleurone-color inhibiting factor, I, on chromosome 9; the pericarp is "dingy" or distinctly colored red, brown, or variegated. The gene for dingy pericarp color also affects the color of the glumes when various alleles at the Tu-tu locus are present. It may turn out to be a "wild" allele at the P locus on chromosome 1, the many other alleles at this locus being mutant forms, arising during domestication. Mid-cob color is often present; the plants are short with tall tillers. Other characteristics are early maturity, leaf sheaths strongly colored and pilose, good adaptation to high altitudes, sparsely branched tassels with prominent central spikes, conical ear shape; low chromosome knob number in the less contaminated collections, and a cross-sterility gene, Ga, often present.

These characteristics are listed in Table 9.1, which shows to what extent they are known to be present in certain races of the countries of this hemisphere. No more than a brief glance at this table is needed to show that *all* of these characteristics occur in a single race only in Mexico. In all other countries some of the characteristics, but never all, are found in certain races resembling Palomero Toluqueño of Mexico.* This suggests either (1) that the Toluca Valley of Mexico is the original home

*Although not included in the table, certain races of popcorn grown in the United States have some of the characteristics of the Mexican Palomero Toluqueño. The variety "Tom Thumb" has the sparsely branched tassels and the pilose leaf sheaths. The "Rice" types of popcorn have the pointed imbricated kernels. "Japanese Hulless" has both features.

of this race and from here it has spread through the hemisphere; (2) or that the race has only recently been introduced to this valley and has not yet become contaminated. Since Palomero Toluqueño has participated in the evolution of a number of races of Mexico (see Wellhausen et al., 1952) the first alternative seems the more plausible of the two.

The reader may ask how it is that despite obvious admixtures so many characteristics of this race still persist. The answer is that many genes and perhaps all of this corn's ten chromosomes are involved in the inheritance of this complex of characteristics. Kernel-row number, for example, is a quantitative character involving multi-factorial inheritance. Pointed kernels involve at least two different hereditary factors (Hayes and East, 1915). Flinty endosperm differs from floury by at least one gene on chromosome 2 and from dent corn by many genes. White endosperm color involves a locus on chromosome 6; the inhibitor of aleurone color is a gene on chromosome 9; dingy pericarp color is probably governed by an allele at the P locus on chromosome 1; mid-cob color and colored leaf sheath involve alleles at the R locus on chromosome 10. The cross-sterility gene, Ga, has its locus on chromosome 4. Linkage studies by Paxson (1953) indicate that pilose leaf sheaths involve genes on at least two chromosomes, 3 and 9. Thus at least six of corn's ten chromosomes are known to be involved in about half of the characteristics of Palomero Toluqueño. We know little about the inheritance of spreading spikelets, inverted kernels, tall tillers, sparsely branched tassels, and adaptation to high altitude, but the indications are that their inheritance is multi-factorial. Thus it is that this race may become strongly introgressed by other races even to the extent of having the majority of its chromosomes replaced by those of the introgressing race. Yet, if it retains even a minority of its original chromosomes it will still exhibit certain of its original morphological characteristics. This fact must be taken into consideration in examining the data on chromosome-knob numbers and positions discussed later in this chapter.

I see Palomero Toluqueño as one of the Ancient Indigenous races of Mexico originating from a different geographical race of wild corn than the Chapalote-Nal-Tel com-

plex of the Tehuacán Valley (Mangelsdorf et al., 1964). The latter descended from a wild corn adapted to intermediate altitudes, the former from a high altitude race. That there was such a high altitude race of wild corn is shown by the fossil pollen found at depths of more than 200 feet below the present site of Mexico City, which is at an altitude of about 7,500 feet. The evidence suggest that this race spread at an early date to regions of high altitudes in South America. It may or may not be significant that the Peruvian Indians call it *Chili-Sara*, meaning "old corn" (Grobman et al., 1961).

Ears with pointed imbricated kernels occur among the prehistoric specimens of Peru and are also represented in the prehistoric ceramics. The Indians of Peru, Bolivia, Ecuador, and Argentina have similar words for the pointed popcorn, *Pisankalla*, and variations of it. The resemblance between the Peruvian pointed popcorn, Pisankalla, and the Mexican pointed popcorn, Palomero Toluqueño, are numerous and apparent. They have been pointed out by Grobman et al., who, however, suggest their common origin to have been the central Andes. I consider it much more likely that their origin is Mexican.

This race has left its mark in varying degrees on many other races of the countries of Latin America. Wellhausen et al. postulated that it is one of the parents of Conico, the most commonly grown race of the Mexican plateau and has influenced Conico Norteño and Pepitilla. The latter is in turn related to the "gourd-seed" corn of the southern United States and may have contributed some of the genes responsible for the high kernel-row numbers in this corn; gourd-seed in turn is involved in the ancestry of the world's most productive race of corn, the Corn Belt dent, of the United States (Wallace and Brown, 1956).

In Peru the characters of this race are also found especially in the races Confite Puntiagudo, Paro, Huancavelicano, Rabo de Zorro, Ancashino, Maranon, and probably Piricinco. The "interlocking" of the kernels of adjacent rows of Piricinco and its Bolivian counterpart, Coroico, described by Cutler (1946), may be an extreme form of the spreading of the members of a spikelet pair characteristic of Palomero Toluqueño. These two races, Piricinco and Coroico, have in turn left their mark on many other races in South America.

The Sweet Corns

Second only to the race of the pointed popcorns in the range of its distribution throughout the countries of this hemisphere is the race of sweet corn. Sweet corn was collected in Mexico, Colombia, Ecuador, Peru, Bolivia, Chile, and Argentina. Although not collected in Guatemala in these studies, it must have occurred there, since Breggar (1921) referred to a Guatemalan sweet corn used in breeding experiments in Puerto Rico. Sweet corn was also grown by numerous Indian tribes in the southwestern, north central, and northeastern parts of the United States. It apparently was not grown in Venezuela, Brazil, and the West Indies.

Sweet corn's wide range of distribution is somewhat surprising, since it actually represents a genetic defect in metabolism. The conversion of sugar into starch in the endosperm of the kernels does not proceed to completion as it does in starchy types of maize: dent, flint, and pop. Thus the storage material in the endosperm is composed of sugars—glucose and sucrose—and of intermediate polysaccharide products generally classified as dextrins. Such starch grains as are formed are quite small compared to those of nonsweet varieties. As the kernels mature and dry, they become wrinkled and translucent like dried syrup and are easily distinguished from those of all other types of maize. Genetically the sugary condition of the kernels is the product primarily of a single recessive gene on the short arm—near the centromere—of chromosome 4, but, as pointed out later in this chapter, cultivated races of sweet corn involve many additional genes affecting this characteristic.

The sugary condition not only reduces the total amount of food stored in the endosperm (Mangelsdorf, 1926, Kiesselbach, 1926) but also renders the kernels more susceptible to damage by molds and other fungi. Sweet corn kernels generally need more favorable conditions for ripening and for germination and growth than do the pop, flint, or dent corns but are not greatly inferior in this respect to flour corn, which is also, to some degree, an inherited defect in metabolism.

Since sweet corn is a mutant form representing a metabolic defect it seems unlikely that it ever occurred in nature as a wild race. Consequently, it should probably be considered a product of domestication. Since

TABLE 9.1 Characteristics of the Pointed Popcorns of the Countries of Latin America Showing Their Relationships to the Mexican Race Palomero Toluqueño

Characteristics	Mexico	Guatemala	Colombia[a]	Colombia[b]	Venezuela	Ecuador[c]	Ecuador[d]	Peru	Bolivia	Chile	Argentina	Brazil
1. Kernels pointed, imbricated	+	+	+	−	−	+	+	+	I−+	I−+	+	+
2. Kernels relatively long, slender	+	I	+	I	I	+	+	+	+	+	+	+
3. Kernel pairs spreading	+	+	+	I	I	I		+				
4. Basal kernel rows irregular	+	I	I	−	−	I		+	+		+	
5. Inverted kernels present	+										+	
6. Endosperm corneous	+	+	+	+	+	+	+	I	+	+	+	+
7. Endosperm white	+	+	+	+	+	+	+	+	+	I	+	+
8. Aleurone colorless	+	+	+	+	+				+	+	+	I
9. Pericarp dingy or colored	+	+	+	−	−	+			+	+		+
10. Mid-cob color present	+	+	+			+	+					
11. Adapted to high altitudes	+	+	+	I	I	+	+	+	+	+	+	
12. Short plants	+	−	I	I	I−+	+	+	+				
13. Early maturity	+	I	+	I−+			I	+				
14. Strong plant color	+	I	+	I	I	I	I	+	−			
15. Strongly pilose	+	I−+	+	I	I	I		+	I			
16. Tall tillers	+			−	I	+	+	−	I			
17. Prominent central spike	+	I			+	+	+	I				
18. Few tassel branches	+	−		I	I	+	+	+			I	
19. Ear shape conical	+	+	+	−	−	+	+	I	I	I	I	I
20. Low chromosome-knob number	+	+	I			+			+	+		
21. Cross-sterility gene present	+	+								+[e]		

[a]Imbricado, [b]Pira, [c]Canguil, [d]Canguil Grueso, [e]Nelson, 1951.

+, character present; −, absent; I, intermediate.

TABLE 9.2 Diameter = Length Ratios and Kernel-Row Numbers of the Native Races of Sweet Corn of the Countries of America, from North to South

	Characteristic			
Country, race, and source of data	Length	Diameter	Ratio	Kernel row no.
U.S.; Pawnee Sweet; Will and Hyde, p. 154	4.5	1.0	.22	10.0
U.S.; Ponka Sweet; Will and Hyde, p. 121	4.4	1.1	.25	10.0
Mexico; Dulcillo; Wellhausen et al., fig. 131	6.2	1.3	.21	11.6
Mexico; Maiz Dulce; Wellhausen et al., table 15	13.7	4.5	.33	14.5
Mexico; Maiz Dulce; Wellhausen et al., fig. 27	6.8	2.8	.41	14.0
Mexico; Maiz Dulce; Kelly and Anderson, table 1	16.6	4.9	.30	17.1
Colombia; Maiz Dulce; Roberts et al., fig. 18	7.1	2.2	.31	14.7
Ecuador; Chulpi; Timothy et al., table 6	12.6	5.0	.40	16.3
Ecuador; Chulpi; Timothy et al., fig. 35	7.1	3.2	.45	16.0
Peru; Chullpi; Grobman et al., table 5A	8.6	5.9	.69	18.0
Peru; Chullpi; Grobman et al., fig. 76	8.0	4.5	.56	18.0
Bolivia; Chuspillo; Ramirez et al., table 8	11.7	5.1	.44	19.7
Bolivia; Chuspillo; Ramirez et al., fig. 46	7.7	3.5	.45	18.0
Chile; Chulpi; Timothy et al., table 6	10.1	4.7	.47	17.4
Chile; Chulpi; Timothy et al., fig. 11	4.8	2.4	.50	16.0
Argentina; Chulpi; Brieger et al., table 12	10.4	4.5	.43	15.5
Argentina; Chulpi; Brieger et al., fig. 39	10.6	3.9	.37	14.0

sweet corns in general are less productive than nonsweet types they must offer some special advantage to the grower which compensates for their shortcomings. That advantage undoubtedly lies in the sweetness of the kernels. Although man has no need for sugar as such in his diet, what is commonly known as a "sweet tooth" appears to be almost universal in the human race, and people in all parts of the world have discovered and utilized sugar-rich substances of both plant and animal origin. In the Old World honey was one of the most highly prized substances if its frequent mention in the Bible is any criterion. Other sweet substances of Biblical times included manna, now believed to have been the secretion of an insect; dates, which, when ripe and dry, contain up to 50 percent of sugar; and St. John's bread, the fruits of a leguminous tree which were the "locusts" on which John the Baptist subsisted in his years in the wilderness and which, like dates, contain up to 50 percent of sugar. Not mentioned in the Bible is the plant destined to become the world's most important sugar plant, sugarcane. The natives of New Guinea and other parts of that region of the world discovered long ago how to identify, and to enjoy by chewing, the sweet stalks that occurred occasionally as mutant forms among the starch-filled stalks of the wild cane.

The ancient people of the New World also had their sweet tooth and a variety of ways to satisfy it. Mexican Indians prized a kind of honey made by a species of stingless bee that does not separate the cells dedicated to the storage of honey from those designed for the larvae, so that consuming the one usually means partaking also of the other. The Indians of the region now the southwestern United States made use of the pods of the leguminous mesquite tree, boiling them with their corn to produce a sweet pudding. And the Indians of the region now the northeastern United States and southeastern Canada not only drank the spring sap of the maple tree but had learned to reduce it to sugar, a very palatable sugar flavored with delectable impurities.

Not the least of the sources of sugar of the American Indians was the corn plant. We know from the prehistoric remains from various caves that the Indians chewed the tender young ears before the kernels had developed very far. The stalks from which the ears are

removed at an early stage become quite sweet because the sugar manufactured in the leaves, having no other place to go, accumulates in the stalks as it does in sugarcane. There is abundant evidence of stalk-chewing in the quids occurring in the vegetal remains from once-inhabited caves.

I make this digression on the prehistoric use of sugar-rich substances to show that the discovery of a new source of sugar, a mutant corn with more sugar and less starch in its kernels, would have been a highly significant event. We can be certain that the sweetness of the kernels of the mutant type did not escape the notice of the Indians, whose powers of observation were remarkably keen—and still are—when not dulled by the distractions of civilization. In Mexico and among the Indians of the upper Mississippi sweet corn was the basis of *pinole*, a confection prepared by grinding the mature seeds (Kelly and Anderson, 1943). In Peru and the adjoining highland regions of South America, kernels of sweet corn were parched to produce a favorite food, *kancha*. In that region sweet corn also found a special niche in the preparation of the native beer, *chicha*, to which by virtue of its greater sugar content it imparted a higher alcoholic potency, a characteristic prized no less in primitive societies than in modern civilizations. It subsequently found a much larger niche in the United States, where as a green corn or "roasting ears" and as canned and frozen corn it has become one of the most popular and widely grown vegetables. Strangely enough the American Indians, although they prized young corn—the Navajo's even had a highly ritualized "green corn" dance—did not generally employ sweet corn for this purpose. It was said to be too "gummy"; but cut from the cob and dried and as mature corn, sweet corn had a place in the cultural practices of many Indian tribes.

Has the mutation from starchy to sugary endosperm occurred more than once? There is no doubt that sugary is a recurring mutation. Singleton once discovered a mutation from starchy to sugary and estimated that it must have occurred once in about 2,000 gametes. Lindstrom (1935) found one sugary mutation involving about 18,000 gametes. Stadler (1942) in controlled experiments on natural mutation rates found mutations to sugary endosperm occurring at the rate of 2 per million gametes. Langham (1945), while

working in Venezuela, found no sweet corn being grown in that country and hoped to develop a sweet-corn variety by looking for a mutation to sugary in a field-corn variety; he was successful. Bear (1944) observed mutations to both sugary and waxy endosperm in inbred lines of dent corn. While working in Texas I noted a sugary mutation in an inbred line of the variety Mosshart Yellow dent.

Assuming the data from Stadler's controlled experiments on the mutation rates to provide the most reliable estimates available on the change from starchy to sugary endosperm, we can calculate that with about 257 million acres of corn grown in the world each year, there should be about one mutation for every 17 acres of corn, or about 15 million mutations annually in the world's total acreage. In view of these circumstances it is not surprising that some authorites on sweet corn, notably Erwin (1942), have concluded that sweet corn, at least that grown in the United States, is a post-Columbian development and the product of a recent mutation or mutations.

Despite the fact that the sugary endosperm of corn is known to be a recurring mutation and despite the fact that I have generally been a proponent of "independent invention" rather than of "diffusion" of cultural traits, I have concluded from my study of the description and illustrations set forth in the publications on races of maize of the countries of Latin America and elsewhere, that the sweet corns occurring in these countries are all related to each other and have had a common origin. This conclusion is based in part on a number of significant resemblances, in part on the statements of the authors of the publications, and in part by what I have learned through experience about the nature of sweet corn as a cultivated race. Cultivated races of sweet corn involve more than the single gene, *su*, for sugary endosperm on chromosome 4. They involve a whole constellation of modifying factors contributing to the expression of this principal gene. I found this out the hard way when working in Texas in the twenties. I attempted to develop earworm-resistant sweet-corn varieties for the southwest by introducing through repeated backcrossing into two locally adapted field-corn varieties, Surcropper and Mexican June, the single gene for sugary endosperm. After several backcrosses to the field-corn varieties,

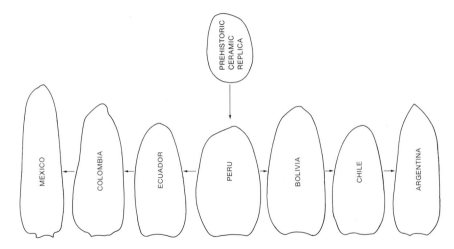

FIG. 9.3 The Peruvian sweet-corn race Chullpi is almost certainly the ancestral form of the sweet corns grown by the Indians in the countries of Latin America from Mexico to Chile. Its genealogical descendants may also include the Indian sweet corn varieties of the U.S. Southwest, the upper Missouri, and the Northeast. If so, virtually all sweet corn varieties now grown carry at least some germ plasm from this race. Chullpi is regarded as the domesticated version of a more primitive progenitor illustrated in Figure 16.11. (One-half actual size; from Grobman et al., 1961.)

FIG. 9.4 Outline drawings showing the shapes of the ears of the races of sweet corn of the countries of Latin America. Note that the ears are alike in being tapered at both ends but become longer and more slender as the distance both north and south from Peru increases. The fact that the ears of sweet corn varieties of some of the Indian tribes of the upper Missouri resemble those of the Mexican sweet corn has led to the conclusion that virtually all modern sweet corn is descended from the Peruvian sweet corn race Chullpi.

the recessive sugary kernels could no longer be readily distinguished from the starchy kernels on the same segregating ears. It was only by selection for modifying factors accentuating the sugary characteristic that I finally succeeded in producing varieties in which the mature kernels were wrinkled and translucent as they are in the sweet corn varieties of the northern United States. Had the backcrossing to the Texas varieties been carried on for several additional generations this selection for modifier complex might not have succeeded. Another experience illustrating the importance of the modifying complex was concerned with the discovery of a mutation to sugary endosperm in the Texas variety, Mosshart Yellow dent. Here, I thought, was an opportunity to produce a new, well adapted, yellow-seeded, sweet corn in one generation by growing the recessive sugary seeds. The sugary mutant proved to be, however, virtually a lethal in the field; not a single one of the sugary seeds germinated. These two experiences convinced me that virtually all cultivated races of sweet corn have modifier complexes that accentuate the expression of the sugary condition on the one hand and

on the other counteract the highly deleterious near-lethal effects of the sugary mutation. I now strongly suspect that the majority of them receive at least part of their complexes from a single original source, the sweet corn of Peru and Bolivia.

Wellhausen et al. (1952) included the Maíz Dulce of Mexico as one of the four Pre-Columbian Exotic races which they thought had been introduced from farther south in prehistoric times. Kelly and Anderson (1943) had earlier noted that the ears of sweet corn in Jalisco are shorter, thicker, and more rounded than those of the field-corn varieties of Jaliscan maize and that they "bulged in the middle" rather than tapering sharply and evenly like most Mexican maize. They regarded cob shape as one of the best criteria of relationship since it depends upon a number of genes, and they noted that the cob of Jaliscan sweet corn differs in shape from that of most other Mexican maize but has a shape similar to one common in South America.

Roberts et al. (1958) considered the Máiz Dulce of Colombia to be related to the Chullpi of Peru. Timothy et al. (1963) regarded the sweet corn of Ecuador to be "probably related" to the Peruvian and Bolivian forms of Chuspillu. Ramirez et al. (1960) pointed out that none of the ears of the Bolivian Chuspillu are as spherical as the classic Peruvian extreme. Timothy et al. (1963) considered the

Chilean sweet corn to be related to the Chuspillu of Bolivia but recognized that its ears were less "ball-shaped" than the sweet corn of Bolivia and Peru. Grobman et al. (1961) are the only authors who did not consider the sweet corn in their country to have been derived from that of adjoining countries—to them there was no doubt that the sweet corn originated in the south central Andes. The following analysis shows that their conclusion is probably correct.

The Peruvian sweet corn race, Chullpi, has a number of features in addition to sugary endosperm. Its ear shape is ovoid; a prehistoric ceramic replica is almost globular in shape. Accompanying this shape are many kernel rows, usually somewhat irregular, and long, slender kernels. The corn is adapted to high or intermediate altitudes. The plants have a low chromosome knob number; in the Peruvian form about half of the plants studied had knobless chromosomes, about half had small knobs in their long arm of chromosome 7, and about half of those had an additional small knob on the long arm of chromosome 6. Of five Bolivian plants of this race examined, all had a medium to small knob on the long arm of chromosome 7 and four had a very small knob on the long arm of chromosome 6. The average knob number of the Mexican plants of Maíz Dulce studied was five.

Table 9.2 shows the average width-length

ratio of the ears illustrated in seven publications in which local forms of the sweet corn race are described. Outline drawings of their shapes are illustrated in Figure 9.4. The data based on the illustrations in Kelly and Anderson (1945) on the sweet corn of Jalisco are also included. The greater diameter of the ears in relation to their length, the higher the ratio. The highest ratio of 0.56 was in Peru. The prehistoric replica illustrated in Grobman et al., is still higher, 0.69. The ratio tends to decrease in the countries both northward and southward from Peru more or less in proportion to the distance from Peru. In the sweet corn from Ecuador, Colombia, and Mexico the ratios are 0.45, 0.31, and 0.33 respectively. In Bolivia, Chile, and Argentina they are 0.45, 0.50, and 0.37 respectively. The same kind of cline occurs with respect to kernel-row number (Table 9.2). The average kernel-row number of the Peruvian Chullpi is 18.0; of the sweet corn of countries northward, Ecuador, Colombia, and Mexico: 16.3, 14.7, and 14.7 respectively; of the countries southward, Bolivia, Chile, and Argentina: 19.7, 17.4, and 15.5 respectively. The fact that kernel-row number is higher in the Bolivian than in the Peruvian sample is probably not significant; ears with kernel-row numbers of 20 or more are not uncommon in Peru. Although not many data are available for the sweet corn varieties of the North American Indian tribes, those that have been published suggest that this tendency to high row numbers prevails. The sweet corn of the Mandans, Pawnees, and Ponkas is described as 10-rowed, 10–16-rowed, and 14-rowed respectively; the kernel row number of the starchy varieties of the same tribes is predominantly 8.

In Peru approximately 40 percent of the ears of Chullpi have pericarp color, brown, red, variegated, or mosaic. Red pericarp color in this race occurs also in the countries northward: Ecuador, Colombia, and Mexico, and among the Indian varieties of the United States. It is not reported in Bolivia, Chile, and Argentina.

There appears to be no doubt that the race of sweet corn had its origin, as Grobman et al. have postulated, in the south central Andes, probably in Peru but possibly in Bolivia. The suggestion of these authors that the race has descended from a popcorn ancestor, Confite Chavinense, is probably also correct. The ears have the same globular

shape as the prehistoric ceramic replicas illustrated in Grobman et al. The ovoid shape of the ancestral form has persisted in modern varieties, becoming less conspicuous in proportion to the distance from Peru and an accompanying increase in contamination by other races.

Are the sweet-corn varieties of the region now the United States, like those of other parts of the hemisphere, related to the Peruvian Chullpi? There is evidence to indicate that they are. The evidence that many of the American Indian tribes possessed sweet corn is extensive. Summarized by Carter (1948), it shows that sweet corn has been collected from the Indians in northwestern Mexico; from the Pima and Papago Indians, some of whom said that it had been introduced from Mexico; from the Pueblo region, where the Hopi, Zuñi, Acoma, Laguna, and San Felipe Pueblos all possessed sweet corn; where it was used ceremoniously by the Hopi; and where it has been found archaeologically. In the upper Missouri, sweet corn was grown by the Hidatsa, Mandan, Cheyenne, Omaha, Pawnee, and Ponka tribes and was used for special purposes. All investigators of the Iroquois food and agriculture mentioned sweet corn and considered it part of the aboriginal crop complex. It was from the Iroquois that sweet corn was apparently introduced to the New England colonists under the name "papoon" by one Lieutenant (later Captain) Richard Bagnal, who participated in General Sullivan's expedition against the Six Nations. Erwin (1934) has questioned the authenticity of this account published in 1822, some 43 years after the alleged introduction in 1779. But Carter (1948) makes an excellent case for its validity based on both internal and concurring external evidence. At least there is no doubt that there was an expedition against the Six Nations, a confederation of Iroquoian people, that one Richard Bagnal participated in it, that one of its purposes was to discover and destroy the Indians' cornfield, and that the Iroquoi grew and used at least two kinds of sweet corn.

In any case it was this New England sweet corn, presumably obtained from the Indians, that became one parent of the earliest named varieties of sweet corn: Darling's Early and Old Colony, new varieties consciously developed by crossing sweet corn with field-corn varieties (Galinat, 1971). Tapley et al. (1934)

are probably correct when, in discussing the origin of the numerous varieties described in their useful monograph, they conclude: "From these native sweet corns and by crosses of these with the original Indian varieties of field corns have come the hundreds of named forms contained in this volume." Thus there is a continuous record of the use of sweet corn by the Indians from west central Mexico to northwestern Mexico, southwestern United States, the upper Missouri, and finally the northeast.

Was this all the same sweet corn or did each tribe or region independently discover sugary mutants and independently develop its own sweet corn varieties? The evidence, though not conclusive, points to the first alternative. I have already shown that in the countries of Latin America there are clines both southward and northward from Peru with respect to the width-length ratios of the ears. The northward cline continues northward and eastward from west central Mexico as is shown in Table 9.2 in which the data are based on photographs of sweet corn in northern Mexico (Wellhausen et al. and Will and Hyde, 1917). The ratios are as follows: west central Mexico 0.33, northwestern Mexico 0.22, Pawnee 0.21, Ponka 0.29, and Mandan 0.25. More significant than the decline in ratios, which are more or less in proportion to the distance of sweet corn from the center of Maíz Dulce in Mexico, is the fact that the sweet corn of the tribes of the upper Missouri has consistently thicker ears than the starchy corn from the same tribe. The starchy corn of the three tribes has width-length ratios of 0.16, 0.20, and 0.19 respectively; the sweet corn of the same tribes 0.21, 0.29, and 0.25 respectively. The data are consistent with the hypothesis that the sweet corn of these tribes traces back to Maíz Dulce of Mexico, which in turn is descended from the Chullpi of Peru. As the distance from Peru increases, the sweet corn is more and more contaminated by corn of other types but has never completely lost all of the genes for thick ears. One possible reason for this is that the sugary gene is located near the centomere on chromosome 4. Such regions generally exhibit less crossing over than more distal regions of the chromosome. Thus the sugary gene, even in the corns of the upper Missouri, may still be closely linked with genes for thickness of the ear inherited from the Peruvian Chullpi thousands of miles and hundreds of generations removed.

In conclusion, it now seems reasonably certain that the sweet corn races of all of the countries of Latin America including the Maíz Dulce of Mexico are descended from the Peruvian race Chullpi. It is also highly probable that the sweet corn of the North American Indians is related to the Mexican Maíz Dulce. Only slightly less certain is the origin of the named commercial varieties of the United States. Some of the oldest of the modern varieties, Darling's Early and Old Colony, were produced by crossing field-corn varieties with a northern sweet corn, probably the one introduced into New England from the Iroquois Indians by Richard Bagnal. If there was an independent domestication of sweet corn anywhere in this hemisphere except in the south central Andes there is at present no evidence of it. The most plausible conclusion is that, despite the frequent mutations to sugary endosperm that must be occurring every year, essentially all of the world's sweet corn is descended from the Andean race.

Related to the sweet corns of this hemisphere, but descending from a collateral line having the same wild progenitor and bearing starchy instead of sugary seeds, are a number of races of the Andean countries of South America with somewhat globular ears. These include Confite Puneño, Paro, Granada, Huayleño of Peru; Confite Puneño and Paro of Bolivia; Marcame, Capio Chico and Capio Grande of Chile; and Capia Amarillo and Capia Variegata of Argentina. The genes for globular ear shape seem not to have gone northward from Peru except as they were part of the race of sweet corn.

10 The Concept of Lineages

If, in attempting to summarize the eleven publications on the races of maize of the countries of this hemisphere, I had hoped to classify the more than 300 races into a smaller number of "superraces," I was disappointed. What I have discovered, however, and this may prove to be even more useful, is that many of the races, perhaps even the majority, can be assigned to a limited number of lineages. Webster's dictionary defines lineage as "descent in a line from a common progenitor," and this is the sense in which I use it here. The concept is by no means new, except as we now apply it to races of corn; lineages have long been recognized in human genealogies and the pedigrees of pure-bred domestic animals. Applied to maize, the idea is somewhat revolutionary, and I have no doubt that it will meet with some degree of skepticism. Its most revolutionary aspect is that, carried to a logical conclusion, it postulates at least six different races of wild corn from which all present-day races have descended.

The idea of a multiple origin of cultivated maize is not new. Reeves and I discussed this possibility at some length in one of our 1959 papers. Randolph in the same year independently suggested that "more than one species of wild corn was involved in the origin of cultivated maize." McClintock (1960) on the basis of her studies of chromosome knobs was led "to consider the possibility that cultivated maize may have had several independent origins." Still more recently Galinat and I (1964), as the result of our experiments on dissecting and reconstituting the tunicate locus, concluded that there must have been at least two races of wild corn differing in their components of this locus. Grobman et al. (1961) postulated an independent origin of certain Peruvian races. Brandolini (1970) in his review of the problem of corn's origin concluded: "It looks highly probable that the origin of cultivated maize has been polycentric." What is new is the idea that there might have been as many as six different geographical races of wild corn and that primitive races stemming from these still exist.

Once having recognized, from the evidence on fossil pollen in Mexico and Panama and archaeological wild corn in Mexico discussed in Chapter 15 that there must have been at least three different geographical races of wild corn in Mexico and Central America, and from the evidence in the previous chapter at least another, the ancestor of all sweet corns, in South America, it was not difficult for me to postulate at least six ancestral forms involved in the present-day races of the world. Actually there may have been even more, but following the principle of parsimony I have kept the number as low as is consistent with our present knowledge of races.

Lineages are obviously not clear-cut taxonomic entities and may not be especially useful in classifying races. Their usefulness lies in showing relationships. Thus any corn that has one or more of the distinctive characteristics of the pointed popcorn Palomero Toluqueño of Mexico—slender, pointed seeds, or spreading spikelets, or dingy pericarp color—is probably in a line of descent from that progenitor, although it may not have received more than a fraction of its germ plasm from that race.

Obviously any modern race may belong to more than one lineage, and the more complex its genealogy the more likely is this to be true. The Corn Belt dent of the United States is an example. Through its Northern flint ancestry it traces one line of descent to the Peruvian popcorn Confite Morocho; through its Southern dent ancestry it brings in characteristics from the Mexican progenitors, Nal-Tel and Palomero Toluqueño. The brown aleurone of some Corn-Belt varieties may trace back to the Peruvian Kculli. I see no evidence of relationship to the progenitor of the sweet corns or of the orange-seeded tropical flint corns. Corn breeders might do well to consider introducing germ plasm from these two lineages into their hybrids. Two of the lineages that I postulate have already been described in Chapter 9, the remaining four are considered in the following pages.

Eight-Rowed Corn

A third race of maize which has spread almost throughout the hemisphere is an eight-rowed type to which Galinat and Gunnerson (1963) have given the name Maíz de Ocho. It includes the eight-rowed flint corns of New England and Canada which Brown and Anderson (1947) have called Northern flints. It includes also the eight-rowed flour corns of the Indian tribes of the upper Missouri described by Will and Hyde (1917) as well as the eight-rowed flour corns of western Mexico described by Wellhausen et al. (1952) under the name Harinoso de Ocho.

Examining both archaeological and modern historical evidence, Galinat and Gunnerson have shown convincingly that this eight-rowed type, which first appeared in western Mexico about A.D. 700, spread into the region now the southwestern United States and from

FIG. 10.1 The Peruvian primitive popcorn race Confite Morocho is considered to be the ancestral form of a lineage that includes most of the eight-rowed corns of South, Middle, and North America: Cuzco Gigante of Peru, Cabuya of Colombia, Harinoso de Ocho and Tabloncillo of Mexico, and the Northern flint corns of the United States. Also included in this genealogy are the highly evolved races Jala of Mexico and Corn Belt dent of the United States. (One-half actual size; from Grobman et al., 1961.)

there it followed a biological life zone northward and eastward. In Mexico and the southwestern United States it was adapted to relatively high altitudes, in the northeastern United States and Canada to high latitudes. This is only one of many examples showing how in the adaptation of cultivated plants latitude can substitute for altitude. A number of the common ornamental flowers grown in New England at little above sea level came originally from high altitudes in regions much nearer to the equator.

When we attempt to trace the earlier history of this eight-rowed corn the facts are somewhat less clear. Wellhausen et al. (1952) included Harinoso de Ocho as one of the four Pre-Columbian Exotic races of Mexico which were assumed to have been introduced into Mexico from Central or South America in prehistoric times. Roberts et al. (1957) suggested that the eight-rowed Colombian flour corn, Cabuya, might be the counterpart of the Mexican Harinoso de Ocho. Consistent with this suggestion is the fact that some of the ears of the eight-rowed flour corns of the Mandan Indians, which apparently came originally from Mexico, are almost exact counterparts of some of the ears of Cabuya grown in Colombia. However, there are other important differences. Cabuya as it grows in Colombia has highly colored and strongly pubescent leaf sheaths and lacks tillers. The

eight-rowed North American flours lack these first two distinctive characteristics and have a tendency to produce tillers.

Perhaps also related to Maíz de Ocho is a Guatemalan high-altitude, eight-rowed subrace of Nal-Tel described by Wellhausen et al. (1957). Other eight-rowed corns somewhat resembling Maíz de Ocho and perhaps in some degree related to it are the races Huevito of Venezuela, Kcello Ecuatoriano and Huandango of Ecuador, Uchuquilla and Niñuelo of Bolivia, Araucano of Chile, and Amarillo de Ocho of northern Argentina. Since the eight-rowed Northern flints are known from historical evidence (Anderson and Brown, 1952) to represent one parent of the Corn Belt dent of the United States, their origin becomes a matter of more than academic interest and perhaps justifies the speculation in which I am about to engage.

In the first place it should be recognized that the eight-rowed condition of Maíz de Ocho is only one of its characteristics. Others are straight kernel rows; paired rows sometimes separated from each other, a condition quite the opposite from that in Palomero Toluqueño in which the members of the same pair tend to spread apart; ear shape almost cylindrical; a slender rachis sometimes slightly to strongly flexible; kernels about as wide as long, relatively thick, apically rounded. The only primitive race so far discovered that exhibits the majority of these characteristics is the popcorn race Confite Morocho of Peru. I think that it is quite possible, as Grobman et al. have suggested, that what we now call Maíz de Ocho had its ultimate origin in that primitive race.

Kculli

A fourth kind of maize that has left its mark widely throughout the hemisphere is the Peruvian race Kculli. This is one of the most distinct of the primitive races. The ears are short, slightly ovoid or conical, with relatively large kernels, usually rounded but sometimes slightly pointed, arranged in irregular rows. The endosperm is white and floury, the aleurone sometimes purple or brown or both, the pericarp a deep cherry red. The corn is used by the Indians as a dye corn and for the preparation of a colored beverage, *Chicha Morada*, which is made in both fermented and unfermented forms.

FIG. 10.2 The Peruvian race Kculli is characterized by its short, thick ears; large, round, floury kernels and a combination of pericarp and aleurone colors that could scarcely have been assembled by primitive plant breeders with no knowledge of genetics. This race is considered to be the ancestral form of all races that have strong aleurone colors, often in combination with pericarp colors, floury endosperm, and thick or basally thickened ears. Included in this lineage are many Peruvian races, the Interlocked Soft Corns of eastern Bolivia and Brazil, Güirua of Colombia, and its Guatemalan counterpart Negro de Chimaltenango, the Conico Elotes of Mexico, and many of the multicolored flour corns of the region now the United States. (One-half actual size; from Grobman et al., 1961.)

That Kculli is an ancient race is shown by a prehistoric replica in the Archaeological Museum of Cuzco (see figure 40, Grobman et al., 1961) and perhaps also by the representation on a vase of the Mochica Period (Grobman et al., figure 34) and by the occurrence of cherry pericarp in prehistoric collections from several sites (Grobman et al.). In relatively pure form Kculli occurs only at high altitudes in the Andean countries Peru and Bolivia, but its several distinctive characteristics have been transmitted in varying degrees to other races in all of the countries considered.

Table 10.1 lists 16 of the more distinctive characters of this race. Especially significant are the genetic factors for color. The most conspicuous genetic character is cherry pericarp, a dark, almost black, color that involves the interaction of three factors: A on Chromosome 3, Pl on chromosome 6, and r^{ch} on chromosome 10. The strong purple plant color of Kculli in Peru and Bolivia involves an additional color gene, B on chromosome 2. When Kculli becomes outcrossed to other races still other color genes are expressed, P for pericarp color on chromosome 1 and Pr

for purple aleurone on chromosome 5. The latter to be expressed must interact with the *C* factor on chromosome 9. Purple aleurone color masks another series of aleurone colors: bronze, orange, and brown (the last a misnomer, actually pale yellow) involving various alleles at the *Bn* locus on chromosome 7. Thus this race has genes involved in the expression of colors in the plant, the pericarp, and the aleurone on seven of corn's ten chromosomes. I pointed out some years ago (1961) that since several of these genes have no visible effect except in combination with others it is unlikely that they were brought together in one genotype by conscious hybridization and selection on the part of Indian plant breeders. It seems more probable that they represent an assemblage of genes characteristic of one race of wild corn. If so, this race is quite distinct in its genotype from certain other races such as Palomero Toluqueño of Mexico, which has an entirely different assemblage of distinctive characteristics.

To what extent has the Peruvian race Kculli influenced the development of other races? I once would have said "only slightly," but a study of the descriptions, data, and photographs appearing in the eleven monographs on the races of maize now convinces me that various components of the Kculli genotype have been transmitted to one or more races in almost every country of the American mainland.

The data that I have assembled from these publications, set forth in Table 10.1, show Kculli to have the maximum number of distinctive characteristics in Peru, and this suggests that it occurs there in its purest form. It is a fact that Peruvian Indian farmers who grow this race usually exercise extraordinary pains to keep it pure. In traveling through Peru at harvest time one sees crops of Kculli laid out in the sun to dry; the ears are remarkably uniform. "Rogue" ears, deviating from the desired type in shape or color, have been sorted out and piled to one side; these are in some instances remarkably few in number. The Indian plant breeder has done an excellent job of preserving this particular genotype; for the reasons mentioned above it seems highly unlikely that he had much to do with creating it.

The Kulli of Bolivia is only slightly less pure than that of Peru. The Racimo de Uva of Ecuador, the Culli of Argentina, and the

TABLE 10.1 Characteristics of Races with Pericarp or Aleurone Colors, Showing Their Possible Relationships to the Peruvian Race Kculli

Countries and races	Tapering ear	Slender shank	Thick kernels	Slight pointing	Irregular rows	Floury endosperm	Cherry pericarp	Other pericarp colors	Red-blue aleurone	Brown aleurone	Short plants	Early	Purple sheaths	Glabrous	Low knob no.	High altitude
Peru																
Kculli	+	+	+	+	+	+	+	+	I		+	+	+	+	+	+
Granada	+	+	+		+	+	−	+			+	+	+	I	+	+
Huancavelicano	+	+	+	+	+	+	−	+			+	I		−		+
Piricinco	+	−	+	−	+	+	I	−	I	+	I	I	−	I	+	−
Shajatu	+	+	+	+	+	+	−	+	+	−	I	−	−	+	+	I
Pisccorunto	+	+	I	+	−	+	−	+	+		+	+	+	I		+
Marañon	+	−	+	+	+	+	−	+		+	−	−	−	−	I	I
Cuzco Morado	+						+		+				+			
Tumbesino	+		+	−	−	+	−			+	−					
Bolivia																
Kulli	+	+	+	+	+	+	+		+		+		+		+	+
Aysuma	+	+	+	+	+			+			+		+			+
Checchi	+	+	+	I	I	+	−		+		+		+		+	+
Morado	I	−	I	−	I	+	+	+			−		+	−		−
Pojoso Chico	+	I	I	−	I	+	−		+		I		−	−		−
Coroico	+	−	+	−	+	+	−			+	−	−	−			−
Ecuador																
Racimo de Uva	+	+	+		+	+	+		+		+		+			+
Candela	I	I	I	−	−	+	−	−	−	+	−		+			−
Chile																
Negrito Chileno	I	I	+	+		+	+									+
Capio Negro Chileno	+	I	+	+	I		−		+							+
Argentina																
Culli	+	+	I	+		−	+						+			+
Oke	+		+	+	+		−	−	+							+
Colombia																
Güirua	+	−	+		I	−	−	+	+		−	+	−			+
Cariaco	+	+	+	−	I	+		+		+	+	+		−	I	−
Sanbanero Bronce	+	−	+	−	I	+	−	+		+	I	I		I		+
Cacao	+	+	+	−	I	+		+	+	+	−	+		+	−	I
Negrito	I	+	I	−	I	+		−		+	+	+		I	−	−
Venezuela																
Cariaco	+	+	I	−	I	+		+		+	I	I		−		−
Cacao	+		+		+	+		+		+	I	I		−		−
Negrito	I		+	−	I	+		+	+		I	+		I		−

+, character present; −, absent; I, intermediate.

TABLE 10.1 (continued)

Countries and races	Tapering ear	Slender shank	Thick kernels	Slight pointing	Irregular rows	Floury endosperm	Cherry pericarp	Other pericarp colors	Red-blue aleurone	Brown aleurone	Short plants	Early	Purple sheaths	Glabrous	Low knob no.	High altitude
Brazil																
Interlocked Soft Corn	+		+	−	+	+			+	+						−
Maisirará	+		+	−	+	+			+	+						−
Guararé	+		+	−	+	+				+						−
Itudoné	+		+	−	+	+		+		+						−
Central America																
Negro de Chimaltenango	+	I	+	−	−	+			+		−	−		I	+	+
Negro de Tierra Fria	+	−	+	−	I	+			+		I	+		−	+	+
Negro de Tierra Caliente	+	I	+	−	−	+			+		I	+		+	−	−
Panama no. 1861						+			+	+				+		
Mexico																
Conico Elotes	+	+	I	−	+	+	+		+							
United States																
Indian Flour Corns	+		+		+	+			+	+						

+, character present; −, absent; I, intermediate.

Negrito Chileno of Chile deviate more widely but all have the distinctive cherry pericarp involving at least three genes: *A*, *Pl*, *r^{ch}*, and usually a fourth, *B*. All exept the Argentinian race have floury endosperm.

If we can assume that any modern race of maize that exhibits a significant number of the distinctive characteristics of Kculli—floury endosperm and cherry pericarp, or floury endosperm and bronze, orange, or brown aleurone color, or purple aleurone with other characteristics—has received at least part of its germ plasm from Kculli then the data in Table 10.1 indicate that the majority of the races listed there, if not all, may be regarded as descendants of Kculli and consequently are *to some degree related to each other* and thus form a lineage of which Kculli is the ancestral form.

The latter fact is especially significant since it helps us to understand more clearly certain relationships of which we were only vaguely aware before. For example, Wellhausen et al., (1957), recognized that the blue-kerneled Guatemalan race Negro de Chimaltenango, is almost identical in size and shape of its ears to the blue-kerneled Colombian race Guirüa. They recognized further that the Guatemalan subraces Negro de Tierra Fría and Negro de Tierra Caliente, are part of this complex. It is now apparent that the Elotes Cónicos of Mexico with their tapering ears, slender shanks, thick kernels, floury endosperm, cherry pericarp, and blue aleurone color are also a part of this complex, and it is probable that other races which are relatively pure for aleurone color e.g., Negrito of Colombia and Venezuela and Capio Negro Chileno of Chile, are likewise related to this complex.

I have recently examined a collection of corn varieties of the Indian tribes of the United States to determine whether this com-bination of pericarp color, aleurone color, and slender shanks reached this region and if so how far northward it was diffused. An ear of Navajo corn from New Mexico with cherry pericarp also has purple aleurone and a very slender shank. Another Navajo ear has red pericarp and purple aleurone but a thick shank. Two ears from the Chickasaw tribe and one from the Arapaho, both of Oklahoma, had both pericarp and aleurone colors, and two of the three ears had shanks more slender than the average for Indian corns.

Pericarp and aleurone colors are also common among the corn varieties of the Indians of the upper Missouri described by Will and Hyde (1917). I strongly suspect, though I cannot now prove, that these colors had their origin in the race Kculli of Peru. I have already presented evidence in the preceding chapter showing that the sweet corn of these Indians is probably related to the Peruvian sweet corn. An even more impressive instance of relationship is that of the flour corns with bronze, orange, and brown aleurone color. In Peru, Bolivia, and Brazil these aleurone colors are associated with interlocking of the kernels to produce ears with odd numbers of rows such as nine, eleven, and thirteen. In my earlier years before this phenomenon was discovered by Cutler I used to tell students that an ear of corn with an odd number of kernel rows is botanically impossible because the pistillate spikelets occur in pairs and twice any number is an even number.

Grobman et al. postulate that this race, in Peru designated as Piricinco, has originated from a cross of Rabo de Zorro with Enano, a popcorn assumed to have been introgressed by *Tripsacum*. Rabo de Zorro in turn is postulated to be the product of hybridization between the popcorn races Confite Morocho and Confite Puntiagudo. I regard both of these postulates as valid. I have myself produced ears resembling those of Rabo de Zorro, with slender cobs and pointed kernels by crossing Confite Morocho with a derivative of Palomero Toluqueño. It is the latter race which probably contributed the characteristic of interlocked kernels to Piricinco and its Bolivian and Brazilian counterparts. But the bronze-orange-brown aleurone complex and floury endosperm must have been contributed by Kculli. Thus Piricinco, its Bolivian counterpart Coroico, and its Brazilian coun-

FIG. 10.3 An ear of the Mexican primitive popcorn race Chapalote. The most distinctive features of this race are its brown pericarp color and dorsally flattened kernels. The early archaeological corn from caves in the Tehuacán Valley considered to be wild corn has been identified as Chapalote, as have also early archaeological remains from caves in northwestern Mexico and the U.S. Southwest. The lineage represented by Chapalote includes the Mexican races Reventador, Tabloncillo, Jala, Celaya, and Cónico Norteño and the Peruvian races Pardo and Cuzco Gigante. (One-half actual size; from Wellhausen et al., 1952.)

terparts, "the interlocked soft corns" of Brieger et al. (1958) have three of the six primitive races of maize—Palomero Toluqueño, Confite Morocho, and Kculli—in their ancestry. They also have some germ plasm from *Tripsacum*, as is shown by the chromosomes with tripsacoid effects that I have extracted from the race Coroico and which are discussed in an earlier chapter. It may be noted in passing that this race of flour corn, which is widely grown in the tropical lowlands of Peru, Bolivia, and Brazil, might offer some unusual and useful genes for the improvement of corn in other parts of the world. More on this subject in another chapter. Related to the Piricinco–Coroico–Brazilian interlocked complex are the races Cariaco and Cacao of Colombia and Venezuela. These, too, are lowland corns with floury endosperm and aleurone color of the bronze-orange-brown series. They have, however, lost the characteristic of interlocked kernels.

The Chapalote–Nal-Tel–Pollo Complex

A fifth lineage is represented by two of the

Mexican races of maize classified by Wellhausen et al. (1952) as Ancient Indigenous: Chapalote and Nal-Tel. These authors also recognized that these two races are related in some degree and have many characteristics in common. Our studies of the prehistoric wild and early cultivated corn of the Tehuacán Valley in Mexico showed that in the early stages of their evolution the two races could be distinguished only by their pericarp color, a characteristic which when considered alone may be somewhat superficial, since the orange pericarp color of Nal-Tel involves an allele of the brown pericarp color of Chapalote. To the early agriculturalists neither color would have had any particular advantage over the other. Studies of archaeological maize from a number of sites described in Chapter 14 indicate that the race with brown pericarp which we now call Chapalote moved westward and northward from the Tehuacán Valley, while the race with orange pericarp color, now called Nal-Tel, moved eastward into Yucatan and Campeche and eastward and southward into Guatemala, where it became one of the predominating races in that country (Wellhausen et al., 1957).

There is some evidence that Nal-Tel also spread further eastward and southward into Colombia and Venezuela to become the primitive Colombian race Pollo. Roberts et al. (1957) pointed out that Pollo showed resemblances to the Mexican Nal-Tel, but it also has some characteristics in common with the Peruvian Confite Morocho. Grant et al., (1963) mentioned the resemblance of the two Venezuelan races Pollo and Aragüito to Nal-Tel and to other somewhat similar Guatemalan races. These authors conclude, however, that it is at present "quite impossible" to distinguish between Pollo as "a very ancient relic or merely a vestige of a more recent cultural interchange between the highland Maya of Central America and the Chibchan-Timotean tribes of the northeastern Andes" (Grant et al., 1963).

A study of early archaeological maize from Venezuela, uncovered by Dr. Mario Sanoja O., although showing similarities between the prehistoric maize and modern Pollo, does not distinguish between the several possibilities with respect to Pollo's origin (Mangelsdorf and Sanoja, 1965). Considering all of the facts now available I am venturing to suggest that the most plausible hypothesis regarding the

FIG. 10.4 Closely related to Chapalote is the Mexican primitive popcorn race Nal-Tel, which has orange-red instead of brown pericarp color. Cobs of this race have been found in archaeological sites in northeastern Mexico and in Guatemala. Nal-Tel has been involved in the ancestry of the Mexican races Dzit-Bacal, Zapalote Chico, Zapalote Grande, Bolita, Vandeño, and Tuxpeño. Through Tuxpeño it has probably contributed to the germ plasm of the U.S. Corn Belt dent. (Seven-tenths actual size; from Wellhausen et al., 1952.)

origin of Pollo is that it is a direct descendant of the Nal-Tel of Central America and consequently an integral part of the Chapalote–Nal-Tel complex. I base this suggestion largely on the shape of the ears of modern specimens and the shape of cobs of the prehistoric ones. In following the spread of the race of sweet corn Chullpi from Peru through this hemisphere, I became impressed with the validity of the conclusion of Kelly and Anderson that shape of the ear may be one of the most useful diagnostic characters, since it usually involves a number of genes. In their shape, tapering and rounded at both ends, the ears of Pollo from both Colombia and Venezuela are quite similar to the ears of the Mexican Nal-Tel illustrated by Wellhausen et al. and are also similar to the cobs of prehistoric wild corn from the Tehuacán Valley. They are less similar to the ears, especially the slender flexible ones, of the Peruvian Confite Morocho. Complying with the maxim of parsimony that the simplest hypothesis that will explain all of the facts is the one to be preferred, I conclude that Pollo is descended from the Mexican race Nal-Tel and not from an indigenous Colombian race of wild maize.

I believe that all of the facts presently available can be explained without postulating a hypothetical wild race of maize in Colombia's ancestral form of Pollo. If there was once such a race it has become extinct without creating an extensive lineage.

If we accept Pollo as a part of the Chapalote–Nal-Tel complex, then to the Mexican and Central American races which have been influenced by these two races we must now add the South American races that have been influenced by Pollo. Wellhausen et al. regard Chapalote as one of the parents of Reventador and through it participating in a development of other Mexican races, notably, Conico Norteño, Bolita, Tabloncillo, and Jala, the last, the world's largest-eared corn. Grobman et al. consider Tabloncillo to be the ancestor of the Peruvian race Pardo, which they regard as one of the parents of Cuzco Gigante, the world's largest-kernelled corn. Wellhausen et al. have Nal-Tel contributing certain of its characteristics to its derivative races Bolita, Dzit Bacal, Zapalote Chico, Zapalote Grande, and Vandeño and being involved in some manner not yet clear in Tuxpeño, one of the world's most productive corns.

In the countries of Central America Nal-Tel became mixed with numerous other races but seems not to have become the direct progenitor of any new races (Wellhausen et al., 1957). Approximately the same situation obtains with respect to Pollo in Colombia and Venezuela. There is evidence of reciprocal introgression of Pollo and Sabanero in Colombia, and the latter has influenced many other races, but there is no clear-cut evidence of Pollo becoming the direct progenitor of any new race.

Pira Naranja

The sixth and last of the ancient primitive races of maize that I shall postulate as the progenitor of a lineage is the Colombian popcorn Pira Naranja, described by Roberts et al. (1957). The chief distinctive characteristics of this race are its rounded, flinty, orange-colored kernels. The orange color of Pira Naranja is located in the endosperm and not, as in Piricinco and similar races, in the aleurone layer. If Pira Naranja did not exist it would have to be invented to account for the origin of one of the most widely grown races

FIG. 10.5 Pira Naranja of Colombia is, as its name suggests, a popcorn with orange-colored kernels, the color residing in the endosperm. This race is regarded as the ancestral form of a lineage that includes the Cateto races of Brazil, Uruguay, and northern Argentina, Camelia of Chile, Costeño and Montaña of Colombia, and Olotón and Jala of Mexico. (One-half actual size; from Roberts et al., 1957.)

of South America, the Cateto of the coastal regions of Brazil, Uruguay, and northern Argentina. Roberts et al. (1957) recognized resemblances between Pira Naranja and Cateto but did not postulate that the former is the progenitor of the latter, probably because of what then appeared to be a wide geographic gap in their distributions. More recent studies have shown a race, called Cateto, with orange-colored kernels and characteristics intermediate between those of Pira Naranja and the Brazilian Cateto, to occur in eastern Bolivia bordering Brazil (Ramirez et al., 1960). A similar race, Camelia, has been collected in Chile bordering Argentina (Timothy et al., 1961). Thus the difficulty of visualizing the spread of a corn from southwestern Colombia to coastal Argentina, Uruguay, and Brazil has to a large extent disappeared.

The inheritance of orange endosperm color is complex, involving, according to Graner (1950), at least three genetic loci. Since, as in the case of the color factors of the race Kculli, it is unlikely that Indian plant breeders would

consciously have brought these loci together from diverse sources, it is probable that the widely grown Cateto is a direct descendant of the Colombian popcorn Pira Naranja. Cateto in turn appears to be involved in some way in the ancestry of the lowland race Costeño of Venezuela and Colombia and is also related to the orange-yellow flint corns of the Caribbean.

In addition to being the progenitor of Cateto, Pira Naranja may also be one of the parents of Montaña, the largest-eared race in Colombia (Roberts et al., 1957). Montaña is the Colombian counterpart of the Mexican Oloton, a race which has probably entered into the ancestry of the Mexican race, Jala, the world's largest-eared corn (Wellhausen et al., 1952). The races of the Pira Naranja lineage have so far been little used by corn breeders in the United States. They would seem to offer promising material for the improvement of corn in this country.

Chromosome Knobs of Races of Maize

All of the American Maydeae, maize, teosinte, and *Tripsacum*, are represented by races or species, some of which bear deeply staining bodies called "knobs" on some or all of their chromosomes. These vary in number, size, shape, and position on the chromosome, some being terminal and others internal, some quite small others very large. Longley (1937) appears to have been the first to employ chromosome knobs in studying relationships. He found that the Mexican teosintes are in general more maize-like in their chromosomes than the Guatemalan teosintes, which have terminal knobs like those of *Tripsacum*. Reeves and I (1939) regarded Longley's observations as evidence supporting our hypothesis that teosinte is a hybrid of maize and *Tripsacum* and is intermediate in the number and position of its chromosome knobs as it is in many other characteristics. We also concluded that the knobs of maize chromosomes might trace their origin to the hybridization with *Tripsacum* and the subsequent repeated hybridization of maize with teosinte, and we postulated that we might well expect to find among South American varieties, especially those of the Andean regions (presumably free of introgression from teosinte), varieties having no knobs on their chromosomes. Testing this postulate, Reeves studied the chromosome knobs of 40 varieties from the countries of South America and found 15 of them, all from the Andean region of Peru, to have essentially knobless chromosomes.* The confirmation of our prediction that maize with knobless chromosomes would be found in the Andean region of Peru we regarded as strong evidence for our conclusion that Peru includes the primary center of maize domestication and that the knobs of maize are evidence of previous hybridization with teosinte and ultimately *Tripsacum*.

In the thirty years since our monograph was published many new facts have come to light. We now know that Peru is an independent or secondary instead of the primary center of domestication; yet of the six primitive races that I have postulated three are Peruvian in origin. It is now reasonably certain that wild maize was a plant of high and intermediate altitudes and not a native of the lowlands, as Reeves and I concluded. There is now good evidence that some South American species of *Tripsacum* have no chromosome knobs, so that the absence of knobs in certain races of maize is no longer reliable evidence that hybridization with teosinte or *Tripsacum* has not occurred. However, chromosome knobs continue to be useful in classifying the races of maize and in attempting to discern relationships, and they have been employed as part of the racial descriptions in the majority of publications on the races of this hemisphere. In addition, there have been two special studies of chromosome knobs, one by McClintock, of which the details are still not completely published, although some of the data are fortunately included in the publications on the races of maize of Ecuador, Bolivia, and Chile, and the other by Longley and Kato (1965). Both of these studies have gone beyond the earlier ones involved in the publications of races of maize—in which only numbers of knobs were recorded—and have identified the chromosomes bearing the knobs, positions on the respective chromosomes, and the sizes and shapes of the knobs. McClintock (1959) suggested that the data from these studies "would be a useful adjunct to the observations of morphological and

*McClintock, in later studies of similar Andean races from Ecuador, Bolivia, and Chile, has identified a small knob on chromosome 7. Reeves probably considered this not a knob but a large chromomere.

physiological properties." This indeed they have proved to be.

One of McClintock's most important findings is that all but 2 of the 32 high-altitude races from Ecuador, Bolivia, and Chile are amazingly similar with regard to knob constitution. In these 30 races she identified a small knob at one particular location in the long arm of chromosome 7. In some but not all of the plants of these races, she found a very tiny knob present in the long arm of chromosome 6. Included among these 30 races are 3, Confite Puneño, Chuspillu, and Kculli, that I have designated as ancestral forms of extensive lineages. A similar situation prevails with respect to a fourth high-altitude Peruvian race, Confite Morocho, in which Grobman et al. found one small knob on the long arm of chromosome 7 and in about one-fourth of the plants another small knob on the long arm of chromosome 6. Thus four of the six primitive races that I have postulated have virtually knobless chromosomes. The following data indicate that a fifth race, Palomero Toluqueño, is also in this category; Wellhausen et al. found plants of Palomero Toluqueño to have a low average chromosome-knob number, 1.2, with some plants having knobless chromosomes. Longley and Kato later found the overall average knob number to be 2.1, and, although they did not report finding plants with completely knobless chromosomes, concluded that in the formation of this race a knobless type of maize was the "basic material." In the Peruvian counterpart of this race, Confite Puntiagudo, Grobman et al. found the knob number to be low, some plants apparently having knobless chromosomes. The knob situation in the Pisankalla of Bolivia and Canguil of Ecuador, however, is quite different. In the former, McClintock identified knobs on all chromosomes except 3 and 10, and in the latter on all chromosomes except 1, 2, and 10. These findings are consistent with the data I have assembled in Table 9.1 which show that Pisankalla of Bolivia and Canguil of Ecuador, although undoubtedly related to Palomero Toluqueño of Mexico, have lost, presumably through admixture with other races, some of the characteristics of the original race. I am inclined to agree with Longley and Kato that the basic material of this race was a form with knobless or virtually knobless chromosomes.

Since the 22 characteristics of Palomero Toluqueño listed in Table 9.1 must involve loci on most if not all of corn's ten chromosomes, it is not difficult to see how the chromosomes of a primitive race with knobless chromosomes could be altered or replaced through hybridization with the knobbed chromosomes of other races. However, if this replacement is not complete some of the characteristics of the primitive race will persist and the relationship of the modified race to the original one may still be recognizable.

The constitution with respect to chromosome knobs of the two remaining races postulated as primitive and ancestral to extensive lineages, Pira Naranja and Chapalote–Nal-Tel, is not clear at this time. Roberts et al. (1957) reported average chromosome knob numbers of the former to be 7.0, but apparently only two plants were examined. For the Mexican Chapalote and Nal-Tel, Wellhausen et al. reported average numbers of 6.0 and 5.5 respectively, but again only a small number of plants were examined. For these same races, Longley and Kato reported average knob numbers of 11.7 and 10 respectively. These high knob numbers might appear to support the suggestion that there may have been two kinds of primitive races of maize, one with virtually knobless chromosomes, the other with numerous chromosome knobs. This possibility cannot be ruled out. On the other hand it is undoubtedly true that the present-day collections of Chapalote and Nal-Tel are far from representing pure races. Our studies of the prehistoric corn from caves in the Tehuacán Valley in Mexico show admixture to have begun at an early stage (Mangelsdorf et al., 1964).

McClintock (1960) speaking of Nal-Tel stated that the chromosomal constitution of the majority of the selections studied reflected the predominant knob complexes found in plants of other races growing in the same general region. Also, when we recall that the Pisankalla of Bolivia had knobs on eight of its ten chromosomes and yet is undoubtedly related to Palomero Toluqueño of Mexico, which has some plants with knobless chromosomes, it is not difficult to visualize the present-day mixed collections of Chapalote and Nal-Tel as having descended from a "pure" race with knobless chromosomes. From the data presently available this strikes me as being somewhat more plausible than the suggestion that there were once two kinds of primitive races, one with knobless chromosomes the other with high chromosome knob numbers. I prefer the former partly because it involves only known entities. Four of the six postulated primitive races do have virtually knobless chromosomes; no primitive race free of admixture with other races and having a high chromosome knob number is known. However, corn's closest relative, teosinte, which crosses readily and repeatedly with corn, does have numerous chromosome knobs, and there is little doubt that at least some of the knobs of modern corn have come from teosinte. Others have perhaps been derived directly from *Tripsacum*, some species of which have knobs on every chromosome. These subjects will be discussed further in later chapters. Here it will suffice to point out that the data from chromosome knob studies are in general consistent with the conclusions set forth in this chapter on races.

Summary

In this chapter I have postulated that the more than 300 races of maize described in the eleven publications on the races of maize of the countries of this hemisphere can be assigned to six lineages, each one descended from a wild race of maize. Two of these are Mexican, one Colombian, and three Peruvian. Proceeding from north to south the still-living ancestral races are:

1. Palomero Toluqueño, the Mexican pointed-seeded popcorn.

2. The Chapalote–Nal-Tel complex of Mexico.

3. Pira Naranja of Colombia, the progenitor of the tropical flint corns with orange-endosperm color.

4. Confite Morocho of Peru, the progenitor of eight-rowed corns.

5. Chullpi of Peru, the progenitor of all sweet corns and related starchy-seeded forms with globular ears.

6. Kculli, the Peruvian dye corn, the progenitor of all races with complexes of pericarp and aleurone colors.

Hybridization between these races and their lineages and with corn's relatives, teosinte and possibly also *Tripsacum*, has been a principal factor in creating the diversity of corn varieties that exists today. This is discussed in detail in the following chapter.

11 The Role of Hybridization in Corn's Evolution

> The work of botanists has shown that distinct species of plants, when they inhabit the same territory, occasionally cross and produce fertile offspring. Of course crossing within a species remains more frequent than that between species. Interspecific crosses, however, cause a diffusion, or introgression, of genes from one species to another. The gene transfer augments the genetic variability and furnishes great opportunities for natural and artificial selection. Dobzhansky, 1955.

One of the most important factors involved in the evolution of maize under domestication has been hybridization. This has apparently been of three kinds: interracial, interspecific, and intergeneric.

Interracial Hybridization

Since corn is a naturally cross-pollinated plant, producing a profusion of pollen grains that are easily carried by the wind, it is inevitable that whenever two races of corn are grown in proximity there will be some degree of interracial hybridization. Also, since man in his peregrinations often carries his cultivated plants with him and since cultivated plants are also spread by trade and warfare, it is inevitable that distinct races of maize will sometimes be brought together in the same area and will cross. A classic example is the evolution of the Corn Belt dent corn of the United States.

Employing evidence from archaeology, history, cytology, and genetics, Anderson and Brown (1950, 1952a, 1952b) have shown convincingly that Corn Belt dent, one of the world's most productive and widely grown races, is the product of interracial hybridization. One parent is the race comprising the long, slender Northern flints, which had dominated corn culture in the region now the eastern United States for at least several centuries preceding the discovery of America. The other parent is the race comprising the Southern dents, which are clearly related to the dent corns of Mexico. The two races were brought together in what is now the Corn Belt when, with the construction of the railroads, agricultural migrants from the southeast and the northeast, carrying their corn varieties with them, converged in the Middle West. Hybridization between the Northern flints and the Southern dents, followed by both natural and artificial selection, produced a new race that is intermediate between its parents in many characteristics, including chromosome knob number. As a final touch to their story of the origin of Corn Belt dent corn, Anderson and Brown crossed its putative parents, and practising selection in several subsequent generations produced a synthesized race quite similar to the modern Corn Belt dent.

When working on the classification of Mexican maize, Wellhausen, Roberts, and I were aware of the conclusion of Anderson and Brown on the hybrid origin of the Corn Belt dent, and we made a search for similar evidence of hybrid origin in Mexican races. We soon found a clear-cut example of it in the race Cónico, the predominating race of the Valley of Mexico. This race seems without much doubt to be the product of the hybridization of one of the Ancient Indigenous races of Mexico, Palomero Toluqueño, and one of the Pre-Columbian Exotic races, Cacahuacintle. Cónico has its center of distribution in the Central Mesa, where both of its putative parents are still to be found. It resembles one or the other of its two parents or is intermediate between them in practically all of its characteristics. Inbreeding of Cónico results in segregates, some of which closely resemble Palomero Toluqueño; others approach but never quite duplicate Cacahuacintle. Finally Wellhausen produced a synthetic Cónico by crossing Palomero Toluqueño and Cacahuacintle. The ears of this hybrid are practically identical with those of Cónico, except that they segregate for endosperm texture. The genetic evidence of the hybrid origin of Cónico is about as conclusive as it is for the hybrid origin of Corn Belt dent; only the historical evidence is lacking.

Stimulated by the clear-cut example afforded by the race Cónico, we looked for and found similar though usually less conclusive evidence of hybrid origins of other Mexican races. The result was that the team of Wellhausen et al. assigned to a category called Prehistoric Mestizos 13 of the 25 races which they recognized. This category includes races which were thought to have arisen through hybridization of four Ancient Indigenous races with four Pre-Columbian races and the hybridization of both with teosinte. Four additional races thought to have arisen from hybridization in historical times were assigned to a category called Modern Incipient. That a race is the product of hybridization seems probable—indeed reasonably certain—when the following four kinds of evidence are available.

1. The race is intermediate between the two putative parents in a large number of characteristics.

2. The putative parents still exist and have geographical distributions that make such hybridization possible and plausible.

3. Inbreeding of the suspected hybrid race yields segregates which approach in their

characteristics one or the other of the two putative parents, in some cases both.

4. A population quite similar in its characteristics to the race in question can be synthesized by hybridizing the two putative parents.

Wellhausen et al. (1952) presented all four kinds of evidence for the hybrid origin of a number of the present-day Mexican races; they presented similar but less complete evidence for the remainder. The authors were well aware, however, that some of their conclusions were tentative. In the foreword to the English edition of the Races of Maize of Mexico they stated: "The conclusions on origins and relationships presented in this monograph are not to be regarded as final. They represent no more than the best interpretations which the authors were able to make from their observations and data. Some have substantial evidence in their support and have already proved useful; others are necessarily still speculative and will require modification as new evidence from the maize of other countries is brought to bear upon the problem. Indeed one of the most useful purposes which this monograph could serve is to focus attention on the need for similar studies in other parts of America."

As mentioned in a previous chapter the monograph did indeed focus attention on the need for similar studies in other parts of America. Soon after moving from Mexico to Colombia Lewis Roberts began the collection, study, and classification of the races of that country, a project in which he invited me to participate. In classifying the races of maize of Colombia both Roberts and I, having been impressed by the role of hybridization in the formation of the Mexican races, looked for and found similar evidence of hybridization in the development of Colombian races. We—the team of Roberts et al.—concluded that a number of Colombian races were the product of racial hybridization, and we attempted to identify the putative parents, fully recognizing as we did so that some of our identifications could be little more than tentative.

When a few years later I joined the team of Grobman and his bright young *Maiceros* in Peru, we naturally followed the procedure that seemed to us to have been effective and fruitful in both Mexico and Colombia.

Authors of the remaining publications on races of maize have been more cautious in reaching conclusions regarding hybrid origins, and some have expressed skepticism about the hybrid nature of races. Brieger et al. (1958), for example, have been critical of some of the conclusions of Wellhausen et al. and others regarding the hybrid nature of races, especially of the evidence involving the intermediate nature of hybrids between their putative parents. In this connection they state that "since the putative parents were selected especially to differ equally in opposite directions from the supposed synthetic race, there is really no possibility of any differences between the F_1 hybrid and the supposed synthetic race. The argumentation and the so-called 'proof' present a perfect circle in reasoning."

Two answers should be made to this criticism. (1) The putative parents were never selected to differ *equally* in opposite directions from the supposed hybrid race. Indeed this would generally have been impossible; there are not that many races to choose from. They were selected because they seemed to possess certain characteristics which might have contributed to the hybrid race. (2) Races were considered putative parents only when their geographical distributions were such as to make the suspected hybridization possible and plausible. This circumstance places definite restrictions on the number of races that can be considered as putative parents of other races.

Despite their criticism, Brieger et al. concede that some races of corn are hybrids (although they prefer the term "synthetics") of other races and that the extension by Wellhausen et al. of this principle first recognized by Anderson and Brown "was and remains of very great importance." It can be said, therefore, that among the students of the origins and relationships of races of maize, there is general agreement that interracial hybridization has been an important factor in corn's evolution. Such differences of opinion as exist are largely concerned with the extent to which putative parents can be reliably identified.

To those of us who have attempted such identifications it is apparent that some of the more highly evolved modern races are the product of repeated hybridization and thus

have complex genealogies. The Corn Belt dents, for example, are hybrids of the Northern flints and the Southern dents. The latter undoubtedly had their origin, as Anderson and Brown (1952) recognized, in Mexico and are related to such Mexican races as Tuxpeño. Wellhausen et al. found Tuxpeño to be intermediate in a large number of characteristics between two other Mexican races, Olotillo and Tepecintle, both of which appear to be products of hybridization with teosinte. Thus there is evidence that the Corn Belt dents are the product of repeated hybridization. The same is true of the Mexican Jala and of the Peruvian Cuzco Gigante. The genealogies proposed for these two races by Wellhausen et al. and Grobman et al. respectively may seem speculative in some details and may require modification as additional facts become available. They are probably valid, however, in showing the races to be the product of successive hybridizations as well as hybridization with teosinte and in some cases with *Tripsacum*.

Hybridization of Maize and Teosinte

No less important than interracial hybridization as a factor in corn's evolution is the interspecific hybridization of corn with its close relative teosinte, *Zea mexicana*. For this there is convincing evidence both archaeological and modern. Let us first consider the latter.

Teosinte grows as a weed in and around the maize fields in parts of Mexico and as a feral plant in pure stands often near maize fields in the Rio Balsas region in Mexico and in Guatemala. Since both species are monoecious and wind pollinated and since in these countries they flower at approximately the same time, abundant opportunities for hybridization exist, and it would be strange indeed if hybridization between the two species were not constantly occurring. Every student of the problem from Harshberger on has been aware of the natural hybridization of corn and teosinte (Harshberger, 1898; Collins, 1921; Kempton and Popenoe, 1937; Mangelsdorf, 1952; Weatherwax, 1954; Randolph, 1955; Wilkes, 1967). Randolph has considered the hybridization between maize and teosinte to be rare and, in relation to the apparent opportunities for its occurrence, perhaps it is. There are undoubtedly partial genetic barriers to the free hybridization of corn and teosinte under natural conditions. Certainly the number of recognizable hybrids found in localities where the two species are growing together in the same field and flowering at the same time is less than would be expected if there were no such barriers. And yet hybrids do occur. Randolph himself counted 45 F$_1$ hybrids in five days of travel in a limited region near the village of Nojoyá and San Antonio Huixta in Guatemala. If this small sample is representative, there must be thousands of new hybrids produced each year and this has been going on in countless localities for many centuries.

Maize-teosinte hybrids are especially common in Central Mexico where teosinte grows as a weed. In 1943 I obtained some data on the extent to which hybridization occurs near the village of Chalco, where teosinte is a common weed in and around the cornfields. In a field where almost half of the plants seemed to be those of teosinte, I obtained the owner's permission to tag and harvest 500 consecutive plants. Of this number I identified 288 as maize, 209 as teosinte and 3 as F$_1$ hybrids. Of the plants classified as maize, 4 showed definite evidence of previous contamination with teosinte. In addition, in an adjacent row not part of the sample of 500 plants, I found one plant similar in its characteristics to a first backcross to teosinte.

In the most comprehensive study yet made of teosinte populations, Wilkes (1967) found high frequencies of F$_1$ hybrids in maize fields near Los Reyes, Chalco, and Amecameca in the Valley of Mexico. In the Amecameca field, approximately 10 percent of the plants that at first glance seemed to be those of teosinte proved on closer examination to be F$_1$ hybrids or backcrosses of such hybrids to teosinte. The occurrence of maize-teosinte hybrids in the maize fields of the Valley of Mexico is recognized by the Mexican cultivators, who, engaged in conversation, will often volunteer the opinion that in three years or so the hybrids will become maize.

Wilkes found a high frequency of hybrids also in the vicinity of Nobogame in northern Mexico. There teosinte is not limited to the cultivated fields but occurs also in dense stands along the streams and in areas on the surrounding hills protected from grazing. Most of the hybrids occur not in maize fields but in the thick stands along their margins or in the willow thickets along the streams. Wilkes found Nobogame teosinte, when grown in Massachusetts, to be one of the most maize-like teosintes in its physiological characteristics. This is consistent with the conclusion mentioned in Chapter 4 that Rogers and I reached many years ago in our experiments in Texas: that Nobogame teosinte differs from maize by only three principal blocks of genes and not by four as do the other teosintes that we studied.

In all other populations that he studied, those of the Central Plateau and Rio Balsas regions of Mexico and Huehuetenango and the southeastern parts of Guatemala, Wilkes found evidence of hybridization of maize and teosinte often accompanied by evidence of backcrossing to one or the other or to both parents. In view of his extensive observations, there can scarcely remain any doubt that hybridization between maize and teosinte is occurring regularly throughout the latter's range.

Evidence of Gene Flow

Randolph (1955), who has questioned the extent to which maize and teosinte hybridize, has also been skeptical about the extent to which there is any substantial gene flow following such hybridization. That there has been some exchange of genes between the two species there can be no doubt. This would be expected on the basis of the following considerations: (1) F$_1$ hybrids of corn and teosinte are usually vigorous and highly fertile and are easily backcrossed to either parent to produce fertile progeny. (2) The chromosomes of the two species are morphologically similar and synapse more or less normally in the hybrids (Kuwada, 1919; Longley, 1924, 1937; Beadle, 1932a, b, 1936). (3) The arrangement of the gene loci, although not identical in the two species, is certainly similar (Emerson and Beadle, 1932). (4) Crossing over between linked loci in maize and teosinte is, with few exceptions, of the same order as that in corn (Emerson and Beadle, 1932).

In view of these well-established facts it is difficult to see how once hybridization has occurred some degree of gene exchange could be prevented, and there is ample evidence that it has not been. There is a constant flow of genes between the two species—a flow which moves in both directions. One result is that the teosinte of Mexico, where it grows as a weed in the maize fields and where the opportunities for hybridization are at a maximum, has become quite maize-like in virtually all of its characteristics except the pistillate spikes and the structures enclosing the kernels. In the vicinity of Chalco, for example, Collins (1921) noted many years ago that teosinte plants have the same plant coloration and the same pilose leaf sheaths as the predominating maize of the area. The conspicuous coloration involves at least one gene, perhaps two: *B* on chromosome 2 or *R* on chromosome 10, or both, and pilosity at least two (Paxson, 1953). These distinctive characteristics are well illustrated by a color photograph reproduced in Wilkes' monograph.

In a collection of seeds of individual plants of teosinte from this same vicinity that I made in 1943, I found both yellow endosperm, and purple aleurone, colors unusual in teosinte. In being transferred from maize to teosinte, the genes for these have undoubtedly carried with them blocks of closely linked genes, and this accounts at least in part for the fact that the teosinte of Chalco is in many respects one of the most maize like of all of the varieties known.

Today in the Chalco area even the keen eyes of the native Indian maize growers can no longer distinguish teosinte plants from maize plants growing in the same field until they flower and the nature of their pistillate inflorescences becomes apparent. Since the teosinte seeds are not usually harvested but fall to the ground, where they remain dormant until the onset of the next rainy season, teosinte plants may reach a high frequency in the maize fields, sometimes exceeding the maize plants in number. Recently Johnson (personal communication) has reported that in the area near Amecameca, a village about 18 kilometers northeast of Chalco, one-third to two-thirds of the plants in the field are obviously not "pure" corn. Even in areas where some of the farmers have adopted the practice of cutting out the off-type plants in their maize fields, they do this only after the plants have flowered and some pollen has been shed. Since the hybrid plants are profuse pollen-shedders, the opportunities for gene flow between the two species has not been drastically curtailed by this practice. In situations such as this there is some question whether teosinte should be accorded even the specific rank, *Zea mexicana*, that Reeves and I assigned to it some years ago (1942). A more realistic classification would say that here maize and teosinte represent a single dimorphic species in which one component is preserved by man and the other by nature.

As a result of the widespread hybridization between maize and teosinte which occurs in the Chalco area and in other parts of Mexico, the teosintes of that country are in general more maize-like than the teosintes of Guatemala, where hybridization between the two species is less frequent than it is in Mexico. The differences between the Mexican and Guatemalan teosintes are not only morphological (Kempton and Popenoe, 1937; Reeves, 1944; Wilkes, 1967) but also genetic (Rogers, 1950a, b, c) and cytological (Longley, 1924, 1941a, b). Longley has explained the cytological differences as a product of "gradients," but the explanation is not tenable (Mangelsdorf and Cameron, 1942; Rhoades, 1955). A much more simple and plausible explanation is that the Mexican teosintes have, on the average, undergone more admixture with maize than have the Guatemalan varieties.

Since some varieties of teosinte have obviously been modified by admixture with maize it is to be expected that some varieties of maize have been modified by admixture with teosinte. There is both morphological and cytological evidence that this has occurred. Of the twenty-five races of maize that they described, Wellhausen et al. (1952) recognized teosinte introgression in twenty-two, primarily on the basis of the induration of the rachis and lower glumes of the cobs. This is a subjective criterion, but their scores for teosinte introgression proved to be strongly correlated with chromosome knob numbers and later were shown by others (Cervantes et al., 1958) to have a remarkable correlation with resistance to a virus disease of corn called "stunt." Here may be a situation similar to that reported by Venkatraman and Thomas (1932), who, observing an infestation of *Aphis maidis* on sorghum while sugarcane and sugarcane-sorghum hybrids were free, concluded that insects are sometimes able to discover more quickly than botanists the real nature of a population. In the case of stunt disease in corn it may be the virus, rather than the insect vector which spreads it, that recognized the degrees of teosinte introgression in the Mexican races and showed that the estimates made by Wellhausen et al., although subjective, are reasonably accurate.

Cytological Evidence of Teosinte Introgression

As previously mentioned all varieties of annual teosinte have deeply staining heterochromatic regions in their chromosomes called "knobs." Their number varies from fourteen to fifteen in the least maize-like race of teosinte from Guatemala to four in the most maize-like teosinte from Mexico. Many modern varieties of maize also have chromosome knobs varying in number from one to sixteen. We concluded some years ago (1939) that the chromosome knobs of present-day maize are the product of admixture with teosinte and, if teosinte is a hybrid of maize and *Tripsacum*, are derived originally from *Tripsacum*. We assumed that varieties of maize not contaminated by teosinte would have knobless chromosomes, and as mentioned in a previous chapter, we predicted that these would be found in South America, especially in the Andean region. And they were. At the time, this appeared to be virtually conclusive evidence that our assumptions with respect to the origin of chromosome knobs in modern maize was valid. It has since been shown, however, that some forms of the South American *Tripsacum*, *T. Australe*, also have knobless chromosomes (Graner and Addison, 1944), so that the absence of chromosome knobs in a corn variety can no longer be regarded as evidence that it has not undergone admixture with *Tripsacum*. This does not, however, invalidate the assumption that the presence of knobs indicates teosinte introgression, and this assumption still has much evidence to support it.

We—James Cameron and I—showed some years ago (1942) that in the maize of western Guatemala the number of chromosome knobs is associated with several characters that might have been derived from teosinte or produced by hybridization with it. These include denting of the kernels, fibrous seminal root systems, and resistance to shattering, lodging, and smut infection. Brown (1949) found high knob numbers to be associated with high kernel-row numbers, denting, absence of husk leaf blades, many seminal roots, and irregular rows of kernels, all of which are characteristics of Southern dents, which have affinities with the Mexican Tuxpeño, a race which we regard as having teosinte in several of its ancestral lines. Brown concluded, however, that more data are needed before chromosome knobs can be regarded as reliable indicators of *Tripsacum* germ plasm.

Those Mexican and Guatemalan races of maize that were classified as primitive and that showed little morphological evidence of admixture with teosinte had a lower average chromosome knob number than the highly evolved hybrid races, which appeared to have marked admixture (cf. Wellhausen et al., 1952; 1957). In Mexico, as mentioned above, chromosome knob number was correlated with resistance to the virus disease "stunt."

The question has been raised whether the chromosome knobs of teosinte can be transferred to maize through hybridization. It would be strange indeed if they could not, since the inheritance of knobs follows the same general pattern as the inheritance of other chromosomal loci. Also, there is direct evidence bearing on the question. Cytological studies by Ting (1956, 1957) of the modified strains of A158 which I developed by introducing chromosomes or parts of chromosomes from teosinte show that knobs were introduced from chromosome 4 of Nobogame teosinte and from chromosomes 1, 2, 3, 5, 8, and 9 of Durango teosinte and from not yet identified chromosomes of teosinte from Guatemala and Honduras.

Another chromosome characteristic that seems to have been transferred from teosinte to some populations of maize is an inverted segment on the short arm of chromosome 8. Ting (1964) has identified this inversion in four varieties of teosinte: Chalco, Xochimilco, Durango, and Nobogame. McClintock (1960) has found the same inversion in four races of maize from central Mexico and one race from Bolivia. It is also found in some cultures of North American maize (Ting, 1964). Since the inversion is common in teosinte and much less so in maize it seems probable that the transfer has been from teosinte to maize.

The cytological evidence supporting the assumption of introgression between teosinte and maize has been well summarized in the following succinct statement by McClintock (1960), who, although still regarding maize and teosinte as distinct genera, recognized that there has been interchange of chromatin and heterochromatin between them: "Since the over-all organization of the chromosomes is much alike in *Euchlaena* and maize, and since the two genera may be crossed readily, interchange of segments of chromosomes, including knob-forming regions, undoubtedly has occurred between them."

Anatomical Evidence of Teosinte Introgression

Employing a method developed by Galinat for preparing median longitudinal sections of mature cobs, Sehgal and Brown (1965) have shown that various inbred strains derived from Corn Belt maize are similar in the internal anatomical characteristics of their cobs to strains of the inbred A158 that have been modified by the substitution of chromosomes or parts of chromosomes from varieties of teosinte. In their pistillate spikelets and their associated rachis internodes, some Corn Belt inbreds are comparable to the A158 strains modified by chromosomes 4 and 9 from teosinte; others are comparable to the A158 strains modified by chromosomes 3 and 4 of teosinte. Some of the Corn Belt inbreds, although having a genetic background different from the experimentally introgressed A158 strains, still resemble them closely in the anatomical details of their pistillate spikelets and rachis segments. Still other strains appear to be even more tripsacoid than the teosinte-introgressed derivatives of A158. One of these, 334, is also one of the best strains from the standpoint of general combining ability in hybrids.

The authors also point out that if Corn Belt maize possesses various degrees of teosinte germ plasm in its genetic constitution, as all the available evidence suggests, then it should be possible to recover segregates comparable to the experimental teosinte-introgressed derivatives by inbreeding the open-pollinated varieties. The authors have observed such segregates in the inbred progeny of the varieties Krug, Lancaster, and Midland. Some of the extreme segregates have many tripsacoid features of both plant and ear, and a few exhibit even a tendency toward

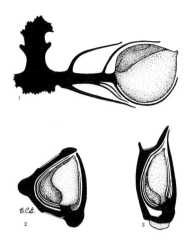

FIG. 11.1 Diagrammatic longitudinal section of mature fruits of teosinte (2) and *Tripsacum* (3) compared with a spikelet of the prehistoric Bat Cave maize (1) drawn to the same scale. In both teosinte and *Tripsacum* the spikelets are sessile, the rachillae are strongly condensed, and the axes of the kernels are parallel to that of the spike. The lower glumes are thickened compared to those of the pure maize and like the rachis segments are highly indurated. These differences, as well as others, permit the recognition of introgression, both prehistoric and modern, into maize. (Drawing by W. C. Galinat.)

single spikelets, one of the botanical characteristics that distinguishes maize from its relatives teosinte and *Tripsacum*.

Comparable but somewhat more detailed observations were made by Johnston (1966) on twelve of the commonly used Corn Belt inbreds in comparison with strains of A158 modified by the introduction of chromosomes or parts of chromosomes from Florida teosinte. Almost all of the strains showed the influence of teosinte introgression in one or more of their anatomical characteristics. As measured by the sum of the numerical values of the characteristics which Sehgal (1963) had previously shown to be especially useful—the angle of inclination of the rachilla and the hardness of the rachis—some of the inbreds were more tripsacoid than the A158 strains modified by chromosomes 1, 3, or 9 from Florida teosinte. None was quite as tripsacoid as A158 modified by Florida chromosome 4.

To determine the position on the chromosomes of the polygene segments responsible for the tripsacoid effects, Johnston crossed all twelve inbreds with a multiple-gene tester stock having recessive genes on nine of corn's ten chromosomes. In the F_2 generation of

these crosses, linkage between the polygene segments and the marker genes were found most commonly on chromosomes 1 and 4 and least frequently on chromosome 7. Examining combining ability in relation to the results of the linkage analysis, Johnston found in inverse relationship between the yields of the hybrids and the degree of homozygosity of the introgressed segments. The significance of this finding for corn improvement is obvious and will be discussed in greater detail in Chapter 20.

The studies of Sehgal and Brown and of Johnston leave little doubt that the modern corn of the United States has sometime during its evolutionary history undergone introgression from teosinte or *Tripsacum* or both. That such introgression has been going on for many centuries is shown by the archaeological evidence discussed in the following section.

Archaeological Evidence of Teosinte or *Tripsacum* Introgression

The botanical descriptions in earlier chapters show that maize differs from its relatives teosinte and *Tripsacum* in numerous characteristics. Among these are the features of the pistillate spikelets which are illustrated by the diagram in Fig. 11.2. These show that the kernels of maize are borne on a relatively long stem, a rachilla which is set at right angles to the main stem, the rachis. The kernels are partly enclosed (completely so in some forms of pod corn) by floral bracts which botanists call glumes. In both teosinte and *Tripsacum* the rachillae are quite short and are parallel to the axis of the spike. The kernels are enclosed in specialized fruit cases consisting of segments of the rachis and lower glumes. Both of these structures are relatively thick and highly indurated.

We know from numerous genetic studies that both teosinte and *Tripsacum* and their hybrids with maize transmit this characteristic of the induration of the tissues in varying degrees to their progeny, and they do this in exactly the way in which a morphologist familiar with the characteristics of the three species would expect them to: they cause the structures in maize that are homologous to those affected in teosinte—rachises and lower glumes—to become indurated. We know from our extensive genetic experiments (Rogers, 1950b, and mine) that a number of

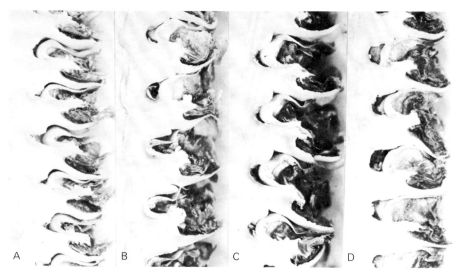

FIG. 11.2 Anatomical evidence of the introgression of *Tripsacum* into maize is provided by these longitudinal sections of cobs of the inbred strain A158 and three strains modified by chromosome substitutions. *A*, the original strain; *B*, modified by the introduction of chromosome 4 of teosinte; *C*, modified by an extracted chromosome from the Bolivian race *Coroico*; *D*, modified by an extracted chromosome from the Argentinian race *Maiz Amargo*. Note that *B*, *C*, and *D* are quite similar and differ from *A* in the wider spacing of the spikelets, the thickening of the floral axis, and the shortening of the rachillae on which the kernels are borne. Since teosinte is unknown in South America the tripsacoid characteristics of *C* and *D* are attributed to introgression from *Tripsacum*. Sections and photographs by S. M. Sehgal (× 3.5).

different chromosomes of teosinte, but especially no. 4, are involved in producing this induration of the rachis and glumes. And it is because of our extensive experimentation with maize-teosinte hybrids over a period of some forty years that we are able to recognize some of the effects of teosinte introgression, both prehistoric and modern, in maize.

When, in collections of prehistoric maize, we see cobs with highly indurated tissues of the rachis and lower glumes, we at once suspect introgression from either teosinte or *Tripsacum*, since there is no other known source from which these characteristics could have been derived. When we find specimens that not only are highly indurated but also have other teosinte characters such as single spikelets or a distichous arrangement of the spikelets, we can be almost certain that there has been such introgression. Many collections of prehistoric maize have now been studied, and several of these are described in detail in Chapters 14 and 15. Here I wish only to present briefly the archaeological evidence for the prehistoric introgression of teosinte, or *Tripsacum*.

The earliest cobs of maize from all of the older archaeological sites so far studied from

Mexico and southwestern United States have polystichous spikes, paired pistillate spikelets, relatively long pedicels, and soft tissues of the rachis and glumes. The cobs from later levels of these same sites or from later sites usually include specimens with indurated rachises and glumes similar to certain segregates of maize-teosinte hybrids. Some of the more extreme types also exhibit other characteristics of teosinte, including single spikelets and a distichous arrangement of spikelets.

These prehistoric cobs from most sites are without kernels—the prehistoric Indians have done a thorough job of shelling. The few kernels that they have missed are centuries old and no longer viable, so we cannot grow them and prove by progeny tests that these specimens with tripsacoid* characteristics actually represent the effects of introgression from teosinte. But what we can do and have

*This useful term, invented by Anderson and Erickson (1941), describes in a single word any combination of characteristics that might have been derived from teosinte or *Tripsacum*. Several students of maize have recently objected to this term on the grounds that there is no conclusive evidence that *Tripsacum* has played any part in the evolution of maize, and they suggest that the

done quite effectively is to employ a counterpart of one of the techniques of the organic chemist, who by synthesizing a compound gains a clear understanding of its molecular structure. We can in our experimental cultures of maize-teosinte hybrids produce specimens which closely match the prehistoric ones. Although this does not *prove* that the ancient specimens are the product of teosinte introgression, it creates a strong inference that they are, especially since there is no other way known to produce modern specimens that match the ancient ones.

Tripsacoid cobs resembling segregates of maize-teosinte hybrids have now been found in caves in the Tehuacán Valley of southern Mexico (Mangelsdorf et al., 1964), in La Perra and Romero's caves in northeastern Mexico (Mangelsdorf et al., 1956, 1968), Swallow, Tau, Slab, and Olla caves in northwestern Mexico (Mangelsdorf and Lister, 1956), Richards and Tonto caves in Arizona (Galinat et al., 1956), and Bat Cave and Cebollita Cave in New Mexico (Mangelsdorf and Smith, 1949; Mangelsdorf et al., 1967; Galinat and Ruppé, 1961). Additional collections not yet completely analyzed but obviously containing tripsacoid cobs have been recovered from sites in Colorado, Nevada, and Oklahoma. Highly tripsacoid cobs can also be recognized in illustrations of prehistoric specimens from Tularosa Cave in New Mexico studied by Cutler (1952) and from the Hueco Mountain caves in Texas reported by Cosgrove (Galinat et al., 1956).

The dates at which the introgression of teosinte (or *Tripsacum*) into maize became well established can now be quite accurately determined by the well-dated remains from two Mexican caves: San Marcos Cave in the state of Puebla in southern Mexico and Romero's Cave in Tamaulipas in northeastern Mexico. In the former, teosinte introgression was quite evident in the Ajalpan phase, dated at 1500–900 B.C. In the latter, it became common in the later part of the Guerra phase, dated at 1500–1200 B.C. The

term "euchlaenoid" be used instead. This suggestion is not consistent with their contention that teosinte is wild corn and hence a species, not of *Euchlaena*, but of *Zea*. It also overlooks the fact that tripsacoid characteristics occur in certain South American races that have not, so far as the present evidence goes, been exposed to hybridization with teosinte.

vegetal remains from the Tamaulipas site are especially interesting and valuable because they include specimens of both teosinte and *Tripsacum*. Since teosinte is unknown in this part of Mexico today, one wonders whether the ancient cultivators who occupied these caves may have planted teosinte in their cornfields to "improve" their corn, as apparently did the Indians of western Mexico, who planted "maizillo" (probably teosinte) in their cornfields for this purpose (Lumholz, 1902). Among the artifacts found in Romero's Cave was a leather pouch containing teosinte seeds. This indicates that teosinte was highly regarded for some purpose or for some supposed usefulness.

In summary it may be said that the evidence of hybridization with teosinte as a factor in corn's evolution is extensive and comprises observations on modern hybridization, archaeological evidence of prehistoric hybridization, and evidence from genetics, cytology, and anatomy. This part of our tripartite hypothesis thus seems to be well established and has in fact now been generally accepted. The evidence of the direct introgression of corn's other relative, *Tripsacum*, although much less extensive, is substantial.

The Evidence of Introgression from *Tripsacum*

Corn crosses with its more distant relative, *Tripsacum*, less readily than with its closest relative, teosinte, but crosses can be made (see Chapter 5). The tripsacoid characteristics of some races of maize have been attributed to *Tripsacum* introgression. Roberts et al. (1957) for example, in their studies of Colombian maize assumed that the unusual race, Chococeño, is the product of the hybridization of maize and *Tripsacum*. They state:

Chococeño is one of the most unusual races of this hemisphere, both in its characteristics and in the primitive way in which it is grown. Its culture is largely confined to the humid coastal region of western Colombia, where rainfall sometimes exceeds 400 inches annually. The maize is grown without cultivation. The fields are prepared by cutting down the small trees and brush. The seed, which is broadcast and not covered, germinates on the surface of the soil. The plants grow up through the branches of the cut vegetation.

FIG. 11.3 When the inbred 4R–3 (*A*) is crossed with Florida teosinte (*C*) the F_1 hybrid ears (*B*) are maize-like in having four-ranked ears, some paired spikelets and partially naked kernels. When a strain of 4R–3 that has been modified by the introduction of chromosome 3 of Florida teosinte (*D*) is crossed with Florida teosinte (*F*) the F_1 hybrid spike (*E*) is almost identical with teosinte in being two-ranked with single spikelets and the kernels enclosed in shells. The modified strain (*D*), although resembling maize in its principal characteristics, obviously carries cryptic genes for several teosinte characteristics.

To succeed under these primitive conditions the maize must have unusual characteristics. Chococeño is highly tripsacoid. It has tough, slender stalks with tillers, narrow, dropping leaves and pendulous tassel branches. It has the general aspect of certain segregates from maize-teosinte or maize-Tripsacum hybrids. Since teosinte does not occur in this region, and Tripsacum is common, it has been assumed that Chococeño is the product of the hybridization of maize and Tripsacum.

Likewise Grobman et al. (1961) attribute the tripsacoid characteristics of the race Enano and its descendants to hybridization with *Tripsacum*, and they postulate that *Tripsacum* introgression has been involved in the evolution of a number of Peruvian races. The fact remains, however, that a natural hybrid of corn and *Tripsacum* has never been discovered, and since hybridization between the two genera is at best rare, it may be a long time before a natural hybrid is found. In the meantime, the evidence for such hybridization must continue to remain circumstantial. I wish now to describe an experiment which adds an increment to the growing body of such evidence.

Cryptic Genes for Tripsacoid Characteristics in Latin-American Races of Maize*

The experiment reported here was an attempt to determine whether chromosomes extracted from modern Latin-American varieties which

*The following section of this chapter was originally published in the *Boletin Sociedad Argentina de Botanica* (Mangelsdorf, 1968) as part of a memorial to my good friend, the late Professor Lorenzo R. Parodi, Argentina's most distinguished botanist and a long-time student of maize and other New World cultivated plants.

127   The Role of Hybridization in Corn's Evolution

single spikelets (Figure 11.3). This simple experiment demonstrated that the strain which had been modified by teosinte introgression carried concealed genes, not expressed in the strain itself, for these two teosinte characteristics. This fact suggested the possibility of detecting cryptic genes for two of the principal botanical characteristics which distinguish maize from its relatives teosinte and *Tripsacum*.

The experiment was basically simple. I had for some years been developing modified strains of A158 by introducing from Latin-America varieties, chromosomes which affected induration of the tissues of the rachis and lower glumes. I now selected nine of these and crossed each one with the pollen from a single plant of the Guerrero race of teosinte. A brief description of the nine Latin-American sources of the substitution chromosome follows:

Mexico 1077. A yellow flint corn from Chihuahua apparently related to the race Onaveño of Wellhausen et al.

Cuba 394. A typical Cuban flint with orange-yellow endosperm related to the race Argentino of Hatheway (1957).

Honduras 1639. A maize with purple aleurone related to the race Negro de Tierra Caliente of Guatemala (Wellhausen et al., 1957).

Nicaragua 501. Like the preceding race, a race with purple aleurone related to Negro de Tierra Caliente.

Venezuela 1536. A yellow flint corn resembling the race Costeño of Colombia and Venezuela (Roberts et al., 1957; Grant et al., 1963).

Bolivia 1157. (Figure 11.4). A flour corn with bronze aluerone and odd numbers of kernel rows, related to Pirincino of Peru (Grobman et al., 1961) and Coroico of Bolivia (Cutler, 1946; Ramirez et al., 1960).

Brazil 1691. A fairly typical ear of the Brazilian race, Southern Cateto (Brieger et al., 1958).

Paraguay 333. *Mais tupi* collected near Concepción. A white flint corn similar to Cristal Paraguay (Brieger et al., 1958).

Argentina 1807 (Figures 11.4 and 11.5). Maíz Amargo, a grasshopper-resistant corn grown in the province of Entre Rios (Rosbaco, 1951).

The last-named race is quite unusual in several respects which deserve mention. This

FIG. 11.4 Ears of inbred strains of A158 which have been modified by the introduction of chromosomes from Latin-American varieties. *Left*, modified by Bolivia 1157· *right*, modified by Argentina 1807; *center*, modified by both Bolivia 1157 and Argentina 1807. Note that the center ear has a prominent staminate spike. Although neither parent exhibits this feature, both may carry cryptic genes for it. ($\frac{1}{2}$ actual size.)

affect induration of the tissues of the rachis and lower glumes also carry genes for other characteristics of teosinte or *Tripsacum*, such as distichous spikes and solitary pistillate spikelets, which cannot be phenotypically expressed in genotypes in which maize germ plasm predominates. That such cryptic genes might exist and de tected was suggested by an observation that I made some years ago (Mangelsdorf, 1952) in which I compared two maize-teosinte hybrids, one having the inbred 4R3 as its maize parent, the other a strain of 4R3 modified by the substitution of a chromosome of teosinte for the corresponding chromosome of maize. The F_1 hybrid ears of these two crosses differed strikingly in several characteristics. The control, the hybrid of teosinte and the unmodified 4R3, had four-ranked spikes and paired spikelets; the hybrid with the strain of 4R3 carrying a teosinte chromosome had two-ranked spikes and

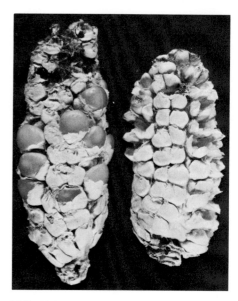

FIG. 11.5 Ears of *Maiz Amargo* from Entre Rios, Argentina. Note the prominent indurated lower glumes. These ears resemble some of the segregates from maize-teosinte and maize-*Tripsacum* hybrids. (Actual size.)

maize is called "amargo" (the Spanish word for bitter) because it is somewhat resistant to the attacks of grasshoppers, presumably becuase it is unpalatable to them. Horowitz and Marchioni (1940) suggested some years ago that this resistance might be the result of previous hybridization with *Tripsacum*. My own observations are consistent with this suggestion. The cobs of this race, which were sent to me by Ing. Urbano Rosbaco, are the most tripsacoid that I have encountered in a modern variety and are quite similar to the cobs of certain segregates from our experimental maize-teosinte and maize-*Tripsacum* hybrids. Two of the most tripsacoid ears of those sent to me by Ing. Rosbaco are illustrated in Figure 11.5. A much larger number are shown in his own article. Brieger et al. (1958) regard all types of Maíz Amergo as "extracts" of Calchaqui White. What is meant here by "extracts" is not clear, but it is conceivable that Maíz Amargo is a segregate resulting from inbreeding a population of Calchaqui White Flint, which is itself a highly tripsacoid race having, according to these authors, "very hard cobs which are highly sclerenchymatized, including the rachis flaps and the horny part of the glumes." These, of course, are among the characteristics that we regard as tripsacoid and that I have employed in identifying chromosomes producing tripsacoid effects. As mentioned later in this section, open-pollinated varieties are usually heterozygous for such chromosomes. Inbreeding should produce some progenies that are homozygous and consequently very tripsacoid. Sehgal and Brown (1965) report such segregates in progenies resulting from inbreeding United States varieties. If Maíz Amargo is an "extract" (a segregate resulting from inbreeding) of Calchaqui White, it provides good evidence that the parent variety is heterozygous for chromatin that produces tripsacoid effects. It is equally possible, however, that Calchaqui White is a descendant of Maíz Amargo, which has lost some of the tripsacoid characters of the original race.

Another significant characteristic of Maíz Amargo is that some of its plants are cytoplasmically male sterile (Gini, 1939). Lewis (1941) regards cytoplasmic sterility as the most common form of male sterility in species hybrids. Most of the cases of cytoplasmic sterility described in Edwardson's excellent review (1956) were derived from intergeneric, interspecific, or interracial crosses. Duvick (1965) considered it reasonable to hypothesize that the two types of cytoplasmic sterility then known in corn may be evidence of interspecific or intergeneric hybridization in the past. Thus even before the present experiment was conducted we had reason on several grounds to suspect that the highly tripsacoid Maíz Amargo is the product of hybridization of maize and *Tripsacum*.

In developing the modified strains of A158 by introducing chromosomes or parts of chromosomes from Latin-American varieties we noted that, in the majority of first generation hybrids of A158 and the tripsacoid races, more than half of the ears exhibited the tripsacoid characteristics of the parent but less than half showed these characteristics to the same degree as the parent. This situation is usually repeated in subsequent generations. One obvious explanation is that plants in open-pollinated varieties are usually heterozygous for chromosomes with tripsacoid effects and thus transmit such chromosomes to only about half of their progeny. Also crossing over with their homologs results in gametes that transmit only part of the chromosomes or polygene segments involved. Data illustrating the counterpart of this

situation in maize-teosinte hybrids have been presented in Chapter 4.

The linkage relations of the introduced chromosomes involved in these substitution strains developed from these nine varieties have not yet been determined. However, in six of the nine strains the red cobs of A158 have been replaced by white cobs. Since teosinte and *Tripsacum* lack the *P* gene on chromosome 1 for red cob this means that the introduced chromosome is no. 1.

The inbred strains employed in the experiment discussed here have involved five lots: (1) A158, the original strain, and the control in this experiment; (2) substitution strains of A158 modified by chromosomes or parts of chromosomes introduced from Florida, Durango, and Nobogame teosintes; (3) substitution strains of A158 modified by chromosomes or chromosome segments introduced from modern varieties of Central America and the Caribbean; (4) similar strains involving modern varieties from South America; (5) similar strains into which chromosomes or chromosome segments had been introduced from two different Latin-American varieties.

The F_1 hybrids of these five groups with Guerrero teosinte were grown in hills of five plants each at the Waltham Feld station in 1965. Since previous experience had shown that maize-teosinte hybrids often do not flower in this latitude before frost, all of the plants were subjected to short-day treatment until floral initiation had been induced. The treatment was stopped when a small tassel was visible to the naked eye in a longitudinal section of a stalk that was sacrificed for the purpose of determining floral initiation.

The data set forth in Table 11.1 are concerned with characteristics of the lateral spikes and represent averages of five spikes from each cross. The length of the staminate and pistillate portions of the spikes are based on actual measurements; the remaining data represent arbitrary grades on the number of ranks, frequency of single spikelets, and fragility of the rachises. A score of 0 represents the most maize-like condition and a score of 4 the most tripsacoid. The final figure "total tripsacoid score" is the sum of the three separate preceding scores.

The data set forth in Table 11.1 establish the following facts: (1) There is a strong correlation between the data involving actual measurements and those based on arbitrary

FIG. 11.6 Lateral spikes of hybrids of Guerrero teosinte with A158 (*left*) and with a strain of A158 modified by the introduction of a chromosome from Argentina 1807, *Maíz Amargo* (*right*). Note that the specimen at right is distichous, has solitary pistillate spikelets, and a prominent staminate spike. It is concluded that *Maíz Amargo* carries cryptic genes for these characteristics and that these may have been derived originally from *Tripsacum*.

TABLE 11.1 Characteristics of the Lateral Spikes of Crosses of Guererro Teosinte on the Inbred A158 and on Strains of A158 Modified by the Substitution of Chromosomes from Teosinte or from Latin American Maize Varieties

Seed parent of the cross	Length of lateral spikes, cm.				Scores of *Tripsacum* influence			
	Male part	Female part	total	percent male	rank	pairing	fragility	total
A158 (control)	8.6	9.8	18.4	47	2.2	0.6	0.6	3.4
Florida 1, 3, or 9	12.2	9.4	21.6	56	2.2	1.4	0.2	3.8
Florida 3	9.6	8.8	18.4	52	4.0	2.2	1.2	7.4
Florida 4+	13.0	7.4	20.4	64	4.0	2.6	3.0	9.6
Durango 1, 7, 9	12.8	6.0	18.8	68	3.0	2.4	3.2	8.6
Nobogame 4A	4.4	7.6	12.0	37	2.8	1.8	3.6	8.2
Nobogame 4B	2.6	1.2	3.8	68	2.4	2.8	3.8	9.0
Av. teosinte derivatives	9.1	6.7	15.8	58	3.1	2.2	2.5	7.8
Mexico 1077A	11.4	7.8	19.2	59	1.6	1.2	2.0	4.8
Mexico 1077B	10.1	7.8	17.9	56	2.8	1.8	3.8	8.4
Honduras 1639	20.6	8.2	28.8	72	4.0	3.0	2.0	9.0
Nicaragua 501	13.6	6.8	20.4	67	2.2	3.6	2.0	7.8
Cuba 394	10.0	8.6	18.6	54	2.8	2.4	2.0	7.2
Av. middle American	13.1	7.8	21.9	63	2.7	2.4	2.4	7.4
Brazil 1691	11.4	2.4	13.8	83	4.0	3.0	3.0	10.0
Paraguay 330	12.4	6.0	18.4	67	4.0	2.6	2.4	9.0
Argentina 1807C	14.4	8.0	22.4	64	3.0	3.0	2.0	8.0
Bolivia 1157	9.2	9.6	18.8	49	2.6	2.0	3.0	7.6
Av. South American	11.9	6.5	18.4	65	3.4	2.7	2.6	8.7
Mexico-Venezuela	6.6	9.4	16.0	41	2.0	1.8	2.2	6.0
Honduras-Nicaragua	19.2	7.4	26.6	72	4.0	3.2	2.4	9.6
Nicaragua-Brazil	10.0	7.2	17.2	58	3.6	3.0	3.8	10.4
Bolivia-Argentina	15.2	7.2	22.4	68	4.0	3.2	2.4	9.6
Av. double crosses	12.8	7.8	20.6	60	3.4	2.8	2.7	8.9

grades, which indicates that the latter, although to some degree subjective, are also to some degree reliable. (2) Every hybrid involving a modified strain of A158 is more tripsacoid than the control in one or more characteristics. Seventeen of the nineteen crosses are more tripsacoid in both actual measurements and arbitrary grades. (3) The chromosomes or chromosome segments introduced into A158 from Latin-American varieties are about as strong in their effects as those introduced directly from races of teosinte. (4) The chromosomes or segments introduced from South American varieties— those from Bolivia, Argentina, Brazil, and Paraguay—are at least as strong in their effects as those introduced from Mexican and Central American varieties. The cross involving Maíz Amargo is illustrated in Figure 11.6. (5) Several of the crosses in Group 5, involving chromosomes or parts of chromosomes introduced from two different varieties, are more tripsacoid than the average of the crosses involving only one parent.

From these facts we may conclude that modern varieties of Latin-America, like those of the United States, contain chromosomes or chromosome segments producing tripsacoid effects which came originally from teosinte or *Tripsacum* or both. Those extracted from Mexican and Central American varieties may have come originally from teosinte, but chromosomes or chromosome segments extracted from South American varieties could scarcely have come from teosinte, since this species does not occur in South America (except rarely in small patches planted for fodder that are usually cut before the plants flower). Tripsacoid characters of certain South American races, for example, the race Pardo of Peru, could be explained as introgression from teosinte, since Pardo is probably a race introduced into Peru from Mexico in post-Conquest times. The tripsacoid characteristics of South American races that have no counterparts in Central America must be explained in other ways. To me it seems highly probable that the chromosomes with tripsacoid effects extracted from these varieties have come directly from *Tripsacum*. The fact that they have some of the same effects as those of teosinte chromosomes is consistent with the idea that teosinte is itself a hybrid of corn and *Tripsacum*. It is also consistent, however, with the hypothesis that

Tripsacum is a hybrid of teosinte and *Manisuris* (Galinat, 1970). In either case teosinte and *Tripsacum* would have genes in common.

Goodman (1965) states that "the general feeling seems to be that teosinte but not (tripsacum) has contributed to the evolution of maize." My own opinion, based on extensive studies of archaeological maize and my own experimentation and that of my students over a period of some forty years on maize-teosinte hybrids and their derivatives, is that the evidence of teosinte's introgression as a factor in the evolution of maize is so conclusive as to leave little doubt; with respect to the role of *Tripsacum* the evidence is less conclusive but still considerable. It can at least be said that all of the facts now available are in general agreement with the assumption of hybridization of this genus with maize and they fit this hypothesis better than they fit any other that has so far been proposed.

12 Mutations

Although the immediate basis of the variation which makes evolution possible is genetic segregation and recombination, its ultimate source is mutation. This fact is implicit in the definition of mutation, as accepted by most modern geneticists.
G. Ledyard Stebbins, Jr., 1950

Under domestication, as in nature, the building blocks of evolution are mutations. To what extent has this been true in the evolution of maize? It is quite probable that some of the more successful domesticated species have been derived from wild species that were, while still in nature, more mutable than species in general. There are several reasons for suspecting that this might be true. (1) Species that are well adapted to their environments tend to have low mutation rates, and these rates are kept low by the pressure of natural selection. Such species are not likely to be amenable to domestication which may involve drastic changes in environment. (2) A species to be successful under domestication must be genetically plastic. It must be capable of being shaped by selection, initially by natural selection in a man-made environment, subsequently by artificial selection practiced by man. Such plasticity can be the product of either genetic recombination following hybridization or of mutability or of both. In the preceding chapter I discussed the important role of hybridization in the evolution of maize. This chapter is concerned with the role of mutations. I hope to show that these two evolutionary factors are interrelated; that hybridization can result in genetic recombination only if there has previously been mutation; that hybridization can also cause mutations.

Whether maize in nature was indeed a mutable species it is now impossible to determine. What we do know is that modern *Zea Mays* contains many mutant forms, hundreds of which have been recognized and described by geneticists. Indeed it is its abundance of inherited mutant forms that has made maize so valuable—for many years second only to the fruit fly, *Drosophila*—as a subject for genetic experimentation.

We know also that modern maize is a mutable species or that at least it contains some highly mutable loci. Stadler some years ago (1942) assembled data on the spontaneous mutation rates for specific loci in maize. These varied from 0 per million gametes for the *Wx* locus to 492 per million for the *R* locus. Later Singleton (1954) found much higher mutation rates for several of these same loci. The spontaneous mutation rates for some loci in maize are of the same order as rates of X-ray induced mutations at the white-eye locus in *Drosophila* (about 510 per million) and these in turn are about 60–70 times the spontaneous rates at the same locus. Thus, compared to the fruit fly, until recently a truly wild species, cultivated maize appears to be a mutable species.

Some of the mutants that occur in maize are in the direction of making the plant more useful to man, usually at the expense of its ability to survive in nature. Mutations at the pod-corn locus discussed in Chapter 8 furnish a classic example; the sugary endosperm of sweet corn discussed in Chapter 9 is another. The floury endosperm characteristic of many Indian varieties is a heritable defect that makes the kernels easier to grind but reduces their ability to germinate in cold, wet soils. A modern example of man's exploitation of single-locus mutation is the development within the past quarter of a century of hybrid corns with waxy endosperm employed for their peculiar chemical composition, and the most recent example is the development of varieties of maize in which the recessive mutant, opaque endosperm, has been introduced to enhance the protein quality of the grain, especially the content of the amino acid, lysine.

These are examples of mutants with effects so conspicuous that they are easily recognized by man and, if useful, employed by him. For every such mutant there must be numerous others whose individual effects are less easily recognized but whose combined effects may be considerable. It is these mutations that make maize so responsive to selection, a fact illustrated especially well by the famous Illinois experiments on selection for high and low protein and high and low oil content.

The Mutagenic Effects of Hybridizing Maize and Teosinte

In a previous chapter I described experiments involving the introduction of chromosomes or parts of chromosomes from various races of teosinte into two inbred strains of maize, initially the Texas strain 4R3 and later the Minnesota strain A158. The original purpose of these experiments was to determine, by comparing the modified strains with the original strains, what kind of characteristics are governed by the genes of the introduced teosinte chromatin and whether the different races of teosinte are similar or alike in the genic constitution of their chromosomes. As shown in Chapter 4 all of the races of teosinte proved to be alike in the fact that their fourth chromosome produces conspicuous effects on the characteristics of the tissues of the rachis and glumes.

As so often happens in biological experimentation, unexpected results occurred which are more significant than those for which the experiment was designed. Spontaneous mutations began to appear in the modified strains at a phenomenal rate. Both of the inbred strains used in our studies, 4R3 and A158, are stable strains that have never produced visible mutations in the many years that we have grown them. Their derivatives, in which teosinte chromatin has been introduced, have mutated at a rate so high that as many as three mutations have sometimes appeared in a single small population comprising the plants of a single nursery row, usually twelve in number. During the years 1950 to 1954 inclusive I observed 42 separate mutations in the maize-teosinte derivatives. During this same period the maximum population of plants grown, based on perfect stands, could not have been over 1844. Thus visible mutations occurred in at least 2.3 percent of the plants grown, or 1.15 percent of the gametes involved. Since perfect stands are seldom attained the actual rate must have been somewhat higher than this.

The kinds of mutations that occurred in these modified inbred strains run the gamut of the recessive characters that appear when open-pollinated varieties of maize are inbred. They include loci affecting (1) the gametophyte, thereby causing disturbances in Mendelian ratios; (2) the endosperm, the majority causing retardation in overall development but one, brittle endosperm, involving a conspicuous change in chemical composition, has also occurred; (3) chlorophyll defects including albinos, virescents, and pale green; (4) dwarfs of various kinds; (5) male sterility; and (6) chromosome translocations.

Some strains that have mutated once have continued, following outcrossing, to mutate at phenomenal rates. For example, a strain in which chromosome 4 from New teosinte had been introduced first mutated to a dwarf. This same stock has subsequently mutated to defective seeds (eight times), albinos, virescents, two other types of chlorophyll deficiencies, and a gametophyte factor affecting the Mendelian ratios on ears segregating for sugary endosperm. The position of several of the loci involved in these mutants has been determined and found to be on the short arm of chromosome 4. During the years in which these mutations, 14 in all, occurred, the total

number of plants of this stock grown (based on perfect stands) did not exceed 195. Consequently more than 7 percent of the plants have shown mutations.

A similar situation occurred in a strain in which chromosome 4 from Florida teosinte was introduced. In a population of not more than 85 plants there were four mutations to defective seeds and one each to dwarf, virescent, yellow-green seedlings, and a gametophyte factor. In this strain mutations occurred in more than 9 percent of the plants, a rate comparable to that resulting from treatment with massive doses of X-rays. The mutations do not occur at random but apparently involve a limited number of loci. Dr. Angelo Bianchi, who made a special study (1958) of the mutants affecting the development of the endosperm, found that of the 29 that he studied one had occurred nine times and others five, three, and two times respectively. The mutants appear to fall into two broad categories with respect to their stability. Some are quite stable, have a fairly constant expression on different genetic backgrounds, and segregate more or less normally in outcrosses with unrelated stocks. Others are quite unstable in their expression even on a uniform genetic background and do not segregate at all in outcrosses with certain stocks.

Stable Mutations

The mutations in the first category usually occur, so far as their positions have been determined, on the chromosome pair of which the introduced chromosome from teosinte is one member. For example, in the first two mutations that I observed, defective seeds designated as de^{f1} and de^{f2} occurred in strains in which chromosome 4 had been introduced from the teosinte varieties Florida and Nobogame respectively. Through linkage tests Bianchi (1957) found that the loci for these two characters are on chromosome 4, with about 14 percent of crossing over between them. Segregating in the "repulsion" phase they form a system of balanced lethals, which Bianchi and Salamini (1963) employed in determining linkages with other genes on chromosome 4. Mutations involving the short arm of chromosome 4 have enabled us to extend the map of this region beyond its previous limits. In the course of our experiments in which one of my former graduate students,

Surinder Sehgal, participated we obtained the following data on crossing over.

Loci	Crossover percentages
$V–De$	28
$V–Ga$	32
$V–Su$	47
$De–Ga$	30
$De–Su$	40
$Ga–Su$	38
$Ga–Gl_3$	47
$Su–Gl_3$	36

Although the data were not all obtained from the same segregating populations and so may include variations in crossing over resulting from differences in the environment and genetic backgrounds, they suggest the following sequence of loci: $V–De–Ga–Su–Gl_3$.*

I should add that the gametophyte factor in this series, a gametic lethal not usually transmitted through the pollen, is not the Ga factor characteristic of many popcorn varieties which confers a selective advantage on gametes that carry it. It may, however, be the same gametophyte factor linked with the defective seed factor that Jones and I described many years ago (1925) and on which we based our conclusions about the sequence of genes on chromosome 4. If so, the conclusions that we drew then about the sequence of genes on this chromosome are still valid, although they may have been based on an erroneous premise, i.e., that the gametophyte factor that we found linked with defective seeds is the same as that occurring in popcorn varieties.

The fact that mutations have occurred at such a high rate following outcrossing of a strain in which at least one mutation had already occurred suggests that mutability is associated with heterozygosity and may be the product of crossing over. Perhaps the

---

*It may be noted in this connection that the mutations involved, v, de, and ga, represent a segment of chromosome 4 at least 32 crossover units in length, a substantial part of the short arm of that chromosome. The suggestion sometimes made that the effects of introduced teosinte chromosomes may be due to single genes with pleiotropic effects is, in view of these results, scarcely tenable.

simplest explanation for the occurrence of these mutants is that they are due to a form of unequal crossing over similar to that involved in the phenomenon of bar eye in *Drosophila*. It is not difficult to suppose, indeed it would be strange if it were not true, that the chromosomes of teosinte, although apparently homologous to those of maize so far as cytological configurations in their hybrids are a criterion, are not, in fact, precisely and completely homologous throughout their entire lengths. Crossing over in regions in which the chromosomes are not exactly homologous could cause rearrangements of various kinds, including minute deficiencies and minute duplications both too small to be detected by cytological techniques. Deficiencies in some cases may be expressed phenotypically by simple recessive Mendelian characters usually deleterious in their effects. The duplications may be impossible to detect individually but in the aggregate may play a major role in corn's evolution under domestication.

The following model shows how a minute extra locus in a teosinte chromosome might cause unequal crossing over and the production of deficiencies and duplications.

Teosinte
chromosome $A'B'C'D'X'E'F'G'H'$
Maize
chromosome $ABCDEFGH$

If crossing over occurred between E and F in the maize chromosome and between X' and E' in the teosinte chromosome, the crossover products should be:

$A'B'C'D'X'FGH$
$ABCDEE'F'G'H'$

The first of these crossovers lacks the locus E, and instead has the locus X. To the extent that X is incapable of performing the functions of E, this represents a deficiency that might be expressed phenotypically as a defective seed, an albino, a virescent seedling, or some other recessive defect in development. The reciprocal crossover has two E loci, and if they are identical this is a simple duplication. However, if E and E' are different, a new compound locus has been created in which the two components may perform slightly different functions. Following the terminology employed by Brink and his associates to designate certain unusual types of mutations, these "duplications" might be called "paraduplications." They represent compound loci or "supergenes," whose components are somewhat different in their functions; but they have been created suddenly by hybridization instead of by a long and slow process of evolution.

Once there has been unequal crossing over of the type illustrated above, crossover chromosomes carrying the paraduplication when paired with "normal" chromosomes can furnish additional opportunity for the production of still more "deficiencies" and "duplications." The pairing of $ABCDEE'F'G'H'$ with $ABCDEFGH$ could result in unequal crossing over at any point to the right of E' in this model. Perhaps this explains the fact that in our modified inbred strains an initial mutation is sometimes followed by a burst of mutations. Perhaps it also explains the fact that although we have never had a mutation in the inbred strain A158, we have had several in a stock in which the Tu locus, which we now know to be compound, has been introduced and maintained in a heterozygous state for many generations.

The Tunicate Locus an Example of Paraduplication. By dissecting the tunicate locus and separating its components, Galinat and I have shown that this locus on Chromosome 4 is a compound one whose components are not identical. By recombining its components we have shown that together the two loci produce an effect at least as great as, if not greater than, the sum of their separate effects. If the reader, interested in this aspect of the problem, will turn to Table 8.2 and plot the weights of tassels against the number of components of the tunicate locus he will find the resulting curve to resemble a typical logarithmic one. This shows that the effect of an additional component is greater at the higher levels than at the lower ones. This is probably true for many characters involving measurements of size; it may be true also of characters involving quantities as in chemical composition. Its importance in this connection is that once the process of duplication begins its effects, as the process continues, can be exponential in nature.

Another phenomenon that may also occur is one known as the "threshold" effect. This means that when a certain physiological threshold is reached there is an effect that does not occur below this level. In the case of the components of the tunicate locus there is a threshold for the proliferation of spikelets which occurs rarely in genotypes with less than four components and commonly in those with four or more. Although there is no apparent selective advantage in proliferation of spikelets in the tassel—indeed the average corn breeder would consider it an abominable monstrosity—it illustrates one of the possible effects of the multiplication of components of a locus. There are, however, undoubtedly other instances of multiplication that result in selective advantage, especially in species evolving under domestication.

At least two other loci in maize, the A locus on chromosome 3 and the R locus of chromosome 10, have been shown by the researches of Laughnan (1952), Stadler (1951), and Stadler and Emmerling (1955), respectively, to be compound. There are undoubtedly many other compound loci not yet easily identified and not yet studied. Loci such as these, which are counterparts in a diploid organism of loci in a polyploid, have undoubtedly played a prominent part in corn's evolution under domestication.

The example that I presented showing how such a duplication might arise is based on the assumption of unequal crossing over between maize and teosinte chromosomes. The same kind of unequal crossing over would be expected to occur when two distinct geographical races of maize, whose chromosomes are homologous but not in perfect register, hybridize. Indeed in our paper (Mangelsdorf and Galinat, 1964) on the dissection and reconstruction of the tunicate locus we postulated that the extreme, somewhat monstrous, form of pod corn is the product of hybridizing races of maize derived from two distinct wild ancestral races. And I suggested in Chapter 10 that all living maize represents at least six "lineages," each one descended from a separate wild maize. Intercrossing between these may have created numerous opportunities for unequal crossing over and the production of compound loci. Later the process could have been accentuated by the hybridization of some of these lineages with teosinte or *Tripsacum* or both. Seen in this light it would appear the modern corn plant is a diploid organism possessing some of the characteristics of an allopolyploid.

Incomplete register may also be the explanation of a phenomenon that has long

FIG. 12.1 *Left*, an ear segregating for an unstable defective seed mutant. The mutant is inherited as a recessive, but the percentage of visible defectives is usually less than 25 percent because some of the homozygous defectives are not distinguishable, by inspection, from normal kernels. *Right*, an ear homozygous for the unstable defective-seed mutant. Note that the development of the kernels varies from extreme defective to almost normal. This is one of several unstable mutants that have occurred in modified strains of A158 in which chromosomes from teosinte or chromosomes with tripsacoid effects from Latin-American races have been introduced.

puzzled corn breeders; segregation for recessive characters in the second, third, or even fourth generations of inbreeding in lines that did not segregate in the first. So long as the lines remain heterozygous for two chromosomes that are not in perfect register, there remains the opportunity for unequal crossing over and the production of deficiencies that may be expressed as defective seeds, albinos, virescents, and the like. However, once such lines have become completely homozygous they may be quite stable.

Unstable Mutations

The mutations in the second category have three principal characteristics: (1) they are highly unstable; (2) some have appeared in strains that have been inbred for many generations and would ordinarily be highly homozygous; and (3) their loci are, in some cases at least, in chromosomes other than the one introduced from teosinte. These unstable mutants probably have less significance as factors in evolution than the stable ones, which are assumed to comprise undetectable duplications as well as detectable deficiencies. They are of interest, however, in suggesting the probable origin of the mutation system studied by McClintock and by Brink and his associates and the unstable mutations that have appeared in inbred lines employed in the commercial production of hybrid corn. The unstable mutations may affect any one of several characteristics. I shall describe three examples.

Defective Endosperm. The unstable mutant that we—Angelo Bianchi and I—have studied more than any other is one affecting the endosperm. This organ is sensitive to certain chromosomal and other irregularities and indeed, as Brink and Cooper (1947) have shown, is an effective isolating mechanism in preventing hybridization between species. Since the endosperm is one of the first stages in the development of the plant in which inherited characters having deleterious effects can express themselves, it is not surprising that endosperm defects are among the most common of those occurring naturally in open-pollinated varieties of maize. When breeders in the early twenties began to inbreed corn on an extensive scale as the first step in the production of hybrid corn, they frequently encountered ears segregating for various types of defective seeds. As a graduate student

during this period I studied 14 of these and found 12 of them to be genetically distinct. Defective seeds were also the most common mutations in the inbred lines that we produced by substituting chromosomes of teosinte for those of maize.

The unstable defective seed that we studied most intensively, designating it by the symbol de^{t5}, appeared first in a modified strain of A158 in which chromosomes or parts of chromosomes had been introduced from Florida teosinte (Fig. 12.1). The defective seeds appeared in one ear in 1952 following the second generation of selfing; it bore 363 seeds, of which 45 were classified as defective. In 1953 I planted the defective seeds of this ear. The self-pollinated ears (S_3) obtained from this planting bore mostly defective seeds, but these varied greatly in the degree of their development, ranging from thin seeds containing almost no endosperm tissue to well-developed seeds scarcely distinguishable in their appearance from normal seeds. It seemed obvious to me that we were dealing with an unstable mutant affecting the development of the endosperm, and it occurred to me that this mutant might provide unusually favorable material for making a quantitative study of the phenomenon of instability because the variation in the expression of this unstable mutant can be easily measured by simply weighing the individual kernels.

The following year I divided the kernels from one of these ears (now S_4) into five grades on the basis of size and planted those of each grade separately. There proved to be some correlation between the size of the seeds planted and the kernels on the selfed ears, but it was far from complete. As in the previous generation, there was considerable variation in kernel size on any one ear. To determine if there was any consistent pattern of variation we weighed the kernels individually and plotted a frequency polygon for each selfed-ear population using as ordinates the percentage of the total number in each class.

Although there is marked variation in the patterns of the resulting polygons, some having their modes at the right, others at the left, and others having two or more modes, the number of patterns is finite and there clearly are "repeats" (Mangelsdorf, 1958). In 17 ears that we studied in 1954 we recognized seven distinct patterns, three of which occurred once and one each two, three, four, and five times respectively. We assumed that each of these seven patterns represents a distinct genotype. Examples of "repeats" were illustrated in my earlier article. In 1955 we selected for planting from among these 17 ears the one that appeared to show the greatest variation in kernel weights. From this planting we obtained 84 self-pollinated ears, of which we weighed the individual kernels on 40. Again we found some correlation between the weight of the seed planted and the average weights of the kernels on the progeny ears, but again the correlation was not complete. In this generation (S_5) we recognized 10 different patterns in the frequency polygons, 7 of which we had encountered in the previous generation. These are illustrated in Figure 12.2. Bianchi (1959) has illustrated similar frequency polygons involving this and other defective-seed mutants.

One of the characteristics of this unstable endosperm defective is that it "disappears" in crosses with certain inbred strains. This was first discovered by my associate, Angelo Bianchi, who crossed a mutant with one of our multiple-gene tester stocks in order to determine its linkage relations. There was no segregation for the mutant in the F_2 generation. A possible explanation of this phenomenon is presented later in this chapter.

In a previous paper (1958) I suggested that this defective seed mutant had arisen when a certain block of teosinte genes—a segment of teosinte chromatin—had become inserted into a maize chromosome where in the homozygous condition it produced an inhibiting action upon the development of the endosperm. This, as already mentioned, is an early stage in the ontogeny of the plant in which deleterious factors can be expressed. I postulated that variations in the length of this block of genes would be reflected in variations in the degree of inhibition of endosperm development and that "mutations" from one state to another might be explained as the product of crossing over between segments of unequal length. This interpretation is still tenable. We have not, however, been able to subject it to really critical tests. The postulated blocks of genes are not visible in cytological preparations; my former associate, Dr. Ting, was unable to find any cytological irregularities at meiosis in F_1 hybrids of the mutant defective with an inbred strain of Wilbur's flint. In F_1 hybrids with another strain he found occasional asynapsis in chromosomes 1, 5, and 9. The significance of this is not clear.

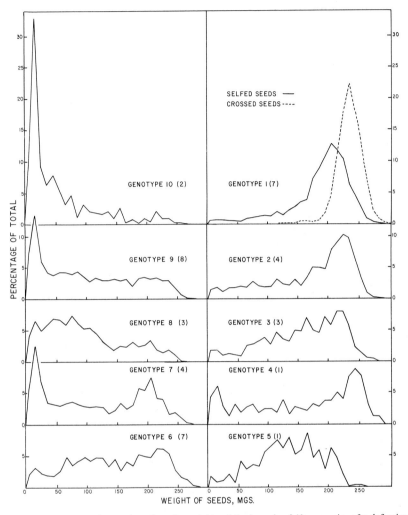

FIG. 12.2 Frequency polygons based on the weights of the kernels of 40 progenies of a defective seed mutation. Ten more of less distinct patterns occurred varying from that shown in the upper left in which the inhibition of seed development is virtually complete to that in the upper right in which the majority of the seeds approach the average of normal seeds in weight. The locus responsible for this defective-seed mutant is obviously unstable. Numerals in parentheses represent the number of individual populations assigned to the genotype.

In a study of the early cell divisions of the endosperm, Ting found no aberrant chromosome behavior similar to the conspicuous chromosomal irregularities that sometimes occur in the development of the endosperm of Paeonia.

There may be significance in the fact that one of the common mutations occurring in strains modified by the introduction of extracted chromosomes from Latin-American varieties is the unstable defective de^{t5}. This shows that these Latin-American varieties with tripsacoid characteristics have something in common with inbred strains derived from maize-teosinte hybrids.

An Unstable Dwarf. A close counterpart of the unstable defective seed is an unstable dwarf that appeared in the S_5 generation of a modified line of A158. This dwarf is inherited as a simple recessive, but the dwarf plants are extremely variable in height. The dwarfness of this mutant is definitely associated with tripsacoid characteristics of the cob; the more extreme the dwarf the more tripsacoid its cob. The variation in tripsacoid characteristics is also apparent in plants heterozygous for the dwarf locus. A dwarf of intermediate height crossed on an inbred strain of Wilbur's flint produced an F_1 population of 18 plants, the cobs of which were either weakly or strongly tripsacoid, indicating that the parental dwarf was heterozygous for several "states."

An Unstable Gametophyte Locus. A third example of an unstable locus involves the functioning of the male gametophyte, which in Angiosperms is concerned primarily with the development and growth of the pollen tubes. The gametophyte, like the endosperm, represents a sensitive stage in the ontogeny of the plant and one in which many irregularities such as chromosomal deficiencies, duplications, and extra chromosomes are commonly screened out. Loci affecting the expression of the gametophyte are detected by disturbances in Mendelian ratios of phenotypic characters that are linked with the gametophyte factor.

The unstable gametophyte factor described here is one of the 14 recognized mutants, mentioned earlier in the chapter, that appeared in a population not exceeding 195 plants. I first discovered it on a selfed ear segregating for sugary endosperm on which

the number of sugary seeds appeared by inspection to be much lower than the 25 percent expected in simple Mendelian inheritance and proved by actual count to be only 15 percent. The progeny of this ear included one ear segregating normally and five ears, designated as "low sugary," in which the percentages of sugary seed ranged from 9.2 to 18.9, with an average of 14.2 (963 seeds total).

Growing the starchy seeds of a cross of this stock with an unrelated sugary inbred and selfing their plants, we obtained nine "normal-sugary" ears (of which six bore 1,552 seeds, 25.8 percent of them sugary) and two "high-sugary" ears (37.4 percent sugary in 447 seeds). The results indicate that the aberrant segregation in the original low-sugary ear was the product of a deleterious gametophyte factor linked with the gene for sugary endosperm. The one normal-sugary ear in the progeny of the original low-sugary ear and the high-sugary ears in the out-crosses are assumed to be crossovers; they comprise 17.6 percent of the ears tested. Plants bearing low-sugary ears when selfed produced only 26.1 percent of sugary seeds (total 721) when their pollen was applied to a homozygous sugary stock; when pollinated by sugary they produced 49.4 percent of sugary seeds (total 987). The marked deviation from a 1:1 ratio in the crosses involving low-sugary plants as pollen parents and the close fit to this ratio when the same plants served as seed parents show that the aberrant segregation is largely if not completely the product of selective elimination in the male gametophytes.

In a series of backcrosses to a sugary stock of plants heterozygous for both sugary endosperm and the gametophyte factor, there was marked variation in the ratios of starchy and sugary seeds. The percentages of sugary seeds differ significantly not only from the 1:1 ratio expected in normal backcrosses involving one-factor segregation but also from each other. This variation is greater than can easily be accounted for by fluctuations in crossing over between the gametophyte and sugary loci, and it suggests that as in the case of the unstable defective seed and the unstable dwarf, we are dealing with a number of different "states" involving various degrees of inhibition of the normal functioning in this case of the male gametophyte. The inhibiting effect of this factor, like those of mutant defective seed and the mutant dwarf described

below, tends to "disappear" in certain crosses.

Disappearance of Unstable Mutants in Out-crosses

As mentioned earlier in this chapter, when Bianchi crossed the mutant defective seed with a multiple-gene tester stock, WMT277, the mutant which had been segregating as a recessive character failed to reappear in the F_2 generation. To determine how general this situation might be I crossed the defective seed with a number of inbred strains including the original A158. There was no segregation whatever in crosses with two strains, B10 and WMT277, the strain employed by Bianchi. In F_2 populations of crosses with inbreds C20, Oh28, and WMT275, some of the ears did not segregate and the remainder had only a few defective seeds. In crosses with P39 and an inbred strain of Wilbur's flint only part of the F_2 populations segregated, but these had quite a number of defective seeds. In the crosses with inbred Hy all F_2 populations segregated but had only a few defective seeds. Only in the cross with the original A158, the parent of the modified strain in which the mutant had first been observed, was the segregation in the F_2 approximately normal.

The cross with B10, one of those in which there was no segregation in F_2, was grown into an additional generation. In the F_3 eleven populations did not segregate; one had a few partial defective seeds.

In a similar experiment we crossed a number of inbred strains by a number of mutant dwarfs that had appeared in various strains of A158 modified by teosinte introgression. In all crosses the F_1 plants were normal. Especially interesting is the segregation in the crosses involving an extreme dwarf, one that was about one-eighth the height of normal plants in the same line. Four of the crosses, those made with inbreds C103, C106, Wilbur's, and WMT275, had no dwarf plants. The remaining three populations, those in which the parents were A158, P39, and WF9, all had more dwarfs than would be expected from one-factor Mendelian segregation.

The dwarfs that reappeared in the three segregating populations were in some cases quite different from the dwarf plant which was the pollen parent of all of the crosses. All of the dwarfs segregating out of the cross with A158 were tiny, ragged plants, about half of which died in the seedling stage. The dwarfs from the cross with WF9 were two types, about half being quite short, the remainder intermediate in height. In crosses with all other dwarfs irrespective of the inbred strains used as parents, there was a deficiency of recessives in the F_2 populations except in the cross of P39 by an intermediate dwarf.

When segregation for the mutant dwarf is compared with that for defective seeds no clear-cut pattern emerges, partly because the two experiments had too few inbred strains in common. Both the defective seed and the extreme dwarf reappeared in high frequency in the F_2 generation of crosses with A158. This is not surprising, since it was in modified strains of A158 that both mutants first appeared. Presumably this strain provides a milieu in which the mutants can be expressed. With respect to other inbred strains the results are not consistent. However, the fact that the defective seed does reappear in crosses with the Wilbur's flint inbred and WMT275, and the extreme dwarf does not, indicates that these two mutants have a somewhat different genetic basis.

To determine whether various types of dwarfness, like defective seeds, are associated with visible tripsacoid manifestations—especially the induration of the tissues of the rachis and lower glumes—we repeated plantings of 14 F_2 populations in which the parental dwarfs had failed to appear or to segregate normally. The behavior with respect to segregation of the dwarfs was the same in the second planting as it had been in the first; there were few or no dwarfs. We found, however, that the ears of the F_2 generation of plants could be classified with respect to tripsacoid manifestations in three categories: normal, intermediate, and tripsacoid. The numbers in these categories were 152, 330, and 165 respectively, a good fit to a simple 1:2:1 Mendelian ratio.

It is clear from these results that the particular block of genes responsible for a particular form of dwarfness is inherited in a normal Mendelian fashion. Whether the block of genes is expressed as a form of dwarfness depends upon the genetic milieu in which it occurs.

Significance of the Phenomenon of Disappearance. We are dealing here apparently with a phenomenon that is the counterpart in human

genetics of variations in "penetrance." Diabetes, for example, has been shown to be primarily the product of a single recessive locus, yet many individuals who are homozygous recessive for the locus never succumb to the disease. The word "penetrance," although useful in describing variation in the expression of a genetic character, does little to explain the variation itself. Lerner (1954) has given particular attention to some of the factors that might be involved in such variation: homozygosity versus heterozygosity for the genotype as a whole, canalization, buffering, and the specificity of the unbalance. There is no doubt that in the case of our unstable mutants we are concerned with serious disturbances in homeostasis. There seems little doubt that these disturbances are more severe in genotypes that are relatively homozygous, such as inbred strains, than in those that are much more heterozygous, such as F_2 populations, but there are obviously still other factors involved. What we can be certain of is that we have introduced into an inbred strain various blocks of genes from teosinte which normally do not occur there and that in certain genetic milieus these have deleterious effects while in others they apparently do not. Whether the mutants have expressed themselves in one genotype because it is already "loaded" with teosinte genes or has failed to do so in another because it is relatively free of such genes or, having them, is also strongly buffered against disruptive effects are questions still unanswered.

The Tripsacoid Nature of Other Variable Mutants

Because the mutants arising in maize-teosinte derivatives are often variable and difficult to classify, it occurred to us that some of the variable mutants occurring spontaneously in maize or appearing after inbreeding may have arisen in the same manner and may be tripsacoid in their cob characteristics. Accordingly we selected from the characters described in Emerson, Beadle, and Fraser (1935) a number of those described as "variable" or "difficult to classify." Plantings of these were made in 1962, and the ears produced were scored for tripsacoid manifestations. The following characteristics proved to be associated with tripsacoid cobs: albescent, brevis, narrow leaf-1, pale-green seedling-2, rootless, silky-1, zebra-1, 2, 3, and 4. In several segre-

gating populations the mutants were tripsacoid and the normal plants not. Only one character of those studied, adherent-1, was not associated with tripsacoid effects.

It may be that many of the recessive characters that maize geneticists have studied and have employed as chromosome markers in linkage studies and other genetic experiments have had their origin in the mutagenic effects of teosinte introgression. For some years we have been introducing marker genes for each of corn's ten chromosomes into our standard inbred A158. Several of the stocks so produced are distinctly more tripsacoid than the original strain. These include bm_2 on chromosome 1, lg_2 on 3, la on 4, v_3 on 5, and wx on 9. The A158 stock carrying v_3 has mutated to a defective seed similar to de^{t5}, a mutation common in maize-teosinte derivatives of A158 but never found in A158 itself. The multiple-gene tester that I developed by bringing nine recessive genes together in one line is one of the most tripsacoid stocks in our collection.

Mutations in Long-Inbred Strains of Maize

It is possible that some of the mutations occurring in long-inbred strains of maize are the counterparts of those occurring in maize-teosinte derivatives and are the result either of unequal crossing over or of the transposition of chromatin or some other kind of change in polygene segments originally from teosinte. There is little doubt that there has been teosinte introgression into the principal commercial field corn varieties in the United States. This would be expected from a knowledge of their history. We know that the Corn Belt dents are the product of hybridizing the Northern flints with the Southern dents. We know that the Southern dents are related to certain dent corn races of Mexico such as Tuxpeño, in which the occurrence of teosinte introgression is well established (Wellhausen et al., 1952). We know from the studies of Sehgal and Brown (1965) and of Johnston (1966) that many well-known inbred strains used in the production of hybrid corn are tripsacoid, some of them in the characteristics of their cobs closely resembling modified strains of A158 into which teosinte chromosomes have been incorporated.

Of particular interest in this connection are the mutants described by Singleton (1943a, b) which occurred in the inbred strain of sweet corn, Indiana P39. This strain, one of the parents of the widely grown Golden Cross bantam, was for many years the most extensively used sweet corn inbred in the United States. But it also had a reputation for being somewhat unstable, because it sometimes "broke down," to use the terminology of practical horticulturists to describe the phenomenon of producing conspicuously off-type plants.

When Singleton crossed a mutant dwarf that occurred in this inbred back to its mother line, there was a noticeable heterotic effect in the hybrid, indicating that the mutant line differed substantially from the mother line in its germ plasm. When he crossed the mutant with another inbred, new variations including germless seeds, brittle endosperm, and virescent seedlings appeared in the cultures. The similarity of these results to those that I have described earlier in this chapter are striking. All of the kinds of mutations occurring in P39 and its outcrosses that Singleton reported have occurred in our maize-teosinte derivatives, and one, brittle endosperm, is probably identical with his mutant brittle.

Of equal interest and perhaps of even greater significance from the standpoint of some of the mechanisms involved in heterosis in maize are the mutations in long inbred lines of maize studied by Jones (1945, 1954, 1957). In his first paper (1945) he described degenerative mutants: narrow leaf, dwarf, blotched, late flowering, and pale top-crooked stalks. The late-flowering and pale top-crooked stalk mutants appeared simultaneously and seemed at first glance to represent a single mutant with several effects. In another line five different variations appeared at about the same time. When the mutants were crossed back to their mother lines the hybrids were more vigorous than the normal parent.

Since the degenerative mutants appeared to differ from their mother lines by single genes, the results seemed to support the concept of heterosis due to "overdominance," the heterozygous genotype, Aa, being superior to either homozygote, AA, or aa. In a later paper, however, Jones (1952) concluded that the differences involved in the degenerative changes were not due to single genes because homozygous recessives and homozygous dominants extracted from the crosses of mutants with their parent lines were more vigorous than the parental plants, indicating that the

visible changes had been accompanied by or preceded by other changes capable of affecting growth rates.

In the final paper on this subject Jones (1957) showed conclusively that in the case of the pale top-crooked mutant both the recessive and the normal genotypes recovered from crosses of the mutant with the parental normal line were significantly taller than their parents. These results together with the finding that the condition pale top could be separated from that called crooked stalk provided convincing evidence that this mutant, at least, involves more than a single locus. At about this same time Schuler (1954) made a similar study of mutants found in inbred lines, some of which had been inbred for less than four generations and others for ten generations or more. Sixteen separate mutations, including eleven different dwarfs, were identified. All of these except two, reddish pericarp and brown midrib, have occurred in our maize-teosinte derivatives.

Making a series of comparisons between the mutants and their parental lines and hybrids between them Schuler found the mutant genotype aa to be usually the least productive and the hybrid Aa to be generally more productive than the original line AA. Comparing mutants recovered from selfing the F_1 or backcrossing it to the parental mutants, Schuler found the recovered mutants to be significantly more variable than the original ones, indicating that the lines still contained "relic heterozygosity." These results led Schuler to suggest that what other investigators had labeled single-gene heterosis may be due to multigenic diversity, a conclusion similar to that subsequently reached by Jones. Another significant feature of Schuler's experiments is that in some crosses the mutants failed to reappear in the F_2 and in others the F_2 Mendelian ratios were quite distorted. The similarity of these results to those that I have described earlier in this chapter is obvious.

In another series of experiments Schuler and Sprague (1956) crossed eight mutants: crinkled leaf, narrow leaf, brachytic, green stripe, two dwarfs, and two small seeds, and their mother lines on three inbred strains produced by the diploidization of haploid plants. Such lines should initially, at least, be completely homozygous. The hybrids were compared in a number of characteristics including yield of shelled grain.

This experiment provided a comparison between the two genotypes, homozygous dominant AA and heterozygous Aa, on a genetically heterozygous but homogeneous background. Without exception there was no evidence of superiority in the yield of the Aa genotypes. Variance ratios, however, indicated that the assumption of genetic homogeneity was not valid for all mutants. The authors attributed the heterogenity to concealed heterozygosity in the original lines. Another possible explanation of the results is that the mutants are similar to the unstable defective seed and dwarf in our maize teosinte derivatives described earlier in this chapter. If the mutations in long-inbred lines are indeed the counterparts of those produced in our cultures by teosinte introgression then the experiments on single-locus heterosis in which they are employed are not critical in distinguishing between overdominance and complementary action as the causes of heterosis.

Introgression and Mutation Systems

Finally I must point out the possibility that the mutation systems studied so intensively by McClintock (1965) and by Brink and his associates may owe their origin to teosinte or *Tripsacum* introgression. McClintock's *Ac* tester stock, which is homozygous for the *Ds* factor, is quite tripsacoid, and a chromosome with effects on the induration of tissues of the cob can be extracted from it. Her *Ac* factor in the presence of the *Ds* factor causes variegation in the aleurone color due apparently to areas of inhibition of the normal expression of the *C* factor for aleurone color on chromosome 9. When we crossed one of her stocks homozygous for aleurone color and the *Ds* factor but lacking the *Ac* factor with one of our maize-teosinte derivatives a certain proportion of the F_1 seeds, usually about 1 to 2 percent, were variegated. From one of these crosses involving chromosome 4 of teosinte we developed a stock breeding true for variegation in aleurone color resulting from inhibition of the expression of the *R* factor on chromosome 10. In this case teosinte chromosome 4 is somehow associated with the creation of an unstable locus on chromosome 10. In other pollinations on this same stock we identified 42 mutations from *R* to *r* in a population of 13,033 seeds and 19 mutations from *Su* to *su* in 23,654 seeds. Mutations from

Pr to *pr* on chromosome 5 and from *Y* to *y* on chromosome 6 were also noted, but data on their frequency are not available.

Brink and his associates (Brink and Nilan, 1952; Brink, 1954) have shown that variegated pericarp in maize is the product of interaction between the *P* locus on chromosome 1 and a modulator factor, probably identical with McClintock's *Ac* factor, which may occupy a number of different positions sometimes simultaneously. It may be more than a coincidence that all of the pericarp patterns that have been described in maize from any part of the world occur in two Peruvian races, Huayleño and Ancashino, in the Department of Ancash (Grobman et al., 1962). Variegated pericarp is, however, unknown in the prehistoric maize from early Peruvian sites, although it occurs in the later levels—after contamination with teosinte was evident—of Bat Cave in New Mexico.

Obviously something happened in Peru to create mutability in the *P* locus. Grobman et al. postulate that Ancashino has had the race Confite Puntiagudo in its ancestry and that the latter, is the product of *Tripsacum* introgression. Whether or not this is true, there is at least no doubt that there has been an explosive evolution of pericarp color in Peru that occurred sometime after domestication began. The Peruvian Department of Ancash has become the center of diversity of expression at the *P* locus.

Both McClintock and Brink have suggested that the phenomena that they have studied may represent the effects of heterochromatin. This is a plausible explanation, since heterochromatin has been held in other organisms, notably *Drosophila*, to be responsible for instability in adjacent or nearby loci. Lewis (1950) has divided position effects into two types: *V* (variegated) and *S* (stable). *V* types result from transposing genes to new positions in the vicinity of regions containing heterochromatin, which in *Drosophila* are concentrated near the centromeres. The mutation systems described by McClintock and by Brink, as well as the unstable defective seeds and dwarfs described in this chapter, might all be considered to represent types of variegation caused by proximity to heterochromatin. This interpretation is not, however, inconsistent with our conclusion that instability in some instances is caused by the introduction of chromatin from corn's relatives teosinte and *Tripsacum*. Indeed we know from evidence presented earlier that replacing corn chromosomes with teosinte chromosomes or chromosomes with tripsacoid effects from Latin-American varieties sometimes introduces heterochromatic knobs. It is even possible that the "foreign" heterochromatin of teosinte and *Tripsacum* is more mutagenic in its effect than the "native" heterochromatin of maize itself.

Some years ago Anderson (1953) suggested that a chromosomal segment derived through introgression from another species may become reduced in length through crossing over to a point where crossing over within the segment occurs rarely; when it does occur the result may be regarded as a mutation. To Anderson's suggestion we may add the possibility that even very small segments of introgressed chromatin may introduce not only foreign euchromatin but also heterochromatin, both of which might create instability in other loci.

Viewed in this light the variegated pericarp of maize may be a close counterpart of the variegation in tulips caused by a virus infection. The two patterns of variegation are so similar that color photographs illustrating only areas of variegated surface without showing the structure in which the variegation occurs—the pericarp of the maize caryopsis in one case and the petal of a tulip in the other—are virtually indistinguishable. The virus of the tulip does essentially what the modulator locus in maize does: it sporadically inhibits the development of color, thus producing variegation.

The analogy has been carried to the molecular level by Jacob and his associates (cf. McClintock, 1961), who regards the phenomenon described by McClintock and by Brink as counterparts of the situation in bacteria where nucleic acid introduced by phage may become incorporated into the bacterial chromatin through which it is then transmitted from generation to generation. Such a transmitted foreign component, which Jacob calls an "episome," has inhibiting effects upon some of the normal activities of the bacterium. The parallelism between the effects of introduced nucleic acid in bacteria and of introduced chromatin from other species in maize is striking.

Summary

Perhaps second only to hybridization as a factor in the evolution of maize is the phenomenon of mutation. There is also interaction between these two evolutionary forces, since hybridization can produce neither genetic recombination nor heterosis unless there has previously been mutation. It now seems clear that hybridization can also produce mutation. Whether wild maize was a mutable species cannot now be determined. There is no doubt, however, that modern maize is mutable. This accounts for the numerous inherited mutant forms that have made maize one of the most useful of subjects for genetic experimentation, and it accounts also for its marked response to selection. Some of the single-locus mutants such as sugary, floury, waxy, and opaque endosperm have proved useful to man. The substitution of teosinte chromosomes for maize chromosomes in a long-inbred strain of corn has resulted in numerous mutations. Some of these are quite similar to mutant forms previously known and are quite stable in their inheritance. Since they are defects in development and metabolism they may represent minute chromosomal deficiencies resulting from unequal crossing over, too small to be detected by cytological techniques. This same process should also result in minute duplications which may create compound loci, rendering the modern corn plant a diploid with some loci characteristic of polyploids. The tunicate locus which we have dissected and reconstituted furnishes an example of this process.

The unstable loci, of which three types, defective seeds, dwarfs, and a gametophyte factor, have been subjected to intensive study, may be less important in evolution, but they are useful in explaining the mutations that occur in long inbred strains and in showing that these cannot be considered as single-gene loci in testing theories of heterosis. The unstable mutants, the products of introgression from another species, may also prove to be related to the mutation systems described by McClintock and by Brink and his associates which Jacob regards as counterparts of "episomes" in bacteria, the product of introducing a foreign nucleic acid into the bacterium's chromosomes.

13 Genetic Drift and Selection

The assumption implicit in the Hardy-Weinberg theorem is that the population consists of so many individuals that chance variations in gene frequencies are negligible. Strictly speaking, this condition could obtain only in ideal, infinitely large populations. In reality, all populations are finite, and some populations are small.
Dobzhansky, 1955

The great power of this principle of selection is not hypothetical. Charles Darwin, 1859

A third factor involved in corn's evolution is "genetic drift," a term applied to changes in gene frequencies resulting from the random sampling of gene pools.* Sewall Wright has given particular attention to the mathematics of this phenomenon. Population geneticists are not in complete agreement on the importance of this factor in evolution in nature; some authorities regard its role as minor. Yet there can be little doubt that under some circumstances genetic drift plays a role. Populations that expand and contract greatly in numbers are especially subject to the influence of drift. Some genes are lost and others increase in frequency when a breeding population becomes greatly reduced in size, as do many insect populations during over-wintering.

Genetic drift must almost certainly be an important factor in the evolution of species of cultivated plants in which populations are broken up rapidly and repeatedly. The late Nikolai Vavilov, Russia's distinguished student of diversity in cultivated plants, recognized the phenomenon—although he did not call it drift—and he employed it systematically as one means of determining centers of origin. He concluded from his world-wide studies of cultivated plants that recessive genes having low frequencies at the center of a plant's origin may attain high frequencies at the periphery of its spread simply as the

*I am indebted to Cambridge University Press for permission to reprint here, with minor changes, a discussion of genetic drift and selection which appeared in a chapter on maize that I contributed to *Essays in Crop Plant Evolution*, 1965, edited by Sir Joseph Hutchinson.

result of random sampling. This process, which Vavilov called "the emancipation of recessives," is certainly one form of genetic drift.

Genetic drift occurs in corn whenever a new population is created from a limited sample of the gene pool in the original one. Experiments by Sprague (1939) have shown that from ten to twenty plants are required for adequate representation of genetic diversity in an open-pollinated corn variety. Since the number of ears saved for seed by Indian maize cultivators with only small plots of land at their disposal is often smaller than this and indeed since new maize populations are sometimes established by growing the progeny of a single ear, it follows that there must often have been genetic drift—changes in gene frequencies resulting from the creation of small breeding populations.

A striking example of genetic drift in maize is the occurrence in parts of Asia—China, Burma, Assam, and the Philippines—of varieties with waxy endosperm (Collins, 1909, 1920; Kuleshov, 1928; Stonor and Anderson, 1949). Waxy endosperm is a simple Mendelian recessive character that affects the chemical composition of starch, which in waxy varieties is composed exclusively of amylopectin and in nonwaxy varieties of a mixture of amylose and amylopectin. Waxy endosperm is the counterpart in maize of the "glutinous" character in rice. It also occurs in sorghum and millet.

Waxy Maize

Varieties of corn pure for waxy endosperm are unknown among the races of maize of

America, but the waxy character itself has been discovered in non-waxy varieties: in a New England flint corn (Mangelsdorf, 1924) and in a South American variety (Breggar, 1928). Bear (1944) reports that waxy endosperm is not an uncommon mutant in Corn Belt dent varieties, he having found three separate mutations to waxy in three consecutive years in a total population of some 100,000 selfed ears.

The fact that waxy maize occurs so commonly in a part of the world that also possesses waxy varieties of rice, sorghum, and millet can be attributed to artificial selection. The people of Asia being familiar with waxy varieties of these cereals and accustomed to using them for special purposes recognized the waxy character in maize after it was introduced into Asia following the discovery of America and purposely isolated varieties pure for waxy endosperm. But the fact that waxy endosperm came to their attention in the first place is probably due to genetic drift. The gene for waxy endosperm, which has a low frequency in American maize, apparently attained a high frequency in certain samples of Asiatic maize. Indeed the practice reported by Stonor and Anderson (1949) of growing maize as single plants among other cereals would result in some degree of self-pollination and in any stock in which the waxy gene was present would inevitably lead in a very short time to the establishment of pure waxy varieties with special properties that people accustomed to the waxy character in other cereals could hardly fail to recognize.

The selection and preservation of waxy maize in Asia exemplifies one of the most

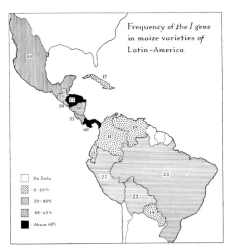

FIG. 13.1 The *Pr* gene for aleurone color has no visible effect except when the genes *A*, *C*, and *R* are also present. This gene has its highest frequency in Peru and Bolivia and in the adjoining countries Ecuador and Colombia. In Mexican maize from which the commercial corn of the United States is thought to be derived, the frequency of the *Pr* gene is 67 percent; in the commercial inbred strains of the United States, it is 98 percent. The difference may be the result of genetic drift.

FIG. 13.2 The *I* gene, which inhibits the development of color in the aleurone, has a low frequency in the Andean countries of South America and a high frequency in Mexico. In the commercial varieties of the United States, the frequency of this gene is zero. The United States varieties, if derived from Mexican races, apparently do not represent a random sample with respect to this gene.

FIG. 13.3 The *Ga* gene, which is responsible for the cross-sterility of certain popcorn varieties, probably once an isolating mechanism protecting maize from excessive hybridization with its relatives, has a high frequency in Mexico and in the other countries of Middle America and a low frequency in Peru and Bolivia. In the commercial inbred strains of the United States its frequency is zero, probably as the result of genetic drift. The low frequency of this gene in Peru and Bolivia lends some support to the suggestion set forth in Chapter 10 that Andean maize may be descended from different races of wild corn than the Mexican maize.

important features of genetic drift: the fact that it may interact with selection to confer greater adaptive plasticity than the species would otherwise possess. This interaction may even produce genotypes that, being inherently inferior to the norm, must, to survive, find new niches in which, through the accumulation of modifying factors, they may eventually become quite well adapted (Dobzhansky, 1955).

The development of waxy varieties in Asia illustrates these steps to perfection. Waxy endosperm is inherently a defect in metabolism, and its low frequency in most maize populations in the face of recurring mutations inidicates that it is acted against by natural selection. In the small breeding populations sometimes employed in Asia, the waxy gene became, to use Vavilov's expression, "emancipated" and reached a high frequency in certain samples. Because the people had special uses for cereals with waxy starch there was immediately at hand a niche for the new type of maize in which it survived and flourished. More recently this niche has been greatly expanded by the development in the United States of new industrial uses of waxy

maize and the establishment of a special market for it. Apparently there was no such niche in the Indian cultures of Pre-Columbian America. Otherwise the American Indians, like the Asians, would have preserved the new type when it occurred as a mutant which they could hardly have failed to see and to recognize as something different.

Flour and Sweet Corns

Other types of maize that differ from the "wild" genotype primarily by single genes probably also had their origins through the interaction of genetic drift and artificial selection. These include both the flour and the sweet corns. Flour corn, the expression of a locus on chromosome 2, found a special niche because it was much more easily chewed when parched and more easily ground than flint corn, the "wild" type. Sweet corn, which involves a gene on chromosome 4, found a special niche in South America in the preparation of the native beer, *chicha*; because of its higher sugar content it is preferred to the starchy corns for this purpose. Sweet corn found another niche, in Mexico and parts of the region now the United States, in the preparation of the native confection, *pinole*.

And, like waxy corn, sweet corn eventually found a niche much larger in the United States for use as green corn or "roasting ears," and as canned or frozen corn it has become a highly popular vegetable which is grown on several million acres annually.

Both floury and sweet corn are, like waxy, inherently defective. The sugary endosperm that sometimes occurs as a mutant in field corn varieties is virtually a lethal character. All of these metabolic defectives exist as cultivated varieties today because, having found new niches in which they can survive, their deleterious effects have been counteracted by the accumulation of favorable modifying factors. A fourth metabolic defective, opaque endosperm, may be going through this process at the present time; more about this mutant in a later chapter.

Gene Frequencies in Mexico and United States

A somewhat different example of genetic drift is furnished by certain gene frequencies of the corns of the United States compared to those of Mexico and other Latin-American countries. Some years ago John Edwardson and I testcrossed a number of varieties of the countries of this hemisphere for the genes *Pr* and *I*, which are concerned with the expression and inhibition of color in the aleurone and for the cross-sterility gene *Ga*. The frequencies of these three genes in the countries of Latin America are shown in Figures 13.1, 13.2, and 13.3. In Mexican maize, from which the commercial corn of the United States has been derived, we found the frequencies of *Pr*, *I*, and *Ga* to be 67, 50, and 56 percent respectively. Among inbred strains isolated from commercial varieties of the United States we found the percentages of these same genes to be 98, 0, and 0. It is clear that the corn of the United States, the world's most extensive center of corn production, is not at all typical of maize in general with respect to these particular genes and indeed represents an almost unique sample. With respect to the loci under consideration it is an example of the decay of variability, one of the consequences of genetic drift. Hence it is altogether appropriate for corn breeders now to give serious consideration to utilizing the germ plasm of corn varieties from other parts of the hemisphere, a subject treated in more detail in a later chapter.

Natural and Artificial Selection

A fourth factor involved in the evolution of corn under domestication is selection, which as already mentioned often interacts with genetic drift. Selection as it applies to corn may be natural or artificial or both. There can be little doubt that natural selection acting in a man-made environment has been an important force in the evolution of maize under domestication. This has probably been especially true since maize became highly heterozygous as the result of hybridization between races and between maize and its relatives teosinte and *Tripsacum*.

I can think of a number of characteristics of maize that are probably the product of natural selection: cold resistance of the high-altitude Peruvian maize; resistance to fungus diseases—to the rust fungus in high-altitude varieties of Mexico and to *Helminthosporium* in tropical varieties of Colombia; the high chromosome knob numbers of lowland tropi-

cal maize,* the long mesocotyls of certain Hopi Indian varieties that have been subjected for many generations to deep planting (Collins, 1914). All of these characteristics are probably the result of natural selection, and it is doubtful that man participated to any substantial degree in their development except as he exposed maize to the environments in which these traits contributed to survival.

Natural selection, sometimes abetted by human selection, may be responsible for certain other characteristics: the early maturity of Canadian varieties; the long, tight shucks providing protection against earworm and weevil damage of some varieties of the southern United States; the large pollen grains of varieties selected for their long ears (Galinat, 1961). These may be examples of the interaction between natural and human selection.

The effectiveness of natural selection is well illustrated by the rapidity with which open-pollinated varieties become adapted to new environments. More than forty years ago Kiesselbach and Keim (1921) showed that in Nebraska, which had then been settled only about a century, the corn varieties of the state had become so different in their adaptation that varieties collected from only a few hundred miles away were less productive than those in which the yield tests were made. There are many other cases cited in the literature, but less precisely documented, of corn varieties becoming rapidly modified in new environments.

At some stage in the development of Indian agriculture, artificial selection began, and it is widely practiced today. The better Hopi maize growers never save seed from an ear that shows evidence of mixture (Wallace and Brown, 1956). Among the pure-blooded, non-Spanish-speaking Indians of Guatemala rigid selection for a type of seed is often practiced (Anderson, 1947). This probably accounts for the fact that in Guatemala there is a high correlation between the percentage of *indigenas*, pure Indians, and the number of well-defined races of maize (Wellhausen et al., 1957). In Peru selection has become an art, and many different races are maintained in a state of uniformity, at least with respect to

---

*The selection would probably have been not for chromosome knobs as such but for genes associated with them.

color, by rigorous selection. I have already mentioned the careful selection that Peruvian Indians practice in keeping the race Kculli pure for its distinctive characteristics. The large-seeded flour corns of Peru, like the sweet corns already mentioned are the products of human selection interacting with hybridization, mutation, and genetic drift.

A number of well-known modern maize varieties of the United States owe their origin to rigorous selection for a specific type. Those which have furnished some of the inbred strains most widely used in the production of hybrid corn are of particular interest. In developing the famous Reid's Yellow dent, Robert Reid of Illinois selected for an early corn with cylindrical ears bearing 18 to 24 rows of kernels. George Krug, also of Illinois, selected for ears borne on good stalks and heavy for their size, with lustrous kernels. Isaac Hershey, the originator of Lancaster Surecrop, mixed together a number of varieties and selected for earliness and freedom from disease (Wallace and Brown, 1956).

The response, often spectacular, of modern corn to selection is nowhere better illustrated than in the famous Illinois experiments on selection for chemical composition. In fifty generations of selection, oil content, initially 4.7 percent, has been raised to 15.36 percent and reduced to 1.01 percent in the high and low lines respectively. Protein content, initially 10.92 percent, has been raised to 19.45 percent in the high-protein line and reduced to 4.91 percent in the low-protein line. That there is still variability for these characteristics in the population is shown by two generations of reverse selection which has been effective in three of the four lines (Woodworth et al., 1952). Other examples of selection, including an especially effective type known as "recurrent selection," are discussed in Chapter 20.

Summary

One of the four principal factors involved in corn's evolution under domestication is "genetic drift"—changes in gene frequencies resulting from random sampling of gene pools. Recessive genes with a low frequency at the center of a cultivated plant's origin may attain high frequencies at the periphery of its spread. Genetic drift may interact with selection to adapt genotypes to new environmental niches; varieties of corn with waxy endosperm in parts of Asia are an example; the development of varieties of sweet corn and flour corn are similar examples. The differences in the frequency of *Pr*, *I*, and *Ga* genes in the commercial corn of the United States and the corn of Mexico may be the product of genetic drift.

Natural selection operating in a man-made environment may be responsible for the cold resistance of high-altitude corn varieties and the disease resistance of others. Natural selection combined with artificial selection is probably responsible for the early maturity of Canadian varieties, the long shucks of ear-worm and weevil-resistant varieties of the southern states and the large pollen grains of long-eared varieties. The spectacular response of corn to continued selection is well illustrated by fifty generations of selection for high and low oil and protein content.

14 Archaeological Evidence of Corn's Evolution

It is clear that when we have archaeological records about a given species, like those of the Egyptian monuments, or of the Swiss lake-dwellings, these are facts of remarkable accuracy. A. de Candolle, 1886

In our 1939 monograph, commenting on the evidence from archaeology bearing on the origin of corn, we (Mangelsdorf and Reeves, 1939) pointed out that archaeologists have had almost as great an interest in the problem of the origin of maize as have botanists, for it has long been evident that maize was the basic food plant of all ancient American civilizations and advanced cultures. Yet at that time archaeological investigations had thrown but little direct light on the problem of origin. This, fortunately, is no longer true. On the contrary, prehistoric vegetal remains uncovered in archaeological excavations have contributed as much to solving the problem of corn's origin and evolution as have the results of any other line of attack and have contributed much more than all of the speculation and controversy combined. In the absence of facts debate can go on indefinitely—sometimes heatedly—but it is difficult to argue with the solid evidence provided by well-preserved archaeological specimens.

The evidence from archaeology began to assume increased importance in connection with the problem of the origin and evolution of corn when some of the younger archaeologists began to give attention to the accumulated refuse in once inhabited dry caves. In such sites there was little prospect of finding any precious gold work, beautiful fabrics, or artistic pottery, which many of the earlier diggers had sought; instead there might be found evidence of the beginnings of agriculture, and perhaps this would turn out to be even more important. One of the first of such excavations was made in 1948 by Herbert W. Dick, then a graduate student in anthropol-

ogy at Harvard University, in an expedition sponsored by the Peabody Museum of Harvard in cooperation with the University of New Mexico.

Bat Cave, New Mexico

The site was a once-inhabited but long-since abandoned rock shelter known as Bat Cave. This is located in Catron Country, New Mexico, on the edge of the plains of San Augustin. Bat Cave was produced by the action of waves of an ancient lake which existed during a wetter period in geological history and which is now virtually extinct. The cave is about 165 feet above the level of the lake floor at an altitude of about 7,000 feet. It was occupied for some 3,000 years by a people practicing a primitive form of agriculture and an even more primitive pattern of sanitation. During the centuries of their occupation the cave dwellers allowed trash, garbage, debris, and excrement to accumulate to a depth of about six feet, thus producing exactly the kind of site that enterprising young archaeologists of this generation delight to dig into. Dick and a group of students from the University of New Mexico spent an interesting summer in uncovering the artifacts and vegetal remains which were later to provide the basis for Dick's doctoral thesis and the first archaeological evidence of an evolutionary series in corn.

Anticipating that significant vegetal remains would be uncovered in the excavations, Dick persuaded another Harvard graduate student, a botanist, C. Earle Smith, Jr., to join the expedition. Smith, in turn, approached me, as Director of the Botanical Museum,

seeking funds for the purchase of a round-trip bus fare to New Mexico and a sleeping bag. In return for these he proposed to deliver to the Botanical Museum all of the prehistoric specimens of plants which the expedition might turn up. After some argument with our then comptroller, who insisted that paying out money for something not yet delivered was highly irregular if not clearly improper, Smith was provided with his bus fare and a sleeping bag at a cost to the Museum of about $150. Seldom in Harvard's history has so small an investment paid so large a return.

From the standpoint of the history of maize, the collection of prehistoric specimens uncovered by Herbert Dick and his associates represents a landmark in at least five respects. (1) It contained the most ancient maize which up to that time had been discovered. (2) The most ancient maize was the most primitive which had by then been turned up. (3) The collection represented the longest sequence of maize remains so far uncovered. (4) It was the first collection to reveal a clear-cut evolutionary sequence. (5) This was the first prehistoric maize in which age was determined by the revolutionary new technique of radiocarbon dating. The collection represents a landmark in still one more respect. Our analysis of the prehistoric specimens created so much interest among archaeologists that there soon began a flow of archaeological material into the museum which enabled us to study prehistoric maize from many sites throughout the hemisphere. It was directly responsible for me becoming associated with Richard S. MacNeish and working with him on the prehistoric corn from La Perra Cave in eastern Tamaulipas, Mexico, two caves from Infiernillo Canyon in southwestern Tamaulipas, and finally the caves in the Tehuacán Valley in southern Mexico in which prehistoric wild corn was found.

In removing the accumulated refuse in Bat Cave in arbitrary levels of 12 inches each, the diggers found cobs and other parts of corn in all of the levels. More significant, however, was the fact that there was a correlation between the size of the cobs and the depth at which they were found. In other words, the cobs from the lower to the higher levels of refuse show a distinct evolutionary sequence with respect to size. This was the first such evolutionary series turned up by archaeologists.

At the bottom of the refuse, resting on the cave's floor were some small cobs, which because of their relatively long glumes we considered to be forms of pod corn. And because of the small size of the kernels which they must once have borne we considered them also a kind of popcorn. These specimens seemed to support the conclusion of Sturtevant (1894) and our more recent one (Mangelsdorf and Reeves, 1939) that primitive corn probably was both a pod corn and popcorn. They gave convincing support to the hypothesis that the ancestor of cultivated corn was corn and not its closest relative, teosinte. Yet many of the later cobs showed evidence of contamination with teosinte, thus lending support to another of our 1939 conclusions that much of the variation in modern maize is the product of introgression of teosinte which occurred as a later step in the evolution of maize. Seeds of beans and squashes were also present. Since the region, a very dry one, was not considered suitable for maintaining plants of these species in the wild state, we concluded that the vegetal remains of the three species were the product of cultivation and that agriculture was being practiced by the cave occupants.

Radiocarbon and Other Dating

The assertion that the Bat Cave maize was the most ancient which up to that time, 1948, had been discovered perhaps needs elaboration. The first estimate of the ancient prehistoric maize from Bat Cave was based on geological considerations. Dr. Ernst Antevs

estimated that the cultural deposits containing the maize had their beginning not later than 2000 B.C. His estimate was based on the fact that the refuse rested upon a layer of wind-blown sand and dust which represents a period decidedly dryer than the present. Antevs considered the layer to have been deposited between 5000 and 2000 B.C. The top level of the maize-bearing deposit was dated by the resemblance of the pottery to other dated sites at A.D. 500–1000. Thus the deposit containing the remains of maize was thought to cover a span of not less than 3000 years.

Within less than two years other estimates of the age of the maize were obtained by radiocarbon dating. For the reader not familiar with this technique, which we owe to the brilliant work of Willard F. Libby and his associates, a brief description may be in order. A radioactive isotope of carbon ^{14}C, occurs in measurable amounts in the atmosphere, where it is formed by the action of cosmic rays. This is incorporated into plants and animals from the carbon dioxide of the air. In living tissues a fixed concentration of ^{14}C is maintained by the balance between assimilation and radioactive decay. When life stops the concentration of ^{14}C runs down at a known rate, half of it being gone in 5,570 years, half of the remainder in the next 5,570 years, etc. Analysis of the residual activity in ancient specimens of wood, charcoal, cloth, bones, etc., by incredibly sensitive instruments permits age of samples sometimes as old as 15,000 years to be accurately determined.

Now that radiocarbon determination of the ages of a great variety of ancient materials has become routine it is difficult to convey the sense of excitement associated with this revolutionary new technique, especially our own at having samples of the prehistoric corn from Bat Cave included in the early determinations that were made in Libby's laboratory. The first radiocarbon determinations on the vegetal remains from Bat Cave employing cobs and wood dated the most recent material at $1,750 \pm 250$ years and the oldest at 3,000 to 3,500 years. These estimates differ somewhat from those based on geological criteria and pottery correlations but are still of the same general order. They lend additional support to the conclusion that the earliest maize from this site is the most ancient maize heretofore discovered.

The Remains of Maize

The maize remains isolated from the refuse comprised a total of 766 specimens of shelled cobs, 125 loose kernels, 8 pieces of husks, 10 of leaf sheaths, and 5 of tassels and tassel fragments. Most of the specimens, even the oldest from the lowest levels, were extraordinarily well preserved; some of them with long plant hairs on the epidermis of the floral bracts still intact. Indeed most of the vegetal material was in such an excellent state of preservation that we might have been inclined to question the estimates of its age had not the specimens of maize from the different levels exhibited, as mentioned above, a distinct evolutionary sequence.

The Cobs. The average length of the intact cobs increased from 7.3 cm. in the lowest level to 9.3 in the topmost. Even more striking is the increase in maximum length, 9.5 to 16.5 cm. There was a comparable increase in the diameter of the cobs, 14.8 mm. to 17.5 mm., and this was accompanied by an increase in average kernel-row number from 10.7 to 11.4. Beginning in the third level from the bottom, there was strong evidence of contamination by teosinte, the closest relative of maize. Our estimates of teosinte contamination, although subjective, are based on a familiarity with maize-teosinte hybrids and their derivatives which we have studied in extensive experiments over a period of years. Beginning in the late twenties, I had by 1948 grown numerous first- and second-generation as well as back-cross populations of maize-teosinte hybrids and had examined and classified thousands of cobs resulting from such crosses. When cobs similar to those in our experimental cultures were turned up in archaeological collections, I had no hesitation in concluding that they represent contamination by teosinte, especially since no plausible alternative explanation to account for certain of their characteristics has been adduced.

One of the most common effects of the contamination of corn by teosinte is an increase in the induration of the tissues of the cobs, especially the lower glumes which partly enclose the kernels, and the rachis, the central stem of the cob to which the kernels are attached. Employing arbitrary scores of teosinte introgression based upon the induration of the tissues of the cob, we found almost no evidence of contamination in the cobs of the two lowest levels. In the four succeeding levels, however, 27, 57, 91, and 73 percent respectively of the cobs were classified as showing contamination by teosinte.

Since teosinte does not now occur in the vicinity of Bat Cave and since there were no remains of teosinte in the cave, we assumed that the admixture with maize, which was evident in some of the archaeological specimens, had occurred in Mexico where corn and teosinte have been hybridizing for many centuries.

The Kernels. The people of the primitive culture (anthropologists ordinarily do not speak of primitive people, only primitive cultures) who grew the maize, the remains of which we found in Bat Cave, were evidently extremely efficient in their corn-shelling operation. Not a single cob among the 766 recovered from the digging bore even one normal kernel. However, a total of 125 loose kernels were uncovered in the refuse, and some were found in each of the six levels. All were reasonably well preserved, some so much so that they were later lost to the depredations of mice, which for some reason preferred them to grains of modern corn available in the same storage room in the Botanical Museum. The kernels in each of the two lower levels were small and corneous and were undoubtedly originally capable of popping when exposed to heat. Thus the primitive maize of Bat Cave-was, as mentioned above, a popcorn.

The kernels, like the cobs, showed a progressive increase in size from level to level, especially in width, from an average of 62 mm. in the lowest level to an average of 84 mm. in the uppermost. Even more striking than the increase in average kernel widths is the increase in maximum widths from 65 in the lowest level to 108 mm. in the highest. Also significant is the fact that although there was an increase in kernel size from level to level the small kernels with which the series began did not disappear from the scene when new types with much larger kernels appeared. What actually occurred was a substantial increase in the range of variation with respect to kernel size. This led to a general assumption that increases in variability may be factors of particular significance in the evolution of cultivated plants.

Evolution of the Husks. Several specimens of husks were found in the refuse, one of which proved to be of special interest. It is about 24.5 cm. long and includes four leaf sheaths. Its shank or stem is quite slender, and the husk showed no evidence of having once been firmly shaped around an ear. We concluded that a husk of this type must once have borne a small ear with a slender stem and that husks in the early stages of corn's evolution were involucres subtending the ears but not completely enclosing them at least at maturity. Similar husks uncovered in a second excavation of Bat Cave and described later in this chapter show that husks of this type may have enclosed not one ear but a cluster of several small ears.

The Tassels. Five specimens of tassels were found. All of these are similar to those of modern varieties in their general characteristics. One from level V—next to the lowest—has spikelets smaller and more delicate than those of most varieties of modern maize but is clearly recognized as a staminate spike of maize, thus providing additional evidence that the ancestor of cultivated corn was corn. We made no attempt to classify the cobs, kernels, or tassels of Bat Cave with respect to race. Indeed in 1948 we could not have done so because little more than a beginning had been made in classifying corn's living races. The recognition of the principal types of corn occurring in Bat Cave had to await the classification and description of the races of maize of Mexico (Chapter 9) and an analysis of prehistoric remains turned up in a second expedition to Bat Cave made in 1950 and described below.

Bat Cave Revisited

The conclusions that we drew from the prehistoric maize uncovered in Herbert Dick's 1948 excavation of Bat Cave (Mangelsdorf and Smith, 1949) gained substantial support and were also amplified by the remains resulting from a second excavation of the cave made in 1950 (Mangelsdorf et al., 1967). The earliest maize from this expedition is more primitive than any of the specimens turned up in the first; it is definitely related to the modern Mexican race Chapalote; it has many of the characteristics of a form of pod corn; it definitely is a popcorn. The evidence of teosinte contamination is even more convincing than it was in the 1948 material.

Since the 1948 expedition had excavated only one section of Bat Cave, Dick made a second expedition in 1950 in the hope of finding still earlier remains of maize as well as charcoal suitable for radiocarbon determinations. The 1950 expedition comprised four members, all from the University of Colorado: Herbert Dick, who directed the excavation, Francis Olson, Allen Olson, and Dr. Paul Maslin. As had been done in the 1948 excavations, the refuse was removed in arbitrary levels of 12 inches each, down to a layer of sterile sand on which the cave deposits rested.

Various Ways of Dating

As mentioned above, one purpose of the second expedition was to obtain material for accurate dating of the vegetal remains. Since charcoal was considered to be especially useful for this purpose, Dick made a special point of obtaining charcoal from undisturbed parts of the deposits from the upper to the lower levels down to the sterile sand. The radiocarbon determinations ranged from $1,610 \pm 200$ for charcoal found at 11 to 15 inches to $5,605 \pm 209$ for charcoal in the 48- to 60-inch level. I accepted these dates at the time—there was no compelling reason to doubt their accuracy—and I mentioned them in several of my articles. We are now reasonably certain, however, that the date of 5,600 years for the 48- to 60-inch level is too early by about 1,600 years, at least as it applies to the prehistoric corn. The date may be quite accurate for the charcoal, which could have been the product of itinerant campers' fire built centuries before maize-growing people occupied the cave. We should have given more consideration than we did to this possibility.

Other methods of dating—geological, correlation with cultural phases in other sites, and correlations with the prehistoric maize of other sites—dated by radiocarbon determinations, all provide more recent dates for the earliest Bat Cave material than do the radiocarbon determinations of the charcoal. I have already stated early in this chapter that the geological estimate of Dr. Ernst Antevs dates the beginning of the cultural deposits containing maize at not later than 2000 B.C. Considering the earliest Bat Cave maize to belong to an archaeological period known as Chiricahua, Willey (1966) estimates its age at

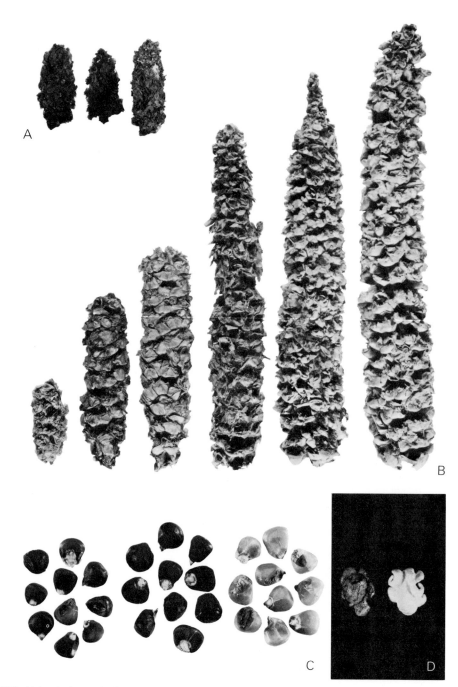

3000 to 2000 B.C. The Chiricahua site in Arizona has been dated by radiocarbon at 2056±270 B.C. On the basis of the characteristics of the cobs, especially those which represent evidence of teosinte contamination, we have dated the earliest Bat Cave maize at 2300 to 1500 B.C. These three estimates, geological, archaeological, and botanical are of the same general order and are also in agreement with the radiocarbon date of the Chiricahua site.

The Cobs. On the whole the cobs from this collection, numbering 444 specimens in addition to one fragment bearing kernels, are quite similar to those from the earlier excavation, with one important difference. Some of the cobs from the lowest levels are smaller and more primitive than any of those previously turned up at this site (Figure 14.1). My associate, Walton Galinat, and I spent a solid week dissecting one of these cobs and measuring all of its parts. So that we could hand slides back and forth between us and see features under identical magnifications, we purchased a second dissecting microscope exactly like the one that I had been using. On the basis of our joint measurements, Galinat prepared the diagrammatic longitudinal section illustrated in Figure 14.2. We concluded that the tiny kernels which this cob must once have borne could easily have been those of a popcorn, a type in which the kernels are small and hard and capable of exploding when exposed to heat. This conclusion was later amply confirmed by finding among the prehistoric grains several specimens of popped corn. These are described later in this chapter.

Whether the earliest Bat Cave corn was also a pod corn depends upon how pod corn is defined. Those botanists whose conception of pod corn is based entirely upon modern pod corn, an often monstrous type in which the kernels when present are usually completely enclosed by long glumes and other floral bracts, object to calling our prehistoric specimens of this type pod corn. There is no denying, however, that the earliest Bat Cave corn has some of the important characteristics of pod corn. The cobs have long floral bracts; lemmas and paleas surround the kernels but do not completely enclose them; the central stem, the rachis, is slender, and the secondary stems, the pedicels and rachillae upon which the kernels are borne, are relatively long.

FIG. 14.1   *A*, three cobs from the lowest level of the 1950 excavation of Bat Cave. These are dated at *ca*. 2300 B.C. or later. The diagrammatic longitudinal section illustrated in Figure 14.2 is based on one of these. There is now reason to think that these may not be intact cobs but the terminal portions of somewhat longer cobs. *B*, a series of Chapalote-type cobs, the shortest (perhaps not intact) from the 48–60 inch level, the longest from the 12–24 inch level. *C*, typical kernels with brown (left), red (center), and colorless pericarp. *D*, a prehistoric popped kernel from the 36–48 inch level compared with a popped kernel produced by exposing a prehistoric kernel from the same level to heat.

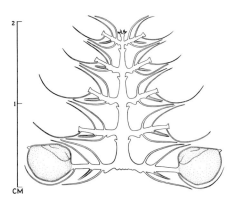

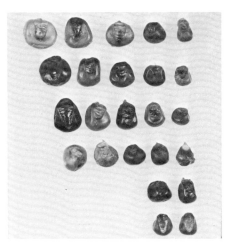

FIG. 14.2   Diagrammatic longitudinal section of one of the early Bat Cave cobs based on measurements of dissected parts. The tiny kernels show that this was a popcorn; the slender rachis (central stem), the long pedicels and rachillae on which the spikelets and kernels were borne, and the long floral bracts almost enclosing the kernels all indicate that this was a form of pod corn. The stump at the apex may be that of a staminate spike broken off. The shape of the specimen suggests that it may be, not an intact cob but the terminal portion of a longer one. (× 2.2. Drawn by W. C. Galinat.)

These are all features of weak forms of pod corn such as those developed in our experimental cultures, all of which involve components of the tunicate locus, *Tu*, on chromosome 4.

The cobs of the earliest Bat Cave corn are related to the modern Mexican race, Chapalote, one of the four Ancient Indigenous races of Mexico described by Wellhausen et al. (1952). They taper at both ends as do the cobs of the modern race. The earliest kernels, like those of modern Chapalote, have brown pericarp color; there is no other modern Mexican race with this color. We had found brown kernels among the remains of the 1948 excavations, but we supposed that these were discolored because of age. Some of the brown kernels fom the 1950 excavation are so well preserved in other respects that we could not doubt that the color is a natural one that has been preserved through the centuries.

The Kernels. Whatever argument there may be about the earliest cobs being pod corn, there is no doubt that the earliest Bat Cave kernels are those of popcorn. Of 299 kernels found in the refuse, 6 were partly or completely popped. We were able by a simple experiment to demonstrate that the early Bat

FIG. 14.3   Variation in kernel size in the six successive levels of the 1948 Bat Cave excavation. The kernels, like the cobs, increased in mean size from level to level, but the smallest size persisted throughout the series. This suggests that selection by man for larger kernels either was not practiced or was not effective.

Cave kernels are still capable of popping. To increase their moisture content we placed ten of the smaller brown kernels from the 36- to 48-inch level in a petri dish with a piece of moist paper towel for 48 hours. When dropped into hot oil all of these popped in varying degrees, providing dramatic proof that primitive corn was a popcorn, as Sturtevant (1894) and Reeves and I (1939) had postulated. One of the popped-prehistoric kernels is illustrated in Figure 14.1, where it is compared with a prehistoric-popped kernel from the same level.

I was not particularly astonished when these ancient kernels popped because I knew from long experience that the small hard kernels of most cereals will pop when exposed to heat if they have the right moisture content. But a young Argentinian botanist working with me on a grant from the National Research Council of Argentina was so excited at seeing these prehistoric kernels explode that he later reported to the council that studying archaeological corn can be an emotional experience.

The Evidence of Teosinte Contamination

The corn from the 1950 expedition, like that from the 1948 one, showed a well-defined evolutionary sequence with respect to rachis diameter, number of kernel rows, increase in the size of the kernels, and above all in total variation. The increased variability is the product, at least in part, of contamination by

FIG. 14.4   Two tripsacoid cobs from the 36–48 inch level of Bat Cave. Both have indurated tissues indicative of introgression by teosinte. About half of the spikelets on the larger specimen and at least one on the smaller are solitary, a teosinte characteristic. (× 2.4).

teosinte. The evidence of teosinte contamination was especially conspicuous in the cobs from the 36- to 48-inch level dated by radiocarbon determinations of associated charcoal at 2,048 ± 170 years. Several of the specimens having indurated tissues and bearing solitary pistillate spikelets are illustrated in Figure 14.4.

Among some 300 well-preserved kernels found in the refuse, 13 from the 36- to 48-inch level had variegated pericarp. Modern maize with variegated pericarp is the product of a mutation system which has been intensively studied by Alexander Brink and his associates at the University of Wisconsin. It involves the interaction of genes at the *P* locus on chromosome 1 with a genetic factor called "modulator" which may occupy various positions on chromosomes 1 or on other chromosomes.

Since we have experimental evidence discussed in Chapter 12 that contamination by teosinte has mutagenic effects we suspect that the prehistoric mutation system represented by the kernels with variegated pericarp was the product of teosinte introgression.

New Evidence on the Husks

The specimens from the 1950 excavations, for the most part confirmed the conclusions that we had drawn from the 1948 remains, but with one important exception. This concerns the nature of the husks. In the earlier collection we had found a long husk which showed no evidence of having been shaped around an

FIG. 14.5   Fragments of tassel branches from the 1948 excavation of Bat Cave. The vertical rows from left to right represent levels VI, V, IV, IV, and IV respectively. The spikelets of the oldest specimen are similar to the earliest of the Tehuacán Caves (Fig. 15.8) in being small and delicate and quite different from the staminate spikelets of teosinte. It was from this tassel fragment that the pollen grains measured by Barghoorn et al. and designated as "Bat Cave Early" (Chapter 15) were obtained. (Actual size.)

ear and we had concluded that this might represent an involucre of leaf sheaths subtending the ear but not completely enclosing it. Additional evidence from the 1950 collections suggests that a long husk of this type probably enclosed, not a single ear, but a cluster of ears, each of which had its own much shorter husks. A similar situation occurs in some of the plants of our genetically reconstructed wild corn described in Chapter 8.

Conclusions Drawn from the Bat Cave Corn

In summary then, the prehistoric specimens from the Bat Cave excavations represent the first collection of archaeological corn showing an evolutionary sequence. The earliest corn dated at *ca.* 2000 years B.C. was a popcorn and probably also a form of pod corn. It is related to the modern Mexican race Chapalote, and, like this modern race, has brown pericarp color. In the later corn there is definite evidence of teosinte contamination in the induration of the tissues of many of the cobs and the occurrence of solitary pistillate spikelets. Contamination by teosinte is undoubtedly a factor involved in the great increase in variation from the lower to the higher levels of the cave deposit, and it may also be responsible for creating a mutation system of which variegated pericarp is one of the products. For personal reasons the prehistoric maize of Bat Cave proved to be especially important because our analysis of it led to many other archaeological collections coming to us. Each one of these taught us something about the nature of primitive corn and its evolution under cultivation and one of them, the collection from Tehuacán, revealed the nature of wild corn.

La Perra Cave, Northeastern Mexico

It was our published analysis of the prehistoric maize from the 1948 excavation of Bat Cave that led directly to my being asked by Dr. Richard S. MacNeish, then of the National Museum of Canada, to undertake a similar analysis of the prehistoric corn that he had uncovered in his excavation, in 1949, of La Perra Cave in eastern Tamaulipas, Mexico. MacNeish had previously offered this material to another expert on maize, who turned it down stating that, in his opinion, there was little to be learned from these prehistoric cobs, however well preserved they might be. At one time I might have expressed a similar opinion, but I had already become convinced some years earlier that the final solution of the mystery of corn might well come from the prehistoric remains uncovered in archaeological excavations. Also by the time that I met MacNeish I had already studied the prehistoric corn from the 1948 Bat Cave excavation and had taken a preliminary look at the corn from the 1950 excavation. Furthermore, and this proved to be of at least equal importance, I had participated in the classification and description of the races of Mexico with the result that I could immediately see in the prehistoric La Perra corn a relationship to one of the still-living Mexican races.

The Site

La Perra Cave, like Bat Cave, is a rock shelter. It is located about eighteen miles northeast of the town of Los Angeles on a steep side of a dry arroyo some 150 feet above the bed. The cave, which is situated in Cretaceous limestone, is about 23 feet deep, 40 feet wide at the mouth, and 48 feet wide at 15 feet back from the mouth. The cave's ceiling reaches to a maximum height of about 15 feet.

Unlike the excavations in Bat Cave, which were based on 12-inch levels, the digging in La Perra Cave attempted to follow a recognizable stratigraphy. This comprised several more or less distinct zones, designated as A, B, C, and D. In analyzing the vegetal remains a distinction was made between low B and high B because this zone appeared to involve several different levels of occupation. The earliest cultural remains came from zone D and consisted largely of fragments of deer bones, projectile points, and a wide variety of scraping, chopping, and cutting tools, all of which suggest that the subsistence of the first occupants of La Perra Cave was based largely on hunting. Zone C was a sterile layer probably formed during a dry period when the cave was unoccupied. The second complex of artifacts as well as preserved vegetal remains made their appearance in zone B. The remains from the lower part of zone B were dated by radiocarbon determinations (Libby, 1952) at $4,445 \pm 180$ years.

There was an abundance of preserved remains of food plants (2,504 specimens), but only a small number of these represented cultivated species, the remainder being those of wild food plants. Fragments of animal bones,

projectile points, and slip loops indicate some dependence on the hunting and snaring of small animals. A few specimens of squash, *Cucurbita Pepo*, and of maize cobs and husks suggest an incipient agriculture.

The foodstuffs in zone A indicate that the La Perra people of this period, estimated at 1,800–2,200 years ago, were primarily agriculturalists, growing not only maize but also beans, squashes, gourds, manihot, and cotton. In summary the archaeological remains seem to comprise three cultural phases; first a nomadic hunting culture with no agriculture; second a semisedentary hunting and food-gathering culture in which an incipient agriculture was practiced; and third a sedentary food-growing culture. The remains of maize found in the cave are confined to the last two cultures.

The Maize

Because the maize remains from La Perra Cave represent a transition from an incipient to a well-established agriculture they are of particular interest. They are also of interest because of their age, because the earliest maize has its affinities in the south, and because the specimens represent a well-defined evolutionary series in a race of maize that still exists in Mexico. The collection comprises 177 specimens, of which the majority are cobs (87) or small cob fragments (31).

The Cobs. Of the 87 cobs which could be studied in detail, 78 can be said without much question to belong to a single race that is clearly related to the living Mexican race Nal-Tel. This race is described by Wellhausen et al., 1952, as one of the Ancient Indigenous races of Mexico and as relatively primitive when compared to modern races of Mexican maize. Nal-Tel is grown most commonly in the states of Yucatán and Campeche and sporadically in Guerrero, Oaxaca, and San Luis Potosí. Early forms of it, among the prehistoric remains of La Perra Cave, indicate that it was once more widely distributed in ancient times.

Of the 78 cobs recognized as belonging to the race Nal-Tel, only 18 are counterparts of the modern Nal-Tel of the Yucatán peninsula. These cobs have approximately the same size and shape and the same kernel row number as those of the living race. The remaining 60 cobs, which resemble Nal-Tel, represent a

more primitive form of this race. The cobs are shorter and more slender; their central stems, the rachises, are smaller in diameter and, probably as a consequence, the ears have a lower kernel-row number. The glumes are fleshy and, in relation to the slender rachis, are prominent. We considered these cobs to represent a form of pod corn resulting from the expression of lower alleles at the *Tu-tu* locus. One exceptional cob had quite prominent glumes. There was no evidence of teosinte contamination in cobs of Early Nal-Tel. There was such evidence, however, in cobs described below of two races which made their appearance in zone A.

Nine cobs in the collection were definitely distinct from those of Nal-Tel. Six of these appeared to be related to one of the races described by Wellhausen et al. (1952), Dzit-Bacal, which has approximately the same range of distribution as Nal-Tel, especially in the states of Yucatán and Campeche, where both are common and where there is introgression of each into the other. The remaining three cobs, all large, cannot be definitely assigned to any race described by Wellhausen et al. They seem, however, to be related to a modern race found in the state of Tamaulipas which Wellhausen (unpublished) has called Breve de Padilla.

The fact that these nine cobs, of the modern races Dzit-Bacal and Breve de Padilla, were found in a zone dated at 250 B.C.–A.D. 150 does not necessarily prove that these races were already in existence at this early date. Actually these few cobs may have been left in the cave by iterant campers long after the last occupation of the cave. They do, however, complete an evolutionary series with respect to size of an early form of Nal-Tel to a large-eared modern variety.

The Husks. The La Perra collection contains 56 well-preserved specimens of husks, some comprising only a single sheath, others entire husk systems still attached to their shanks. The husks were rather uniform in length. Twenty-one specimens were intact or almost so with respect to length and were much longer than the cobs of Nal-Tel occurring in the same zones. The husks, which average 140 mm. in length, were 2.3 times as long as the corresponding cobs, which averaged 60 mm. In this respect the husks resemble those found in Bat Cave, one specimen of which described earlier in this chapter was 2.7 times as long as a corresponding ear. Some of the husks give the impression of having flared open, exposing the ear.

A single specimen of a young ear shoot partly chewed showed clearly that at the time of emergence of the silks the young ear is tightly enclosed by the husks. Consequently, if the husks flare open, this must occur later as the ear matures. On one of my trips to Mexico, shortly after analyzing the La Perra prehistoric maize, I made a special effort to determine whether the flaring open of the husks at maturity is a characteristic of modern Nal-Tel. I found that many of the plants of this race do have this feature. The ears droop down, and the husks spread open, providing a kind of hood or canopy over the ear that protects it to some extent from birds but still allows the kernels to be exposed to the air, thus deterring the development of molds and other destructive fungi. It was a common practice at one time in the Corn Belt of the United States for farmers to select for this type of ear in order to insure adequate ripening before frost and to facilitate the husking operation. We now have evidence from specimens in the second Bat Cave excavation described earlier and from a site in South America that long husks of this kind found in La Perra Cave may have enclosed not one ear but a cluster of two or more.

Flag Leaves and Ligules. Husks which in modern corn surround and usually tightly enclose the ear are modified leaf sheaths. Sometimes these are terminated by short leaf blades known as "flag leaves," and when these are present there is a recognizable ligule at the point of attachment of the blade to the sheath. Of nine completely intact husks in this collection, not one had either a flag leaf or a ligule.

The Shanks. The husks, like the cobs, had relatively slender shanks or peduncles. Twelve specimens had an average diameter of 6.7 mm. at the point where the ear had been removed. The average shank diameter of modern Nal-Tel reported by Wellhausen et al. is 7.1 mm. These same specimens also show how the husks which arise at the nodes are spaced on the peduncle. In the La Perra maize the internodes of the peduncle are extremely

short. In the 12 specimens that we measured we found the average length of the region on the peduncle from which the husks arise to be only 8.6 mm. This figure suggests that the ears were borne high upon the stalk, since on plants bearing several ears there is an inverse correlation between the relative height of the ears and the lengths of the shank. Galinat (1954), for example, found that the shank lengths of 12 ears borne on a single stalk of Argentine popcorn varied from 19 mm. in the uppermost ear to 134 mm. in the lowermost. The data of Wellhausen et al., (1952), show that in modern Nal-Tel the average number of leaves above the ear is 4.3 and the internodes above the ear are relatively short. Prehistoric Nal-Tel may have borne its uppermost ear in an even higher position.

Primitive Uses

The remains of corn from La Perra Cave tell something about how prehistoric corn was used as well as what it was like. There were quids composed of thoroughly chewed young ears. After seeing these we tried chewing young ears enclosed in husks and found them tender, succulent, and sweet. Consuming maize in this way is a quick and simple method of obtaining a little sugar; no equipment of any kind is required. But from the standpoint of efficient utilization of the maize plant the method is quite wasteful, since for a morsel now, it sacrifices a full meal later. Also if the practice were general, there would be no production of seed for planting the next year's crop.

The second use of prehistoric maize is as green corn in the milk stage. Several of the cobs have the basal parts of kernels remaining within the glumes, the upper parts having been removed either by chewing or by cutting with a coarse blade. Ears of green corn called *elotes* are a popular delicacy in modern Mexico; apparently they were also relished in ancient times. The evidence from Bat Cave that one use of corn was by popping has been mentioned. We found no actual remains of popped corn among the specimens from La Perra Cave, but many of the cobs were slightly scorched or charred, showing that they had been exposed to heat. Not much equipment is needed to produce popcorn. A bed of hot coals, a pointed green stick thrust into the base of the ear, the ear held over the coals and slowly rotated, so that the kernels explode

while on the ear and are easily picked off and eaten. The glumes are slightly charred during this process but are not completely carbonized. We have simulated this procedure with modern popcorn ears substituting an electric hot plate for glowing coals.

A more sophisticated use of corn is by grinding followed by baking. On many of the cobs the glumes were battered and broken as though the ear had been beaten or forced across a rough surface. This would indicate a shelling operation and the use of shelled grain. The stone mortars found in the cave suggest grinding to produce a meal. There is no evidence to show how the meal was used; no remains of any kind of bread were found in the cave.

What the La Perra Corn Showed

The prehistoric remains of corn from La Perra Cave in northeastern Mexico, like those from Bat Cave in New Mexico, showed that early corn was both a popcorn and a form of pod corn. The La Perra remains are of particular interest in being the first to be studied in which the majority of prehistoric specimens could be related to a still-living Mexican race Nal-Tel (Mangelsdorf et al., 1956).

The Caves of Infiernillo Canyon, Mexico

While excavating La Perra Cave in eastern Tamaulipas, MacNeish made some preliminary soundings in several caves in southwestern Tamaulipas which persuaded him that still earlier corn, perhaps even prehistoric wild corn, might be found in the lower levels of the refuse of these caves. Financed by modest grants from the American Philosophical Society, the Botanical Museum, and the DeKalb Hybrid Corn Company, Mac-Neish, with the assistance of David Kelley, then a graduate student in anthropology at Harvard, in 1954 excavated two caves, Romero's and Valenzuela's, and sampled the refuse in a third cave, Oja de Agua, all located in Infiernillo Canyon. The vegetal remains uncovered in the refuse of these once inhabited caves are of particular interest in containing specimens of all three of the American *Maydeae*: corn, teosinte, and *Tripsacum*, as well as hybrids of maize and teosinte.

The Site Described and Dated

The Canyon Infiernillo is located in the Sierra Madre Oriental Mountains in southwest

Tamaulipas in the northern part of the Municipio Ocampo. It is about 75 miles southwest of La Perra Cave, which is located in the Sierra de Tamaulipas Mountains and has a somewhat different environment. The canyon is quite dry and has a xerophytic vegetation in which maguey, cactus, mesquite, and chaparral predominate. The three caves are high above the canyon floor at the base of limestone cliffs. They are quite dry, and the refuse they contain has been almost perfectly preserved. All three caves had stratified occupational layers which not only revealed a long cultural sequence but also yielded many botanical specimens which could be brought to bear upon the problem of early agriculture and subsistence activities.

Radiocarbon determinations and correlations of artifacts with those of other sites permitted MacNeish to recognize six more or less distinct cultural phases beginning at about 2350 B.C. These phases and their ages are:

San Antonio	A.D.	1450–1800
San Lorenzo	A.D.	1050–1450
Palmillas	A.D.	200–800
Mesa de Guaje		400–1200 B.C.
Guerra		1200–1800 B.C.
Flacco		1850–2350 B.C.

The maize specimens from only one of the three caves, Romero's, have been analyzed in detail, but an early cob from the Flacco phase of Valenzuela's Cave has been included in the analysis as well as several specimens of teosinte and *Tripsacum*. The collection of prehistoric remains of maize and its relatives described here comprise 12,014 specimens and include virtually all parts of the plant: pieces of stalk, leaves, husks, cobs and cob fragments, tassels, and tassel branches. There were also a large number of quids of chewed stalks, young ears, and tassels as well as several specimens of teosinte and *Tripsacum*.

Classification of the Cobs

All of the intact and nearly intact cobs were classified on the basis of their resemblance to the existing races of corn described by Wellhausen et al. (1952). Nine different races or subraces were identified. The great majority of the cobs, about two-thirds of the total, were assigned to the race Chapalote and its precursors or derivatives. This race is found today only in western Mexico (Wellhausen et al., 1952) but it was once much more widespread. Prehistoric ears, kernels, and cobs of this race have now been identified from archaeological sites in northwestern Mexico and in the United States in Arizona, Utah, Colorado, Nebraska, Oklahoma, and Texas.

The fact that Chapalote is the predominating race in the prehistoric cobs from the Canyon Infiernillo caves while the early corn from La Perra Cave, only some 75 miles away, is exclusively Nal-Tel (Mangesldorf et al., 1956) suggests that races of cultivated maize, like other artifacts, are characteristic of the cultures to which they belong. The people who occupied La Perra Cave were evidently related to lowland agriculturists in eastern Mexico, while those of the Canyon Infiernillo caves had their affinities westward and northward.

Pre-Chapalote. Following the nomenclature employed in our earlier description of Mexican prehistoric maize (Mangelsdorf et al., 1956; Mangelsdorf and Lister, 1956) we called the earliest corn uncovered in Romero's Cave Pre-Chapalote. The cobs of this race have the same tapering shape as those of modern Chapalote but are much smaller. None are as small, however, as the prehistoric wild corn from Tehuacán Valley described in the next chapter. We assumed that even the earliest corn from these Tamaulipas caves was cultivated corn. This subrace made its first appearance, a single cob, in the Flacco phase of Valenzuela's Cave dated at 2350–1850 B.C. It appears last in the Mesa de Guaje phase dated at 1200–400 B.C. The total number of cobs of this subrace is 17.

Early Chapalote. The cobs of this subrace, comprising 133 specimens, are intermediate in size between those of Pre-Chapalote and modern Chapalote. In other respects the cobs are quite similar to those of modern Chapalote. Cobs of this subrace appeared first in the Guerra phase dated 1850–1200 B.C. and last in the San Lorenzo phase, A.D. 1050–1450.

Tripsacoid Chapalote. Slightly more than half of all identified cobs, 1,546 specimens, were assigned to a subrace which we have called Tripsacoid Chapalote. The cobs are similar to those of Chapalote except for the induration of their tissues, especially the rachises and glumes, which are highly indurated. Cobs of this race are probably the product of hybridization of Chapalote with corn's closest relative, teosinte. Specimens of teosinte and of corn-teosinte hybrids described later in this chapter were found among the prehistoric vegetal remains. The Tripsacoid Chapalote appeared first in the Guerra phase but at a later level than the Early Chapalote. A single cob occurred in level 5, and the subrace was well established (126 cobs) in level 4b, which represents the end of the Guerra phase, probably about 1500–1200 B.C. This corn became the predominating type in the two succeeding phases, 1200 B.C.–A.D. 800 and thereafter was gradually replaced by other races, persisting, however, as a prominent component in the complex until A.D. 1800 after the arrival of the Spaniards.

Chapalote. Cobs of the type quite similar to those of modern Chapalote appeared first in the Mesa de Guaje phase, 1200–400 B.C. They increased in frequency in the Palmillas phase (14.5 percent of all cobs) and continued to maintain approximately this frequency until the end of the series. A total of 361 cobs were assigned to this category. Wellhausen et al. (1952) collected modern Chapalote in only two states in Mexico, Sinaloa and Sonora, but as pointed out above it must at one time have been much more widespread.

Breve de Padilla. Making its first appearance only slightly later than the modern Chapalote —level 4a of the Mesa de Guaje phase—is a race called Breve de Padilla by Wellhausen (unpublished). The origin of this race is not definitely known. Its cobs are longer and thicker than those of Chapalote. It may be the product of hybridization between Chapalote and Harinoso de Ocho. The fact that the suspected hybrid appears before one of its putative parents may not be significant, since Harinoso de Ocho has a low frequency at all levels and its absence in the Mesa de Guaje phase may represent nothing more than a sampling error.

Whatever its origin, Breve de Padilla appears to have been a productive race which soon replaced other races. Starting with an initial frequency of 4.6 percent, the cobs of this race increased in succeeding phases to 23.5, 35.1, and 43.1 percent. In the San Antonio phase, A.D. 1450–1800, this was the predominating race; altogether 836 cobs were assigned to it. Cobs identified as those of Breve de Padilla had previously been found among the prehistoric specimens in La Perra Cave (Mangelsdorf et al., 1956) but only in the uppermost levels and in low frequency. Its high frequency in Romero's Cave and its low frequency in La Perra Cave may be regarded as evidence that this race is western in origin and that the caves of the Sierra Madre and those of the Sierra de Tamaulipas, although only 75 miles apart, represent different peoples with respect to, their agriculture.

Minor Races. A total of 112 cobs were found which could not be assigned to any of the races described above, but all seemed to be related to others of the existing races described by Wellhausen et al. Among these was Harinoso de Ocho the putative parent of Breve de Padilla, which first appeared in the lower levels of the Palmillas phase A.D. 200–850. At no time, however, did it attain a high frequency, and the total number of cobs assigned to this race was only 32, or about one percent of the total identified cobs.

A total of 40 cobs were assigned to the race Nal-Tel, which is the eastern lowland counterpart of Chapalote and was the predominating race in La Perra Cave. It is not even certain that all of these cobs are actually those of Nal-Tel, since the two races are quite similar, differing slightly in their cob shapes and also in their pericarp color, Chapalote having brown pericarp color, Nal-Tel, orange red. Since refuse in the Sierra Madre caves was virtually devoid of kernels, pericarp color could not be determined.

A total of 37 cobs were assigned to the race Palomero Jaliscience, which is described by Wellhausen et al. as a subrace of Palomero Toluqueño, one of the four Ancient Indigenous races of Mexico. Among the living races of Mexico it has been collected only in southern Jalisco. Three unusually large cobs were found which showed some resemblance to Tuxpeño, the predominating modern race of the lowlands of eastern Mexico. It is possible, however, that these are not cobs of Tuxpeño but unusually large cobs of Breve de Padilla resulting from better than average growing conditions.

Other Parts of the Corn Plant

A total of 8,525 specimens of other parts of the corn plant were identified. These included stalks, leaves, husks, tassels, and quids of chewed parts. Pieces of stalks, 47 in number, showed like the cobs, an evolutionary sequence with respect to size, the earlier ones being on the average more slender than the later. Nine leaves or leaf fragments among the specimens add no significant information, since they are similar to the leaves of modern corn. The husks, a total of 219 pieces, show an evolutionary sequence with respect to length, the earlier ones shorter, on the average, than the later.

An amazingly large number of tassels, tassel branches, and tassel fragments, 8,099 specimens in all, were found among the vegetal remains. There is great variation among these; the later ones are indistinguishable from those of modern races of maize, but some of the earlier ones have smaller and more delicate spikelets. Although some students of maize (Anderson and Cutler, 1942) have considered the tassel to be an especially useful organ in classifying maize, we have not yet been able to discover any clear-cut diagnostic characters that will allow us to assign the tassels, as we have the cobs, to recognized races. We are, however, preserving all the specimens of tassels in the hope that some future student of prehistoric maize may see in them more than we have so far been able to discern.

The most puzzling aspect of the tassels is why they should have been preserved in such large numbers by the occupants of the caves. One of the specimens among the quids is that of a chewed young tassel. Since corn pollen is known to be rich not only in a number of amino acids but also in vitamins and minerals, it may be that tassels were sometimes chewed for the nutritive value of their pollen grains, these serving as a kind of vitamin-mineral supplement. However, the majority of the prehistoric tassels branches are those of ripe tassels which have shed their pollen and lost their anthers.

The refuse contained a large number of chewed quids, some of which, a total of 151, were identified as those of corn. Three types were recognized: chewed tassels mentioned above, chewed young ears, and chewed stalks. All three were probably chewed more for their sweetness than for the few calories which they added to a none-too-adequate diet. It is a fact, however, that a growing stalk from which the young ears have been removed accumulates sugar in concentrations comparable to those of sugar cane, and the chewing of such stalks might have provided an appreciable amount of energy. The chewing of both young ears and stalks must have been at the expense of subsequent grain production. Perhaps being provident had not yet become a virtue in this stage of culture and was seldom practiced.

Teosinte

We identified nine specimens of teosinte and three of maize-teosinte hybrids. Although cobs considered to be the product of teosinte contamination have been found in the remains of every archaeological site in Mexico and North America that we have studied, these are the first and only prehistoric remains of teosinte so far discovered. The earliest specimen, the fragment of a fruit case, occurred in feces in one of the lower levels of the Guerra phase dated at 1850–1200 B.C. Other specimens occurred in the two succeeding phases. Several of these comprise clusters of spikes. The specimens that we identified as maize-teosinte hybrids differed from those of teosinte in having thicker stalks and less fragile rachises. We cannot be certain whether these are F_1 hybrids or segregates occurring in subsequent generations.

The presence in the caves of remains of teosinte is puzzling. The seeds of teosinte are nutritious, having a higher protein content than those of corn (Melhus et al., 1953), but they are enclosed in hard, bony shells from which they are difficult to remove. Although this can be done by popping if the moisture content of the seed is right (Beadle, 1939) there is no evidence from the prehistoric remains that teosinte was used in this way. On the contrary, Dr. E. O. Callen, who has made a study of the feces from these caves, has found in them a number of teosinte fruits with their hard, bony shells unchanged. Since it is unlikely that the consumption of the intact teosinte fruits provides any satisfaction or nutritional benefits to the consumer, there must be some other reason for their use. Hernandez states that teosinte seeds are a cure for dysentery (cf. Wilkes, 1967) and although there is little reason to believe this, the inhabitants of the caves may well have believed them to be efficacious as a remedy for this or other ailments. Or perhaps the Indians of this region practiced the custom of planting teosinte in their cornfields to improve the corn, as did those of western Mexico (Lumholz, 1902) and of some parts of Guatemala (Melhus and Chamberlain, 1953). The finding of several fruits in a leather pouch suggests that they were regarded as valuable. The occurrence of maize-teosinte hybrids suggests that teosinte grew in or near the cornfields in prehistoric times, although it is unknown in Tamaulipas today.

Tripsacum

Slightly less puzzling, since it still grows in Tamaulipas, is the presence of *Tripsacum* in the refuse of the Canyon Infiernillo caves. Five specimens were found, and these included both staminate and pistillate parts of the spikes. Seeds of *Tripsacum*, like those of teosinte, are nutritious but are difficult to remove from the bony shells that enclose them. Although not at all promising as a source of food, they must have sometimes have been gathered for this purpose. Gilmore (1931) found fruits of *Tripsacum* in the prehistoric refuse of a cave in the Ozarks.

Three American Maydeae from One Site

The prehistoric refuse of the Canyon Infiernillo caves is of particular interest in containing the remains of all three of the American Maydeae: maize, teosinte, and *Tripsacum* as well as maize-teosinte hybrids. The remains, which are well dated by radiocarbon determinations and correlations with other sites, show that the earliest maize (2350–1850 B.C.) is related to the modern Mexican race Chapalote. Teosinte made its first appearance at 1850–1200 B.C., and teosinte-contaminated maize appeared at 1500–1200 B.C. and became the predominating type through the period 1200 B.C.–A.D. 800, when it was replaced by a new hybrid race, Breve de Padilla. The races of maize found in the caves and the presence of teosinte indicate that the occupants of the cave had their affinities with western rather than eastern Mexico.

Swallow and Other Caves, Northwestern Mexico

Soon after completing our analysis of the prehistoric remains of maize from La Perra Cave in northeastern Mexico, we had the opportunity to make a similar analysis of the remains from five caves in northwestern

FIG. 14.6 (*Top*), prehistoric teosinte from Romero's Cave in Tamaulipas, Mexico, dated at 900–400 B.C. In its triangular rachis segments having blunt apices, this teosinte resembles the modern races Nobogame and Central Plateau described by Wilkes (1967). Since teosinte is not known in Tamaulipas today, this specimen may have resulted from a practice, once common in western Mexico, of growing teosinte with corn to "improve" it. (*Bottom*), a spike considered to be that of a segregate from a maize-teosinte hybrid. This differs from teosinte in having a solid rachis and herbaceous instead of indurated glumes. The soft glumes are probably the expression of one of the alleles at the *Tu–tu* locus. (About actual size.)

Mexico. These caves, known as Swallow, Slab, Tau, Olla, and Dark, were excavated in 1952, 1953, and 1955 by Robert H. Lister, then of the University of Colorado.

These caves are located in the Sierra Madre Occidental of northwestern Chihuahua and northeastern Sonora, Mexico. Most of the Sierra Madre in this area is a plateau into which rugged canyons have been cut by stream action. The average elevation of the plateau is somewhat more than 5,700 feet, with a maximum of 8,200 feet. The climate is temperate, with an annual precipitation of 24 inches, three-fourths of which falls during July, August, and September. Four of the caves from which maize specimens were collected are located in the canyon of the Rio Piedras Verdes in Chihuahua, known to local inhabitants as "Cave Valley" because of the numerous caves which occur there. Many of these were occupied by the original peoples of the area, whose agricultural lands were located in the wider sections of the canyon bottoms, tributary arroyos, and on adjacent slopes. Numerous rock retaining walls thought to have been associated with agricultural practices still stand along arroyos at steeply sloping areas of cultivatable land. Dark Cave, the fifth of the caves yielding specimens, is located in Sonora approximately 35 airline miles northwest of Cave Valley in a narrow canyon known locally as Arroyo el Concho.

Swallow Cave, the largest of the four located in Cave Valley and the one producing the best archaeological record and the most significant sequence of maize specimens, is situated about 100 feet above the canyon floor. The cave extends into the cliffs to an average depth of 30 feet and has a mouth about 160 feet wide. The accumulated cultural deposits, which in one part of the cave reached a depth of 96 inches, contained maize at all but the lowest levels. The lower levels were lacking in potsherds but contained flakes of stones.

On the basis of the correlation of the cultural remains with those of other sites, Lister concluded that the most recent potsherds might be dated at A.D. 1000–1100. Slightly earlier artifacts were assigned to the Mogollon 4 period, dated at approximately A.D. 1000. Still earlier remains were assigned to the late Mogollon 3 period, dated at A.D. 900. How much earlier than this is the preceramic maize uncovered at 30 to 36 inches below the earliest Mogollon remains we do not know. The botanists and archaeologists agreed that these early maize remains were too precious to be sacrificed for radiocarbon determinations.

The Maize

Once again the botanist is indebted to the archaeologist for furnishing in the form of prehistoric specimens tangible and highly significant evidence on the evolution of maize. The archaeological maize from these five caves in northwestern Mexico is of unusual interest for five reasons. (1) The majority of the specimens are related to a primitive race of maize, Chapalote, which is still grown in Mexico. (2) The earliest maize appears to be a precursor of this modern but primitive race. (3) There is evidence of the introduction of an eight-rowed flour originally from South America. (4) There is evidence of hybridization with teosinte. (5) These several entities are finally blended into a modern race of maize, Cristolina de Chihuahua, which is grown in Chihuahua today and which has affinities with modern maize of the southwestern United States.

The botanist is also indebted, as he was in the case of the prehistoric maize from La Perra Cave and that of Infiernillo Canyon, to the corn breeders of Mexico, who from strictly utilitarian motives collected, classified, and described the living races of maize of that country. Their monograph on the subject (Wellhausen et al., 1952) with its detailed descriptions and excellent illustrations has furnished the clues to the identity of all of the different types of maize uncovered in the five caves.

The prehistoric cobs from these caves comprise five recognizable types, four of which still occur in the states of Sonora or Chihuahua. The fifth type is a now-extinct precursor of one of the four living types. Brief descriptions of the five types, all of which were found in Swallow Cave, follow.

Pre-Chapalote.
The earliest intact cob from Swallow Cave comes from level 13 (about 78 inches below the surface, since the cave was excavated in arbitrary six-inch levels). This is a carbonized specimen 3.5 cm. long, having twelve rows with an average of nine functional spikelets per row. In shape this cob is quite similar to the earliest cobs from Bat Cave in

New Mexico, now dated at about 4,000 years. This early maize from Swallow Cave also shows a resemblance to the modern Mexican race Chapalote, which, as mentioned earlier, occurs in the western part of Mexico and which has been collected in northwestern Mexico in the states of Sinaloa and Sonora (Wellhausen et al., 1952). The prehistoric cob is much shorter than that of modern Chapalote, but it has the same general shape, tapering at both ends, the same row number, 12, and prominent cupule rims, also known as "rachis flaps." Of the living races of maize in Mexico, Chapalote seems to be the only one which could be the modern counterpart of the earliest Swallow Cave corn which we therefore designated as "Pre-Chapalote."

Early Chapalote. The next recognizable element in the Swallow Cave cobs is even more clearly related to Chapalote. This type has the characteristic shape of Chapalote, tapering at both ends; it has the same predominating row number, 12, prominent glumes perhaps representing a weak allele of tunicate, and prominent cupule rims. Cobs of this type designated as "Early Chapalote" were found in several of the caves.

Further evidence that Chapalote or something very much like it was grown in this region is provided by the extensive collection of kernels from Dark Cave comprising about 3,000 kernels from levels 5 and 6. Almost all of these are similar to the kernels of modern Chapalote in their size, shape, and brown pericarp color. Chapalote is the only modern race in Mexico which has brown pericarp color.

Tripsacoid Maize. Beginning with level 2 in Swallow Cave there is evidence of maize which has been modified by teosinte introgression. Included in this type designated as "tripsacoid" are small cobs with strongly indurated glumes and occasional single spikelets or partially aborted pedicellate spikelets. Some of the specimens from the surface layer have strongly indurated curved lower glumes which are set at right angles to the rachis like the teeth of a coarse wood rasp. Some of these archaeological specimens, like those described by Galinat et al. (1956) can be almost exactly duplicated by modern specimens from our experimental cultures of maize-teosinte hybrids. The combination of

indurated glumes and solitary or partially aborted spikelets leaves little doubt that there has been introgression of teosinte into the maize of Chihuahua in prehistoric times. Teosinte is fairly common in western Mexico and has been collected in the state of Chihuahua (Wilkes, 1967). Lumholtz, cited earlier in this chapter, stated that the Indians of western Mexico practiced the custom of interplanting maize and *maizillo* (probably teosinte) for the purposes of improving their maize. Thus there is both historical and botanical evidence of the introgression of teosinte into maize in western Mexico.

Harinoso de Ocho. Also first becoming evident in level 2 of Swallow Cave are cobs of an eight-rowed, large-seeded corn similar to the race Harinoso de Ocho described by Wellhausen et al. (1952), who postulated that it was introduced into Mexico from South America in Pre-Columbian times. A race similar to it has subsequently been found in Colombia, where it is known as "Cabuya" (Roberts et al., 1957). Still later Galinat and Gunnerson (1963) showed that an eight-rowed corn, to which they gave the general term *Maiz de Ocho*, spread from Mexico through southwestern United States, the upper Mississippi Valley, and finally to the region now New England.

Cristalino de Chihuahua. The fifth race of corn in Swallow Cave found mainly in levels 1 and 2 is larger than any of the preceding types, is more or less cylindrical, and represents the blending of the characteristics of the three preceding entities: Chapalote, Tripsacoid, and Harinoso de Ocho. The cobs of this maize are counterparts of a modern race grown in Chihuahua and described by Wellhausen et al. under the name of Cristalina de Chihuahua. The prehistoric cobs are slightly shorter than the typical cobs of the modern race, but in other respects they are quite similar. Also the prehistoric kernels from several of the caves resemble the kernels of the modern race in being thick, flinty in texture, sometimes slightly dented, and predominantly white. This type of corn has affinities with the modern corn of some of the southwestern Indians: Zuñi and Navaho.

The Evolutionary Sequence

The sequence of steps in the evolution of

maize in northwestern Mexico is reasonably clear. All of the prehistoric maize, whether from Swallow Cave in Chihuahua or from Dark Cave in Sonora, is related to the living but still primitive Mexican race, Chapalote. The earliest maize of this type (from levels 13 and 14 of Swallow Cave) is smaller and more primitive than modern Chapalote and is a precursor of it. During the period (perhaps a very long one) represented by the difference between levels 13 and 14 and levels 1 and 2 in Swallow Cave there was little change in this race except for a slight increase in size. In this respect the situation is similar to that in La Perra Cave in northeastern Mexico, where another Ancient Indigenous race, Nal-Tel, remained remarkably constant during a long period of time.

This gradual evolution within a single race was suddenly interrupted when two new entities, an eight-rowed maize originally from South America and a tripsacoid maize the product of hybridization with teosinte or *Tripsacum*, became involved in the evolutionary sequence. With almost explosive rapidity these three elements merged to produce an entirely new and highly advanced race similar in its characteristics to the modern race Cristalina de Chihuahua. This spectacular evolutionary spurt can be accounted for by three genetic forces: genetic recombination, heterosis, and the mutagenic effects of teosinte introgression. Of these three the last, discussed in detail in Chapter 12, may have been the most important. In any case the maize of northwestern Mexico in a short period of time, not more than several centuries at the most, was almost completely transformed.

Approximate Date of Evolutionary Changes

None of the maize specimens from these Chihuahua and Sonora caves has been dated by radiocarbon determinations. Correlations of the cultural manifestations mentioned earlier in this chapter suggest that the most recent of the archaeological specimens should be assigned to A.D. 1000–1100. The sudden changes beginning in level 2 probably represent a slightly earlier date. Cutler (1952) in his study of the maize from Tularosa Cave in New Mexico noted a decrease in average kernel-row number which he attributed in part to the introduction from outside of the area of varieties with lower row numbers. The

FIG. 14.7 Prehistoric cobs from Swallow Cave, Chihauhua, Mexico, illustrating an evolutionary series. *A*, pre-Chapalote from level 13 (72–78 inches). *B*, early Chapalote from level 12 (6–12 inches). *C*, Tripsacoid cob considered to be the product of introgression from teosinte and exhibiting hybrid vigor. *D*, an eight-rowed race introduced into Mexico from South America considered to be the product of introgression from *Tripsacum* and exhibiting hybrid vigor. *E* and *F*, cobs of the modern race Cristalino de Chihuahua, which may represent four kinds of heterosis: (1) maize × maize; (2) maize × teosinte; (3) maize × *Tripsacum*; (4) teosinte × *Tripsacum*. (About three-fourths actual size.)

FIG. 14.8 Two cobs in which the lower portions were immersed for 24 hours in sulfuric acid. The nontripsacoid cob (*left*) with its soft tissues was so eroded by this treatment that only a delicate framework of the rachis remained. The tripsacoid cob (*right*) had lost its floral bracts, but the indurated rachis was still intact. (Actual size.)

change was most sudden between the Georgetown and San Francisco phases (*ca.* A.D. 700). On the basis of these several estimates we concluded that the very marked change in the maize of northwestern Mexico might have occurred at about A.D. 750 ± 250.

We now know that Chapalote, the principal early corn of the Chihuahua and Sonora caves, was diffused rather widely and was the principal if not the only race of maize of the early cultures in northwestern Mexico and the southwestern United States. Prehistoric remains of Chapalote or something closely related to it have been found by Kelley in the states of Jalisco and Sinaloa in Mexico (Anderson, 1944), by Anderson (1947) in the material from Painted Cave, and by Hurst and Anderson (1949) in the maize from Cottonwood Cave in Colorado. Although we did not recognize it initially, our later studies of the Bat Cave remains showed the earliest corn from that site to be related to Chapalote. The pre-pottery maize from Tularosa Cave illustrated by Cutler (1952) may be an early form of Chapalote. The earliest maize from Romero's Cave in Tamaulipas, Mexico, de-

scribed earlier in this chapter is definitely related to Chapalote. Because all of the five types of maize uncovered in Swallow Cave and the other caves of Chihuahua and Sonora have also been found as archaeological remains or living varieties in the southwestern United States, we concluded that the highlands of northwestern Mexico served as a corridor for the northward diffusion of maize.

Richards' Caves and Tonto Monument, Arizona

Through the cooperation of Mr. Lloyd Pierson of the National Park Service, Galinat and I came into possession of a large number of well-preserved cobs along with kernels, husks, shanks, and tassel fragments from two sites in Arizona: Richards' Caves and Tonto National Monument. These remains are neither very ancient nor do they show an evolutionary sequence. Their value lies in their number (3,342 cobs), which enabled us for the first time to subject prehistoric specimens to a statistical analysis (Galinat et al., 1956).

As my participation in classifying and describing the races of maize of Mexico proved to be invaluable in helping me to recognize precursors of living races in the prehistoric remains of La Perra, Swallow et al., and Romero's Cave, so years of experimentation with maize-teosinte hybrids, involving thousands of individual plants of first, second, and later generations as well as backcrosses to both parents, proved to be useful in analyzing material from these two Arizona caves. Many of the cobs from these sites resembled closely those which occur in F_2 and backcross generations of maize-teosinte hybrids. In fact it was possible to match many of the archaeological specimens feature for feature with modern cobs of segregates of such hybrids.

As mentioned earlier, one of the conspicuous differences between maize and its two relatives teosinte and *Tripsacum* is in the induration of the tissues of the fruits, especially those of the rachis and cupules and the lower glumes. In both teosinte and *Tripsacum* as they are described in Chapters 4 and 5, the caryopsis is enclosed in a hard, bony case consisting of a segment of the rachis and a lower glume. These structures become highly indurated as the fruit matures. Extensive experiments with maize-teosinte hybrids has

shown that the genes responsible for this induration are most frequent on chromosome 4 but also occur on several if not all of the other nine chromosomes of teosinte.

Estimating the Degree of Teosinte Introgression

In analyzing the cobs from the Arizona sites we sought for an objective method of estimating the degree of teosinte introgression. We hoped that a fairly direct estimate might be made by determining the specific gravity of the cobs. This proved for various reasons— the presence or absence of pith—not to be the case. There was a significant correlation between specific gravity and degree of induration, but it was not strong enough to be useful in estimating the amount of teosinte introgression.

We also experimented briefly with treating the archaeological cobs with sulfuric acid, which dissolves cellulose and leaves lignin. We found that when cobs of teosinte derivatives from our experimental cultures were immersed in sulfuric acid for 24 hours they lost all of their tissues except the rind of the rachis, the cupules, and lower glumes. When archaeological specimens are subjected to the same treatment the cobs considered to be tripsacoid remained almost intact while the nontripsacoid cobs eroded to the extent that only a delicate framework of cupule margins remained. This method of estimating the degree of teosinte introgression proved, however, to be too cumbersome for the purpose of analyzing more than 3,000 cobs, and we finally resorted to making subjective estimates based on five arbitrary grades, grade 1 designating the nontripsacoid, "pure" maize and grade 5 approaching maize-teosinte F_1 hybrids. The chief criterion in scoring the specimens was the degree of induration. Also considered was the presence of solitary pistillate spikelets, since these are characteristic of teosinte and are common in maize-teosinte hybrids but are rare in typical maize.

In the cobs from both Richards' Caves and Tonto there was a significant correlation between our estimates of teosinte introgression and kernel row number, the row number decreasing with an increase in teosinte introgression. This correlation was especially strong with respect to row numbers lower than eight. Indeed virtually all cobs with four or six rows were highly tripsacoid. With

FIG. 14.9 Resemblances between archaeological cobs from Richards' Caves showing introgression from teosinte and their matching counterparts derived from experimental maize–teosinte hybrids. The modern specimens (right in each pair) have been boiled in oil to simulate the dark coloration characteristic of the prehistoric specimens. (About six-tenths actual size.)

respect to cob diameter and teosinte introgression, there was no correlation in the cobs from Richards' Caves but a highly significant one, 0.984, in the cobs from Tonto. The reasons for the difference are not clear. With respect to cob length and teosinte introgression the correlation was strongly curvilinear, in the cobs from Richards' Caves both the longest and the shortest cobs being the most tripsacoid. In the Tonto specimens teosinte introgression was accompanied by a decrease in cob length.

My associate, Dr. Galinat, who made the statistical analyses of the cobs from the two Arizona sites, suggested that the curvilinear correlation in the cobs from Richards' Caves might be explained by assuming that the short tripsacoid cobs are homozygous for genes from teosinte and the long tripsacoid cobs are heterozygous for such genes and thus to some degree heterotic. Since there is abundant evidence from our experimental cultures of maize-teosinte hybrids that blocks of genes from teosinte may be beneficial when heterozygous and deleterious when homozygous, this explanation seems quite plausible.

Matching Prehistoric Cobs with Maize-Teosinte Derivatives

The Arizona collections were the first to come to us in which the number of cobs was large enough to justify statistical analysis. They were also the first in which we attempted to match a number of archaeological specimens with cobs from our experimental cultures of maize-teosinte hybrids. Four pairs of such matched specimens are shown in Figures 14.9 and 14.10. Since the archaeological cobs are discolored by age and the modern cobs quite fresh in appearance, we simulated the effect of aging by boiling the modern cobs in oil to produce the dark coloration characteristic of the prehistoric specimens. In some matched pairs the prehistoric and modern cobs are so similar in their coloration and in other characteristics that only experts can distinguish them.

Since all of the prehistoric cobs have had their kernels removed and since these would in any case have been incapable of germination, it is impossible to grow progenies of prehistoric corn to determine whether any of them are indeed the product of teosinte introgression. However, the fact that many of the prehistoric cobs can be matched almost exactly by modern cobs from our experimental cultures of maize-teosinte hybrids is persuasive evidence that both are the products of the hybridization of maize with its closest relative, teosinte. I know of no other way in which the prehistoric cobs could be so closely matched by modern ones.

Cebollita Cave, New Mexico

Shortly after completing the analysis of the prehistoric cobs from Richards' Caves and Tonto in Arizona we received from Reynold J. Ruppé, then a graduate student in anthropology at Harvard, a collection of prehistoric remains of maize which he had uncovered in his excavation of a once inhabited rock shelter, Cebollita Cave in New Mexico. Like the collections from the Arizona sites, this one contained a large number (2,375) of cobs, but it differed from the Arizona material in coming from a stratified site. Thus the Cebollita Cave material was the first large stratified (five levels) collection to become available to us for statistical treatment. Since Galinat had been so successful in applying statistical techniques to the prehistoric cobs from the two

FIG 14.10 Additional examples of resemblances between archaeological cobs from Richards' Caves and their matching counterparts selected from experimental maize–teosinte hybrids. Note the distichous arrangement of the spikelets in one pair. We have seldom found a prehistoric cob that we considered to be the product of teosinte introgression that could not be matched in its characteristics by cobs selected from our experimental maize–teosinte hybrids. (About six-tenths actual size.)

Arizona sites, I was glad to give him complete responsibility for analyzing the Cebollita Cave material. The conclusions that he and Ruppé (Galinat and Ruppé, 1961) reached added another significant chapter on the role of teosinte introgression in the evolution of maize. But first a word about the site.

Description of the Site

Cebollita Cave is located in the Cebolleta Mesa* area in Valencia County, New Mexico, about twenty miles south of the town of Grants. The cave was in the Upper Sonoran climatic zone at an elevation of about 7,000 feet and is located in a vertical sandstone cliff. It faces south and looks out on a broad valley which, before channel cutting had commenced, must have been an ideal floodfarming area. The climate is semiarid, with an average annual rainfall of about 11 inches. Within the cave is a 14-room pueblo which

*According to the principal maps of the area the name of the mesa is spelled Cebolleta, while the name of the cave which contained the archaeological maize is spelled Cebollita. The latter is the correct spelling of the Spanish word meaning "little onion."

furnished some protection to the inhabitants from the hazards of rock fall from the roof of the cave. An enormous block of sandstone falling from the roof at one time apparently caused a temporary abandonment of the pueblo. When it was reinhabited it was by a group that had a culture slightly different from that of the previous occupants. The entire occupation of the pueblo was encompassed within the Early Pueblo III period dated by ceramic typology at about A.D. 1050 to 1200. Although the cave is not far removed geographically from Bat Cave, the remains of corn which it contained differ in several respects from the latter.

The Prehistoric Maize

Most of the prehistoric maize was found in Room B of the pueblo, which contained a deposit of about four feet of refuse. Level 5, which begins with a hard-packed adobe floor, marks the earliest occupation of the pueblo. Lying on the floor were more than 500 charred ears as well as concentrations of charred shelled kernels and heat-warped potsherds. Before the fire, which caused the charring, the shelled kernels had probably been stored in pots and the loose ears may have hung from the roof beams.

The maize ears from level 5, the lowest stratum, are almost perfectly preserved by carbonization. The ears taper slightly at both ends. Small kernels are rounded on the top and are nearly isodiametric in their dimensions. Prominent glumes may protrude between the kernels. In all of these characteristics the ears resemble those of the modern Mexican race Chapalote, one of the Ancient Indigenous races of Mexico. This, it will be recalled, is also the race to which we assigned the early corn of Swallow and other caves in Chihuahua and Sonora, Mexico, and Romero's Cave in southwestern Tamaulipas. The termination of the level 5 occupation was marked by a fire which carbonized or charred all of the original Chapalote ears and which probably also caused a large rock fall from the ceiling and a temporary abandonment of the cave.

Level 4 represents the first reoccupation of the cave after the fire and was accompanied by a drastic change in the corn. Whereas all of the corn from level 5 was Chapalote, 85 percent of the cobs from level 4 were tripsacoid, and 3 percent of these were almost

exact counterparts of modern F_1 maize-teosinte hybrids or segregates from later generations of such hybrids. These cobs were two-ranked for part or all of their length and had highly indurated, upward curved lower glumes. Such cobs were scored grade 5 in our system of estimating teosinte introgression on the basis of the same five arbitrary grades employed in the earlier study of cobs from the Arizona site.

At the other extreme, 15 percent of the level 4 cobs were scored as grade 1 because they had long, soft glumes which were structurally similar to the carbonized ears of the Chapalote race found in the previous level 5. The teosinte introgression which marks the cobs of level 4 was accompanied by a reduction in the average size of the cob. Cob diameter was reduced by 10 percent below that of the cobs in level 5. Kernel row number was reduced by 11 percent and cob length by 22 percent.

The correlation between scores for teosinte introgression and cob length was curvilinear as it was in the cobs from Richards' Caves in Arizona, both the longest and the shortest cobs being on the average the most tripsacoid. The cobs of intermediate length tended to resemble the original Chapalote corn in having soft glumes.

Beginning in level 3 there was a progressive reduction in teosinte introgression, or more accurately perhaps a modification in its expression. The average scores for teosinte introgression declined while the averages for kernel-row numbers, cob diameters, and cob lengths increased. Most pronounced, however, was the increase in variability as reflected by the standard deviation. For levels 5 to 1 the standard deviations for cob lengths were respectively 1.01, 1.89, 2.27, 2.39, and 1.86 cm.

Galinat interpreted the curvilinear correlation, as he earlier had a similar one in the cobs from Richards' Caves, as homozygosity for teosinte gem plasm in the short cobs and heterozygosity accompanied by hybrid vigor in the long ones. The apparent decline in teosinte introgression he interpreted as the result of a selective elimination and/or a buffering against the more detrimental effects of teosinte contamination while retaining the beneficial effects of maize-teosinte heterosis. The increase in variation he attributed in part to genetic recombination and in part to the

mutagenic effects of introducing teosinte germ plasm into the maize, a phenomenon which I had discovered several years earlier (Mangelsdorf, 1958) and which is discussed in some detail in Chapter 12. We were able, as we had been with the Arizona specimens, to match some of the most tripsacoid cobs with modern counterparts from derivatives of our experimental maize-teosinte hybrids. In summary, the prehistoric cobs of Cebollita Cave revealed an evolutionary sequence in which a "pure" maize related to the Mexican race Chapalote was replaced by a highly tripsacoid maize in which the expression of the teosinte introgression gradually declined in successive levels of the refuse while the average cob diameter, kernel-row number, and cob length gradually increased and the variation as measured by standard deviation increased by a marked degree.

Spread of an Eight-Rowed Maize from Mexico and the Southwest

An especially fruitful case of collaboration between a botanist and an anthropologist occurred when Walton Galinat joined forces with James Gunnerson, then spending his final year as a graduate student in the Peabody Museum of Archaeology and Anthropology. Gunnerson and Galinat first analyzed the prehistoric maize which Gunnerson had uncovered in the Snake Rock site in Sevier County, Utah, then analyzed a number of earlier collections stored in the Peabody Museum, and finally they reviewed published papers to determine the percentages of eight-rowed maize in sites throughout the United States. From these data they traced the spread of an eight-rowed corn, which they called Maíz de Ocho, northward and eastward from the southwestern United States. Their joint contribution showed how an eight-rowed corn, perhaps originally from South America, spread from Mexico into the southwestern United States and from there, following mainly the Transition Life Zone, moved into the region of the Upper Mississippi and eventually into the northeastern United States.

These authors also prepared a useful map in which the prehistoric distribution of Maíz de Ocho is compared with the historic distribution of the eight-rowed type as determined by a survey made by Plumb (1898). The two distributions agreed closely in showing Maíz de Ocho to occur mainly in the Transition Life Zone and only sporadically in the warmer Upper Austral. Prehistorically it also occurred in several sites in the southern states but is no longer grown there. Their map shows a considerable gap in its occurrence between the region represented by Arizona, New Mexico, and Utah and the region of the Upper Mississippi. This gap has subsequently been partly filled by the prehistoric maize from a site in southeastern Colorado excavated by Robert G. Campbell and analyzed by Galinat (Galinat and Campbell, 1967). Thus their conclusion that Maíz de Ocho spread from the southwest to other parts of the United States has received substantial new support.

The "eastern" influence which Carter (1945) postulated and which Brown and Anderson (1947) apparently accepted as a factor in the development of the corn of the southwest proves actually to have come from northwestern Mexico, and the resemblances between the southwestern corn and those of the northeast are the result of a spread northward and eastward from the southwest. Galinat and Gunnerson suggested that it was the productiveness of this type of maize that may have been responsible for the expansion of the Pueblo II culture into areas not previously occupied by agriculturalists and that this may have been a more important factor in the spread than a period of more favorable rainfall. These authors also show, in a genealogical chart illustrated with photographs of actual ears, how the hybridization of Chapalote, the universal early corn of western Mexico and southwestern United States, produced the prehistoric dent corn which they called Freemont dent and which has modern counterparts in the Maiz Blando de Sonora (Wellhausen et al., 1952) and the corn of the Pima-Papago Indians. It is this corn that Anderson (1959) considered to be related to the Zapalote Chico of southern Mexico, and although there is a superficial resemblance between the two types, Galinat and Gunnerson concluded that it is unlikely that Zapalote Chico is involved in the origin of Freemont dent. Another product of this hybridization, Cristalino de Chihuahua, of northwestern Mexico has a close counterpart in modern Pueblo corn.

The conclusions of Galinat and Gunnerson on the spread of Maíz de Ocho represent a major contribution in elucidating the origin of the Corn Belt dent corn of the United States. It is reasonably certain that this, probably the world's most productive corn, is the product of hybridization between a gourd-seed type of dent corn from the southern states and an eight-rowed northern flint type from New England. It now appears that both of these putative parents came originally from Mexico. The eight-rowed parent may have come from still farther south, either from Guatemala or from South America. The genealogy of the Corn Belt dent is discussed in greater detail in Chapter 11.

Summary

Prehistoric corn from twelve once-inhabited dry caves in Mexico and the United States Southwest comprise some 20,000 specimens representing various parts of the plant: cobs, kernels, husks, leaf sheaths, and tassels. Slightly more than half of the specimens are cobs. The earliest of these, dated at about 2400–2000 B.C., are all identified as precursors of the modern but primitive Mexican popcorn race Chapalote or the closely related race Nal-Tel. These cobs have the principal botanical characteristics of modern corn; they show no evidence of having evolved from teosinte. Specimens from later levels, however, show strong evidence of the introgression of teosinte, some of them closely resembling certain segregates from maize-teosinte hybrids. Seeds of teosinte and specimens identified as maize-teosinte hybrids occurred in a cave in Tamaulipas, Mexico, but not in the earliest levels. A larger-eared, eight-rowed corn first made its appearance in the region about A.D. 750 ± 250 and spread northward and eastward, becoming the predominating prehistoric corn in the region now comprising the northern United States and southern Canada.

15 Prehistoric Wild Corn and Fossil Pollen

I dare not hope that maize will be found
wild, although its habitation before it was
cultivated was probably so small that botanists have perhaps not yet come across it.
The species is so distinct from all others, and
so striking, that natives or unscientific
colonists would have noticed and spoken of
it. The certainty as to its origin will probably come rather from archaeological discoveries. A. de Candolle, 1886

Remains of Prehistoric Wild Corn

Even more exciting than the archaeological
maize described in the previous chapter are
the still older prehistoric remains of all parts
of corn plants that were turned up in the
refuse of once-inhabited caves in the Tehuacán Valley of Mexico. In analyzing this highly
significant collection our studies of archaeological maize reached a climax, for we concluded that the earliest cobs from this site,
dated at about 5000 B.C., were those of wild
corn. Like the prehistoric corn of La Perra
and Romero's Caves in northeastern Mexico
described in the preceding chapter, the cobs
of wild corn came to us through Richard S.
MacNeish, who has accomplished more in
his search for the beginnings of agriculture
in this hemisphere than any other archaeologist.

On the basis of his excavations in Tamaulipas and the discovery of the fossil corn pollen
in the Valley of Mexico, MacNeish concluded
that the evidence for the earliest domestication of maize and the beginnings of agriculture in America must be sought farther
south. A reconnaissance that he made in
Honduras and Guatemala in 1958 yielded no
results of promise. Excavations in 1959 of
Santa Marta Cave in Chiapas in southern
Mexico uncovered corn and other vegetal remains including pollen but none older than
that which had already been found farther
north. Turning northward again, MacNeish
made a reconnaissance of sites in Oaxaca and
Puebla which led him to the conclusion that
the Tehuacán Valley of southern Puebla and
northern Oaxaca might, because of its dry

climate and ever-flowing springs, offer the
most promising area so far discovered for
seeking prehistoric wild corn and the beginnings of agriculture. A preliminary sounding
in 1960 in one of the numerous caves in the
cliffs surrounding the valley uncovered tiny
cobs which MacNeish brought to me in Cambridge and which after a microscopic examination, I pronounced as probably those of wild
corn. We celebrated the momentous discovery
in appropriate fashion that evening. Full-
scale excavations which MacNeish conducted
the following season confirmed this first
conclusion.

It was not until the fall of 1962, however,
that Galinat and I with our wives went to
Mexico to make a study of the prehistoric
corn which MacNeish and his field staff of
some 80 diggers had turned up. By this time
all of the corn had been screened out of the
refuse of five caves and the specimens were
packed in shoe boxes—favorite containers
among archaeologists—each box carefully
labeled with the designation of the cave from
which it had come, the location within the
cave, and the zone within the refuse. By this
time numerous radiocarbon determinations
of the prehistoric materials had been made
under the direction of Frederick Johnson,
and as we studied the specimens of corn we
knew their approximate age. The reader can
scarcely imagine what a thrill it was for me to
see spread out on the floor and tables an evolutionary sequence of prehistoric corn beginning at 5200 B.C. and ending about A.D. 1536,
a period of almost 7000 years. Here was corn
telling us its own history. Here, for me, was
the culmination of more than forty years of

study, experimentation, and speculation on
the origin and evolution of corn. Here was the
material that would prove or disprove our
own previous theories. Galinat, of course, was
equally thrilled, as was also my wife, who,
although fastidiously cleanly, was so fasci-
nated by the story of corn's evolution spread
out before our eyes that she enthusiastically
and effectively participated in the task of
measuring thousands of cobs, many of which
were well coated or impregnated with prehistoric offal, including a substantial component of excrement.

Before describing this epoch-making collection of prehistoric remains of corn, a
survey of the physical features of the Valley
of Tehuacán may be in order. A detailed description has been published elsewhere by
MacNeish; here I will mention only the most
salient features related to corn. At first glance
this valley, with its predominantly xerophytic,
drought-resisting vegetation consisting of
cacti and thorny leguminous shrubs, might
not appear to be a suitable habitat for wild
corn. Maize is not noted for its drought resistance; in order to thrive it requires a steady
and adequate supply of water. It was for this
reason among others that Reeves and I in
1939 had postulated that wild maize, if still
extant, should be found in the humid regions
of the tropics and subtropics. Actually I had
visited Tehuacán Valley a number of times
beginning in 1941, and at no time had it occurred to me that wild corn might once have
grown there. Now, however, the abundance
of prehistoric remains from the caves demanded a closer examination of the valley's
environmental features, and this suggested

that the habitat furnished by this arid valley may, in fact, have been almost ideal for wild corn. The average annual rainfall in the valley is low—approximately 500 mm. per year at the valley's center—but more important than the total rainfall is its seasonal distribution. About 90 percent of the annual precipitation occurs during the growing season from April to October. It reaches its peak in midsummer, which is the most critical period for corn, when it would normally be silking, shedding pollen, and nourishing the rapidly developing young kernels.

The remaining months are quite dry—in midwinter the valley is virtually a desert—and comprise a period during which the seeds of wild maize and other annual plants could have lain dormant, ready to sprout with the beginning of the summer rains, and never in danger of germinating prematurely only to succumb to the vicissitudes of winter. Seeds of many wild plants have devices which provide against this danger, mechanisms of various kinds for delaying germination, but corn— modern corn at any rate—lacks such a mechanism, perhaps because it evolved in an environment where none was needed such as the winter desert-like conditions of Tehuacán Valley. The perennial vegetation of this valley, which year after year must survive the dry winter months, is necessarily xerophytic, but the annual vegetation—and we now know that wild corn was an annual—need not be especially drought resistant. Nor need the wild corn have been subjected to severe competition with the drought-resistant perennial plants. The sites which are most suitable for corn, the alluvial terraces and fans of the canyons and barrancas, seem not to be well adapted to the growth of cacti and shrubs. Indeed the deeper the alluvial soil the less likely are cacti and shrubs to be found growing on it. Instead there are grasses and other annual plants among which wild corn might have been quite at home.

Before considering the corn itself we should say a word about the caves in which the remains of maize were uncovered. Of some twelve caves that MacNeish excavated, five, Coxcatlan, Purron, San Marcos, Tecorral, and El Riego, yielded maize in archaeological levels. These caves were situated in three or four different environments which might have had considerable bearing upon the possibility of wild corn having once grown nearby and

the extent to which agriculture was practiced in the immediate vicinity.

Coxcatlan and Purron Caves

A long, narrow rock shelter, Coxcatlan Cave is situated in the southeastern part of the valley in one of the canyons flanking the Sierra Madre mountain range. First found in 1960, it was one of the richest in vegetal remains and was where the first cobs of wild corn were uncovered. The shelter faces north and looks out on a broad alluvial plain covered with grass, mesquite, other leguminous shrubs, and cacti. Augmenting the meager annual rainfall is a certain amount of water drainage from the nearby mountain slopes which would have benefited wild corn in some favored spots and would later have been an important factor in the development of agriculture. Excavations of the cave revealed 28 superimposed floors or occupational levels covering two long, unbroken periods from 10,000 to 2300 B.C. and from 900 B.C. to A.D. 1536. Fourteen of the upper floors, those from 5000 to 2300 B.C. and from 900 B.C. to A.D. 1536, contained well-preserved corn cobs. There is a hiatus in the archaeological sequence in the remains of corn from about 1500 to 900 B.C.

A few miles south of Coxcatlan in the same set of canyons is Purron Cave. This is a somewhat smaller rock shelter, but it contains a long, continuous occupation (25 floors) from about 700 B.C. to A.D. 500. Archaeologically it is much poorer than Coxcatlan Cave, with only the top 12 floors (from 2300 B.C. to A.D. 500) containing preserved remains of food plants.

El Riego Cave

Situated in the north end of the valley only a mile north of the modern town of Tehuacán is a deep recess facing south in the travertine face of a cliff. Under these cliffs and flowing from them are the famous Tehuacán mineral springs, the source of the well-known Agua Tehuacán, the bottled table water available at hotels and restaurants almost throughout Mexico. Because the soils in front of the cave are fertile and well watered the vegetation around the cave is oasis-like. Prehistorically it may have been too lush for wild corn to compete with, but for agriculture it is an excellent area. El Riego Cave contained an

FIG. 15.1 The archaeologist Richard S. Mac-Neish (seated), and the botanist Paul C. Mangelsdorf in San Marcos Cave, from which the prehistoric wild corn illustrated in this chapter was obtained. The cave is not much more than an open shelter providing some protection from the elements. (Photograph by Walton C. Galinat.)

abundance of preserved specimens. Its five archaeological zones with preserved remains, however, do not extend far back in time, covering a period between A.D. 300 and A.D. 1500.

San Marcos and Tecorral Caves

The last two caves, San Marcos and Tecorral, are situated close together, facing east on the steep west wall of a narrow canyon leading into the valley. The shelters look out over alluvial terraces covered by grass and small thorny trees. Plants collected from this canyon include a number of endemics, species not found elsewhere in the valley. The surrounding travertine-covered canyon walls have a vegetation like that of the Sonora desert. The area receives water in the rainy season, and much of it floods the lower terraces, which might have furnished an almost ideal habitat for wild corn. Agriculture would have been possible in the rainly season even without irrigation. Once the practice of irrigation began, for which there is good evidence in the stone-lined ditches that once controlled the movement of water from the upper to the lower reaches of the canyon, agriculture should have been highly successful.

As I stood with MacNeish in the San

Marcos Cave looking down on the alluvial terraces I could see in my mind's eye wild corn plants growing here and there, never in thick stands but in favored spots. I could visualize the prehistoric cave-dweller, primarily a hunter of small game and a food-gatherer, bringing back small ears of wild corn to his shelter, picking off the small, flinty kernels, much too hard for him to chew even with his excellent teeth, and exposing them one by one to the glowing coals of his fire. After absorbing heat for several minutes, the kernels would have exploded, transforming the stony little grains into tender, tasty morsels having about the same nutritive value as whole-grain bread.

Wild corn, although probably by no means the most abundant source of food, may well have been one of the most prized. Never one of the most common plants in the valley, it became even less so as the result of man's inroads upon its natural reproduction. As it approached extinction man, or more likely his women, observed that corn plants growing spontaneously near their habitation from accidentally dropped seeds were far superior to the corn plants growing in the wild. These well-drained sites, free of competition with other vegetation, and well fertilized by excrement and other offal, would be certain to improve the vigor of growth of the stalks. The invention of maize culture may have had two mothers, one necessity, and the other keen observations by the Indians on the behavior of certain plants growing in a man-made environment.

Cultural Zones and Dating

So much for my flight of fancy into the remote past and back to the ancient corn itself. With respect to the prehistoric maize, MacNeish recognized eight cultural zones which were dated by radiocarbon determinations as follows:

Venta Salada	700 A.D.–1536 A.D.
Palo Blanco	150 B.C.– 700 A.D.
Santa Marta	900 B.C.– 150 B.C.
Ajalpan	1500 B.C.– 900 B.C.
Purron	2300 B.C.–1500 B.C.
Abejas	3400 B.C.–2300 B.C.
Coxcatlan	5000 B.C.–3400 B.C.
El Riego	7000 B.C.–5000 B.C.
Ajuereado	10,000 B.C.–7000 B.C.

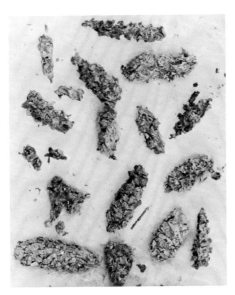

FIG. 15.2 Cobs from the Coxcatlan phase, San Marcos Cave, dated 5000–3400 B.C. Because the cobs are characterized by uniformity in size of the intact specimens, relatively long, soft glumes, and fragile rachises and because there is no evidence from other species that agriculture had been invented and domestication had begun, we concluded that these were the cobs of prehistoric wild corn. They differ from teosinte in being predominantly many ranked, in having paired spikelets, and in lacking induration of the rachis and glumes. (Actual size.)

The Remains of Maize

From several standpoints the prehistoric maize from these caves is the most interesting and significant so far discovered. (1) It includes the oldest well-preserved cobs available for botanical analysis. (2) The oldest cobs are almost certainly those of wild maize. (3) This maize appears to be the progenitor of two of the previously recognized ancient indigenous races of Mexico, Chapalote, and Nal-Tel, of which prehistoric prototypes had previously been found in Swallow and La Perra caves respectively (Mangelsdorf et al., 1956; Mangelsdorf and Lister, 1956). (4) Specimens of all parts of the plant were preserved, and these support the evidence from the fossil pollen described later in this chapter in showing that the ancestor of cultivated corn was corn. (5) The collection portrays a well-defined evolutionary sequence covering a period of about 6,500 years.

In all, 24,186 specimens of maize were found in the five caves; 12,860 of these, or more than half, are whole or almost intact

cobs. In addition to the intact cobs there are 3,941 identified and 3,878 unidentified cob fragments. Among the remaining specimens are all parts of the corn plant: 46 roots, 506 pieces of stalks, 442 leaf sheaths, 282 leaves, 245 inner husks, 706 outer husks, 12 prophylls, 127 shanks, 384 tassel fragments, 47 husk systems, 5 midribs, and 797 kernels. There are also numerous quids, representing 83 chewed stalks or leaves and 140 chewed husks.

When beginning a study of a collection of prehistoric specimens of maize, botanists are usually confronted with the problem of where best to begin. Because we were almost overwhelmed by the sheer volume of the prehistoric remains to be analyzed, we decided to concentrate first on the relatively limited collection from San Marcos Cave, totalling 1,248 specimens. Since the number of specimens was not large, it was possible to spread out the entire collection and view it as a whole. We hoped, too, that the study of this relatively small collection would provide a pattern for studying the much larger collection of over 15,000 specimens from Coxcatlan Cave. This proved to be the case, and it also turned out that the collection from San Marcos Cave tells virtually the whole story, the collections from the other caves adding little more than corroboration. Consequently, I propose to describe the specimens from San Marcos Cave in some detail and those from the remaining caves much more briefly. The descriptions that follow are essentially those that were published in *The Prehistory of the Tehuacan Valley*, volume I, "Environment and Subsistence," (Mangelsdorf, MacNeish, and Galinat, 1967). They are included here with minor changes by permission of the University of Texas Press.

The Nature of Wild Corn

From MacNeish's Zone F, representing the early part of the Coxcatlan phase and dating to nearly 5000 B.C., there is one small cob. Since this specimen is essentially like the cobs from Zone E, it probably represents only a slightly earlier part of the Coxcatlan phase than the later Zone E. The following description of the cobs from Zone E applies also to this single specimen.

The maize from Zone E consists of 26 cobs or fragments of cobs. A number of these are illustrated in Figure 15.2. These cobs are

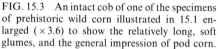

FIG. 15.3 An intact cob of one of the specimens of prehistoric wild corn illustrated in 15.1 enlarged (×3.6) to show the relatively long, soft glumes, and the general impression of pod corn.

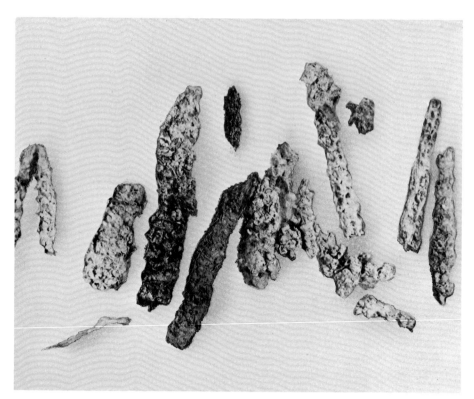

FIG. 15.4 Cobs from the Abejas phase of San Marcos Cave dated 3400–2300 B.C. These are larger and more variable than the earlier cobs and are considered to be representative of early stages of cultivation. They are similar to the wild corn in having long glumes, fragile rachises, and soft tissues. (Actual size.)

remarkably uniform not only in size but also in botanical characteristics. Of eight cobs that are apparently intact with respect to length, the shortest is 19 mm. and the longest 25 mm. The number of kernel rows is usually eight, but one cob is distichous, with four rows, and another has eight rows in the basal and four in the apical region. The number of functional spikelets per row varies from six to nine, and the total number of spikelets (estimated by multiplying row number by number of spikelets per row) varies from 36 to 72. The average number is 55.

The glumes of the spikelets are rather long in relation to other structures and are soft and herbaceous rather than horny and indurated. In some cobs the glumes are folded back, probably as the result of the removal of the kernels. When examined under a dissecting scope these cobs produce an impression of a pod corn from which the kernels have been removed. A photograph of one of these cobs taken at low magnification is reproduced in Figure 15.3.

The spikelets, which are uniformly paired, as they are in modern corn, are attached to cupules which are as long as they are broad or longer, fleshy rather than indurated, and almost glabrous, bearing only sparse short hairs. The rachis consists of individual cupules joined together at their sides and ends. The cross-section of the rachis of an eight-rowed ear thus has the appearance of a square with flaring corners. The rims of the cupules form the structures which in modern maize are sometimes referred to as "rachis flaps." In the earliest San Marcos maize these are quite prominent. The adjoining cupules are not tightly bound to one another, so that the rachis is by no means as rigid as the rachises of modern maize and disarticulates rather easily thus providing a means of dispersal.

The majority of the ears represented by these cobs were originally bisexual, bearing pistillate spikelets below and staminate spikelets above. Of the fifteen apparently intact cobs, ten had recognizable stumps at their tips where a staminate spike had presumably been broken off. Others may also have had staminate spikes, since it is not always possible to distinguish between the stump of a staminate spike and the broken end of a cob which has lost one or more of its terminal pistillate spikelets.

In bearing both male and female spikelets in the same inflorescence, the prehistoric wild corn resembles our genetically reconstructed ancestral form (Mangelsdorf, 1958a) as well as certain races of modern corn regarded as primitive: Nal-Tel and Chapalote of Mexico (Wellhausen et al., 1952), Pollo of Colombia (Roberts et al., 1957), Confite Morocho of Peru (Grobman et al., 1962), and corn's wild relative, *Tripsacum*, which regularly bears pistillate spikelets below and staminate spikelets above on its lateral spikes.

The earliest cobs—27 from San Marcos Cave and 44 from Coxcatlan Cave, some of which date to nearly 5000 B.C.—we regard as being those of wild maize for six reasons. (1) They are remarkably uniform in size and other characteristics and in this respect resemble most wild species. (2) The cobs have fragile rachises as do many wild grasses; these provide a means of dispersal which modern corn lacks. (3) The glumes are relatively long in relation to other structures and must have partially enclosed the kernels as they do in other wild grasses. (4) There are sites in the valley, such as the alluvial terraces below San Marcos Cave, that are well adapted to the growth of annual grasses, probably including wild corn, and which the competing cacti and leguminous shrubs appear to shun. (5) There is no firm evidence from other plant species that agriculture had yet become well established in this valley, at least in the earlier part of the Coxcatlan phase. (6) The predominating maize from the following phase, Abejas, in which agriculture definitely was well established, is larger and more variable than the earliest corn. This combination of circumstances leads to the conclusion—one that to me seemed almost inescapable—that the earliest prehistoric corn from the Tehuacán caves is wild corn. For the purposes of this discussion I have assumed here that it is.

The regular pairing of the pistillate spikelets, and the relatively soft tissues of the rachis and the glumes provide convincing botanical evidence that the wild ancestor of cultivated corn was corn and not one of its relatives, teosinte or *Tripsacum*, both of which have solitary pistillate spikelets and highly indurated tissues. Together with the much earlier fossil pollen from the Valley of Mexico, these prehistoric cobs from the Tehuacán Valley provide virtually conclusive evidence that corn is an American plant and not of Asian origin, as some writers have suggested (see Chapter 17).

Early Cultivated Corn

Cobs

From Zone D, representing the earlier part of the Abejas phase, which extended from about 3400 to 2300 B.C., 102 cobs or fragments of cobs were recovered. All but one of these, described below, were similar in their general characteristics to the 27 specimens from Zones E and F, but the majority were larger. The characteristics of ten intact cobs from this zone are set forth in the accompanying tabulation. A number of the cobs are illustrated in Figure 15.4.

Length, mm.	No. of rows	Spikelets per row	Total spikelets
28	8	8	64
30	8	8	64
32	8–10	10	90
40	8	11	88
44	8	16	120
44	8	17	136
45	8	12	106
51	8	17	136
55	8	18	144
61	10	19	190
Average 43.0	8.3	13.6	113.8

Not originally assigned to Zone D but obviously belonging to it is a collection of 43 cobs from Zone C[1] which are quite similar to those described above. One of these cobs, which is intact with respect to length, is 49 mm. long, has 8 and 10 rows, 15 spikelets per row, and a total of about 135 spikelets. In the structure and texture of the glumes and cupules, these cobs from Zone D, altogether 145 in number, are quite similar to those in the earlier Zones F and E. Their characteristics are illustrated in Figures 15.5 and 15.6. Their larger size indicates, however, that they were grown in a better environment. Since these larger cobs came from levels in which there occurred remains of two species of squashes (*Cucurbita moschata* and *C. mixta*), tepary and common beans (*Phaseolus acutifolius* and *P. vulgaris*), bottle gourds (*Lagenaria siceraria*), chili peppers, avocados, and amaranths, we assumed that they represent the product of an improved environment resulting from the practice of agriculture—that maize had at this stage been domesticated and become a cultigen. Consequently we designated this maize as "early cultivated."

If the maize from Zone D is cultivated maize, it is noteworthy that domestication had effected little change except in its size. In its botanical characteristics the early cultivated maize is virtually identical with the wild maize from which it stemmed. And if the longer cobs that were associated with squashes, beans, and other food plants do indeed represent cultivated maize then it seems all the more likely that the smaller cobs from the lower zones in which there is no evidence of agriculture represent maize grown in the wild.

The one exceptional cob, a fragment mentioned above, must now be accounted for. It is

FIG. 15.5 (*Left*), cob of early cultivated corn, Abejas phase, San Marcos Cave. The slender rachis and the relatively long glumes are characteristic of weak forms of pod corn. (Actual size.)

FIG. 15.6 (*Right*), cobs of early cultivated corn, Abejas phase, San Marcos Cave. *Left*, a rachis of an eight-rowed cob with floral bracts removed showing the long, relatively shallow cupules, which are quite different from those of teosinte. *Right*, intact cob of a similar ear showing the glumes set at right angles to the rachis; in teosinte the glumes are parallel to the rachis. (Actual size.)

FIG. 15.7 Left to right, fragment of a stem, a husk system, a cob, and staminate spike from early zones of San Marcos Cave. (Actual size.)

larger in diameter than the cobs of early culti-vated maize and has strongly indurated tissues in its rachis and glumes. It resembles a type that is quite common in Zone C and will be described below. This cob, however, seems to represent the beginning of a new race, "early tripsacoid," as similar cobs were found in contemporaneous levels of Coxcatlan Cave.

Husks

Associated with this collection was an almost intact husk system which yielded a surprising amount of information. Its shank or peduncle was broken immediately below the lower leaf sheath and has a diameter of 4 mm. The shank has two nodes 4 mm. apart. A basal fragment of an ear bearing the floral bracts of two pistil-late spikelets is attached immediately above the upper node; this shows that the upper leaf sheath of this specimen must have been the uppermost leaf sheath of the original husk system. That the lower leaf sheath of the speci-men was the lowermost husk and therefore the outer husk of the original system is less certain, although it does have the aspect of an outer rather than an inner husk, as the des-cription below suggests. Both husks appear to be intact with respect to length, although they are somewhat frayed at the tips. The lower one is 90 mm. long and the upper 50 mm. Both have conspicuous venation, the lower one having more prominent veins than the upper. The ridges of the lower husks bear short hairs, and the surface and margins of the upper husk have scattered long hairs. Neither husk is ter-minated by a ligule and leaf blade, as are the husks of many modern varieties of corn. The husks are illustrated in Figure 15.7, which also shows a cob and a section of stalk, which might have come from the same kind of plant.

Because these two husks differ in venation and hairiness, we have assumed that one is the outer and the other the inner of a two-husk system. Ears with few husks are usually borne in the upper part of the stalk, sometimes im-mediately below the tassel. The husks protect the young ear before pollination and in the early stages of kernel development but flare open at maturity, allowing the ear to disperse its seeds. Ears with few husks may also occur in lateral inflorescences in which ears are borne at secondary positions at the nodes of the peduncle. We are unable to determine whether this specimen represents the leaf

FIG. 15.8 Staminate spikelets of a tassel frag-ment from the Coxcatlan phase, San Marcos Cave. Floating on water the glumes are seen to be thin and translucent and quite different from the much thicker glumes of *Tripsacum* or teosinte. (Actual size.)

FIG. 15.9 Three types of cobs from the Abejas phase, San Marcos Cave. *Left*, early cultivated corn with long, soft glumes. *Center*, early tripsa-coid with stiff indurated glumes. *Right*, a possible hybrid of the two preceding types (Actual size.)

sheaths of an ear borne high on the stalk or whether it was part of a branched lateral inflorescence.

Since the average length of the longer ears from this zone is about 50 mm., it appears that the inner sheath would have barely covered the pistillate portion of an average ear. If there was a terminal staminate spike, as the stumps at the apices of some cobs indicate that there may have been, the young spike would have been enclosed or partially enclosed by the outer husk, which probably opened before anthesis, allowing the staminate spikelets to shed pollen. We have observed many ears of this type in our extensive cultures of pod-popcorn representing our genetically recon-structed ancestral form.

A Tassel Fragment

Found also in Zone D was a small fragment bearing staminate spikelets. The spikelets were arranged in two ranks, indicating that the spike was a lateral branch from a branched tassel. The spikelets occurred in pairs, one sessile and the other pedicellate; the glumes, which average 7.8 mm. in length, are thin, lacking in distinct keels, weakly veined, and completely glabrous. The branch is illustrated in Figure 15.7. After this photograph was taken, we attempted to clean the specimen by dipping it in hot water. This caused most of the spikelets to separate from the rachis. Floating on the surface of the water, they pro-vided an interesting picture (Figure 15.8), showing that the glumes are thin and trans-lucent, quite different from the much thicker glumes of *Tripsacum* or teosinte. Thus in the staminate spikelets of the tassels as well as in the pistillate spikelets of the cob, the early corn from Tehuacán has exactly the same basic botanical characteristics as modern maize and furnishes additional evidence that corn's ancestor was corn.

Hybridization Plays a Role

Cobs

The cobs of Zone C, representing the Ajalpan phase, which dates from 1500 to 900 B.C., in-cluded a new type similar to the single speci-men described immediately above. Because of the indurated tissue of its rachises and glumes, we have designated this as "early tripsacoid." As mentioned in Chapter 11 the term "trip-sacoid" is one proposed by Anderson and

Erickson (1941) to describe any combination of characteristics which might have been introduced into corn by hybridizing with its relatives teosinte or *Tripsacum*. In both of these species the tissues of the rachis and the lower glumes are highly indurated and the lower glumes are thickened and curved. We regard archaeological cobs showing these characteristics as the product of the hybridization of maize with one of its two relatives.

How this highly tripsacoid maize came into existence is still an unanswered question. There seems little doubt that it is the product of the hybridization of maize with one of its hard-shelled relatives. Where did this occur? Neither teosinte nor *Tripsacum* is known in the Tehuacán Valley today, nor are there remains of either among the archaeological specimens. This does not, however, rule out completely the possibility that one or both species once grew in the valley. Another possibility is that early cultivated corn of Tehuacán was carried into other regions where it hybridized with teosinte or *Tripsacum* and that some of the hybrid progeny was later returned to the Tehuacán Valley. Both *Tripsacum* and teosinte occur widely today in the adjoining state of Guerrero. A third possibility is that wild maize grew in other parts of Mexico, and once domestication began, hybridized with *Tripsacum* or teosinte to produce the new race "early tripsacoid." The fossil pollen from the Valley of Mexico described later in this chapter is evidence that wild maize once existed there. There may be still other possibilities. It is at least clear that as early as the Abejas phase a new element had been introduced into the maize complex in the Tehuacán Valley.

Whatever the origin of the early tripsacoid corn, it evidently hybridized with both the wild and the early cultivated corn of the Tehuacán Valley to produce hybrids which in their cob characteristics were intermediate between those of the parents. First-generation hybrids in turn crossed back to both parents to produce great variation in both the wild and cultivated populations (Figures 15.10 and 15.11).

One of the most conspicuous products of this presumed hybridization was a type which we have called "wild-type segregates." Cobs of these are about the same size as those of the original wild corn, but they possess some of the characteristics, especially the indurated

FIG. 15.10 Various hybrid combinations considered to be the result of the crossing of early cultivated corn with early tripsacoid corn followed by backcrossing to both parents. (Actual size.)

FIG. 15.11 Cobs from a single cache in San Marcos Cave showing the great variation that followed hybridization of early cultivated corn with early tripsacoid corn. The two smaller cobs are classified as wild-type segregates. (Actual size.)

171   Prehistoric Wild Corn and Fossil Pollen

FIG. 15.12   Specimens from the Palo Blanco phase, Coxcatlan Cave, classified as wild-type segregates. At lower left is a tassel branch with basal pistillate spikelets, one of the few such specimens found. Adjacent to it is a distichous spike with single spikelets; these are characteristics of teosinte. (About seven-tenths actual size.)

FIG. 15.13   Leaf sheaths, Palo Blanco phase, San Marcos Cave, showing absence of pilosity. In this respect the prehistoric corn from Tehuacán is quite different from the Mexican highland popcorn race Palomero Tolequeño. (Actual size.)

rachises and glumes, of the early tripsacoid putative parents. Specimens of these are shown in actual size in Figure 15.12. A substantial proportion of the cobs are two-ranked like the spikes of *Tripsacum* and teosinte. These disarticulate readily, and in this respect some of the wild-type segregates may have been better adapted for survival in the wild than was the original wild corn.

Husks

Eleven pieces of outer husks intact in lengths measured 90, 90, 90, 93, 100, 110, 114, 120, 120, 135, and 140 mm. These specimens are quite thick and conspicuously ridged compared to the inner husks. All are lacking in ligules and leaf blades. Most are glabrous, but several have sparse long hairs on the ridges. Four pieces of inner husks have lengths of 90, 90, 120, and 120 mm. The two shorter pieces have profuse long hairs; the longer pieces are almost glabrous.

Leaf Sheaths

Of 20 specimens of leaf sheaths in this zone, 14 are completely glabrous. All of these are narrow, indicating that they had come from slender stems. Two of these are still attached to slender pieces of stalk. Three of the sheaths have sparse long hairs on the ridges, two have short hairs between the ridges, and one, the

broadest of the lot, has profuse long hairs on the ridges and profuse short hairs between. The glabrous condition of the narrow sheaths is probably a characteristic of the wild type. No modern corn, known to us, has leaf sheaths as completely glabrous as the leaf sheaths of the early Tehuacán corn. Several of the sheaths are depicted in Figure 15.13.

Stalks

In addition to the pieces of stalk with attached roots to be described below, there are 14 other short pieces of stalks from this zone. These vary in diameter from 4 to 15 mm., with an average of 9 mm. Since the position of the pieces on the original plants is not known, there is no way of distinguishing the thicker basal pieces from the more slender apical pieces. However, even the thickest of the pieces, which may represent one of the lower internodes, is still quite slender compared to the basal internodes of modern corn.

Roots

Five specimens of basal stalk fragments with roots attached provide some interesting pieces of information. Four of these were intact (Figure 15.14) and have diameters of 12, 11, 17, and 10 mm. respectively. These dimensions are considerably smaller than those of modern corn grown under agricultural con-

ditions. The primitive races of Colombia, Pollo and Pira, described by Roberts et al., 1957, had minimum and maximum stem diameters slightly above ground level of 21.8 to 24.2 and 21.6 to 24.1 mm. respectively. The mesocotyls on two of our specimens were intact and were quite short, measuring 11 and 7 mm. Since the mesocotyl elongates during germination until the emerging seedling reaches the surface of the soil, this indicates that the kernels were barely covered with soil. One of the roots had virtually no mesocotyl, indicating that the kernel from which it grew germinated on the surface of the soil. Such a plant may have been the self-sown progeny of a wild plant.

One of the specimens still had the main root of the primary or seminal root system. Although broken, this was 95 mm. long and had the stumps of numerous broken secondary roots. The seminal root system, which arises from the radicle of the embryo, is in most modern maize a temporary one, maintaining the seedling only until the permanent root system, which develops adventitiously from the lower nodes of the stalk, begins to function. However, in certain drought-resisting varieties grown by the Indians in the southwestern United States, the primary root system makes an early and rapid penetration into

FIG. 15.14 A root and lower internode from the Palo Blanco phase, San Marcos Cave. The extreme shortness of the mesocotyl, indicated by the arrow, suggests that this plant grew from a kernel germinating on the surface of the soil. The absence of scars at the base of the stalk shows that the plant had no tillers. (Actual size.)

Virtually all grasses, wild or cultivated, have the capability of producing tillers. Most grasses are single stemmed only when they are grown under conditions of great stress and are depauperate. Some of the plants of our genetically reconstructed ancestral form develop tillers profusely. It has generally been assumed that wild corn was a freely tillering plant (Weatherwax, 1954; Mangelsdorf, 1958a). Yet none of the basal stalks from San Marcos Cave or from any of the other caves show the scars indicating that the plants once had tillers.

One more bit of information can be gleaned from these five root specimens. The upper regions of the permanent roots have no secondary roots or scars where secondary roots had once been attached. This shows that they probably were not covered with soil. The practice of hilling—piling soil around the base of the plant—which is common in Mexico today, had apparently not yet been invented. All-in-all, these five specimens of basal-stalk internodes with their roots attached provided so much information that at times I felt that we were spying on the ancient inhabitants of San Marcos Cave.

Corn at the Time of Christ

The Zones C and B, representing the Palo Blanco phase of about 200 B.C. to A.D. 700, contained an abundance of remains of all parts of the maize plants except the kernels, of which there were only two. A brief description of the specimens follows.

Cobs

The introduction of the tripsacoid maize into the Tehuacán Valley and its subsequent hybridization with the wild and early cultivated maize resulted in the production of several new types. The most conspicuous of these is a new race which appears to be the ancestral form of two of the ancient indigenous races of Mexico, Chapalote and Nal-Tel, described by Wellhausen et al. (1952). In its brown pericarp color, Chapalote is one of the most distinctive races of Mexico. As mentioned in Chapter 14 it is found today in northwestern Mexico, principally in the states of Sinaloa and Sonora. Nal-Tel, which is closely related to Chapalote (Wellhausen et al.), differs from it primarily in having an orange-colored pericarp. Nal-Tel also tends to have ears somewhat shorter and with slightly fewer rows than those of Chapalote.

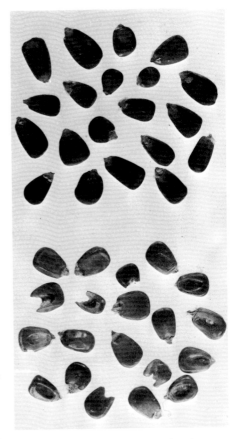

FIG. 15.15 Kernels of the races Chapalote, *top*, and Nal-Tel, *bottom*, from the Venta Salada phase of El Riego Cave. The former are brown, the latter orange-red. (Actual size.)

In other characteristics the two races are quite similar, and it is not generally possible to distinguish them by their cobs. For this reason we hoped to find well-preserved kernels to show us with which race we were dealing. In the last of the boxes containing the hundreds of maize fragments from Zone B there were two small packages each labeled "grano," indicating that each contained one kernel. At this point I said to Galinat, "These kernels should tell us to which race, Chapalote or Nal-Tel, the prehistoric corn is related." Opening the first, we found a kernel with orange pericarp like Nal-Tel; opening the second we found a kernel with brown pericarp like Chapalote. We have since encountered both colors in collections from the other caves in Tehuacán Valley. Kernels from El Riego Cave of the two races are illustrated in Figure 15.15.

Because we were unable to distinguish the cobs of these two races, we combined them

the deeper, moister soils and continues to function throughout the life of the plant (Collins, 1914). Persistence of a primary root system in one of the archaeological specimens indicates that the Tehuacán maize had a similar adaptation to sub-humid conditions. This would have been a useful trait in a wild maize growing in this region. A seedling resulting from germination induced by the first rain does not immediately develop its permanent root system if by the time the seedling emerges the surface soil has become dry. Such a seedling is in a precarious state with respect to its water supply unless its seminal root system can quickly penetrate the deeper moister soils and so maintain the seedling until additional rains moisten the surface of the soil and the permanent root system develops.

Another fact revealed by these five roots is that the plants had single stalks and no tillers; otherwise there would have been visible scars where the tillers had been removed or lost. We found this discovery somewhat surprising.

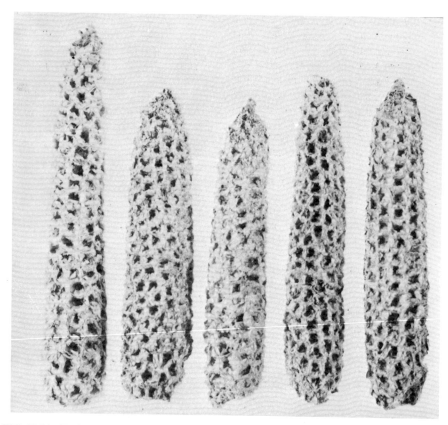

FIG. 15.16 Typical cobs of the Nal-Tel-Chapalote complex from the Palo Blanco phase, San Marcos Cave. Except that they are slightly smaller, the cobs are similar to those of the still-existing Mexican races Nal-Tel and Chapalote. (Actual size.)

and described them in an earlier article (Mangelsdorf et al., 1964) as the Nal-Tel–Chapalote complex, and I shall continue to use this term here. Further study, however, makes it now seem probable that the original wild corn of the Tehuacán Valley was Chapalote rather than Nal-Tel. One evidence of this is that the majority of early cobs are brownish in color. Previously we had considered this to be the product of aging—old cobs often do turn brown. Later it became apparent that even among the well-preserved cobs the lemmas and paleas of some are light and those of others are brown. Of the various alleles at the *A* locus on chromosome 3, which is responsible for plant color, two, A^b and a^p, in combination with the *P* locus for pericarp color on chromosome 1, produce brown color in the pericarp and cob which are dominant to other colors of these tissues (Emerson et al., 1935). Dominant genes are often more common in wild or primitive populations than in highly evolved ones, and in this respect

Chapalote may be regarded as more primitive than Nal-Tel. Also the fact that both *Tripsacum* and teosinte have brown pericarp suggests that this may be a "wild" color. The orange color of Nal-Tel may be a mutation which occurred after maize had hybridized with *Tripsacum* or teosinte; we know that the introgression of *Tripsacum* or teosinte into maize has mutagenic effects (see Chapter 12).

Of 581 cobs in these zones, 133, or 23 percent, could be assigned to the Chapalote–Nal-Tel complex (Figure 15.16). Eight cobs from Zone C (Ajalpan phase) can now also be included in this complex, as can two cobs from Zone IX (Abejas phase) of Coxcatlan Cave. All of these are primitive. The remaining cobs from Zones B and C represent, for the most part, types previously encountered. The introduced tripsacoid had increased in frequency and now comprised 30 percent of the total cobs. Strongly tripsacoid segregates accounted for another 18 percent. These two types together, now called "early tripsacoid,"

were the predominating type, accounting for 48 percent of all identified cobs in Zones B and C[1].

Second in importance numerically, accounting for 38 percent of the cobs in Zone C[1], is the type that we have called "wild-type segregates." These cobs resemble the original wild maize of Zones F, E, and D in size, but the majority have glumes and rachises which are more indurated than those of wild corn. A number are two ranked, and some have single spikelets.

One explanation for these wild-type segregates is that the introduced tripsacoid corn hybridized not only with cultivated corn but also with wild corn still growing in the valley. The first-generation hybrids arising in the wild habitat would then have backcrossed to the wild maize, producing segregating populations containing some individuals which were quite capable of surviving in the wild. Indeed the wild maize may actually have been improved in some respects by the genes for toughness and hardness conferred on it by the tripsacoid maize.

Some of the remaining cobs were classified as intermediate between the introduced tripsacoid and the wild maize. In the frequency polygons in Figure 15.23 all of the tripsacoid types have been combined in the category designated as early tripsacoid. Beside these more primitive tripsacoid cobs there were 108 cobs that showed tripsacoid introgression into the more modern races (Chapalote–Nal-Tel). These are termed "late tripsacoid" and are essentially Chapalote or Nal-Tel with indurated lower glumes.

Data on intact cobs from all zones of San Marcos Cave are presented in the accompanying tabulation. The figures for each zone are averages based on the total number of intact cobs from the zone.

Tc 254 zones	Cobs	Length, mm.	Rows	Spikelets per row	Total spikelets
B	91	55	11.0	14.8	163
C[1]	60	47	10.8	12.4	134
C	2	45	10.0	12.0	120
D	10	43	8.3	13.6	113
E–F	8	22	7.3	7.5	55

Husks

There were many fragments of husks in this zone, but only 45 were intact in length. Thirty-six of these were classed as outer husks and nine as inner. The length of the former varied from 70 to 210 mm., with an average of 145 mm. The majority were fairly close to the average. The husks classified as inner varied from 100 to 160 mm., with an average of 135 mm. All of the husks except two were lacking in flag leaves. One of the exceptions had a rudimentary leaf, the other a well-developed leaf 100 mm. in length. As mentioned below in the description of leaf sheaths, some of the husks, which actually are modified leaf sheaths, were colored. Husks are shown in Figures 15.17 and 15.18.

Prophylls

"In the branch system of grasses, each lateral shoot bears a bikeeled first leaf (prophyll) facing the axillant leaf and addorsed to the main axis" (Arber, 1934). This structure, the prophyll, is especially prominent in modern maize, where it subtends the ear-bearing branch. By the layman it may be regarded as one of the husks. It differs from the husk, however, in being distinctly bikeeled. Three prophylls were found in Zone B, and as was to be expected, all were bikeeled.

Shanks

The stem subtending the ear and from which the husks arise, in botanical terms a peduncle, is commonly called the shank. On six complete husk enclosures, the diameter of the shanks at the base of the lowermost husk averaged 8.3 mm., and the distance between the lowermost and uppermost husk varied from 23 to 74 mm., with an average of 36 mm.

Stalks

The total of 43 pieces of stalks were sufficiently well preserved to be measured with respect to diameter. Thirteen of these had remnants of roots and therefore represent the basal internodes. Their average diameter was 11 mm., which is about half that of the Colombian primitive races, Pollo and Pira, grown under agricultural conditions. The remaining fragments, probably representing a more or less random sample with respect to their original positions on the plant, varied in diameter from 3 to 18 mm., with an average of 7.2 mm.

Leaf Sheaths

Of the 79 leaf sheaths examined, 67 proved to be completely glabrous, and 11 had short

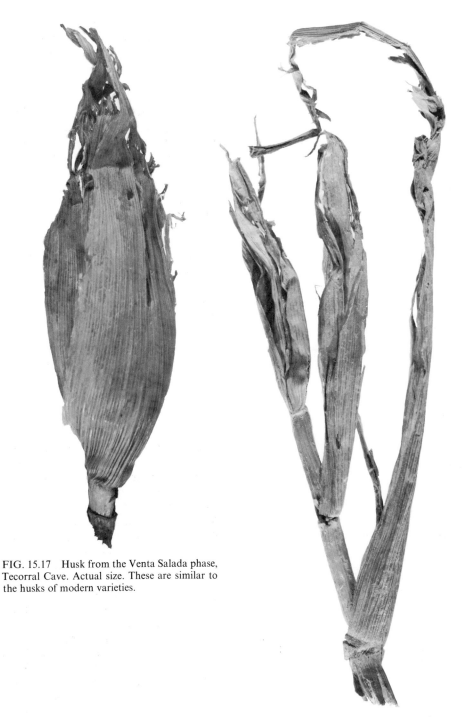

FIG. 15.17 Husk from the Venta Salada phase, Tecorral Cave. Actual size. These are similar to the husks of modern varieties.

FIG. 15.18 Branched lateral inflorescence from the Palo Blanco phase, San Marcos Cave. Actual size. Such plants probably have at least two ears in one husk system.

175   Prehistoric Wild Corn and Fossil Pollen

FIG. 15.19 *Left*, a fragment of a leaf from the Venta Salada phase, Tecorral Cave, showing the typical venation. *Right*, part of a tassel from the Palo Blanco phase, Coxcatlan Cave, showing the central spike. (Actual size.)

hairs in the grooves between the ridges. One had sparse hairs on the ridges. Actually, to the naked eye as well as to the touch, even the 11 short-haired sheaths appeared glabrous. Thus with respect to hairs or bristles, these 79 specimens represent the most uniform lot of sheaths which we have encountered in either archaeological or modern maize. It seems highly improbable that one of the ancient indigenous races of maize of Mexico, Palomero Toluqueño, which has strongly pilose leaf sheaths, could have originated from the populations found in the Tehuacán Valley. This is one of several reasons for supposing that there may have been more than one kind of wild corn.

Practically all of the prehistoric leaf sheaths are brownish or reddish brown. We supposed initially that this represented discoloration from aging. Some specimens, however, had a distinct reddish cast and when soaked in hot water imparted a reddish tinge to the water. It now seems probable that the color of these specimens represents the remains of natural plant color. As mentioned above, some of the husks were also colored.

Leaves

Numerous fragments of maize leaves were found. The majority were too badly broken to yield much information, but four were intact and ten additional ones were intact in width only. The intact leaves had an average length of 47.6 cm., and an average width of 4.1 cm. The leaves intact in width only varied from 2.7 to 5.0 cm., with an average of 3.8 cm. These dimensions are much smaller than those of any modern races of maize in Mexico, even the Ancient Indigenous races, which, compared to some other races, have relatively narrow leaves. Leaves arising from the ear-bearing nodes of Chapalote and Nal-Tel, for example, have average lengths of 80.5 and 65.6 cm. respectively and widths of 7.6 and 8.7 cm. (Wellhausen et al., 1952). The venation index—a measure of the number of veins per unit (cm.) of width—of 17 leaves which were scored for this characteristic varies from 3.8 to 6.7, with an average of 4.5. This is much higher than the venation indices of modern Mexican maize (Wellhausen et al., 1952) and puts prehistoric maize within the range of teosinte with respect to this particular characteristic (Figure 15.19).

Tassel Fragments

Thirty-three tassel branches or fragments of branches were found in this zone. The length of the pedicel on the pedicellate spikelet was measured on one spikelet from each specimen. This varied from 1 to 6 mm., averaging 3.7 mm. The lower glumes on these same spikelets varied in length from 7 to 12 mm., averaging 9.2 mm. There was some variation in the shape of the spikelets, the majority being somewhat flat dorsally, giving the impression of a bikeeled structure. This is characteristic of the glumes of some species of *Tripsacum*. Five of the 33 specimens, however, did not give the impression of having distinct keels, and all of these were shorter than the average and resembled the spikelets in the single fragment found in Zone D and illustrated in Figure 15.8. Four of these five tassel fragments differed also from the majority in lacking hairs on the glumes. In size, shape, and hairiness of the glumes, these four specimens resemble the specimen from Zone D that we considered to be wild or early cultivated corn.

One of the unique characteristics of maize is the central spike of the tassel, which differs from the branches in being polystichous rather than distichous. There have been various theories about the origin of this structure (see Mangelsdorf and Reeves, 1939, for a review of this subject). It is interesting that the polystichous central spike was characteristic of corn in prehistoric times. The several fragments representing central spikes are illustrated in Figure 15.20.

Because our genetically reconstructed ancestral form bore seeds in the branches of the tassel, thus providing a means of dispersal which modern corn lacks, we expected to find some evidence of this characteristic in the tassels of the prehistoric Tehuacán corn. In this we were disappointed. Of the 384 specimens of tassels found in the five caves, only one from Coxcatlan Cave seemed to be a tassel branch bearing pistillate spikelets at its base. Since depauperate corn plants even of modern corn sometimes exhibit this condition, its rare occurrence among the prehistoric specimens is not especially significant. On the contrary, it is surprising that there are not more of these, since both wild and cultivated corn must at times have been exposed to environmental conditions which were con-

FIG. 15.20 Tassel fragments from the Venta Salada phase, San Marcos Cave. ($\frac{4}{5}$ actual size.) All of these have counterparts in modern corn.

FIG. 15.21 Numerous quids found in the vegetal remains from Palo Blanco phase, San Marcos Cave, show that one of the uses of corn was to chew the stalks. The specimen at left shows how the chewing proceeds along the stem. (Actual size.)

ducive to the production of depauperate plants.

Quids

In this zone there were 58 remains of quids, of which 32 were chewed stalks and 26 chewed husks. The latter probably originally contained young ears. We had previously found from studying the quids in La Perra Cave (Mangelsdorf et al., 1956) that young ears enclosed in husks are quite sweet. Also as mentioned in Chapter 14 it is well known that growing corn stalks from which the ears have been removed, or which are barren for other reasons, often accumulate sugars and are

about as sweet as sugar cane—although somewhat less palatable—when chewed. Since a "sweet tooth" is almost universal in the human race, it is not surprising to find that the Tehuacán people made use of the sweetness derived by chewing young ears and stalks of corn. Quids in various stages of maceration are shown in Figure 15.21.

Collections from Other Caves

The maize specimens from the remaining four caves (Coxcatlan, Purron, Tecorral, and El Riego) did little more than to corroborate the conclusions drawn from the collection from San Marcos Cave. But they did so very effectively. Coxcatlan Cave, which was occupied more often and over a longer span than San Marcos, had a large sample, totaling over 15,000 specimens. The majority of the earliest specimens from components of the Coxcatlan phase showed that Coxcatlan Cave's earliest maize has four- and eight-rowed cobs with relatively long glumes and slender rachises. These we assume to be the cobs of wild corn. However, one cob in Zone XIII and three from Zone XI were somewhat larger, and

these are presumably the first products of cultivation. In the Abejas levels of Coxcatlan Cave, cobs of both wild and early cultivated types occur but specimens of the early cultivated far outnumber those of the wild. Also, at this time the first tripsacoid cobs make their appearance. Two tiny cobs of the Chapalote–Nal-Tel complex were uncovered in Zone IX.

The Purron phase was meagerly represented in Zones K and K^1 of Purron Cave by three partly carbonized cobs of early tripsacoid corn. Zone J of Purron Cave supplemented our larger Ajalpan sample from Zone C of San Marcos. Zone J also showed a dominance of early tripsacoid types, with a lesser number of early cultivated types. In the Santa Maria zones of Coxcatlan Cave and Purron Cave there is further evidence of more use of the Chapalote–Nal-Tel complex as well as hybridization of this maize with both wild and cultivated corn to produce new variation on a grand scale.

A New Type, Slender Pop

In the Palo Blanco phase the collections from

FIG. 15.22 Cobs of the type called "Late Tripsacoid" from the Venta Salada phase, Coxcatlan Cave. (Actual size.) The stiff, somewhat curved lower glumes are regarded as indicating previous hybridization with one of corn's relatives, teosinte or *Tripsacum*, probably the former. (About actual size.)

Coxcatlan, Purron, and El Riego caves do, however, reveal a type of maize which, if it occurred in San Marcos Cave, was not clearly recognized. The cobs are more slender than those of Chapalote or Nal-Tel and are more nearly cylindrical. We have designated this type as "slender pop." Remains of occasional kernels indicate that they were quite small, rounded, and orange. This may be the prototype of a Mexican popcorn, Arrocillo Amarillo, one of the four Ancient Indigenous races described by Wellhausen et al. (1952). This race, which is now mixed with many others, occurs in its most nearly pure form in the Mesa Central of Puebla at elevations of 1,600 to 2,000 meters, not far from the Tehuacán Valley and at similar altitudes.

Judged by its cobs alone, the slender pop might be expected to be less productive than Chapalote or Nal-Tel. The fact that it increased in prominence suggests that although the ears are small the stalks may have been prolific, bearing more than one ear. The present-day race Arrocillo Amarillo, to which the slender pop bears some resemblance and to which it may be related, is prolific, usually producing two or three ears per stalk. The maximum yields of corn in the United States are made by prolific types of corn grown under irrigation. Perhaps this was also true in prehistoric Tehuacán.

Cobs of Modern Races

A few cobs which could be assigned to various modern races were found in the upper zones of Coxcatlan and El Riego caves. These included several cobs and kernels resembling the race Conico and one cob each of Zapalote Chico, Tepecintle, and Chalqueño, as well as kernels of an unidentified dent corn. The surprising thing is not that specimens of these modern races were present, but that there were so few of them. These races were thought by Wellhausen et al. (1952) to have been in existence in Mexico centuries ago.

The polygons in Figure 15.23 show when each of the recognized types of corn made its first appearance, reached its highest frequency, and in the case of the wild and early cultivated corn, disappeared. The boundaries between the categories represented in Figure 15.23 are by no means hard and fast. The type called late tripsacoid, for example, might well be included as part of the Chapalote–Nal-Tel complex. It differs from the latter only in the

fact that the cobs are on the average more tripsacoid than those of Chapalote and Nal-Tel. Also some of the cobs are similar to those of the slender pop but somewhat more tripsacoid. In any case it is these three types, Chapalote–Nal-Tel, late tripsacoid, and slender pop, which supported the expanding populations of the Tehuacán Valley from about 900 B.C. to A.D. 1500.

Wild Corn Reconstructed

A well-preserved early cob and an intact husk system consisting of an inner and outer husk from the Abejas phase, together with a piece of staminate spike (actually found in the Ajalpan phase), all from San Marcos Cave, provide the material for the reconstruction of an ear of Tehuacán wild corn. This is illustrated in actual size in Figure 15.24. An ear with only two husks was probably borne either in a high position on the main stalk or in a terminal or near terminal position on a branched lateral inflorescence. In either case the husk would have served primarily to protect the young ear and would have flared open at maturity, permitting the ear to disperse its seeds. The ear, like ears of some modern primitive races, was terminated by a fragile staminate spike bearing spikelets in pairs, one member of each pair pedicellate and the other sessile. The glumes of the staminate spikelets were membranous, glabrous, and lacking in keels.

The wild corn was probably a form of pod corn. Its seeds were partially but not completely enclosed in floral bracts. These are similar in their relative lengths and the extent to which they enclosed the kernels to a genotype which we have produced in our experimental cultures by combining one of the alleles at the tunicate locus, tu^h, on chromosome 4 with a major tunicate modifying gene, Ti, on chromosome 6. The terminal inflorescences, the tassels, were similar to those of modern maize in having a polystichous central spike and distichous branches. Apparently they were completely staminate and in this respect differ from our genetically reconstructed ancestral form, in which many of the tassels bear both staminate and pistillate spikelets. The kernels, which were borne in pairs, were almost isodiametric in their dimensions, were rounded dorsally, and were either brown or orange, probably the former. The specimens from both San Marcos and

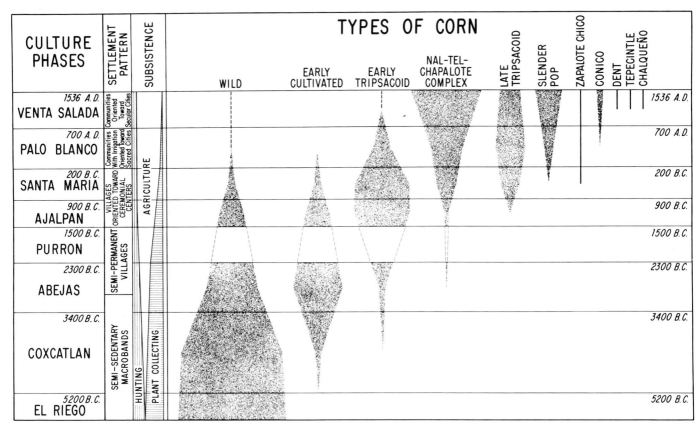

FIG. 15.23   Changes in types of corn in the Tehuacán Valley from 5000 B.C. to A.D. 1500 in terms of percentages of the number of cobs identified. For some reason specimens of corn from the Purron phase are almost completely lacking.

Coxcatlan caves show that the plants apparently had single stalks, although they may have had the capacity to produce tillers under especially favorable conditions, as do virtually all other wild grasses. The leaf sheaths and husks were completely glabrous, and some, if not all, were reddish brown. The ear-bearing lateral branch had as its first leaf a bikeeled prophyll. The root system, like that of modern maize, comprised both seminal roots arising from the radicle of the embryo and permanent roots arising from the first and higher nodes of the stalk. The seminal root system, like that of certain drought-resistant varieties grown by the Indians of the American southwest, may have been more extensive and persistent than it is in most modern varieties.

In all of its essential botanical characteristics, the wild and early cultivated corn of Tehuacán was identical with modern corn but was smaller in all of its parts. Indeed, the wild corn with its tiny ears must initially have been less useful as a food plant than the wild squashes and scarcely more promising than some of the weedy grasses of our gardens and lawns. It was, however, responsive to the improved environment provided by agriculture, and cultivation immediately produced a substantial increase in size. Subsequent hybridization with its relatives *Tripsacum* or teosinte initiated an explosive evolution which resulted in tremendous variability and a manyfold increase in size. Despite the spectacular increase in size and productiveness under domestication, which helped make corn the basic food plant of the pre-Columbian cultures and civilizations of America, there has been no substantial change in 7,000 years in the fundamental botanical characteristics of the corn plant.

Conclusions about Wild Corn Questioned

When we, MacNeish, Galinat, and I, completed our studies of the early prehistoric corn from the Tehuacán caves and the evolutionary series that followed, I felt that the problem of the origin and evolution of corn had been pretty well solved, and on several occasions I gave a lecture entitled "A Botanist's Dream Come True." And the title did seem to me quite appropriate, for the earliest corn appeared to support completely our principal conclusion, Reeves' and mine, published in 1939, that the ancestor of cultivated corn was

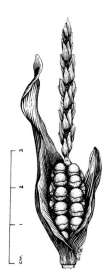

FIG. 15.24 Artist's reconstruction of prehistoric wild corn based on specimens found in the early levels of San Marcos Cave. The small ears, bearing about 50 kernels, were enclosed when young in a husk system that opened up at maturity, permitting dispersal of the seeds. The kernels were round, brown, and partly enclosed by glumes. (Drawing by W. C. Galinat.)

corn and not teosinte. It also appeared to be, as we had postulated, both a pod corn and a popcorn. True, it was not the extreme form of pod corn that geneticists were growing in their cultures, but there was no doubt that the kernels must have been small and were partly enclosed by relatively long, soft glumes. Also we had by that time made considerable progress in our studies of the components of the tunicate locus (Chapter 8), and we recognized the condition in the early Tehuacán cobs as similar to that produced by combining either the tu^h or the tu^d component with the tunicate-inhibiting gene Ti. Our conclusion that this early corn was wild corn was generally accepted by both botanists and archaeologists and even by several of the more ardent diffusionists, whose views on the Old World origin of maize are described in Chapter 17.

It was not until I attended a meeting at Urbana, Illinois, in September 1969 that I learned that the old teosinte theory was being revived. Leading the revival was an eminent geneticist who thirty years earlier had written a laudatory review of our book, concluding it, however, with the assertion that he still held to the teosinte theory because he preferred simple hypotheses. Now retired from active service he planned to dedicate himself to proving that this long-cherished theory was

correct and all others wrong. A revival usually attracts followers, especially when led by an eminent or colorful figure, and this one was no exception. Also revivals are usually marked by a flowering of faith and fervor and in this respect, too, the current one was typical.* When I showed a series of slides illustrating the early corn from Tehuacán and its subsequent evolution there was little reaction or response. The revivalists, having found what they believed to be the truth, were not about to allow themselves to be confused by mere facts.

Eventually, however, persuaded, despite their fervor, that this Tehuacán evidence is critical, they now contend that the early corn dated at 5000 B.C. represents, not wild corn, but an early stage in corn's domestication. This must have been preceded, they postulate, by selection in teosinte practiced by man in a still earlier period arbitrarily estimated at 7,000 to 10,000 years ago. They seem not to be disturbed by the fact that there is no archaeological evidence of the use and domestication of teosinte or even of food-preparing artifacts at this early period. Such evidence, they insist, will be forthcoming when still other sites are excavated.

Naturally I regard such unfounded conclusions as farfetched. The early Tehuacán corn differs from teosinte, not in one or two characters, but in many: its spikelets are paired, not solitary; the majority of the spikes are many ranked, not two ranked. We did find several two-ranked spikes among the earliest specimens, but they showed no other resemblance to teosinte, and although we reported their occurrence we saw no special significance in these, because we know that secondary ears on eight-rowed corns are sometimes two ranked. An especially important difference is that the tissues of the rachises

*The element of faith is apparent in a statement by George Beadle (1972), the eminent geneticist referred to above, who, as a graduate student in the early thirties, had become convinced that teosinte is the ancestor of cultivated corn. Forty years later, despite the considerable body of evidence in conflict with his conviction, he could still assert: "I remain firm in my belief that teosinte is the true and only wild corn."

The fervor of some of the revivalists is well illustrated by the title of a paper that one of them presented at an annual meeting of the Society for Economic Botany: "The Maize Mystique *vs* the Teosinte Truth."

and lower glumes of the early corn are soft and not at all indurated as they are in teosinte. The kernels are not sessile but are borne on rachillae. They are round, not pointed. The axes of the spikelets are at right angles to the axis of the rachis, not parallel with it. The leaf sheaths are glabrous, not pilose like those of the teosinte of the Valley of Mexico, the race postulated as the ancestral form.

To me these numerous differences are convincing evidence that this early corn is not a domesticated teosinte. They involve many different genes: paired spikelets, at least two; many-rowed ears, at least two; soft tissues, at least one major gene combined with numerous modifiers; round kernels, two genes; glabrous leaf sheaths, at least two. The inheritance of the remaining characters has not been precisely determined, but it appears to be multifactorial. Having spent some 15 years in bringing together marker genes on nine chromosomes in maize in developing a multiple-gene tester, I am quite unable to visualize a situation in which man, not yet practicing agriculture and knowing nothing about the principles of inheritance, could have brought all of these characteristics together in one genotype. This would be a challenging task even for a modern plant breeder.

And so the theory that the progenitor of cultivated maize is teosinte, proclaimed as the "most simple" and "direct" has become the most complicated and involved of any now under consideration. Until there is archaeological evidence in its support I shall continue to conclude that the early prehistoric corn from Tehuacán was wild corn and that the ancestor of cultivated maize was maize. What can be more simple and direct than that?

Fossil Pollen

One of the most momentous discoveries concerned with the origin and evolution of maize is the finding of fossil maize pollen at a depth of more than 200 feet below the level of Mexico City. This single discovery pretty well disposes of two questions which students of maize have debated for many years, for it shows beyond a reasonable doubt that (1) the ancestor of cultivated corn was corn and not teosinte or any other of corn's relatives and (2) that corn is an American plant and did not originate in the Old World as some botanists have argued.

This is how this important discovery came about: Paul B. Sears, professor of botany at Yale University and author of the widely read book *Deserts on the March* had for some years made studies of fossil pollen at various sites as one means of charting climatic changes, a subject in which he had long been interested. Among these studies was one including the sediment in a lake bed near Mexico City. When he learned that engineers were taking deep drill cores in two localities, Madero Street and Belles Artes, in Mexico City, in preparation for the construction of the city's first skscraper, he persuaded them to let him have the drill cores for palynological studies.

Pollen, an Indicator of Climatic Changes

The pollen of certain plant species is not difficult to recognize and palynologists assume that changes in pollen frequency, for example, from a preponderance of pines and other conifers to a preponderance of oak and ash, reflects the change from a cool, moist period to a warmer, dry one. The cores which Sears obtained provided the data that he sought for charting climatic changes. They also turned up fossil pollen of a wild maize. Thus the skyscraper subsequently built upon the site from which these drill cores were taken, an impressive 43-story building floating upon the ancient lake-bed sediments of Mexico City, is a monument not only to engineering and architectural progress but also to the solution of several problems concerned with corn's origin and evolution.

In the course of analyzing the pollen contents of the core, Sears' assistant, Mrs. Kathryn Clisby of Oberlin College, observed several unusually large grass pollen grains in sediments close to the 70-meter level in the Belle Artes boring. The size of the grains (75–135 microns in slides prepared by the method known as acetylation) at first appeared to preclude the possibility that they were derived from any native wild grass then extant in the Valley of Mexico. She reported the findings to Professor Sears, who concluded with her that they might be pollen grains of corn's closest relative, teosinte.

It was in the summer of 1951, if I remember correctly, that I received a telephone call from Professor Sears in New Haven in which he stated that Mrs. Clisby and he had made a discovery which seemed to demolish the theory held by Reeves and me that teosinte is a recent hybrid of corn and *Tripsacum* (Chapter 4). To this startling announcement I could only reply that if our theory was about to be exploded I was glad to be the first to be told about the impending denouement.

One of the purposes of the call, it turned out, was to ask me for pollen grains of modern teosinte for comparisons, since neither Sears nor Mrs. Clisby had previously studied such pollen. It was in early September in 1951 that I attended the meeting of the American Institute of Biological Sciences in Minneapolis. There I met Sears and Mrs. Clisby, and in the course of our conversation I learned that, after comparing the fossil pollen grains from Mexico with the teosinte pollen which I had sent them, they were not certain that the pollen grains were those of teosinte. Some of them were too large to be so identified. When I suggested that they might be fossil grains of maize, Sears proposed to send the slides and cores to me for further study. I countered with the proposal to turn them over instead to Harvard's paleobotanist, Elso S. Barghoorn. My reasons for this were twofold: (1) Barghoorn was then already a world-recognized authority on studies of fossil pollen, while my own studies had been largely confined to modern corn using much simpler techniques. (2) Barghoorn could examine the fossil pollens objectively, uninfluenced by any preconceived ideas. I would have tried to be unbiased, but this might have been difficult in view of our theory (Reeves' and mine) on the recent origin of teosinte; and even if I had succeeded there might have been doubt regarding my conclusions on the part of some of my contemporaries, especially those committed to the idea of a lowland South American or an Asiatic origin of maize.

Barghoorn, with the assistance of one of his graduate students, Margaret K. Wolfe, began his studies of the fossil pollen in his usual thorough and systematic way. In order to establish a critical basis for identification of the fossil pollen, he and Miss Wolfe made an extensive study of the size range of the pollen of the varieties of maize and teosinte and the various species of *Tripsacum*. I furnished them with pollen of fourteen varieties of modern maize, three collections of teosinte, and eight of *Tripsacum* for purposes of comparison. To assure uniformity in the data, they prepared all slides of both living and

from 72.0 to 141.7 microns. It was evident that the smaller grains of some of the varieties of cultivated maize fell within the range of teosinte and close to the upper limits of the range of *Tripsacum*. It seemed apparent to Barghoorn that size alone could not always distinguish the pollen of maize from that of its two relatives.

Because of the paucity of structural features and the undistinctive sculpture patterns of the pollen exines of the three species under consideration, as well as of grasses in general, as revealed by conventional microscopy, it became necessary to attempt other means of distinguishing the three pollen types. One possible solution to the problem was to compare the ratios in the size of the pore to the length of the long axis. In order to establish these ratios and to determine their degree of constancy, Barghoorn and Wolfe measured approximately 50 additional pollen grains of each of the lots of maize, teosinte, and *Tripsacum* with respect to these two dimensions. In the case of the fossil grains they measured all of those with intact pores. The measurements were averaged and the ratios for each species computed from the averaged values. The results showed encouraging consistency, the ratio of pore diameter to long axis being an unexpectedly conservative value, and more important to the problem at hand, significantly different among the three species in question.

When compared with respect to their average pore-axis ratio, the three species were now quite distinct. The ranges in the ratios proved to be 1:6.0–1:6.5 for maize; 1:5.1–1:5.4 for teosinte; and 1:3.9–1:4.1 for *Tripsacum*. With respect to the ratios for individual grains there was no overlapping between maize and *Tripsacum*, but the ratios of both overlapped those of teosinte, with no intermediates between the two. However, only 22 of a 113 teosinte pollen grains, 19.5 percent of the total, overlapped those of *Tripsacum*, while 434 of 677 or 68.4 percent of the maize pollen grains overlapped those of teosinte. More important, however, is the fact that 243 of maize pollen grains, 31.6 percent of the total, were beyond the range of teosinte.

The observations of Barghoorn and his joint authors, Miss Wolfe and Mrs. Clisby, showed that fossil pollen grains resembling maize, teosinte, and *Tripsacum* in size and in their pore-axis ratio occurred in the upper

FIG. 15.25   Cobs illustrating an evolutionary sequence from about 5000 B.C. to A.D. 1500. From left to right: wild corn, Coxcatlan phase, San Marcos Cave; early cultivated, Abejas phase, San Marcos Cave; Chapalote, Palo Blanco phase, San Marcos Cave; Chapalote, Venta Salada phase, Coxcatlan Cave; Conico, ibid. (Actual size.)

fossil grains by the same techniques—a modification of the method of Erdtman, 1943—and permanent slides were made with glycerin jelly as a mounting medium. These permanent slides were later to add another chapter to the story of the fossil pollen.

Comparing Pollen of Corn and its Relatives

It became quite clear early in the studies that there is a wide range of size in the pollen grains of each of the three species. In *Tripsacum* the average for the long axes of the pollen grain varied from 41.7 microns in *T. maizar*, a Mexican species, to 57.4 microns in *T. australe*, the principal South American species. The extreme range for the genus as a whole was 33.6 to 64.0 microns. In teosinte, of the three forms examined, the average length varied between 79.3 and 86.4 microns and the extremes ranged from 74.0 to 102.0 microns. In maize the range of the averages was from 87.2 to 122.8 and of the extremes levels of both the Madero and the Belles Artes core. The larger grains, 38 in number, from these upper levels (above 6.0 meters) they considered to be the pollen grains of cultivated corn. Three pollen grains of intermediate size and pore-axis ratio from the 3.3–3.6 meter levels might represent pollen of teosinte which today is a common weed in and around the maize fields in the Valley of Mexico. Three grains conforming to *Tripsacum* occurred in the upper levels of the Belles Artes core, and a fourth was found in the 45-meter level. No pollen grains clearly assignable to teosinte occurred below the 3.6 and 3.3 meter levels respectively in either of the two cores. Although this might indicate that teosinte did not become established in the Valley of Mexico until sometime after maize cultivation had begun, the total number of grains involved is small and differences in levels may be the product of sampling.

Corn Pollen Deep below Mexico City

By far the most significant and important discovery was that of large grains closely resembling the pollen of modern maize in the 69- and 70-meter levels in the Belles Artes core. In all 19 large grains were found, of which 14 were sufficiently well preserved to permit pore-axis measurements. Eight of the 19 were well outside the extreme size of grains for teosinte; of the remaining 11, 4 were outside the extreme pore-axis ratio for teosinte, although within the upper limits for teosinte's long axis dimensions.

With this rather impressive body of data at hand some investigators would have regarded the existence of fossil pollen of wild maize in the Valley of Mexico as completely established. Not Barghoorn. In his usual careful and meticulous manner he first considered various alternative possibilities:

1. The fossil grains are those of a wild grass not related to maize or its relatives. Because no grass pollen approaching this size is known except for the cultivated cereals of the Old World, wheat, barley, rye, etc.—and even these have much smaller pollen than maize—this possibility seemed remote.

2. The grains are the result of contamination occurring in the laboratory. Corn pollen is light and can be carried by the wind for considerable distances. This possibility was eliminated for two reasons (a) the physical, chemical, and optical properties of the fossil grains were quite different from those of modern grains; (b) additional grains were isolated from the cores under laboratory conditions in Cambridge where contamination could be very well ruled out.

3. The grains represent contamination which occurred in Mexico during the core drilling either (a) as atmospheric contaminants* or (b) as stratigraphic contaminants during the drilling operation. The first possibility was excluded for the reason given under 2. The second was excluded on the assumption that if large pollen grains from the upper 6-meter level were carried down into the 69- and 70-meter level, some should have also been found at intermediate levels. Only a single pollen grain of *Tripsacum* was found at these levels.

4. The grains are those of either *Tripsacum* or teosinte which have increased in size and pore-axis ratios as a result of preservation under unusual sedimentary conditions. Although impossible to disprove, this would be wholly inconsistent with previous extensive experience by numerous investigators dealing with Pleistocene and even more ancient Tertiary pollen and other microfossils.

5. The pollen grains are those of ancient maize. After ruling out the alternative possibilities considered, Barghoorn et al. concluded this to be the most reasonable interpretation and from the evidence at hand the only plausible one.

A personal communication to Barghoorn from Paul Sears indicated that the fossil pollen is at least as old as the early stages of the Iowan advance of the Wisconsin Ice Sheet. If this is so, it almost certainly antedates the practice of agriculture in North America and probably precedes the advent of man on this continent. More recent evidence has assigned the earliest fossil pollen to the last interglacial period which ended about 80,000 years ago.

*At a conference on prehistoric maize and fossil pollen held at the Botanical Museum in June, 1972, one of the participants suggested in all seriousness that the maize pollen found at the 6 meter and 69–70 meter levels might represent contamination during two different corn-growing seasons. I do not know whether the drilling operation covered two such seasons but I do know that the Belles Artes site where the fossil pollen was found is in the heart of Mexico City, miles away from any corn-growing areas. Both Professor Sears and I regard the possibility of contamination by modern corn pollen during the drilling as remote indeed.

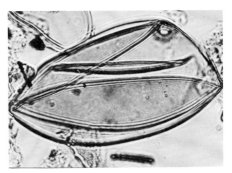

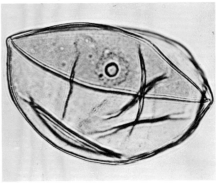

FIG. 15.26 Fossil pollen grain of corn (*upper*) from more than 200 feet below the present site of Mexico City compared with a pollen grain of modern corn (*lower*) at the same magnification. In spite of some 8,000 years difference in their age, these pollen grains are quite similar in their characteristics. The fossil pollen grain shows that (1) corn is an American plant and not of Asiatic origin as some botanists have argued; (2) the ancestor of cultivated corn was corn and not one of its American relatives teosinte or *Tripsacum*. (×450. From Barghoorn et al., 1954.)

The Fossil Pollen Reexamined

I supposed that the fossil pollen grains, some of which seemed unmistakably to be those of maize, had answered once and for all the questions of corn's botanical and geographical origin, but doubts arose again in the minds of some students when Kurz et al. (1960), growing corn in 16 different controlled environments, representing various combinations of temperature, photoperiod, and water supply, found that environmental changes affected both axis length and pore diameter to a marked degree. They concluded from these results that "the use of the axis-pore ratio to differentiate between *Zea*, *Tripsacum*, and *Euchlaena* is inadequate."

To some readers this conclusion seemed to mean that the identification of some of the fossil pollen grains from the deep Belles Artes cores as maize pollen was questionable. These readers overlooked the fact that Kurz et al., although questioning the general adequacy of the axis-pore ratio, did not deny that some of the larger fossil grains studied by Barghoorn et al. were undoubtedly those of maize. Still Barghoorn, one of the most conscientious of scientists, was disturbed at the implication that the conclusions which he and his associates had reached by their meticulous studies might be erroneous, and so with another graduate student he began a reexamination of the pollen with other techniques.

Since the studies of Barghoorn et al. published in 1954, several developments in the studies of pollen had occurred of which one of the most important was the use of phase-contrast microscopy. In specimens that are transparent, or almost so, it is very difficult to see any detail by ordinary transmitted light. In such specimens the altered waves of phase-contrast light produce an image which simulates that of specimens having gradations in density. Using phase-contrast microscopy, Grohne (1957) investigated the pollen of "wild" and cereal-type grasses in Europe and suggested that discrimination between the two was possible on the basis of certain phase changes in the exine pattern. An explanation of these phase changes was offered by Rowley (1960), who found that the grasses of the "wild" type appeared to have three levels of phase retardation while grasses of the cultivated type have only two.

This distinction between cereal and "wild-type" grasses has been shown to be questionable, if not invalid, by more recent studies employing transmission electron microscopy through the use of double replica techniques. In the case of the maize-teosinte-*Tripsacum* complex of pollen types, however, the criteria described below, based on phase-contrast light studies, are not contradicted by electron microscope observations. Further electron microscope study of the pollen of these three genera as well as other cereal grasses is now in progress by Umesh Banerjee, a graduate student working with E. S. Barghoorn.

Henry Irwin, a graduate student working under Barghoorn's direction, undertook an extensive study under phase microscopy of the characteristics of the pollen of maize and its relatives. Galinat and I supplied him with pollen of 18 varieties of maize, a diploid and tetraploid form of *Tripsacum dactyloides*, 11

varieties of teosinte, a maize-*Tripsacum* hybrid, and inbred strains of maize into which various chromosomes of teosinte had been introduced. Irwin and Barghoorn also examined under phase microscopy the fossil pollen from the Belles Artes drill core from Mexico City.

The results of their joint studies, published in 1965, leave no doubt that there are differences in the morphological characteristics of the exine as revealed by phase-contrast light. Under the phase contrast the pollens of corn and its relatives and hybrids appear to be beset with spinules, but the arrangement of the spinules varies. In a large number of the races of maize and especially in those which are regarded as primitive and relatively uncontaminated by teosinte or *Tripsacum*, races such as Chapalote and Nal-Tel of Mexico and Confite Morocho and Puno of Peru, the spinules appear to be evenly spaced on the surface of the ektexine and are strong and dark in appearance. In *Tripsacum* the spinules appear to be distributed irregularly on the ektexine and to occur in clusters. In teosinte the arrangement is similar to that of maize and quite different from that of *Tripsacum*. The authors conclude that the pollen of maize and teosinte can be easily distinguished from that of *Tripsacum*.

In dealing with hybrid races of maize or teosinte, discrimination appears to depend largely on how much germ plasm each species has absorbed from the other. Thus the pollen of the maize-like teosinte Chalco, which grows commonly in the maize fields in Mexico and is constantly crossing with maize, is difficult to distinguish from that of the very tripsacoid Costa Rican race of maize, Huesillo. The inbred strains of Minnesota A158 and Texas 4R3 have fairly strong regular patterns of spinule distribution, while the introduction of teosinte germ plasm causes them to lose the strength and regularity of the pattern.

The authors concluded that pollen morphology is genetically controlled. Perhaps even more important than this conclusion are the results of their reexamination of the pollen grains from the Belles Artes sediment. Employing phase-contrast microscopy Irwin and Barghoorn identified some of those as maize, others as *Tripsacum*. These observations support the previous conclusions of Barghoorn et al. that wild maize existed in

the Valley of Mexico during late Pleistocene time.

Fossil Corn Pollen in Panama

More recently another of Barghoorn's graduate students, Alexandria Bartlett (1969), has discovered maize pollen in sediments from Gatun Lake in Panama dated at *ca.* 6,000–7,000 years. Like the fossil pollen from the Valley of Mexico, it shows under phase-contrast microscopy a regular distribution of spinules. That this is the pollen of a wild maize is indicated by the absence of any evidence of agriculture. Other maize pollen dated at *ca.* 3,000 years is associated with the pollen of the sweet potato, a cultivated plant originating in South America, as well as with drastic changes in pollen frequency indicating deforestation resulting from the practice of agriculture. The fossil pollen from Panama, presumably that of a wild maize, represents a second locality for the occurrence of prehistoric wild maize.

Identification of Fossil Pollen Challenged

The identification by Barghoorn et al. of the fossil pollen from Mexico and of Bartlett et al. of that from Panama has, like our conclusions about the early Tehuacán corn, been challenged. Devotees of the teosinte theory initially dismissed this evidence as "not to be taken seriously." Finally, realizing that it is generally accepted by both botanists and archaeologists, they have attempted to discredit it, not by direct criticism but obliquely by implication, casting doubt on the pollen-preparation techniques and the adequacy of the samples, raising the question of contamination, suggesting that the teosinte pollen with which the fossil pollen was compared may have come from depauperate plants and was not normal (see Galinat, 1971). If the purpose of all this was to obfuscate it has succeeded.

Considering the possibility of contamination, I refer the reader to the statements on this subject presented earlier in this chapter in which Barghoorn et al. give four cogent reasons for concluding that the fossil pollen grains are not the product of contamination. To these I can now add a fifth, that concerned with the early archaeological pollen from Bat Cave. I know this to be uncontaminated because I teased the pollen grains from a partially dehisced anther in a tassel fragment taken from the lowest level of the cave. Since there was no corn—or any other cultivated plant—growing in the vicinity of the cave there is no possibility of contamination occurring during its excavation. How does this uncontaminated early archaeological pollen compare with the fossil pollen? It is quite similar. The average lengths of the grains is 93.1 and 95.6 microns respectively; the pore-axis ratios are 1:6.1 and 1:6.2. Two samples taken from the same population could scarcely be much closer.

The implication that the comparisons may not be valid because of the method of preparation is misleading to say the least. The acetylation method of separating fossil pollen from the material in which it is embedded is standard procedure in palynological laboratories throughout the world. Modern pollen that is to be compared with fossil pollen must, of course, be treated in the same way. Galinat is correct in stating that the data from this method are not comparable to those obtained from other methods such as mounting the grains in lactic acid, which causes the grains to swell. Obviously the fossil pollen cannot be treated in this way.

The implication that the pollen grains of teosinte used in the comparison may have been reduced in size because they were borne on plants subjected to short-day treatment is absurd. One has only to turn to Chapter 3 and look at Wilkes's illustrations of races of teosinte based on plants grown in Massachusetts and brought into flower through short-day treatment to realize that they are not depauperate. If vegetative vigor affects pollen size then the pollen of Chalco teosinte plants grown in Massachusetts that I furnished to Barghoorn is probably larger, rather than smaller, than would have been the pollen of the same race growing in the wild in competition with other vegetation. Also concerning the size of the pollen is Galinat's (1971) assertion that the apparent large size of the fossil pollen is not consistent with the correlation that he (Galinat, 1961) reported between pollen size and ear length, the implication here being that since the wild corn had short ears it should also have had small pollen

grains. The fact is that the correlation reported by Galinat is strongly influenced by two races, Jala and Huesillo, that have unusually long ears and unusually large pollen grains. The correlation among the remaining eight races is by no means as strong, and there are notable exceptions. The highly evolved race Vandeño, for example, has pollen grains of about the same size, 83.9 microns, as that of the primitive race Nal-Tel, 81.2 microns, although its ears are more than twice as long as those of Nal-Tel, 17.2 and 7.9 cm. respectively.

With respect to the adequacy of the samples I do not know whether the implication is aimed at the number of pollen grains measured of *Tripsacum*, teosinte, and maize or at the number of varieties compared. If the first, I can say that the number of pollen grains most commonly measured, 50, is exactly twice that, 25, in some of the varieties of maize included in Galinat's (1961) study of the correlation between pollen size and ear length. If it be argued that more species of *Tripsacum* and races of teosinte and maize should have been included in the comparison, I would suggest that, on the contrary, there should have been fewer. The study would be more sharply focussed if only those entities were included which might have been present in the Valley of Mexico. This would include one of the Mexican *Tripsacum* species, probably *T. lanceolatum*, Chalco teosinte, the race of the region from which the fossil pollen came, and one of the ancient indigenous races of maize, Chapalote, since the wild corn of Tehuacán has been identified as related to this still living race. The data from these three entities follow:

Averages

Kind of pollen	Length of grain, microns	Diameter of pore, microns	Pore-axis ratio
T. lanceolatum	49.9	12.5	1:4.0
Chalco teosinte	81.8	15.9	1:5.1
Chapalote maize	94.9	15.4	1:6.2
Bat Cave maize	93.1	15.2	1:6.1
Fossil	95.6	15.5	1:6.2

The pollen grains of the fossil pollen, the early archaeological pollen, and the Chapalote pollen are remarkably similar in their dimensions, and they differ substantially from those of teosinte and even more from those of *Tripsacum*. I find it difficult to see how an objective examination of the data can lead to any conclusion except that the fossil pollen grains are those of corn.*

Summary

Cobs of prehistoric corn dated at about 5000 B.C. are considered to be those of wild maize, since there is no archaeological evidence of the practice of agriculture in the levels in which they occurred. They show no evidence of having evolved from teosinte. Fossil pollen grains from Mexico and Panama are regarded as those of wild maize, since some of them are much larger than those of modern races of teosinte and are earlier than the beginning of agriculture. If these identifications are correct—and I see no good reason to question them—then we can only conclude that the ancestor of cultivated corn was corn.

*The electron-microscope studies of pollen of maize and its relatives, mentioned in Chapter 4, by Barghoorn and his student Umesh C. Banergee, have now shown clearly that the exine patterns of the pollens of corn and teosinte are quite similar. Thus the principal difference between corn and teosinte pollen is one of size and in this respect there is some overlapping between the species, the smaller pollen grains of corn being no larger than the largest pollen grains of teosinte. Consequently pollen grains of intermediate size, those falling within the range of overlap, cannot be identified with respect to species. However, identification of some pollen grains is still possible; grains larger than the largest teosinte grains can be considered to be corn while those smaller than the smallest corn, but larger than those of *Tripsacum*, can be regarded as teosinte. Kurz et al. (1960) concede that five of the fossil pollen grains studied by Barghoorn et al. are sufficiently large to be classified as maize "with a high degree of reliability". None can be classified as teosinte with the same degree of reliability.

16 Corn in Prehistoric Art

So far as concerns the New World, the above facts tend to support the general proposition, to which the history of advancement in the Old World suggests no exception, that nothing worthy the name of civilization has ever been founded on any other agricultural basis than the cereals. . . . Cereal agriculture, alone among the forms of food-production, taxes, recompenses and stimulates labour and ingenuity in an equal degree. E. J. Payne, *History of the New World Called America*, 1892

Payne's conclusions, reached almost a century ago, seem to be still well supported by the now-available facts. Egypt, Greece, and Rome had their wheat and barley; China, Japan, and India their rice. All of the advanced cultures and civilizations of the New World—the Inca of Peru, the Maya of Middle America, and the Aztec of Mexico—had corn for their basic food plant. And as the cereals of the Old World became the subjects of solemn rites and ceremonies in which various deities concerned with cereals and their production were worshipped so also did maize in the New World become a religious object. And so were deities—remarkably similar in some respects to those of the Old World—invented and revered. The Roman goddess Ceres—from whose name the word cereal is derived—and Demeter, the Greek goddess of grain production, had counterparts in the deities of the ancient civilizations of the New World. And in the other direction, across the Pacific, there were remarkable similarities between the corn-mother myths of Asia and those of America (see Hatt, 1951).

There are those who see in these and related similarities evidence of the diffusion of important cultural traits from the Old World to the New. I regard them, however, as nothing more than the consequence of cereal culture, successfully practiced, leading the way to civilization and arousing in its practitioners an instinct that seems to be almost universal in the human race: a tendency to resort to various forms of mysticism in attempting to explain natural phenomena and to appease by appropriate rites the mysterious natural forces beyond human control.

Related to the religious rites and ceremonies that involved maize was its use as a motif in various art forms. This was especially true in Peru and Mexico, the centers of the most highly developed civilizations of prehistoric America. In Peru maize was depicted in stone, in ceramics, in textiles, and in the two most precious metals, gold and silver. In Mexico the representations of maize were primarily in stone and ceramics.

The extent to which maize was the motif in various forms of artistic expression is, like its role in religion, evidence of the high regard, indeed the veneration and reverence, in which it was held. This aspect of the subject although interesting and intriguing is, however, not the principal theme of the present chapter; here we are primarily concerned with the question: what, if anything, can the prehistoric art tell us about the prehistoric corn?

If the corn is represented only stylistically, as in many cases it is, we cannot learn much about its characteristics, but even in stylistic depiction we can sometimes learn something. For example, many of the carved stone objects from Peru representing corn—called *conopas*—that were customarily placed in the fields when food crops were cultivated to insure that the crops would be abundant were distinctly ovoid in shape (see Mangelsdorf and Reeves, 1939, figure 14). Since no maize resembling these is found in Mexico, they provide at least an indication that the early corn of Peru was different from that of Mexico, a possibility discussed at length in earlier chapters.

But in both Peru and in Mexico there are realistic as well as stylized representations of

maize. Especially significant are the ceramic replicas cast from molds made from the actual ears. These are of particular interest in our study of the origin and evolution of corn, for they are almost as useful as actual prehistoric ears in showing us the size and shape of the ears and of their kernels and the arrangement of the kernels, whether in straight rows or irregular. Compared to cobs from which all of the kernels have been removed—the most common specimens in archaeological sites in Mexico and the American southwest—the ceramic replicas provide vivid pictures of the prehistoric maize that they were fashioned to represent. Thus once the maize of Mexico and Peru and their adjoining countries had been classified and the principal races described, it became possible to identify the maize depicted on various ceramic objects, and since these have been associated with certain archaeological time periods that had been dated by archaeologists employing modern techniques, it is possible to determine within broad limits when certain races were present. A beginning in this direction was made by Wellhausen et al. (1952), who identified several Mexican races depicted on Zapotec funerary urns. Similar studies were carried somewhat farther by Grobman et al. (1961), who identified a number of prototypes of modern Peruvian races on prehistoric ceramic objects, especially funerary urns and other vessels.

Contrary to the opinion of some archaeologists, the realistic representations of corn are often easily distinguished from the stylized ones because of one of corn's most distinctive characteristics. As I pointed out in

Chapter 1 the pistillate spikelets, which bear the kernels, are, like the staminate ones in the tassel, arranged in pairs. This botanical characteristic has two interesting consequences: (1) in practically all races of corn the number of kernel rows is always even, since two times any number is necessarily an even number; (2) in ears that have distinct kernel rows the rows occur in pairs, the kernels within a pair being situated side by side and the pairs of rows alternating with adjacent pairs. Artisans almost invariably overlook this distinctive feature and show the kernels of one row alternating with those of the adjacent rows. One of my favorite possessions is a beautiful carved Chinese ivory ear of corn. The seventeenth-century artist who fashioned it and who proudly put his signature upon it in Chinese characters was meticulously accurate in reproducing all of its botanical details except one, the pairing of the kernel rows. In contrast the pairing of the kernel rows is shown distinctly in the specimens illustrated in Figures 16.2 and 16.9, which were undoubtedly cast from molds made from the actual ears.

The fact that the custom of depicting maize was largely confined to Peru and Mexico raises the intriguing question whether it developed independently in the two centers or originated in one and diffused to the other. There are certain resemblances, the principal one being the casting of replicas from molds made from actual ears. This is a highly sophisticated form of ceramic art that so far as I know was not generally practiced with native plant products in any other part of the world, although stone carvings to represent Old World plants were not uncommon. For example, the portals of the Temple of Bacchus at Bal-Baek in Lebanon, which I visited some years ago, are ornamented with carvings in bold relief of the grape, barley, and poppy plants from which were derived the three principal intoxicating preparations of that part of the world: wine, beer, and opium. The carvings are quite artistic but are by no means as realistic as the replicas of maize in Peruvian and Mexican ceramics. Some of the Peruvian

maize deities have other characteristics—somewhat less conspicuous—in common with the Mexican ones: discs with their centers depressed suspended from the ears common in the Mexican deities and occurring in some of the Peruvian ones. Protruding lips occur in the deities of both countries.

If the custom of ornamenting urns and other vessels with corn had a common origin where did it begin? This is a question that cannot be answered until extensive further studies have been made. On the basis of the evidence presently available it would seem that it is more likely to have originated in Peru and diffused to Mexico than the reverse. The principal reason for suspecting this is that in Mexico the only plant product commonly depicted is corn; others, such as squashes, occur much more rarely. In Peru, however, a great variety of plants are represented. Vargas (1962) in his study of the phytomorphic representations of the ancient Peruvians identified 23 species of cultivated plants, 6 ornamentals, and 11 miscellaneous plants among those depicted. Vargas was quite correct in stating with respect to Peru that "the representation of plant motifs on various kinds of objects in pre-Columbian times reached enormous proportions and attained great authenticity of expression."

If diversity can be considered an attribute of a center of origin then the custom of employing plants as art motifs would seem to have had its center in Peru. Whether Mexico is a secondary center or an independent one is a question still to be answered. Perhaps lending some support to the idea of Peru as the original center is the fact that there was a diffusion of races of maize from South America to Mexico and adjoining countries. Wellhausen et al. (1952) postulated that certain of the Mexican races, those designated as Pre-Columbian Exotic, had been introduced into Mexico from Central or South America in prehistoric times. Four of these, Cacahuacintle, Harinoso de Ocho, Olotón, and Maíz Dulce were recognized. Counterparts of all of these were subsequently found in Colombia and were described by Roberts

et al. (1957) under the names respectively of Sabanero, Cabuya, Montaña, and Maíz Dulce. As was pointed out in Chapter 9 the last-named race is probably of Peruvian origin. An additional race that spread from South America to Mesoamerica is the Colombian Güirua which has its counterpart in Guatemala in the race Negro de Chimaltenango (Wellhausen et al., 1957).

A careful study of the ceramic objects on which corn is depicted realistically may tell us when some of this diffusion occurred. For example, if the vessel illustrated in Figure 16.5 can be accurately dated by archaeologists it may show when a South American corn with thick, dorsally rounded kernels was first introduced into Mexico. Likewise the dating of the vessel illustrated in Figure 16.4 would show at what time the Mexican race Tuxpeño had already come into existence. Since this race is regarded as the product of the hybridization of South American corn with teosinte the dating would provide some estimate of when the South American race was introduced into Middle America.

The problem is, however, complicated by the fact that in Mexico, at least, there has developed in recent times a trade in modern forgeries. Some of the most blatant fakes can be identified by their inconsistencies, for example, the wrong glyphs associated with a particular deity. But others are so well done that they are difficult to distinguish from the authentic objects. Now chemical and photochemical techniques are being developed that will aid archaeologists in distinguishing between the authentic and the forged. When these have been perfected the representations of corn in the prehistoric art can provide valuable information about the times at which certain events occurred in the evolution of corn. In the meantime the illustrations presented here will serve to show some of the possibilities that I hope future research will pursue. Indeed it is gratifying to be able to report that at least one student of archaeology is now engaged in studies of this nature.

FIG. 16.1 The ears ornamenting the headdress of this Zapotec maize deity are quite similar in size, shape, and the characteristics of the kernels to the ears of Nal-Tel, one of the ancient indigenous races of Mexico that still exists. This race is represented more frequently on Zapotec funerary urns than any other. This idol was turned up in the excavation of Monte Albán, Oaxaca, Mexico, and is dated by archaeologists at A.D. 600–800. Note the marked protruding lips of the idol, a common characteristic of the urns on which maize is represented. (Courtesy National Museum of Anthropology, Mexico, D.F.)

FIG. 16.2   An unusually ornate Zapotec funerary urn in which maize is represented, not on the headdress, but by two ears suspended from the necklace. In their size, shape, kernel-row number, and characteristics of the kernels these molded ears are quite similar to the Ancient Indigenous Mexican race Nal-Tel described and illustrated by Wellhausen et al., 1952. (From *Urnas de Oaxaca*, Caso and Bernal, 1952.)

FIG. 16.3 This idol identified as one of the numerous versions of the Mexican Rain God is ornamented with two replicas of ears of corn. The one at the right is of particular interest in showing distinctly at one point the paired arrangement of the kernel rows, thus virtually proving that the replicas were cast from molds made from actual ears. In their shapes, the relatively low kernel-row number, and the number of kernels per row, the replicas are quite similar to the ears of Nal-Tel, one of the Ancient Indigenous races of Mexico still cultivated by certain Indian groups. (From Wellhausen et al., 1952.)

FIG. 16.4 Dent corn molded on a prehistoric Zapotec funerary urn. This urn is of particular interest in illustrating quite distinctly the denting of kernels, a characteristic of the Mexican race Tuxpeño, which is regarded as one of the ancestors of the Corn Belt dent of the United States. If the urn is authentic, as it is generally considered to be, it shows that this kind of corn was already in existence many centuries ago. All of the ears represented on this urn were probably made from the same mold, which was made from an actual ear. (Courtesy National Museum, Washington, D.C.)

191   Corn in Prehistoric Art

FIG. 16.6 This idol is an enigma. It has some of the principal characteristics of the conventional funerary urns of Oaxaca: the protruding lips, the elliptical shape of the eyes, the the discs attached to the ears. But in several other respects, including posture, it is quite atypical, as is also the corn represented by the replicas. The barren tips of the ears are indicative of a lowland rather than a highland race, but no race has been described in either Mexico or Guatemala to which these replicas can be assigned with any degree of confidence. One guess is that the ear from which the replicas were made is a segregate of a hybrid of two recognized Mexican races: Oloton, which has a tapering ear and rounded kernels, and Tabloncillo, which has a cylindrical ear and a low kernel-row number. If the characteristics of the maize can be employed as criteria in determining the age of an idol this one must be suspected of being post-Columbian. (Courtesy Musée de l'Homme, Palais de Chaillot, Paris.)

FIG. 16.7 (*Top*), this Zapotec urn is unusual in the fact that the two ears molded on the headdress are terminated by structures that may have been intended to represent staminate spikes. The early prehistoric corn from the Tehuacán Valley described in Chapter 15 probably had such terminal staminate spikes as the artist's reconstruction in Figure 15.24 suggests. An enlargement of one of the ears molded on the headdress of this idol is shown in Figure 16.8. (Courtesy Botanical Museum, Harvard University.)

Fig. 16.8 (*Bottom*), a close-up view of one of the ears molded on the headdress of the idol shown in Figure 16.7 illustrating the terminal protuberance which may represent a spike bearing staminate spikelets, a characteristic of certain primitive races. The significance of the wavy lines in the figure is not known.

FIG. 16.5 A funerary urn decorated with replicas of ears of corn. Because of their cylindrical shape and low kernel-row number Wellhausen et al. (1952) regarded these replicas as possibly depicting ears of the Mexican race Olotillo. But Olotillo is a lowland race and the urn is presumably highland in origin. Also the distinctly rounded kernels are not characteristic of Olotillo but are more like those of Quicheño, a race commonly grown in northwestern Guatemala in the Departments of Huehuetenango and El Quiche adjoining Mexico and at altitudes similar to those of the Valley of Oaxaca in Mexico. The replicas are remarkably similar to an ear of Quicheño illustrated in Wellhausen et al. (1957). Since this race is regarded as of South American origin this urn is of special interest in suggesting that the introduction of South American races into Central America which these authors postulated had already occurred when the urn was made, probably during the late Classic, A.D. 600–900. The urn is also of some interest in showing with special clarity that all of the replicas were cast from the same mold; all exhibit a slight irregularity of the kernel rows at the bases of the ears.

FIG. 16.9 Another archaeological sherd showing a terminal protuberance of the ear which may represent the staminate tip, a characteristic of certain primitive races. This specimen is also of interest in illustrating the distinctly paired arrangement of the kernels and the alternation of a paired row with adjacent paired rows. This shows that the mold from which this ear was cast was almost certainly made from an actual ear. This sherd was found near Mitla, Oaxaca, by Dr. H. Garrison Wilkes. (Actual size.)

FIG. 16.10 This photograph, kindly sent to me by Dr. Matthew Stirling of the Smithsonian Institution, of a specimen found by him in a small *colegio* in the town of Tlacotalpan in the State of Vera Cruz, Mexico, is of interest in showing both an ear and a tassel. The ear represented in this figure may be stylized, but in its ovoid shape, rounded kernels, and absence of distinct kernel rows it shows a remarkable resemblance to some of the Peruvian replicas illustrated in figure 15 of Mangelsdorf and Reeves, 1939, and figure 41 of Grobman et al., 1961. Note the similarity in design of the stylized tassel branches to the staminate tip illustrated in Figure 16.9. There is also a similarity, which may or may not be significant, to the tassel branches on the Peruvian clay vessel illustrated in Figure 16.18.

FIG. 16.11 Three prehistoric replicas (actual size) of corn ears from the Peruvian highlands. The replica at left may, like that illustrated in Figure 16.12, represent pod corn. The other two replicas, characterized by their ovoid shapes, rounded kernels, and absence of distinct kernel rows, are definitely related to the sweet corn race Chullpi, which is regarded as the progenitor of one of the lineages described in Chapter 10. The fact that there are no races of maize in Mexico with these characteristics is one of the reasons for concluding that there has been a separate origin of maize in Peru. (Courtesy Peabody Museum, Yale University.)

FIG. 16.12 Two views of a prehistoric replica (actual size) of an ear of corn from the highlands of Peru which we (Mangelsdorf and Reeves, 1939) regarded as representing pod corn and as evidence that this kind of corn was known in prehistoric times. Weatherwax (1954) questioned this identification and suggested that the object might represent an ear with pointed, imbricated kernels characteristic of certain Andean races of corn. Because of the flattened dorsal appearance of the individual units and the fact that the basal ones tend to be longer than the apical ones I still favor the pod corn identification. (Courtesy Peabody Museum, Yale University.)

FIG. 16.13 A ceramic vessel of the Peruvian Mochica culture (*ca.* A.D. 500–1000) representing a maize deity and ornamented with replicas in relief of a distinctive corn quite different from any represented on Mexican funerary urns. The long, slender ears with eight kernel rows, the pointed, imbricated kernels, and the tendency for spreading apart of members of spikelet pairs all indicate that this corn is related to the Peruvian race Rabo de Zorro. If this race is a hybrid of Confite Morocho and a pointed popcorn, as Grobman et al. (1961) have suggested and if the pointed popcorn is originally Mexican, as I have postulated in Chapter 9, then the introduction of this Mexican popcorn into Peru must have occurred more than a millenium ago. (Courtesy American Museum of Natural History, N.Y.)

195   Corn in Prehistoric Art

FIG. 16.14 Another Peruvian Mochica vessel representing a maize deity and ornamented with pointed, imbricated kernels similar to those illustrated in Figure 16.13. At first glance this maize may seem to be stylized, but the paired rows of kernels alternating with adjacent paired rows, clearly discernible in the forefront of the vessel, indicate that molds of actual ears were somehow employed. The molds may have been flat clay surfaces on which the maize ears were rolled. (Courtesy American Museum of Natural History, N.Y.)

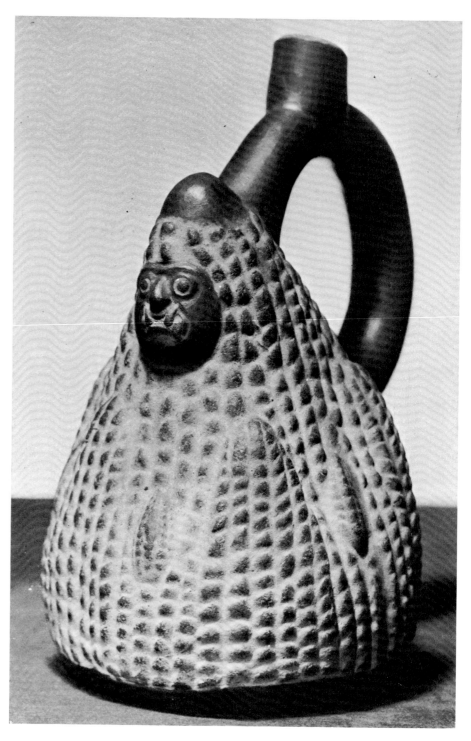

FIG. 16.15 A Peruvian Mochica water vessel depicting a maize deity with *hijos*—replicas in miniature of himself—that are supposed to signify fertility. The vessel is ornamented with replicas in relief of an ear of corn characterized by a tapering shape, irregular kernel rows ten or more in number, and flat kernels that are wider than thick. Although not completely typical the ear from which the mold was made was probably related to the race Cuzco. (Courtesy American Museum of Natural History, N.Y.)

197   Corn in Prehistoric Art

FIG. 16.16 Black Chimu vessel from the Peruvian coast decorated with casts of ears of corn. Unlike the replicas on the Mexican vessel illustrated in Figure 16.5, all of which were cast from the same mold, these casts represent a number of different ears. The lowermost ear, for example, is quite similar to the Peruvian race Confite Morocho. The ear immediately above this, with its straight rows and relatively large kernels, may be a prehistoric version of the race Cuzco. The center ear of the three above the mouth of the vessel is distinctive in its slightly twisted kernel rows and slightly imbricated kernels. Since the diameter of the vessel is $8\frac{1}{4}$ inches, the length of the casts can be easily determined from the photograph. These are found to vary from about 7 to 10 cm. Few typical ears of modern Peruvian races are this short. It is of interest that the Peruvian race Cuzco Gigante, famous for its enormous kernels, is not represented in this or any other known ceramic object; it is apparently a post-Columbian development, one that may have occurred after the large-seeded race Tabloncillo had been introduced from Mexico into Peru. (Courtesy Peabody Museum, Yale University.)

FIG. 16.17 Although obviously somewhat stylized the maize depicted on this Peruvian water vessel has ovoid ears and sharply pointed kernels. These are characteristics of fasciated forms of the Peruvian race Confite Puntiagudo, which, as I have postulated in Chapter 9, is probably related to the Mexican pointed-popcorn race Palomero Toluqueño. This vessel and others similar to it (cf. Grobman et al., 1961, fig. 26) suggest that a corn with pointed kernels, whether of Mexican origin or not, was known in Peru in pre-Columbian times. (Courtesy Berkeley Galleries, London.)

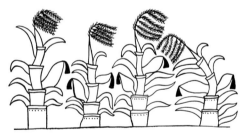

FIG. 16.18 A drawing on a clay vessel from the Valle de Chicame, Peru, which shows that the ancient Peruvians were quite aware of the principal botanical characteristics of the corn plant. The graceful terminal inflorescences, the tassels, are differentiated into central spikes and lateral branches, but no distinction is made between the polystichous arrangement in the former and the distichous condition of the latter. The spikelets are shown alternating on the two sides of the central axis. The ears are enclosed in husks and are terminated by fasicles of silks. The dots on the lower internodes probably represent the rudiments of adventitious aerial roots. Despite a certain degree of artistic license there is no mistaking the plants for anything but corn. (From Lehman and Doering, *The Art of Old Peru*.)

199   Corn in Prehistoric Art

17 Corn's Spread to the Old World

> Maize is of American origin, and has only been introduced into the old world since the discovery of the new. I consider these two assertions as positive, in spite of the contrary opinion of some authors. A. de Candolle, 1886

The first of these positive assertions is as valid today as it was when de Candolle made it more than three-quarters of a century ago. The discoveries of fossil pollen and of prehistoric wild corn in Mexico, described in previous chapters, virtually prove the American origin of corn and rule out the possibility of an Old World origin. Even the most ardent proponents of pre-Columbian contacts between the Old World and the New now seem willing to concede this well-established fact (Heine-Geldern, 1958; Jeffreys, 1967; Carter, 1968). De Candolle's second assertion, that maize was introduced into the Old World only after the discovery of the New, has in recent years been seriously challenged, although no convincing, tangible evidence disproving it has yet been adduced in spite of concerted efforts to discover it.

Most of this challenge has come from the "diffusionists," a school of geographers, anthropologists, and a few imaginative botanists who profess to see in art forms, myths, and other cultural traits, including the use of plants, some remarkable and highly significant similarities between America and the Old World, especially Asia but also Africa and Europe, and who, for reasons that I have never been able to understand, seem determined to prove that many of these traits have diffused from a common center. Most of the diffusionists write well and fluently, and what they have lacked in facts they have made up for in eloquence and volume. The result has been that the literature in which the possibility of a pre-Columbian diffusion of maize is discussed and debated now equals or exceeds that reporting new experimental evidence on

the origin and evolution of the plant. To review this extensive literature in detail would demand more space than, in my opinion, it deserves. Consequently in this chapter I shall treat only what seems to me to be the more salient points.

The question as to which part of the world gave rise to maize is not new, since it is one upon which students of plants have differed for more than four centuries. Sturtevant, a careful student of maize and of the early literature pertaining to it, compiled lists (1879) of names of prominent herbalists and early botanists who had expressed opinions on the geographical origin of maize. Among those who regarded maize as a plant of Old World origin were Bock, Ruellius, Fuchs, Sismondi, Michaud, Gregory, Lonicer, Amoreux, Regnier, Viterbo, Donicer, Tabernaemontanus, Bonafous, St. John, de Turre, Daru, de Herbelot, and Klippart. Equally impressive is the roster of those who believed maize to be an American plant: Dodoens, Camerarius, Matthioli, Gerard, Ray, Parmentier, Descourtilz, de Candolle, Humboldt, Darwin, F. Unger, Von Heer, de Jonnes, Targioni-Tozzetti, Hooker, Figuer, Nuttall, Mrs. Somerville, and Flint.

The opinion of earlier students that maize is a plant of Old World origin was based in part on botanical and in part on linguistic evidence. Wilkes (1967) has pointed out that the Old World plant, *Coix*, Job's tears, was at one time regarded as the closest relative of maize, and since *Coix* was correctly regarded as an Old World plant it was supposed that maize was likewise of Old World origin. This error became apparent when the close rela-

tionship of maize to teosinte, an American plant, was recognized. Among the names applied to maize one of the most common, with variations in different languages, is "Turkish Wheat" suggesting that maize had originated in the Near East. This name, as de Candolle pointed out, apparently first used by Ruellius in 1536, is as erroneous as the English "turkey," an abbreviation of "turkey fowl" given to a bird of unquestionable American origin, but one that may have been introduced into England from Turkey.*

De Candolle's case for the American origin of maize, first presented in 1855 and later reinforced in various editions of his famous work *Origin of Cultivated Plants*, was so convincing and the evidence that he marshaled to support his conclusions so substantial that the possibility of an Old World origin of maize received little consideration from serious students during the ensuing half century. Especially was this true after Ascherson (1875) called attention to the close relationship of maize and teosinte, a plant unmistakably American.

Pre-Columbian Maize in Asia?

I know of only two suggestions within this century, those of Anderson (1945) and Stonor and Anderson (1949), that maize might have originated in Asia, and the first of these was put forward only "*as one possible hypothesis*" (italics his). It has not been taken seriously by students of maize (see Randolph, 1955;

*Another possibility, one preferred by some authorities, is that the turkey was confused with the guinea fowl, which may have reached England from Africa by way of Turkey.

Venkateswarlu, 1962) with the possible exception of Suto and Yoshida (1956), who at least consider it likely to "be proved to be more correct than the Mangelsdorf's hypothesis." The question of a pre-Columbian distribution of maize in Asia has, however, been raised at least twice in this century, first by Collins (1909) and more recently by Anderson (1943, 1954) and by Stonor and Anderson (1949).

It was the discovery in Chinese maize of a peculiar type of endosperm known as "waxy" (cf. Chapter 12) that led Collins (1909) to reopen the question of the possible pre-Columbian occurrence of maize in Asia. This kind of maize was unknown as a cultivated variety in America but was widespread in China, Burma, and the Philippines (Collins, 1909, 1920) from 5° to 45° north latitude (Kuleshov, 1928). However, self-pollinated ears of American varieties segregating for waxy endosperm have been reported by Mangelsdorf (1924) and by Breggar (1928) and mutations to waxy were found by Bear (1944). I have shown in Chapter 12 how a recessive mutation such as waxy endosperm can become the predominating type under certain conditions, as it has in parts of Asia.

Anderson (1943) was struck by what he considered to be the unusual characteristics of a corn collected by Carl and Jonathan Sauer from an Indian village on the Rio Loa in northern Chile at an altitude of about 2,500 meters. Seeing some resemblance between this maize and prehistoric Peruvian maize and in several "peculiar characters" a resemblance to certain Oriental varieties, he concluded: "These facts reopen the entire problem of pre-Columbian contacts between the New World and the Orient." For some diffusionists they did precisely that. We now know, because of extensive studies of the races of maize of the countries of this hemisphere, that the Rio Lao maize is not particularly unusual, since it has close counterparts in some of the high-altitude corn of Peru, Bolivia, and Ecuador. Anderson, an acute observer, did not fail to see this resemblance and to call attention to it, but he was less impressed by this resemblance than that of the Rio Loa corn to certain Oriental varieties.

The diffusionists found even more support in a subsequent report by Stonor and Anderson (1949) describing a group of maize varieties with characters said to be "unusual"

being grown by the hill peoples of Assam. This maize, which the authors designated as "Race A," was said to be unknown in the coastal regions of Asia but was rather widely distributed in central Asia; further, it seemed to resemble certain South American maize also designated as "Race A" which is common archaeologically and has certain characteristics which are still to be found, although rarely, among living South American varieties. These peculiar Asiatic varieties were said to differ profoundly from those of "Race C," which also occurs in both Asia and America but which in Asia are largely confined to coastal regions and are admittedly post-Columbian introductions. These facts were regarded by the authors as "fantastic," and they stated that "any satisfying hypothesis must border on the miraculous." They concluded that maize "must either have originated in Asia or have been taken there in pre-Columbian times."

Knowing that the characteristics of the Assamese maize are not particularly unusual, since they are found also in living varieties of South America which I had grown in my own experimental cultures, I was disturbed at these sweeping conclusions drawn from what appeared to me to be wholly inadequate evidence. But for almost two years I did not challenge them publicly for several reasons: I did not wish to enter into a controversy with Anderson, who as a graduate student had been my roommate at Harvard and subsequently a close friend with whom I had for many years engaged in a fruitful exchange of ideas on maize; I believed that time spent on my own experimental work would contribute more to an eventual solution of the problem than debating a subject already somewhat emotionally charged; discussion of the origin of maize had sometimes in the past reached the point of acrimonious debate, an outcome that I earnestly desired to avoid. However, after being requested by a number of anthropologists to review the article and evaluate the evidence on which it was based, I was finally persuaded by one of my colleagues at Harvard, Douglas Oliver, an anthropologist who contended that my silence was being generally regarded as assent, to join him in writing a critique of the Stonor-Anderson thesis. Merrill (1950), an authority on the botany of Asia and the islands of the Pacific, had already questioned the article's principal

conclusions on general botanical grounds, and Weatherwax (1950), an authority on the botanical characteristics of corn and its relatives, had quite appropriately called attention to certain fundamental botanical differences between maize and its Asiatic relatives. Oliver proposed to treat the ethnological aspects of the subject, and I agreed to examine especially the assertions regarding the uniqueness of the Assamese maize.

Our joint effort was published in April 1951 under the title "Whence Came Maize to Asia?" We showed, to our own satisfaction at least, that the Assamese maize is not unique, since it has close counterparts in certain Colombian varieties; that the uses to which maize are put in Assam are precisely those to which one would expect a cereal to be put. In short we found nothing in either the botanical or the ethnological evidence presented by Stonor and Anderson to justify their conclusion that maize must have either originated there or been taken there in pre-Columbian times.

Our article, written in a somewhat derisive vein, was not calculated to please the diffusionists, and it did not. It was soon attacked in properly indignant terms by Carter (1953): "By such a mass of quibbling, assertions and counter-hypotheses," he stated, "Mangelsdorf and Oliver conclude that there is no tangible evidence of any kind of the existence of maize in any part of the Old World in 1492." He conceded, however, that "The *possibility* that maize is American remains." Our article was also criticized but in more moderate language by Heine-Geldern (1958), who actually expressed agreement with several of our final conclusions. These conclusions, since they appeared in the *Botanical Museum Leaflets*, a journal of limited circulation, may not be easily available to some readers and are here reproduced:

We can find nothing in either the botanical or ethnographic evidence presented by Stonor and Anderson on Assamese maize to justify their conclusion that maize must either have originated in Asia or been taken there in pre-Columbian times. The maize itself is not unique, since it resembles the living varieties of Colombia and thus conforms to the general rule that all Old-World maize has its counterparts somewhere in America. The uses to which maize is put in Assam are exactly those to which one would expect such a cereal to be put when introduced into Asia, and there are no other special circumstances about its utilization, or the traditions connected with it, which indicate a great antiquity in Asia. The fact that maize, if introduced into Asia in post-Columbian times, must have been rapidly accepted by backward people, merely indicates that, like the potato in Ireland, it met an acute and pressing need. Certainly there is nothing in the evidence which is in conflict with the long-established and well-supported opinion that maize is an American plant—one which has perhaps been introduced into Asia twice: once in early post-Columbian times from the west by a land route, and a second time, perhaps somewhat later, when tobacco and the potato were also introduced from the east by seafaring people. There is no factual evidence in conflict with this simple and rational explanation; but there is abundant evidence to support it.

The door is still wide open for hypotheses about pre-Columbian cultural diffusion between the Old World and the New and the problem is an extremely important one which merits the most careful and critical attention on the part of scholars in several fields. The problem is not likely to be solved, however, by putting forward sweeping and sensational conclusions which are based upon inadequate and fragmentary evidence, especially when these are all too likely to be seized upon by other imaginative writers who treat them as "evidence" or, worse still, as "virtually unassailable proof" (Zelinsky, 1950).

Perhaps there has, indeed, been a pre-Columbian, trans-Pacific diffusion of culture and perhaps maize has been involved in it. To speculate upon this possibility certainly can do no harm. But fancy ought not to be confused with fact. The *fact* is, that, at the present time, there is no tangible evidence of any kind—botanical, archaeological, ethnographic, linguistic, ideographic, pictorial or historical—of the existence of maize in any part of the Old World before 1492. Until such evidence is discovered, any case for pre-Columbian, trans-Pacific diffusion must rest on evidence other than maize.

Commenting on these conclusions Heine-Geldern wrote: "They are right in warning about hasty conclusions on the basis of indices that give insufficient proof. Indeed, they even admit the possibility of a pre-Columbian transportation of maize via the Pacific and merely find that, at this time, no evidence for this exists. 'Until such evidence is discovered, any case for pre-Columbian, trans-Pacific diffusion must rest on evidence other than maize.' One can only concur with all this."*

The case for the pre-Columbian occurrence of maize in Asia, never a very convincing one, was still further weakened by the new evidence presented by Ho (1956), an eminent Chinese historian. After a searching study of Chinese historical sources, which, Carter's assertion to the contrary (1953), are quite extensive, Ho concluded that maize was introduced into China early in the sixteenth century arriving there by both overland and maritime routes. He states: "Summing up the introduction of maize into China, we may say that maize was introduced into China two or three decades before 1550; that it was probably introduced by both the overland and maritime routes; that there is little reason to justify Laufer's far-reaching conclusion, especially in the light of the introduction of other New World plants, that in the dissemination of food plants 'a land route is preferred over a sea route as their way of propagation'; and that, barring a sensational discovery in Chinese sources clearly indicating a pre-Columbian introduction, Chinese maize as a topic for anthropological speculation should be closed." It seems strange that Ho's article, which appeared in a well-known and widely circulated journal, has been generally overlooked or ignored by the proponents of diffusion.

To support their argument for an Asian origin or pre-Columbian diffusion of maize, Stonor and Anderson cited the hypothesis of Hutchinson, Silow, and Stephens (1947) which postulated that the New World cultivated cottons are tetraploid hybrids of a wild American diploid, probably *Gossypium Raimondii*, and a cultivated diploid, *G. arboreum*, introduced from Asia by man crossing the Pacific after the invention of agriculture in Asia. This hypothesis was also cited by Carter (1950) and Zelinsky (1950) as evidence of pre-Columbian trans-Pacific diffusion. In the joint article with Oliver, I challenged the hypothesis on genetic and botanical grounds, concluding that: "the case for the trans-Pacific, pre-Columbian diffusion of Old

*Quoted from the translation of Goodman et al., 1964, of the Heine-Geldern article.

World cultivated cottons is no better, in our opinion, than the case for an Asiatic origin or pre-Columbian diffusion of maize. To use the one as evidence in support of the other, is to assume that two guesses have, through some strange alchemy, a greater validity than one." It was scarcely to be expected that this conclusion would please the diffusionists, and it did not do so. Both Carter (1953) and Heine-Geldern (1958) have taken issue with it at some length, Carter branding it as "hardly scholarly." Their criticism turns out, however, to be largely irrelevant. Cytogenetic studies by Gerstel (1953) have shown that the affinities of the American cultivated cottons may be closer with African than with Asiatic species. Both Hutchinson (1962) and Stephens (1963) have had second thoughts about their earlier theory of a recent origin of the American tetraploids, Hutchinson stating that the assumption of their originating under domestication "is no longer tenable" and Stephens likewise ruling out the possibility of a recent origin when writing of the presence of the tetraploid, *G. tomentosum*, in Hawaii. He stated "nor is it likely that this species was derived from an ancestral mainland form prior to the Pleistocene period." Cytogenetic studies by Phillips (1963) do not support the theory of a recent origin of the New World cultivated cottons.

In this connection we should not overlook the historical evidence on the occurrence of maize in China. This early literature is not "sparse" as Carter (1953) has asserted, and it is not true that "economic plants were given little notice." On the contrary the literature is extensive and contains many references to useful plants. Ho (1969) has this to say about one of the most famous of the Chinese works: "Of all the literary works, *The Book of Odes (Shih-ching)* contains by far the most extensive botanical records. Sinologists the world over agree on the authenticity and textual excellence of this ancient work, which illuminates the life of the Chinese from the late eleventh century to the middle of the sixth century B.C. It is true that this anthology of 305 songs and odes collected from the Chou royal domain and the feudal states mentions less than 150 plants, a number that is infinitesimal as compared with the number of species known to botanists today. But when it is remembered that the total numbers of plants known to and mentioned by the ancient

Egyptians, the Bible, Homer, and Herodotus are only fifty-five, eighty-three, sixty, and sixty-three, respectively, *The Book of Odes* is really a mine of information for historians and botanists. For a majority of cases, moreover, *The Book of Odes* states the type of topography in which a plant grows—mountain, plain, wet lowland near water, marsh, pond, or river." And this is only one source. Goodrich (1938) states that local histories sometimes called gazetteers were written in China from A.D. 347 on. By the thirteenth century there were at least 220 of them, and there are now some 6,000. Maize was first accurately described in a work published in the sixteenth century.

New "Evidence" on Prehistoric Maize in India

Despite what would seem to be conclusive evidence against an Asian origin of maize, the question continues to be raised. Several recent discoveries have again aroused interest in the possibility either of an Asiatic origin of maize or of its pre-Columbian diffusion to that part of the world. One of these is a report of a variety of maize with primitive characteristics being grown in Sikkim, a small mountainous state, a protectorate of India, lying in the eastern Himalayas adjoining Nepal. The characteristics of this maize as described by Dhawan (1964), who called it "Sikkim Primitive" are as follows: In its native habitat the plants attain a height of 130 to 200 cm.; each plant has two to four tillers, each main stalk or tiller bears four to six ears and is terminated by a drooping tassel with an average of five primary and seven secondary branches. The basel end of the lowermost one or two primary branches bear from 10 to 20 functional pistillate flowers that when pollinated develop viable seeds. The four to six ears on each main stalk or tiller are borne in successive nodes starting with the node immediately below the leaf subtending the tassel. The uppermost ear is about five cm. long and is terminated by a well-developed staminate spike bearing functional male flowers. Proceeding downward the successive ears increase in length, the lowermost ear being about 7.5 cm. to 10.0 cm. in length. The staminate spike decreases in length and in the lowermost ears is vestigal. The ear diameter ranges from 1.5 cm. to 2.3 cm.; the kernel rows, eight to ten in number, are

irregular; the husks, five to seven in number, tend to open at maturity; the glumes, which are soft, enclose the kernel for about half its length. The kernels are small, round, and hard, and they pop when heated.

This maize has also been described and illustrated by Thapa (1966), a young Sikkimese in charge of maize experimental work in Sikkim which is conducted in collaboration with the Indian Council of Agricultural Research and the Rockefeller Foundation. The author correctly sees some resemblance of this maize to the prehistoric wild corn of Tehuacán Valley (Mangelsdorf et al., 1964), and his illustration shows one of the conspicuous resemblances, the staminate tips of the ears. He does not describe the shape of the kernels, but as nearly as I can determine from the rather poorly reproduced photograph at least the uppermost kernels of the ear are pointed.

Considering its characteristics as a whole it is evident that this Sikkimese maize is related to the Mexican popcorn race Palomero Toluqueño, which, as I have shown in Chapter 9, has counterparts in all of the countries of this hemisphere in which the races of maize have been collected and studied. Since this race in various degrees of purity is widespread in America, it could have been introduced into Asia from almost any part of America. Because of its adaptation to high altitudes it might have almost immediately found a place in high-altitude cultivation in Sikkim.

Through the good offices of Dr. E. W. Sprague and N. L. Dhawan of the Rockefeller Foundation's agricultural program in India, I received in September 1966, 18 packets of seed of various strains of maize from the northeastern Himalayan region. Fourteen of these were originally from north Sikkim. A number of these had the typical pointed kernels of the Mexican race Palomero Toluqueño and its derivatives, and several also had the dingy pericarp color characteristic of this race. Because a new strain of the downy mildew disease had recently been discovered in India, the United States Plant Quarantine Service was reluctant to release this shipment of seeds received from India and did so only after obtaining my assurance that plants of this maize would not be grown in the field. For this reason I am, unfortunately, unable to furnish any additional data on its charac-

teristics. However, Dhawan's description and Thapa's illustration and my own examination of the kernels leave little doubt that the Sikkimese corn is a derivative of the Mexican popcorn race Palomero Toluqueño. Also consistent with this conclusion are the frequency polygons illustrated by Vishnu-Mittre and Gupta (1966) of the pore axis ratios of the pollen of the Sikkimese maize, which are quite similar to those of Palomero Toluqueño. Unless we now assume that there has been an Old World as well as a New World origin of maize—a possibility, but certainly a remote one—the important question is: "When was this American popcorn introduced into Asia?"

Thapa, on the basis of the much discussed illustration of Li Shih-Chen, which is now dated at 1560–1570 but was formerly thought to be earlier, some rather meager linguistic evidence, and the ritual in the eastern Himalayas of offering maize ears to the deity before harvesting, considers it "clear that maize was not unknown in the Old World during the pre-Columbian era." I find this evidence far from convincing. No more convincing, at least to me, is the discovery of subfossil pollen grains of a grass in lake sediments in the Kashmir Valley in several levels. Those found in level "H" representing the topmost part of the pollen sequence are undoubtedly post-Columbian and are not older than 200 to 300 years. Those found in levels E and F, however, are more puzzling since they are presumably older. Ranging in diameter between 70 and 83.7 microns, they are considered too large to be those of other cereals or native wild grasses and though rather small for maize they are within its size range. Since the dating of the sediments is still uncertain, and since the possibility of "intrusions" of pollen grains from the topmost level finding their way into the lower ones has not been completely ruled out, I regard the evidence of a pre-Columbian cultivation of corn in the Kashmir Valley as far from conclusive.

The third bit of "evidence" of pre-Columbian maize in Asia is a potsherd suspected of being prehistoric which bears upon its surface impressions arranged in orderly rows somewhat like the kernels on an ear of corn. A photograph of this specimen was sent to me by Mr. Vishnu-Mittre for my opinion. Although I recognized a superficial resemblance of the impression to those of kernel rows of corn it was clear that they were not made by rolling an ear over wet clay because the characteristic *pairs of rows* alternating with other pairs of rows which results from the fact that the pistillate spikelets of corn occur in pairs was not apparent in the photograph. Dr. Richard MacNeish, then of the National Museum of Canada, to whom I showed the photograph, considered it to be the impression of a piece of basketry or coarse fabric.

Pre-Columbian Corn in Africa?

The controversy that has raged for the past two decades over the possibility of a pre-Columbian diffusion of maize between America and Asia has had a close counterpart in the debate on the possibility of a pre-Columbian diffusion to Africa. This has now become so involved that it would require a long chapter to treat in detail the extensive literature which it has developed. I shall deal here only with what I consider to be the more relevant works, several of which have extensive bibliographies that scholars interested in this aspect of corn's history may find useful.

There is one important difference between the African and the Asian debate. No one, so far as I am aware, has seriously proposed that maize originated in Africa. Even Wiener (1920–1922), who attempted to prove that other well-known American plants, notably tobacco, sweet potatoes, cassava, and peanuts, had their origin in Africa and reached tropical America a few centuries before Columbus made his successful voyage in 1492, did not include maize in that category.

The chief proponent of a pre-Columbian diffusion of maize to Africa is Professor M. D. W. Jeffreys of the University of Witwatersrand, Johannesburg, South Africa. Jeffreys (1953a) first assembled extensive historical references purporting to show that there had been Arab-Negro contacts with the Americas beginning about A.D. 900 and that maize had been introduced into Africa before 1492. When Goodwin (1953) described potsherds from Ife in Nigeria, apparently decorated by rolling maize cobs over wet clay, Jeffreys (1953b) proceeded to date the introduction of maize into the region in Africa represented by Ife at A.D. 1000–1100, a date slightly earlier than the one which he had arrived at on the basis of other evidence. Through the kindness of Dr. W. R. Stanton, then of the West African Maize Research

Unit, I have had the opportunity of examining some of the potsherds in question and I have also since received from him a copy of an article (Stanton, 1963) in which some of the sherds are illustrated. There is no doubt that some of them were produced by rolling cobs or ears over the clay. The pairing of the spikelets so characteristic of the pistillate inflorescences of maize is clearly evident. There is so far, however, no indisputable evidence that the Nigerian pottery on which these impressions were made is pre-Columbian. Unfortunately, no radiocarbon dates of associated vegetal remains seem ever to have been made.

In attempting to show that maize had reached various parts of Africa before 1492 Jeffreys has made extensive use of linguistic evidence, some of which is highly involved. Although such evidence obviously has a place in the search for the origins and spread of cultivated plants it should be employed with caution. Jeffreys' conclusions about the introduction of maize into Western Africa have been severely criticized by Willett (1962). His article has in turn been subjected to criticism by Miracle (1963). I shall not attempt here to pass judgment on these various arguments. I can say, however, that there is certainly a real possibility that maize has sometimes been confused with the cultivated sorghums, which it resembles in many of its vegetative characteristics.

Pre-Columbian Maize in the Phillippines?

Not satisfied with "proving" that maize had reached both western and southern Africa in pre-Columbian times, Jeffreys (1965) attempted to show also that it had reached the Phillippines before 1492. Here, too, it was the globe-circling Arabs who are assumed to have made the introduction. The evidence for the early presence of maize in the Phillippines hinges on the interpretation of the word *miglio*, which Pigafetta mentioned in his account of Magellan's voyage. Merrill (1954) had no doubt that *miglio* referred to a species of millet, but Jeffreys contended that since Pigafetta used the word to refer to maize in Brazil he must have used it for maize in the Phillippines. The majority of Jeffreys' conclusions are strongly based on argumentation involving linguistics. Not being a philologist myself, I find the arguments difficult to follow and I am skeptical of the conclusions which they reach. On this point I find myself in com-

plete agreement with de Candolle (1886), who wrote: "The common names of cultivated plants are usually well known, and may afford indications touching the history of a species, but there are examples in which they are absurd, based upon errors, or vague and doubtful, and this involves a certain caution in their use." And again: "Scholars display vast learning in explaining the philological origin of a name, or its modifications in derived languages, but they cannot discover popular errors or absurdities. It is left for botanists to discover and point them out." And finally: "When we desire to make use of the common names to gather from them certain probabilities regarding the origin of species, it is necessary to consult dictionaries and the dissertations of philologists; but we must take into account the chances of error in these learned men, who, since they are neither cultivators nor botanists, may have made mistakes in the application of a name to a species."

Perhaps the most fantastic recent attempt to show that maize was known to the Old World before 1492 is that of Hui-Lin Li (1960–61). In attempting to identify the locations and natural products described in two Chinese works dated at A.D. 1178 and 1225, he concluded that the localities which certain seafaring Arabs reached might have been "the northern coast of South America, perhaps Maracaibo, the large gulf and lake along the coast of Venezuela, which would afford anchorage for seafaring ships." He identified the large melons "six feet round . . . enough for a meal for twenty or thirty men" as the American pumpkin, *Cucurbita pepo*, and the grains of wheat "three inches long" as the kernels of the large-seeded flour corns of the Andean region.

The conclusion that the "wheat" kernels three inches long represented the large-seeded flour corn of the Andean region overlooks the fact that this type of corn is post-Columbian in origin and is not found among the vegetal remains in archaeological sites nor is it represented on prehistoric ceramics. Li's other conjectures are equally farfetched, and his article would scarcely deserve mention in a serious work on the history of maize were it not that Jeffreys (1967) in his articles on the introduction of maize into southern Africa and into the Phillippines accepts these conclusions and at one point makes the unqualified statement that Li "showed not only that the Arabs had crossed the Atlantic well before A.D. 1100 but had also described maize." These articles by Jeffreys will undoubtedly be cited by other diffusionists—a shorter version of one of them has already been cited by Carter (1968). Thus do highly erroneous conjectures become accepted "evidence" in creating a case for pre-Columbian diffusion.

Pre-Columbian Corn in Europe?

Finan (1950), in a study of the maize illustrated and described in the herbals, concluded that there were two distinct types: the first, characterized by conspicuous prop roots, was probably a tropical form introduced into Europe from the Caribbean area soon after 1492; the second, which lacks prop roots but sometimes has numerous tillers, is similar to the northern flints of eastern North American and appears to have been well known in Europe within 50 years after America's discovery. Both Finan and Anderson (in the foreword to Finan's work) raised the question whether corn could have been introduced into Europe by the Norsemen before 1492.

Suto and Yoshida (1956) have gone even further than this in their unqualified assertions that the so-called Aegean type from which the European maize is assumed to have been derived and which was first described by Anderson and Brown (1953) was, like the so-called Persian type, diffused throughout the Old World before 1492. The suggestion of a pre-Columbian occurrence of corn in Europe has also had the support of Sauer (1960), who contends that the early establishment of maize culture in parts of Italy and the Balkans cannot be attributed solely to the advantages of soil and climate.

It has been generally supposed that maize was introduced into Europe by Columbus on the return from his first voyage, but Weatherwax (1954) in a careful search of the old chronicles failed to find explicit support on this point. He concluded that maize may have been taken to Spain when Columbus returned from the first voyage but that there is no definite record of this. It is known to have been there soon after the return from the second voyage. As early as 1494 there was a reference to a grain which Weatherwax thinks "must have been maize." There is no question about a reference to maize in the 1511 edition of the *Decade* of Peter Martyr, an Italian scholar attached to the Spanish court. The first reference to corn in the botanical literature is in the herbal of Jerome Bock, which Weatherwax dates at 1532. The first illustration of the plant, a very artistic one, is that of Leonard Fuchs, 1542.* During the sixteenth

*Sauer (1960) stated that two earlier illustrations are known but he did not reproduce them.

century maize was described in numerous herbals, representing virtually all of the countries of Europe: Germany, the Low Countries, Italy, Spain, Switzerland, and England (Finan, 1950). In contrast it was not mentioned in a single botanical work in the fourteenth century and only in the last decade of the fifteenth century.

To me one of the most significant facts in corn's history is that actual prehistoric remains of it, which are abundant in the New World, being found virtually throughout the range of its culture, including the wet tropics, are completely lacking in all parts of the Old World. As Edgar Anderson (1947) has pointed out, corn, more than any other plant, documents its own history. Some of the tissues of its cobs are highly indurated and are well designed for preservation under a variety of conditions. Yet not a single corn cob, unmistakably pre-Columbian, has yet been found in any part of the Old World.

It may be, as various students have concluded, that maize reached Europe or west Africa or south Africa or the Philippines or Asia or India or several or all of these places before 1492, but indisputable evidence of this is still lacking, and the two assertions of de Candolle quoted at the beginning of this chapter are, for all practical purposes, still valid.

18 The Prehistoric and Modern Improvement of Maize

Charles Darwin did far more than propound the theory of evolution, the doctrine of natural selection and the descent of man. In 1876 he came out with a book on cross- and self-fertilization in plants, containing ideas which were destined to reach into the heart of the Corn Belt and change the nature of the corn plant for all time. Wallace and Brown, 1956

The improvement of domestic animals and plants through breeding is essentially a form of evolution directed by man. Consequently, understanding the past evolution of a species should be useful to the breeder in formulating his breeding plans or in designing experiments concerned with various aspects of improvement. On this point I cannot agree with my good friend Jay Lush, who once stated that the history of evolution is of little importance to the practical breeder. If this is true with respect to domestic animals, it is far from true of cultivated plants, especially of those species in which interspecific hybridization has been a factor in their evolution. In such species operationally sound decisions must often depend on information about what Lerner (1958) has called the "genetic biography" of a population. In the case of maize at least, I believe that a knowledge of its past evolutionary history has much to offer in aiding the breeder to avoid pitfalls and to improve his breeding techniques. Consequently, I propose in this final part of the book to show how certain features of corn's evolutionary history appear to me to have a bearing upon its future evolution under man's direction and to point out opportunities for improvement which seem to me to be not yet fully appreciated or exploited. First, however, I wish to review briefly the methods of breeding that have been employed in the past and those that are in use now. For more extensive treatment of the conventional methods of modern times I refer the reader to the excellent treatises of Richey (1950), Sprague (1955), and Hayes (1963).

Corn Breeding before Columbus

Several students of maize, especially Kemp-ton (1931, 1936) and Weatherwax (1942, 1954), have regarded cultivated maize as the product of the plant-breeding skills of the American Indian, and it is true that in the hands of Indian cultivators maize had reached a high state of development when America was discovered. All of the principal commercial types of corn recognized today: dent, flint, flour, pop, and sweet, were already in existence when the white man appeared on the scene and, until hybrid corn was developed, the modern corn breeder, for all his rigorous selection, had made little progress in improving the productiveness over the better Indian varieties.

It is doubtful, however, that the Indian was an accomplished plant breeder in the sense of visualizing a new type of maize and selecting toward it. He did have, and in many parts of the hemisphere still has, a deep affection for the plant. A corn plant coming up in a field of any other crop or even in a lawn or public park in Mexico is seldom removed by laborers; it is much too precious to be destroyed. Wallace and Brown (1956) speak of the Hopi as having an intense, purposeful, thoughtful, never-ending interest in corn, and the same statement might be made of the Indians of Peru, of Guatemala, of Mexico, and indeed of almost any area where corn is the Indian's basic food plant. Yet these authors are also correct in stating that "it is not scientifically feasible to demonstrate that *thought*, that *loving* a corn plant, will improve it or its progeny."

Weatherwax (1954) states: "There is no avoiding the conclusion that, regardless of his other merits or faults, the Indian was a good corn breeder. He has, however, failed to pass on to the white man any details as to how he accomplished what he did. It is probable that he had no idea of how he did this or that he even realized what he was doing." In the last analysis the question of whether the Indian was or was not a skilled plant breeder becomes one of semantics. It should, however, be useful to distinguish between the unconscious roles and the purposeful ones that he played in the improvement of corn over a period of 6,000 years or more, and without wishing to diminish his stature as a corn breeder I shall attempt to do this.

There is no doubt that the Indian domesticated corn, and this is in itself a monumental achievement. Yet some of the steps involved in domestication were undoubtedly largely fortuitous. I am unable to follow Sauer (1952) in visualizing domestication as the progeny of leisure, itself the product of an abundant food supply, during which the initial steps in domestication were presumably thought out, planned, and executed. It is much more likely, I think, that the mother of maize domestication, like that of many other inventions, was necessity. The plant was probably never abundant in the wild—never in any region the principal component of the native vegetation. And unlike the sweet potato and cassava, which when harvested renewed the colony from fragments of roots overlooked by the diggers, a colony of maize might be completely extinguished by a single thorough picking. As the Indians found their resources of wild maize diminishing on the one hand and on the other observed how much more productive than wild plants were the "volunteer" maize plants which grew up in the well-fertilized soil around their dwellings from accidentally dropped seeds, it was but a step to the conscious sowing of seed in selected sites. Removal of competing vegetation was an obvious second step or indeed might have actually have been the first, with sowing coming later.

These two steps marked a momentous milestone in the history of corn's evolution under domestication. The plant had entered a new man-made environment in which selection began to operate in new directions, especially to preserve mutant forms which in nature were selected against. Thus the mutation from tunicate to nontunicate, which in corn's evolutionary history may have occurred again and again only to be promptly eliminated, now had survival value, and once it became the predominating type many other changes followed—long husks, for example, to protect the naked seed, and selection by man for larger ears, which, as a secondary effect, tended to reduce the development of tillers.

Pre-Columbian Selection

It is doubtful that in the early stages of domestication artificial selection was practiced to any great extent. The corn of Bat Cave, for example, the product of a primitive form of agriculture, shows no evidence of having been selected by man. During the 3,000 years or more that the cave was occupied, the variability of the maize increased enormously but there is no indication of the fixation through selection of a type. The cave Indian merely accepted the maize that the forces of mutation, hybridization, genetic drift, and natural selection presented to him. Whatever selection he practiced was probably dysgenic. Like his modern counterpart among the Chibcha Indians of Colombia, he probably consumed first the maize that he considered best and used for planting that which remained.

At some stage in the development of Indian agriculture, however, the practice of artificial selection began and has persisted to this day. The Hopi, it is said, never save seed from an ear which shows evidence of mixture. The care and maintenance of particular varieties was, in earlier times, kept largely within families, the responsibility for a variety being handed down within the family from generation to generation (Wallace and Brown, 1956).

In Mexico it is common during the winter months to see ears, selected for planting the following rainy season, hanging in a tree near the dwelling. In tropical regions such ears are hung indoors suspended from the ridgepole. Here they are fumigated each time a meal is prepared by the smoke of the cooking fire, which in the absence of a chimney collects at the peak of the roof, gradually seeping out through the thatch. This practice not only prevents damage by weevils, grain moths, and rodents but also coats the kernels with a film of tar, which probably acts as a deterrent to bird damage when the corn is planted. Whether stored in trees or under the ridgepole, the ears selected for planting are always more uniform than the general run of corn of the region and show evidence of careful selection.

Among pure-blooded non-Spanish-speaking Indians of Guatemala rigid selection for type of seed is often practiced (Anderson, 1947). Guatemalan Indians are also prone to preserve unusual types such as those with fasciated ears of the subrace, Grueso, and with branched ears of the subrace, Ramoso (Wellhausen et al., 1957). The practice of preserving distinct types accounts in part for the fact that in Guatemala there is a high correlation between the diversity of maize and the percentage of Indians in the population. In the nine departments of Guatemala in which the proportion of Indians, "Indigenas," as they are called in the Guatemalan census, is two-thirds or more are found all except one of the Guatemalan races of maize.

The Guatemalan practice of rigorous selection for type, like the Guatemalan maize itself, may be a cultural trait introduced from South America. There, especially in Peru, selection has become an art. I have seen at intermediate altitudes in Peru the harvest of fields of Kculli, the black corn used for dying and for coloring nonalcoholic *chichas* and the puddings, *mazomorras*, made from corn and tapioca flours, laid out in the sun to dry with not a single color deviant in sight. These have already been removed from the drying area and set off in a pile to one side. The pile of off-type ears is usually remarkably small and testifies to the effectiveness of previous generations of rigid selection. In the race Cuzco, seven types differing in endosperm texture or pericarp color are main-

tained in states of relative purity. Several of these involve mosaic or variegated pericarp which is controlled by an unstable locus, and there is little possibility of ever establishing a completely pure race with such a pericarp pattern. Yet varieties are maintained in which the great majority of the ears are of this type.

The Peruvian Indian must be given credit also for developing other special-purpose types of corn such as the sweet corn race Chullpi, which is especially prized for making the native beer *chicha*, its high sugar content contributing to developing a high alcoholic content in the final product. The Peruvian Chullpi is the progenitor of the Colombian and Mexican Maíz Dulce, and as first suggested by Kelly and Anderson (1943), the latter may have played a role in the origin of modern sweet corn varieties. The most spectacular product of selection in Peru, however, is the race Cuzco Gigante, the large-seeded flour corn of the Urabamba Valley of the Cuzco region. Repeated hybridization has undoubtedly played a part in the evolution of this race (Grobman et al., 1961), but human selection for two characteristics, floury endosperm and size of seed, have also been important. The Peruvians have given particular attention to these two characteristics because most of the maize they use for food is consumed as individual, unprocessed kernels to which heat is applied. In contrast, the Mexican Indians, whose staff of life is an unleavened bread, the *tortilla*, make it by steeping the maize in limewater, rubbing off the pericarp, grinding the hulled kernels, and fashioning the dough into round, flat sheets which are baked on hot metal surfaces. The Peruvian method, when used on small corneous kernels of wild maize or maize in the early stages of domestication, causes the kernels to explode or "pop," thus instantly converting the hard, bony seed, too difficult to chew, into a tender, tasty morsel, nutritionally the counterpart of a whole-grain bread.

As kernel size increased, largely because of hybridization between races, the kernels lost their ability to pop and when exposed to heat merely became parched. Parched flour corn is more easily chewed than parched flint corn, and when parching became the predominant method of processing, selection for floury endosperm became a rewarding practice, and, since the individual kernel was still the unit of consumption, selection for large seeds was undoubtedly also practiced. By the time of the Spanish conquest, Peruvian Indians had varieties of maize with grains as large as the seeds of the European broad bean *Vicia faba* (cf. Grobman et al., 1961). Following the creation of new genetic diversity resulting from the hybridization of a large-seeded Peruvian type with a fairly large-seeded race, Tabloncillo from Mexico, several more centuries of selection have culminated in the production of Cuzco Gigante, the corn with the world's largest kernels.

By this time boiling had largely replaced parching as a means of preparing maize for food. I have seen groups of Peruvian Indian farm laborers making their entire noonday meal from the boiled grains, eaten individually, of Cuzco maize. Should the modern corn breeder seek to improve kernel size in maize, he will find the genes needed for that purpose assembled for him in the large-seeded varieties of Peru, the product of centuries of selection on the part of the Indians.

Pre-Columbian Hybridization

A second important factor in corn's evolution under domestication was racial hybridization. To this the Indian contributed unconsciously by bringing distinct races into crossing proximity through his migrations, warfare, and trade. And if my conclusion set forth in Chapter 10 that there were at least six different races of wild corn is correct, then the importance of racial hybridization as a factor in corn's evolution looms very large indeed.

Although it may have been unconscious on his part, the Indian was nevertheless a maize hybridizer on a grand scale. Through the centuries he brought distinct races together repeatedly, with the result that some have very complex pedigrees. The modern corn breeder will wish to examine these to identify, and in some instances to employ for breeding material, those races which have appeared repeatedly in the ancestry of highly successful modern races. The breeder may also profitably give some attention to those races which, highly successful in some parts of the world, seem not yet to have been generally employed in modern breeding operations.

There have been reports of an Indian practice of mixing seed to promote hybridization, but if this was ever done consciously, it was the exception rather than the rule. There is some evidence, considered below, that the Indians planted maize and teosinte together. Cutler reports that the Indians of the South American highlands often grew two distinct types, such as field corn and sweet corn, in the same field, but this is done from necessity and not for the purpose of promoting hybridization.

Hybridization with Teosinte

It is possible that the Indians of Western Mexico have long recognized the beneficial effects of teosinte introgression, for Lumholtz (1902) has reported their practice of interplanting *maizillo* (probably teosinte) and maize for the purpose of improving the latter. Melhus and Chamberlain (1953) were told by one of their informants, a Guatemalan Indian woman 75 years of age, that her husband often planted teosinte in the corn rows to improve the corn. The practice may be an ancient one, for teosinte seeds were found at several levels of a cave, Romero's Cave in Tamaulipas, excavated by Dr. Richard MacNeish. Since living teosinte has never been collected in Tamaulipas or indeed in any part of eastern Mexico, it is possible that these cave-dwelling Indians, whose maize shows affinities with the races of western Mexico, were interplanting teosinte with their maize. Alternative possibilities which should also be considered are that (1) teosinte was once native to Tamaulipas and has since disappeared, and (2) it segregated out of maize-teosinte hybrids introduced as admixtures in maize brought in from the west.

If the Indian of any time or place consciously and purposefully brought maize and teosinte into crossing proximity with the object of improving the maize, he must be regarded as a skilled plant breeder even if he knew nothing about the mechanisms of hybridization. Indeed he was far ahead of his time. The corn breeder of today is only beginning to appreciate the possibility of improving maize through hybridization with its relatives teosinte and *Tripsacum* or in consciously employing the germ plasm from these species that has been incorporated into modern maize through introgression of the past. There is at least no doubt that much of the maize of Mexico has undergone admixture with teosinte and has been improved thereby. This fact should be of some importance to the modern corn breeder.

Corn Breeding in Early Historical Times

As the Indian accepted, with little conscious effort to change them, the races of maize which the forces involved in evolution had delivered to him, so the white man arriving in the New World accepted, with no initial interest in modifying them, the races that he took over from the Indian. It was not until the latter part of the eighteenth century that conscious efforts to improve the corn plant were made. What the white man did to corn between that time and the invention of hybrid corn is well described in several chapters of the interesting and useful book, *Corn and Its Early Fathers*, by Henry Wallace and William Brown. Here I wish to consider only the main factors participating in the evolution of the principal varieties of corn of the United States and in doing so shall divide the period from 1772 to 1918 into four parts: the periods (1) of mixing varieties, (2) of mass selection, (3) of the corn shows, (4) of ear-to-row breeding. Each of these periods has left its mark on present-day corn.

One of the first settlers to practice mixing of varieties of corn or at any rate one of the first to write about it was Joseph Cooper, who farmed in New Jersey across the river from Philadelphia. In the first volume of the *Proceedings of the Philadelphia Agricultural Society*, published in 1808, Cooper described a mixed planting he had made shortly after 1772 between a corn from Guinea (probably a tropical flint) and the "larger and earlier" kinds of corn. He then saved seeds from stalks which produced the greatest number of ears and ripened first and he was "not a little gratified to find its production preferable, both in quantity and quality" to that of any corn he had ever previously planted. Here is the first recorded attempt at combining earliness and multiple ears. Cooper's observations were repeated in several other publications and were undoubtedly widely read. Some seventy years after Cooper's time, farmers continued their attempts to combine the characteristics of earliness and many ears. Some of the prolific corn so highly prized by early nineteenth-century farmers may have received their germ plasm from this kind of mixing and selection

(Wallace and Brown, 1956*). Perhaps the time is ripe for repeating some of these early experiments.

The first record of mixing of the two types which we now know to be the progenitors of Corn Belt dent corn is that contained in the report of John Lorain in a letter dated July 21, 1812, written to the Philadelphia Agricultural Society, in which he stated that yields of corn could be improved by at least one-third "By forming a judicious mixture with the gourdseed and flinty corn" (W. & B.).

Peter Browne, professor at Lafayette College, in "An Essay on Indian Corn" published in 1837 a list of 35 kinds of corn, many of which he said were mixtures of gourdseed and flint (W. & B.). However, a large part of the ancestry of modern Corn Belt hybrid corn traces back to three varieties, all produced by mixing: Reid's yellow dent, Krug's yellow dent, and Lancaster Sure Crop. Reid's yellow dent was developed by Robert Reid, who, on his farm about 30 miles south of Peoria, Illinois, replanted a poor stand of Gordon Hopkins corn, a gourdseed type originally from Virginia, with Little Yellow, an early flint variety. From the cross Robert Reid and his son, James, selected for an early corn with cylindrical ears bearing 18 to 24 rows of kernels. The Reid corn came to public attention when it won a prize at the Chicago World's Fair in 1893. Later it was widely distributed throughout the Corn Belt states (W. & B.). George Krug of Woodford County in central Illinois combined a Nebraska strain of the Reid corn with a variety called Iowa Goldmine. In selecting for ears heavy for their size borne on good stalks and for lustrous kernels with an "oily" appearance, he undoubtedly incorporated into an otherwise somewhat variable variety some degree of resistance to stalk and ear rot (W. & B.). Lancaster Sure Crop was the handiwork of Isaac Hershey of Lancaster County, Pennsylvania. Hershey mixed a late, rough, large-seeded corn with an early flinty corn. To this mixture he added from time to time at least six other varieties of corn. He then selected for earliness and freedom from disease (W. & B.).

From these three varieties, Reid, Krug, and Lancaster Surecrop, have come a number of

*Referred to hereafter as W. and B.

the inbred strains widely used in the production of modern hybrid corn. All three, as their history shows, grew out of mixtures of the gourdseed or southern dent types with the northern flint types. To find out how much gourdseed and how much flint was in these three varieties, Brown (W. & B.) inbred them for six years, self-pollinating every stalk, whether it looked good or bad, generation after generation. The percentage of surviving inbreds which were flint-like was much higher than those resembling the gourdseed type. The inbreds out of Krug and Lancaster, neither of which had been rigorously selected for uniformity or "spoiled by corn shows" were more variable than those of Reid, although these too were quite variable, almost none resembling the kind of "show" corn for which this variety had been selected during the corn-show period. Brown concluded that "you can get anything out of Reid."

We now have some conception of the reason for the productiveness of these Corn Belt varieties and for the fact that almost anything can be gotten out of them by inbreeding. In an earlier chapter I showed that Corn Belt dent corn is a complex hybrid, the product of repeated racial hybridization, of introgression from teosinte, and perhaps also from *Tripsacum*.

The Period of the Corn Shows

Since it would be difficult indeed to improve upon the description of the corn shows which Wallace and Brown have given in one of the liveliest chapters of their book, I quote verbatim with permission of Michigan State University Press several excerpts, the first describing the shows:

Beginning about 1900, an unusual social phenomenon known as the Corn Show sprang up all over the Corn Belt. Certain so-called corn experts would, at harvest time, pick out their best ears of corn. They would put their ears of corn together in a ten-ear sample in which all the ears would have eighteen or twenty rows and none of the ears would be shorter than nine and one-half inches or longer than ten and one-half inches. The kernels would be deep and uniformly keystone-shaped with large germs. "Uniformity" of both ear and kernel type was the objective of nearly all cornbreeders from 1900 to 1920. The agri-

cultural colleges in those days trained both regular students and farmers to judge corn. In a mild way, the Corn Belt was swept by a craze similar to the tulip mania in Holland in the Seventeenth Century. A grand champion ear of corn would sell for one hundred and fifty dollars. So great was the prestige of the corn shows that very few of the corn-show judges trained by the agricultural colleges thought of planting the grand champion ears in comparison with ordinary corn to see how they would yield in competition with each other. And woe betide the corn judge who failed to place first that sample which was most "uniform" for ear-length and kernel-type!*

The second recalling some of the tricks employed:

Here and there some of the old corn-show tricks still hold on. Special prizes are sometimes offered to the men who will bring in the longest ear of corn. One man got the winning quarter of an inch on his ear by soaking it in water overnight. Another man tried to win his prize for the heaviest ear by inserting a metal rod through the center. His deception was found by X-ray. To produce ears with odd numbered rows, special surgical operations have been performed while the young kernels were developing. Sometimes kernels are carefully pasted into place to make the ear seem more perfect. But gone are the days when the corn exhibitors from one state would meet in a hotel room to trade ears to increase the uniformity of their respective ten-ear samples. Corn shows, like all shows, are social occasions. They were never meant to be scientific; it is unfortunate that many

*It is of interest in this connection that Henry Wallace himself as a boy of sixteen had doubts that the prize ears of the corn shows would actually produce a better crop than some of those considered less desirable by the judges, and in 1904 he conducted a test in which he compared the yields of forty individual ears of corn which one of the most prominent corn judges had placed according to the show standards then current. It happened that the row planted from the ear which had been placed first had one of the lowest yields. This boyhood experiment convinced Wallace that the corn shows did more harm than good, and he campaigned against them at every opportunity until they eventually became discredited. In later years he said, "Neither corn nor man were meant to be completely uniform," and pointed out that in the United States 85 percent of the corn is fed to livestock "which is not impressed in the slightest by the appearance of the ears."

farmers were deceived for decades into thinking them practical.

And the third, the continuing influence of the corn-show standards: "Even to this day most breeders of hybrid corn try to combine their inbreds in such a way as to produce a hybrid that will meet corn-show standards. It is not unlikely that this anachronism costs Corn Belt farmers many thousands of dollars a year." The last statement, although true when it was published (1956), is no longer applicable. In developing hybrids adapted to machine harvesting, corn breeders have pretty well abandoned the old corn-show standards. Whatever resemblance still remains between modern hybrids and corn-show types is the result of the continued use of certain outstanding inbred strains developed in the early years of hybrid corn breeding.

Ear-to-Row Breeding

The history of corn breeding would not be complete without some mention of the method known as ear-to-row breeding. Introduced by the Illinois Agricultural Experiment Station in 1896 as part of the now-famous experiment on selection for oil and protein content, it proved successful in developing "high" and "low" strains of both oil and protein. In fifty years of selection the high-oil strain was increased from 4.7 to about 15 percent, the low strain decreased to 1.01 percent. Similar changes were achieved in the high and low protein lines. In another experiment at the Illinois Station involving selection for height of the ears, which is associated with the height of the plant, the high ear strain at the end of six years was approximately twice the height of the low-ear strain.

The theory underlying the method is basically sound: that an ear of corn should be judged not by its appearance, chemical composition, or other characteristics but by its progeny. Actually this is only a slight modification of the method of progeny testing employed by Louis de Vilmorin in France in successfully raising the sugar content of beets from 9.8–11 percent to 16–17 percent. In the case of corn the individual unit is the ear and the progeny test is the row grown in the field from the seed of one ear. The selected line is perpetuated by combining either superior progeny or the remnant seeds from ears that had produced superior progeny.

Since the method was successful in changing chemical composition and height of plant, should it not also be successful in improving yields? Wide variation in the yields of the individual rows suggested that it should be, the higher-yielding rows sometimes producing almost twice as much grain as the low-yielding ones. Yet although the method has been widely tested, it seems never to have lived up to its expectations; no really outstanding variety of corn seems ever to have been developed from it. Selection for low yield, however, in the extensive experiments conducted in Illinois was quite effective. Various reasons have been given for its failure: (1) Combining only the higher yielding rows or the remnant seeds of their parent ears has resulted in a certain amount of inbreeding which has depressed the yields. Attempts were made to avoid this supposed difficulty by detasseling all of the rows in the increased plot and saving the seed only from such rows. (2) A single row is not an adequate test of the progeny. (3) The more productive ears usually were so primarily because they were superior chance hybrids which did not breed true (Richey, 1950).

I do not find any or all of these explanations completely satisfying. Detasseling alternate rows should have reduced inbreeding to a minimum; single rows have proved adequate in identifying low-yielding progenies; the fact that superior rows are hybrids is irrelevant—productive synthetics have been developed by combining superior single crosses. The real explanation in my opinion lies in the fact that the composites produced by combining the superior ear rows derived from a single variety or the remnants of their parent ears do not involve enough genetic diversity.

The History of Hybrid Corn*

Scientific achievements, like royalty and pure-bred livestock, often have long and complex genealogies. Lines of descent usually trace back to more than one distinguished progenitor, and the same progenitor sometimes participates in more than one line. Hybrid corn is no exception to this general rule. Two

*Much of the material on hybrid corn in this chapter is adapted from my article "The Mystery of Corn" that appeared in *Scientific American* in August, 1951. I am grateful for permission to reproduce it here.

distinct lines of descent have converged to make hybrid corn an accomplished fact, and a third line has had a marked influence. If we trace these lines to their recognizable sources they lead us to three famous biological scientists of the nineteenth century: Charles Darwin, Gregor Mendel, and Frances Galton and to four eminent twentieth-century biologists: Johannsen, Shull, East, and Jones.

One line of descent in the genealogy of hybrid corn begins with Charles Darwin, who made extensive investigations on the effects of self- and cross-pollination in plants. Included among his experimental subjects was the corn plant. His were the first controlled experiments in which crossed and self-pollinated individuals were compared under identical environmental conditions. He was the first to see that it was the crossing between unrelated varieties of a plant, not the mere act of crossing itself, that produced hybrid vigor, for he found that when separate flowers on the same plant and different plants of the same strain were crossed their progeny did not possess such vigor. He concluded quite correctly that the phenomenon occurred only when diverse heredities were united.

Darwin's experiments were known even before their publication to the American botanist, Asa Gray, with whom Darwin conducted a regular and lively correspondence (Dupree, 1959). One of Gray's students, William Beal, became, like Gray, an admirer and follower of Darwin. At Michigan State College Beal undertook the first controlled experiments aimed at the improvement of corn for the utilization of hybrid vigor, or as he preferred to call it, "controlled parentage." He selected some of the varieties of flint and dent corn, then commonly grown, and planted them together in a field isolated from other corn. He removed the tassels—the pollen-bearing male flower clusters—from one variety before pollen was shed. The female flowers of these emasculated plants had then to receive their pollen from the tassels of another variety growing in the same field. The seed borne on the detasselled plant, being crossed seed, produced only hybrid plants the following season.

The technique Beal invented for crossing corn—planting two kinds in the same field and removing the tassels of one—proved highly successful and until recently, when replaced by the use of cytoplasmic male sterility, was the method employed in producing millions of bushels of hybrid seed corn. As a device for increasing yields of corn, this method of crossing two open-pollinated varieties—each genetically heterogenous—was not completely effective. Many of the crosses were more productive than their parents but seldom enough so to justify the time and care spent in producing the crossed seed. The missing requirement—the basic principle that made hybrid corn practical—was discovered by George H. Shull at the Carnegie Institution. Yet the line of descent represented by Darwin, Gray, and Beal was to continue, and, as we shall see later, it produced results that complemented those of the Shull line, which also traces back to Darwin through Johannsen and Galton.

Shull's Contribution

Shull's discovery was an unexpected byproduct of theoretical studies on inheritance which he had begun in 1905. His contribution grew from certain earlier studies made by two great European scientists: Darwin cousin and scientific disciple, Francis Galton,* the founder of two sciences, biometrics and eugenics, and the Danish botanist, Wilhelm Johannsen. Galton had recognized that the result of the combination of parental heredity could take two forms an "alternative" inheritance, such as the coat color of Bassett hounds, which came from one parent or the other but was not a mixture of both, and a "blended" inheritance, such as in human stature. He observed that children of very tall parents are shorter on the average than their parents while children of very short parents tend to be taller. These observations, based on extensive measurements, led Galton to formulate his "Law of Regression," which holds that the progeny of parents above or below the average in any given character tend to regress toward the average.

*"I always think of you" wrote Galton in a letter to Darwin dated December 24, 1869, "in the same way as converts from barbarism think of the teacher who first relieved them from the intolerable burden of their superstitions... Consequently the appearance of your *Origin of Species* formed a real crisis in my life; your book drove away the constraint of my old superstition as if it had been a nightmare, and was the first to give me freedom of thought" (Trattner, 1938).

This regression is seldom complete, however, and Johannsen saw in that circumstance an opportunity for controlling heredity through selection in successive generations of extreme variations from the average. He tested the possibility by trying to breed, by repeated selection, unusually large and unusually small beans. His choice of the bean as his experimental subject was to have unexpected and far-reaching results. He found that although selection apparently was effective in the first generation it had no measurable effect whatever in later generations. From these results Johannsen concluded that in self-fertilized plants, such as the bean, the progeny of a single plant represents a "pure line" in which all individuals are genetically identical and in which any residual variation is environmental in origin. He postulated that an unselected race such as the ordinary garden bean with which he started his experiments was a mixture of pure lines differing among themselves in many characteristics but each one genetically homogeneous.

Johannsen's pure-line theory has been widely applied to the improvement of cereals and other self-fertilized plants. Many of the varieties of wheat, oats, barley, rice, sorghum, and flax grown today are the result of sorting out the pure lines in mixed agricultural races and identifying and multiplying the superior ones. We also owe to Johannsen the concept—a working tool of every plant and animal breeder—of the "genotype," which represents the individual's hereditary endowment and the "phenotype," which represents its physical characteristics.

Shull's contribution was to apply the pure-line theory to corn with spectacular though unpremeditated results. His experiments started with the objective of analyzing the inheritance of quantitative or "blending" characteristics, and he chose as an inherited quantitative character suitable for study the number of kernel rows on ears of corn. Since corn is naturally a cross-pollinated plant, he practiced artificial self-pollination in order to produce lines breeding true for various numbers of kernels rows. These lines, as a consequence of the inbreeding resulting from self-pollination, declined in vigor and productiveness and at the same time each became quite uniform. Shull concluded correctly that he had isolated pure lines of corn similar to those described by Johannsen in beans. Then as a

first step in studying the inheritance of kernel row number he crossed these pure lines. The results were surprising and highly significant. The hybrids between two pure lines were quite uniform like their inbred parents, but, unlike their parents, they were vigorous and productive. Some hybrids were definitely superior to the original open-pollinated varieties from which they had been derived. Inbreeding had isolated from a single heterogeneous species the diverse germinal entities whose union Darwin had earlier postulated as the cause of hybrid vigor.

Shull recognized at once that inbreeding followed by crossing offered an entirely new method of improving the yield of corn. In two papers published in 1908 and 1909, he reported his results and outlined a method of corn-breeding based upon his discovery. He proposed the isolation of inbred strains through self-pollination as the first step and the crossing of two selected inbred strains as the second. Only the seed of the first generation cross was to be used for crop production because hybrid vigor is always at its maximum in the first generation.

Shull's idea of maintaining otherwise useless inbred strains of corn solely for the purpose of utilizing the heterosis resulting from their hybridization was revolutionary as a method of corn-breeding. As a genetic system it is comparable to types of adaptive polymorphism described by Dobzhansky (1949) in which natural selection (in the case of hybrid corn, selection acting in a man-made environment) preserves certain chromosomes, not primarily because of their intrinsic worth but because they interact effectively with other chromosomes similarly preserved to produce a highly successful Mendelian population.

That Shull should have been able, from the limited experiments then completed and from unreplicated yield tests which present-day agronomists would consider wholly inadequate, not only to draw valid conclusions regarding the effects of inbreeding and cross-breeding in corn but also to design a new method of corn-breeding based upon the exploitation of heterosis is a creative achievement of the first order. If genius is, as one writer has defined it, "the ability to draw correct conclusions from inadequate data" then Shull must be counted a genius. Yet Shull's method, although he traveled extensively through the Corn Belt to promote it, was not adopted. The crossing of two inbred strains—now known as a single cross—proved impractical as a method of seed production. Because the inbred strains were lacking in vigor and were unproductive, hybrid seed obtained in this way was considered to be too expensive for general use. Hybrid corn did not come into being until the line of descent represented by Shull, Johannsen, Galton, and Darwin converged with the line represented by Jones, East, Beal, and Darwin to produce a kind of genealogical "heterosis" whose results have been nothing short of explosive. One further major development was needed to make hybrid corn practical and the great boon to agriculture which it has become. This contribution came from the Connecticut Agricultural Experiment Station.

The story begins in 1906 when Edward M. East arrived in New Haven from the University of Illinois. There he had participated in two highly significant corn-breeding experiments: one concerned with the effects of inbreeding initiated by two of Beal's former students, Eugene Davenport and P. G. Holden, the other under the direction of Cyril Hopkins an attempt to change the chemical composition of corn by selection. At the Connecticut Station East began a series of studies on the effects of inbreeding and cross-breeding on corn which were to continue for more than fifty years and which have yielded a great deal of information about corn, including the effects of selection upon its chemistry. It was East who called attention to the need for developing a more practical method of producing hybrid seed. It remained for Donald F. Jones, one of East's students who assumed charge of the Connecticut experiments in 1915, to invent the method that solved the problem.

Jones's Contribution

Jones's solution was simply to use seed from a double cross instead of a single cross. The double cross, which combines four inbred strains, is a hybrid of two single crosses; for example, two inbred strains, A and B, are combined to produce the single cross A × B. Two additional inbred strains, C and D, are combined to produce a second single cross, C × D. All four strains are now brought together in the double cross (A × B) × (C × D).

At first glance it may seem paradoxical to solve the problem of hybrid seed production by making three crosses instead of one. But the double cross is actually an ingenious device for making a small amount of scarce single-crossed seed go a long way. Whereas single-crossed seed is produced on undersized ears borne on stunted inbred plants, double-crossed seed is produced on normal-sized ears borne on vigorous single-cross plants. A few bushels of single-crossed seed can be converted in one generation to several thousand bushels of double-crossed seed. The difference in cost of the two kinds of seed is reflected in the units in which they were formerly sold: double-crossed seed was priced by the bushel, single-crossed seed by the thousand seeds. Double-cross hybrids are never as uniform as single crosses, but they may be just as productive or more so.

Ironically, once the use of double crosses established the production of hybrid corn on a highly successful scale, corn breeders found that by developing more vigorous inbred strains than those isolated by the early breeders, and by employing cytoplasmic male sterility to avoid detasseling (see Chapter 20) it was possible to employ single crosses instead of double crosses in the production of hybrid seed corn. In the decade of the sixties much of the hybrid corn grown in the United States was represented by single crosses.

Although the best of the single crosses produce somewhat higher yields than the best double crosses there is a danger inherent in growing them extensively. This stems from the fact that although they are hybrids and thus genetically heterozygous they are also genetically homogeneous, all plants of a single cross being genetically identical. In this respect they are similar to the pure lines of self-fertilized crops such as wheat and are subject to the same hazards.

Writing on this subject some years ago Jones (1958) stated: "These genetically uniform pure line varieties are very productive and highly desirable when environmental conditions are favorable and the varieties are well protected from pests of all kinds. When these external factors are not all favorable the result can be disastrous." The result was disastrous in 1970 when a new mutant strain of the southern corn blight, *Helminthosporium maydis* spread throughout the corn-growing

states, causing a loss in yields for the nation as a whole of 13 percent and much higher losses than this in some states. The almost universal use of the Texas cytoplasmic male sterility was the principal factor involved in the widespread susceptibility of the 1970 crop, but the genetic homogeneity of single crosses was also a factor. Farmers still growing double crosses generally suffered smaller losses than those growing single crosses.

Jones's statement (1958) on the superiority in certain respects of double crosses over single crosses is relevant: "The first double crosses were designed to overcome the handicaps that the single crosses had in seed production. It actually turned out that the genic equilibrium, genetic inertia, genetic homeostasis, or whatever it may be called, is by far the more important. It is the gyroscope that holds the ship steady in a surging sea." Jones made a second important contribution to the development of hybrid corn by presenting a genetic interpretation of hybrid vigor in terms of the hereditary factors of Gregor Mendel and the recently developed chromosome theory of Thomas H. Morgan and his students. Shull and East had suggested that hybrid vigor was due to some physiological stimulation resulting from hybridity itself. Shull was quite certain that something more than gene action was involved; he thought that part of the stimulation might be derived from the interaction between the male nucleus and egg cytoplasm.

Jones proposed the theory that hybrid vigor is the product of bringing together in the hybrid the favorable genes of both parents. These are usually partly dominant. Thus if one inbred strain has the genes *AA BB cc dd* (to use a greatly oversimplified example) and the other has the genes *aa bb CC DD*, the first generation hybrid has the genetic constitution *Aa Bb Cc Dd*. Since the genes *A*, *B*, *C*, and *D* are assumed not only to have favorable effects but to be partially dominant in their action, the hybrid contains the best genes of both parents and is correspondingly better than either parent. Jones's theory differs from a similar earlier theory in assuming that the genes involved are so numerous that several are borne on the same chromosome and thus tend to be inherited in groups. This explains why vigor is at its maximum in the first generation after crossing and why it is impossible through selection in later generations to incorporate all of the favorable genes into a new variety as good as or better than the first-generation hybrid. The ideal combination, *AA DD CC EE*, which combines all of the favorable genes, is impossible to attain because of chromosomal linkage. For example, the genes *B* and *c* may be borne at adjacent loci on the same chromosome and thus be inseparably joined in their inheritance.

Historically, then, hybrid corn was transformed from Shull's magnificent design to the practical reality it now is when Jones's method of seed production made it feasible and his theory of hybrid vigor made it plausible. To attempt to say who contributed most to this epoch-making achievement in applied biology would be difficult if not presumptuous. The important fact is that the combination proved irresistible to even the most conservative agronomists.

Soon after 1917, when Jones's theory of heterosis was published, hybrid corn-breeding programs were initiated in many states. By 1933 hybrid corn was in commercial production on a substantial scale and the U.S. Department of Agriculture had begun to gather statistics on it. By 1950 more than three-fourths of the total corn acreage of the United States, some 65 million acres, were in hybrid corn. This immense achievement stems from the work of many corn breeders, variously associated with the U.S. Department of Agriculture, the state experiment stations, and private industry. Among the pioneers in the breeding of hybrid corn and in conducting critical experiments on various aspects of the new method were Henry A. Wallace, Herbert K. Hayes, Frederick D. Richey, and T. A. Kiesselbach.

The biological phenomenon that underlies the efforts of these and other breeders and that they have consciously and successfully exploited, is *heterosis*, earlier known as "hybrid vigor." This phenomenon has been so important a factor in corn's evolutionary history and is so important a factor in its present and future improvement that it deserves the detailed discussion that I have given it in the next chapter.

19 The Nature of Heterosis

Various mathematical models have been devised for estimating additive genetic variance, degree of dominance, etc. It is well to remember that any such approach is based on certain biological assumptions. To the degree that these biological assumptions fail to conform to reality, the results obtained through the use of such models also will fall short of maximum biological usefulness. Sprague, 1955

It is obvious that hybrid corn, which has literally revolutionized the production of corn in the United States and other parts of the world, is a method of exploiting hybrid vigor or heterosis. Thus heterosis must be considered, along with replication of the hereditary material and photosynthesis, one of the three great biological phenomena underlying the practice of agriculture. What then is the genetic basis of heterosis? And how can it be used more effectively in the breeding of corn? These are questions to which thoughtful corn breeders have been addressing the major part of their attention during the past several decades. To understand the numerous experiments that have been conducted in efforts to answer these questions one must first consider the principal theories which have been proposed to explain heterosis.

Because the phenomenon of heterosis is of overwhelming importance from the standpoint of both theoretical genetics and of plant and animal improvement it has been widely discussed and has been the subject of extensive experimentation. Two comprehensive symposia were held on the subject in the fifties, and papers presented there and elsewhere now comprise a literature so extensive that merely to assemble a complete bibliography on heterosis would be a formidable task and one beyond the scope or purpose of this work. I propose, instead, to discuss briefly the principal theories concerned with heterosis in general and to attempt to ascertain from the wealth of experimental results which theory or theories seems best to explain the phenomenon of heterosis in maize and to

evaluate the various methods of maize improvement in the light of the nature of the heterosis which the maize plant exhibits. For extensive reviews of the literature, the reader is referred especially to East, 1936, Whaley, 1944, Gowen, 1952, and to a series of papers by Mather, Darlington, Pontecorvo, Haldane, Huxley, and others published in the *Proceedings of the Royal Society*, 1955.

Four principal explanations have been offered to account for heterosis: (1) heterozygosity per se; (2) accumulation in the heterozygote of favorable dominants from each parent; (3) allelic interaction, now usually referred to as "overdominance"; (4) nonallelic interaction, commonly referred to as "epistasis." The first theory, because of its vagueness, has never been very satisfactory. Darwin (1876) attributed the vigor often associated with cross-breeding to the sexual elements being in some degree differentiated. Shull (1914) and East and Hayes (1912) thought in terms of the physiological stimulation resulting from heterozygosis. In its original form the theory has been pretty well discredited (see Jinks and Mather, 1955), but newer versions of it still persist. Haldane (1955) has suggested that heterozygosity may promote biochemical diversity, that the heterozygotes may be more versatile than homozygotes and adapted to a wider range of environment, either internal or external. Lerner (1953) has argued that a return to the theory of heterozygosity per se at least in relation to stability and development is overdue. This is essentially an admission that the manifestations of heterosis and inbreeding

depression are not completely explained by any current theory of heterosis concerned only with superior gene content. To the extent that this admission leads to a search for additional explanations it can be useful. However, the theory of heterozygosity per se, if it is indeed a theory, cannot per se explain anything. East and Hayes (1912) made this clear a half century ago when, stating that the heterozygous condition carries with it the stimulus to development, they added "it is clearly recognized that this is a statement and not an explanation." The remaining three theories are alike in explaining heterosis in terms of superiority of gene content but differ in the way in which this superiority is accounted for.

Dominance of Linked Factors

It seems desirable to review the history of this theory in some detail not only because there has been some misinterpretation of the historical facts but also because in order to recognize its relevance to the problem of heterosis, the theory must be clearly understood in all of its implications. The genesis of the theory of favorable dominants is sometimes attributed to Bruce (1910). I consider this a historical error. It may be true, as Richey (1945) and some later writers have stated, that Bruce's contribution is "elegant" and of general applicability. But this is something seen only in retrospect. Little attention was given to the contribution at the time, and it was not a factor in the development of the theory of linked favorable dominants. This theory is the product of applying the concept of chromosomal linkage to the earlier concrete explanation of hybrid vigor by Keeble and Pellew (1910), who explained the increased height of a cross of two varieties of peas over their parents. A quotation from these authors makes clear the nature of their contribution: "The suggestion may be hazarded that the greater height and vigour which the F_1 generation of the hybrids commonly exhibit may be due to the meeting in the zygote of dominant growth-factors of more than one allelomorphic pair, one (or more) provided by the gametes of one parent, the other (or others) by the gametes of the other parent." Two principal objections to this explanation were raised by Shull (1911), East and Hayes (1912), and Emerson and East (1913): (1) it should be possible to combine in one race the favorable dominant factors of both parents; (2) the F_2 distribution with respect to quantitative characters should be skewed.

Both of these objections were overcome when Jones (1918) effectively combined the earlier idea of favorable dominant factors complementing each other with the then emerging chromosome theory of heredity and explained heterosis in terms of Mendelizing groups of genes. His reasoning is clearly revealed by the following quotation from his paper: "Abundant evidence is fast being accumulated to show that characters are inherited in groups. The different theories accounting for this linkage of characters make no essential difference in the use to which these facts will be put here. It is only necessary to accept as an established fact that characters are inherited in groups and that it is these groups of factors which Mendelize. The chromosome view of heredity, as developed by Morgan and others (1915), will be used because it gives a means of representation in a simple, graphical manner."

Since this paper and its significance is sometimes overlooked in current discussions of heterosis, it may not be amiss to quote its summary in full:

1. The phenomenon of increased growth derived from crossing both plants and animals has long been known but never accounted for in a comprehensible manner by any hypothesis free from serious objections.

2. The conception of dominance, as outlined by Keeble and Pellew in 1910 and

illustrated by them in height of peas, has had two objections which were: a. If heterosis were due to dominance of factors it was thought possible to recombine in generations subsequent to the F_2 all of the dominant characters in some individuals and all of the recessive characters in others in a homozygous condition. These individuals could not be changed by inbreeding. b. If dominance were concerned it was considered that the F_2 population would show an asymmetrical distribution.

3. All hypotheses attempting to account for heterosis have failed to take into consideration the fact of linkage.

4. It is shown that, on account of linked factors, the complete dominant or complete recessive can never or rarely be obtained, and why the distributions in F_2 are symmetrical.

5. From the fact that partial dominance of qualitative characters is the universal phenomenon and that abnormalities are nearly always recessive to the normal conditions, it is possible to account for the increased growth in F_1 because the greatest number of different factors are combined at that time.

6. It is not necessary to assume perfect dominance. It is only necessary to accept the conclusion that many factors in the one n condition have more than one-half of the effect that they have in the two n condition.

7. This view of dominance of linked factors as a means of accounting for heterosis makes it easier to understand: a, why heterozygosis should have a stimulating rather than a depressing or neutral effect; b, why the effect of heterozygosis should operate throughout the lifetime of the individual, even through many generations of asexual propagation.

The fact that Collins (1921) later showed that these objections to earlier theories could have been met without applying the concept of linkage is wholly irrelevant except that it illustrates a widespread skepticism of that period to the chromosome theory of heredity. Although it was not confined to plant breeders—witness Bateson's longstanding reluctance to accept the theory—it was perhaps more prevalent among this group than among biologists in general and in some instances persisted longer. As recently as 1950, for example, we see Richey minimizing linkage as a factor in heterosis in the following statement: "As East (1936) has emphasized, Jones' (1917) suggestion was linkage, and not

the interaction of dominant genes. But linkage, by itself, cannot cause hybrid vigor. As Richey (1945b) recently emphasized, then, the dominance theory may be attributed most appropriately to Bruce and to Keeble and Pellew, and the date of placing hybrid vigor on a Mendelian basis should be 1910, that of their papers, rather than the date of Jones' (1917) paper as has been stated by Jones (1942)."

A comparison of this curious statement with that of East is revealing: "Bruce and Keeble and Pellew (1910) are sometimes given prior credit for this conception, but not upon just grounds. The earlier authors based their theory upon independent segregation at a time when linkage was not understood, and their scheme did not and could not fit even the then known facts. The particular use made of the notion that the chromosomes are strings of genes obeying special laws of transfer was what made Jones' theory acceptable, not simply the employment of the words *dominant* and *recessive*."

I present these opposing viewpoints in some detail not merely as excerpts of an academic controversy, although there undoubtedly was that and these quotations illustrate it, but also to emphasize the importance of the concept of chromosomal linkage to any consideration of the problem of heterosis. The failure, almost a half century later, to recognize linkage as a factor in heterosis has caused some experiments to be based on assumptions which are quite unrealistic in their conception. Much time and effort might perhaps be saved if these early papers of Bruce, Keeble and Pellew, Shull, East and Hayes, Emerson and East, and Jones were more carefully studied and better understood by those designing experiments concerned with heterosis or with the improvement of corn. The fact is that Jones's theory of linked dominant factors is, after almost fifty years, still one of the most satisfactory explanations of hybrid vigor which has yet been devised. The only real question which remains is whether it accounts for *all* of the manifestations of heterosis.

Overdominance

Overdominance is a term first used by Hull (1946a) to describe heterosis resulting from the interaction of alleles at a single locus as postulated by East (1936). In its simplest form the theory says that the heterozygous genotype, *Aa*, is superior to either homozygote, *AA* or *aa*. However, if there are more than two alleles at a locus, *AA'* may be superior to either *AA* or *A'A'*. This phenomenon has also been called "superdominance" by Fisher and "double-dose disadvantage" by Huxley (1955). Except for its excessive alliteration Huxley's term is perhaps the best one so far proposed.

There seems to be no doubt that East was the first to suggest interallelic interaction of the type which has come to be known as overdominance, although Rasmusson came close to making such a suggestion. The term overdominance was applied originally to the heterozygote for two alleles at a single locus, but it has also been used when there is heterozygosity for two genetically different but homologous sections of chromosomes, a situation for which Hull (1952) suggests the term "pseudo-overdominance." Since the effects of closely linked blocks of genes are virtually indistinguishable from those of alleles at a single locus, and since what were once regarded as single-gene loci have often proved, when subjected to refinements in analysis, to be compound loci the distinction between overdominance and pseudo-overdominance often becomes meaningless. Crow (1952), for example, cites the augmented areas of anthocyanin formation in the aleurone of maize in heterozygotes for certain alleles at the *R* locus mentioned by Stadler (1942) as an example of overdominance. Since this locus has now been shown to be compound (Stadler and Nuffer, 1953), the example is instead one of pseudo-overdominance if we adhere strictly to Hull's original definition of the term. The fallacy of doing so becomes even clearer when recent studies on the fine structure of genetic loci are considered, for example, those of Benzer (1961) in bacteriophage, Nelson (1959) on the waxy gene in maize, and Mangelsdorf and Galinat on the tunicate locus in maize.

Actually there are very few clear-cut examples of "one-gene heterosis." Pontecorvo (1955) recognized only two: sickling in man and *pab* in *Neurospora*. Perhaps a third example is that mentioned by Haldane (1955) of a heterozygote in man making an agglutinin of higher molecular weight than either homozygote. A more recent example is the interesting case reported by Manwell et al. (1963) of "hybrid" hemoglobins in certain fish crosses. The hybrid hemoglobins have superior blood-gas-transport properties when compared with the hemoglobins of the parental species or with simple mixtures. Not all of the fish crosses studied have hybrid hemoglobins, and it may be significant that the two crosses which do are considered among the best for pond culture. Another example of a molecular basis for heterosis is that described by Schwartz and Laughner (1969), who found that combining in heterozygous alleles an unstable active enzyme with a stable but inactive one produced a hybrid enzyme that was both stable and active.

Overdominance in the broader sense is considered an evolutionary phenomenon which in an outbreeding diploid species results from exposing new gene combinations in a predominantly heterozygous condition. This subject and the extensive literature relevant to it are treated extensively by Parsons and Bodmer (1961). An important fact which seems to have been generally overlooked is that overdominance or its biometrically indistinguishable counterpart, pseudo-overdominance, is inherent in Jones's model of dominant linked factors. For entire chromosomes to exhibit complete or close linkage and to behave as alleles may be rare, but for blocks of genes within the chromosomes to behave in this way is the rule rather than the exception, a fact clearly recognized by Jones and described by him with admirable clarity and succinctness: "It is only necessary to accept as an established fact that characters are inherited in groups and that it is these groups of factors which Mendelize."

Epistasis

The term "epistasis" like the term "overdominance" was originally much more limited in its meaning, being applied to two different genes which are not alleles, both affecting the same part or trait of an organism, the expression of one covering up or concealing the expression of the other. In the area of biometrical genetics it has come to be widely used in referring to any type of nonallelic interaction. Both Fisher (1949) and Mather (1955) have regarded such interaction as an important factor, if not the major one, in heterosis and inbreeding depression. Pontecorvo (1955) points out that physiological genetics provides no crucial arguments to choose between a model of heterosis based on epistatic gene interactions and one based on allelic interactions. He suggests further that advances in our knowledge of gene structure and action have led to the realization that there may be no absolute distinction between the two types of interaction and that the distinction between the two types of models for heterosis no longer has precise meaning, although it may still be useful at certain levels of approximation. I shall argue that the distinction may still be useful in evaluating various methods of corn-breeding.

That nonallelic interaction is a factor in heterosis would be expected on a priori grounds. There is scarcely a unit character or trait which has been studied intensively in any organism which cannot have its expression modified by genes other than the major one which controls it. It was a recognition of this fact that led some geneticists to conclude many years ago that every character of an organism is influenced by all genes and that every gene participates in controlling all characters. This may be an overstatement of the situation, but it is probably nearer the truth than any conclusion that does not recognize epistasis as a factor in heterosis.

The role of nonallelic interaction in heterosis is nowhere better illustrated than by the vigor of allopolyploids. In these the vigor cannot be attributed to heterozygosis per se, since in self-pollinated plants such as the cultivated wheats the majority of phenotypes are highly homozygous and whatever genic interaction is involved must be nonallelic. And here the distinction between *euheterosis*—fitness—and *luxuriance* made by Dobzhansky (1952) begins to lose precise meaning, for under domestication a type of vigor which to the uninitiated appears to be mere luxuriance may, especially in forage and ornamental plants, impart fitness and survival value. Thus what in nature is luxuriance becomes euheterosis under domestication. The important point here is that allopolyploids originating from species hybrids are often more vigorous than their parents in a great variety of measurable characteristics and this type of hybrid vigor must be the product of interaction between genes in different genomes and not of interaction between those within the same genome.

An outstanding example of heterosis in an allopolyploid is the hexaploid bread wheat *Triticum aestivum*. This wheat, which has a haploid chromosome number of 21, is the product of hybridization of three different species, each with seven chromosomes, *T. monococcum, Aegilops speltoides,* and *Aegilops squarrosa.* Only one of these, *T. monococcum,* has ever been cultivated. It is grown on the poor soils of Greece and parts of the Near East, where it yields of grain are said to be "insignificant." The other two species, *A. speltoides* and *A. squarrosa,* have never been considered worth domesticating. Yet the allopolyploid hybrid, which combines the chromosomes of these three unpromising grasses, has become one of the most productive of the world's cereals, capable of producing a record yield of more than 200 bushels per acre. The interaction of the chromosomes of the three component species has created a high degree of hybrid vigor which in turn has greatly increased the genetic potential.

It is of course also true that the genomes of the species which comprise an allopolyploid do have loci in common and if they have descended from a common ancestor some of these loci were once allelic. Consequently, classical genetics, like physiological genetics, provides no absolute distinction between allelic and nonallelic interaction.

Genic Balance a Factor in Heterosis

Although not strictly speaking a theory of heterosis, the concept of genic balance is helpful in understanding some of the manifestations of inbreeding and outbreeding. This is well illustrated by the experiments of Dobzhansky (1952) on inversion heterozygotes in *Drosophila.* The general argument pertaining to genic balance is clearly expressed by Mather. In speaking of the loss of heterosis in F$_2$ of crosses in species of *Drosophila* reported by Vetukhiv (1954), he states: "Thus heterosis must be a property of certain polygenic combinations acting together, being lost when these combinations are broken down by recombination. It is an expression of genic balance." And further: "In outbreeding species, for example, the naturally occurring genotypes will virtually always be partially heterozygous, and natural selection will therefore favour those combinations which combine in homologous pairs to give a good balance. Combinations will, on the other hand, seldom be exposed in the homozygous condition, so that no selection will have been

acting to pick out from the great mass of possible genotypes those which show a good homozygous or internal balance. Inbreeding would thus virtually always lead to a phenotypic depression, reflecting a balance which was poor because it was untested, and vanishing when the tested hybrid balance was restored by crossing."

Heterosis and Genetic Homeostasis

Not a theory of heterosis but useful in understanding the manifestations of heterosis is Lerner's (1955) concept of genetic "homeostasis"—the ability of organisms to utilize and to regulate genic variability. Jones's (1944) term "genic equilibrium" and Darlington and Mather's (1949) term "genetic inertia" involve similar concepts. Jones (1958) considers this ability of organisms to adapt themselves to genetic variability—whatever that may be called—one of the most important factors in the success of hybrid corn and one that offers great promise in the further improvement of many naturally self-fertilized crop plants. He states that in corn, double crosses, which are genetically more variable in composition than single crosses, are also more stable and consistent in performance. Perhaps the present trend in the commercial seed-corn industry of replacing double crosses with single crosses is a step in the wrong direction on two counts: (1) it may lose some of the advantage of genetic homeostasis; (2) it invites disastrous epidemics of new strains of disease to which all plants of a single cross, being genetically identical, might be susceptible.*

Inbreeding Depression in Maize

One of the most puzzling aspects of the problem of heterosis in maize, one to which both Brieger (1950) and Crow (1952) have given particular attention, is the marked depression which accompanies inbreeding in this species. We now have good reason to believe that this is explicable, at least in part, in terms of blocks of genes from teosinte or *Tripsacum.* To the extent that vigor of open-pollinated varieties of maize is the product of heterozygosity of blocks of genes from teosinte or *Tripsacum,* inbreeding depression will reflect

*Since this was written the epidemic in 1970 of the southern corn blight demonstrated in a dramatic fashion the susceptibility of genetically homogeneous populations to a new strain of a pathogenic fungus.

TABLE 19.1 Grain Yields of Modified Strains of A158 Compared with Yield of the Original Strain

Strain	Gm. per plant	% of control
A158 (control)	126	100
Florida 3	138	110
Florida 3, 4, 9	111	88
Florida 4$^+$ A	128	102
Florida 4$^+$ B	162	129
Florida 9	141	112
Durango 1, 7, 9	68	54
Nobogame 4$^+$	118	94
Averages	124	98
Mexico 1077	134	106
Honduras 1639	112	89
Nicaragua 501	94	75
Cuba 394	111	88
Brazil 1691	126	100
Paraguay 333	127	101
Argentina 1807	116	92
Bolivia 1157	107	85
Venezeula 1249	107	85
Averages	115	91
Honduras & Venezuela	61	48
Nicaragua & Brazil	77	61
Brazil & Venezuela	46	37
Argentina & Venezuela	46	37
Bolivia & Argentina	57	45
Bolivia & Mexico	37	29
Venezuela & Nicaragua	22	17
Averages	49	39

the deleterious effects of homozygosity for these blocks. These deleterious effects can be explained in part genetically in terms of Mather's concept of balance. They represent combinations which have been preserved by natural selection only in the heterozygous condition. Inbreeding leads to a phenotypic depression because it reflects the balance which in the homozygous condition is poor because it has been untested by natural selection.

That some chromosomes or parts of chromosomes from teosinte or those with tripsacoid effects extracted from open-pollinated races of corn can have depressing effects on yield when their genes are homozygous is demonstrated by the data in Table 19.1. These

FIG. 19.1 *Left*, a hybrid of inbred strains 701 and 4R–3. *Right*, a strain of 4R–3 modified by introducing three chromosomes or parts of chromosomes from Durango teosinte. *Center*, a hybrid of 701 and the modified strain of 4R–3. This hybrid demonstrates that some modern strains of corn can absorb a considerable amount of teosinte germ plasm without being deleteriously affected so long as the teosinte genes are heterozygous. It also explains why inbreeding some races of maize produces certain lines that are scarcely recognizable as corn.

FIG. 19.2 An ear of the inbred strain A158 compared with its derivative strains which have been modified by the introduction of extracted chromosomes from several races of teosinte. *A*, inbred A158. *B*, a strain modified by chromosome 4 from Florida teosinte. *C*, strain modified by chromosome 4$^+$ from Nobogame teosinte. *D*, strain modified by chromosomes or parts of chromosomes 3, 4, and 9 from Florida teosinte. *E*, strain modified by chromosomes 1, 7, and 9 from Durango teosinte. The yields of these strains in terms of percentages of the original strain are: 100, 102, 94, 88, and 55 respectively (see Table 19.1). The lowest-yielding strain, *E*, was one present in each of the five highest-yielding hybrids tested by Sehgal (1963).

represent the results of a yield test with four randomized replications in which I compared various modified strains of A158 with the original strain. The data show that the introduction of teosinte germ plasm had little effect on yield except when three different chromosomes or parts of chromosomes were introduced. The yield of the modified strain Durango 1, 7, 9 was only slightly more than half of the original.

The depressing effect, when homozygous, of the genes of introduced chromosomes or parts of chromosomes from open-pollinated races of corn is somewhat more general. The yields of six of the nine modified strains were lower than the yield of the original strains, three of these by 15 to 25 percent. Strains modified by the introduction of extracted chromosomes or parts of chromosomes from two different open-pollinated races were consistently less productive than the originals,

and their average yields were less than half of that of the original strains. It is these modified strains that are the counterparts of some of the inbred strains that have been "improved" by the method of convergent improvement.

Photographs comparing a typical ear of A158 with several of its modified derivatives are shown in Figure 19.2.

But part of the inbreeding depression in maize should be explained in other terms: morphological and physiological rather than genetic unbalance. The patterns of development controlled by the maize genes are in some respects quite different from those dictated by the genes received from teosinte or *Tripsacum*. As long as the genes from the two species are heterozygous, there is apparently no conflict in their patterns, but when blocks of foreign genes in maize become homozygous they produce derangements which amount in some instances to insults to devel-

opmental homeostasis. Examples of these have been given in an earlier chapter. The induration of the tissues of the rachis and lower glumes, characteristic of teosinte and *Tripsacum*, may, if it becomes excessive, constrict the fibrovascular bundles to the extent that the kernels are not adequately nourished; a secondary effect may be the accumulation in the stalks and cobs of unused sugars, creating a favorable medium for the growth of smut and other fungi. The result is a marked depression in yield. Inbred strains having such characteristics in extreme form are ordinarily discarded by corn breeders on the assumption, probably a correct one, that any inbred which can be maintained only with difficulty is essentially worthless from the standpoint of commercial seed production. Yet some of these unattractive, unproductive, disease-susceptible strains are quite capable of producing vigorous, productive hybrids.

219   The Nature of Heterosis

FIG. 19.3 Modified strains of the inbred A158 showing how an excess of foreign germ plasm can have deleterious effects. *A* and *C* are ears of strains in which extracted chromosomes have been introduced from Bolivia 1157 and Argentina 1807 respectively. *B* is a strain carrying the extracted chromosomes from both. The yields of the three strains in terms of percentages of the original inbred A158 are, from left to right: 85, 45, 92 (see Table 19.1). The reduction in yield resulting from introducing chromosomes with tripsacoid effects from the races is regarded as the counterpart of the deleterious effects on certain inbred strains of the method of convergent improvement.

They are poor not because they have poor genes but because their good genes are good only in a heterozygous state.

In their effects, which are deleterious when homozygous, these blocks of genes from teosinte or *Tripsacum* are similar to the "bottle-neck" loci postulated by my brother, Albert Mangelsdorf (1952), who, during a lifetime spent in breeding sugar cane, has given much attention to the problem of heterosis. In sugar cane as in maize, interspecific hybridization has played a role in the plant's evolution under domestication. The deleterious effects of blocks of teosinte and *Tripsacum* genes may explain some of the conflicting results reported by Nilsson-Leissner (1927), Jorgenson and Brewbaker (1927), Jenkins (1929), and others on correlations between the yields of inbred strains and their hybrids. All of these investigators found positive correlations

between yield of inbreds and their single crosses, but notable exceptions to the cor-relations occurred in all of the experiments. In Nilsson-Leissner's tests of thirteen lines of dent corn and their single crosses, the second highest-yielding cross had the second lowest-yielding inbred as one parent. Jorgenson and Brewbaker reported that two relatively low-yielding inbreds were consistent in producing high-yielding hybrids when crossed with other selfed lines. In the three groups of lines tested by Jenkins (1929) some of the lower-yielding inbreds in each group produced some of the higher-yielding hybrids, and in one group of 39 lines from late yellow varieties the lowest-yielding line, No. 50, produced the highest-yielding single cross in a total of 282 crosses. Also the mean yield of all of the crosses involving this line as one parent was higher than the mean yield of the crosses in-volving the highest-yielding inbred, No. 67, which had a yield more than six times as great as the lowest-yielding inbred, No. 50.

These striking exceptions to the general correlations between yields of inbreds and their crosses are of particular interest to us here because they suggest that some of the lower-yielding inbred lines included in the tests carry genes such as those contributed by teosinte and *Tripsacum*, which are deleterious in their effects when homozygous but bene-ficial when heterozygous.

If inbreeding depression is often the pro-duct of homozygosity for teosinte and *Trip-sacum* genes then the depression should be less marked in varieties of maize which do not contain such genes. In an article on intro-gression (1961), I stated that it is doubtful whether among varieties of maize now in existence there is any which is completely free of introgression from teosinte or *Tripsacum*. In an earlier work Reeves and I assumed that some of the Peruvian varieties with knobless chromosomes were uncontaminated by teo-sinte, but there is now evidence (Grobman et al., 1961) that even these have undergone some admixture with *Tripsacum australe*, some forms of which have knobless chromo-somes (Graner and Addison, 1944).

Inbreeding Depression in Popcorns

Although it may be true that there is no living maize completely free of contamination with teosinte or *Tripsacum*, it is also true that there

are great differences in races of maize in the amounts of teosinte and *Tripsacum* admixture which they contain. The nearest approach to "pure" maize is probably to be found among the varieties of popcorn, especially those with low chromosome-knob numbers. Although popcorn varieties have never been intensively studied to determine their response to in-breeding, it is common knowledge among those who have worked with them that the depression which they undergo when inbred is not nearly as drastic as it is in the majority of other races.

Brunson, who has had extensive experience in breeding both dent corn and popcorn, has stated (1937) that less loss of vigor from in-breeding is experienced normally in popcorn than in dent corn, and Brunson and Smith (1945) reported that in the production of commercial hybrid popcorn single and three-way crosses were more feasible than with dent corn, since popcorn inbreds are relatively more vigorous and easier to propagate than those of dent corn.

We have had similar experience in our experiments on reconstructing the ancestral form of maize by combining pod corn and popcorn. Cultures containing a large propor-tion of genes from Lady Finger, Argentine pop, and Baby Golden do not suffer con-spicuous depression when inbred, and virtu-ally all inbred strains can be maintained with-out difficulty. Grobman informs me that the high-altitude popcorns of Peru, which have low chromosome-knob numbers and are among the least tripsacoid of living races of maize, do not undergo severe depression when inbred. Although there is a rather exten-sive literature on popcorn, it is concerned largely with problems of production and im-provement, and published data on inbreeding depression in popcorn, as compared to in-breeding in other types of corn, are almost completely lacking. This is one of the con-spicuous gaps in our knowledge on the effects of inbreeding in maize.

Inbreeding Depression in Teosinte

Also almost completely lacking are data on the effects of inbreeding in teosinte. Collins reported some years ago (1918) that teosinte does not lose vigor when it is inbred to the same extent as does maize, and it has been our own experience that quite vigorous selfed lines can be isolated without difficulty from some varieties of teosinte. The possible explanation of this may be that teosinte, although undoubtedly heterozygous for many loci, is homozygous or nearly so for those blocks of genes which enable it to maintain the principal characteristics in which it differs from maize: distichous spikes, single pistillate spikelets, and indurated lower glumes and rachis segments. If inbreeding accentuates these characters in teosinte no harm is done; there is no breakdown of developmental homeostasis.

The teosinte of the Chalco region of Mexico, where teosinte hybridizes constantly with maize and where by absorbing maize genes it has become a mimic in the maize fields, might be expected to be unusually heterozygous and to contain blocks of genes which have been subjected to selection in the heterozygous condition. If so, this teosinte should show more than average depression when inbred and at least some of the lines should be the counterpart of highly tripsacoid maize lines containing instead blocks of maize genes which in teosinte are beneficial when heterozygous and deleterious when homozygous.

Types of Heterosis in Maize

To no group of geneticists is an understanding of the nature of heterosis of greater importance than it is to corn breeders. Consequently a great deal of research has been conducted, much of it of a statistical nature, to distinguish between the several theories explaining heterosis. A discussion of some of the more salient results follows.

On the basis of the evidence for racial hybridization and hybridization of maize with its relatives teosinte and *Tripsacum*, Grobman et al. (1961) suggested that yield heterosis in modern corn may be thought of as having five major components: (1) maize × maize; (2) *Tripsacum* × *Tripsacum*; (3) maize × *Tripsacum*; (4) *Tripsacum* × teosinte; (5) maize × teosinte. Actually, since crosses between different races of teosinte exhibit heterosis and since the introgression to which maize has been subjected in Guatemala and Mexico has involved different races of teosinte, a sixth category of heterosis, teosinte × teosinte, should be added to those listed by Grobman et al.

As part of his thesis research one of my former students, Surinder Sehgal, compared the yields of 109 crosses of modified strains of A158 with other modified strains and with the original strain. A comparison of these crosses with respect to five of the six categories of heterosis mentioned above produced the following results in terms of average yield per plant expressed as percent of the yield of the control (Sehgal, 1963, table 9):

Control (A158)	100.0
Maize (A158) × teosinte derivatives	116.4
Maize (A158) × "Tripsacum" derivatives	87.9
Teosinte derivatives × teosinte derivatives	117.6
"Tripsacum" derivatives × "Tripsacum" derivatives	98.9
Teosinte derivatives × "Tripsacum" derivatives	108.1

The lower than average yields of the maize × *Tripsacum* derivative crosses compared to the higher than average yields of the maize × teosinte derivative crosses suggests that the germ plasm introduced by the extracted *Tripsacum* chromosomes is different from that introduced by the teosinte chromosomes. Why the maize × *Tripsacum* derivative crosses should be less productive than the control is not clear. Sehgal (1963) suggested that perhaps A158 already possesses an optimum assortment of genes similar to the ones in the Latin-American tripsacoid races. Since these races are so different from the strain A158 or the open-pollinated variety from which it was isolated, I doubt that this is the explanation.

Especially interesting to me is a comparison that I have made—using Sehgal's data—of the ten highest-yielding entries. These are shown in Table 19.2. It is highly significant, I think, that the five highest-yielding crosses all had the modified strains Durango 1, 7, 9 as one parent. The strain itself yielded only 54 percent as much as the original (Table 19.1). This is a particularly striking example of blocks of genes—in this case blocks known to be derived from teosinte—having a depressing effect on yield when homozygous and a strongly heterotic effect when heterozygous. It is also significant that all ten of the highest-yielding crosses had teosinte derivatives as one or both parents. I regard this as

TABLE 19.2 Grain Yields in Grams per Plant of the Ten Highest-Yielding Crosses between Inbred A158 and Its Modified Strains

Rank	Cross	Grams per plant
1.	Florida 4[+] × Durango 1, 7, 9	168.0
2.	Durango 1, 7, 9 × A158	156.7
3.	Florida 3A × Durango 1, 7, 9	155.9
4.	Durango 1, 7, 9 × Brazil	147.8
5.	Bolivia × Durango 1, 7, 9	140.6
6.	Florida 4[+] × Nobogame 4A	140.2
7.	Florida 3, 4, 9 × Durango 1, 7, 9	138.2
8.	Nobogame 4A × A158	138.0
9.	Florida 4[+] × Honduras	137.3
10.	Nobogame 4A × Paraguay	136.9
68.	Control (A158)	99.6

Source: Sehgal, 1963.

indicating that the germ plasm of teosinte is quite different from the germ plasm of maize and that this may be considered as additional evidence that corn is not, as some students of maize now contend, a domesticated form of teosinte. Perhaps also significant is the fact that four of the ten highest-yielding crosses have *Tripsacum* derivatives as one parent and teosinte derivatives as the other. This indicates that the germ plasm introduced by the *Tripsacum* chromosomes extracted from open-pollinated varieties is different from that introduced directly from teosinte. This is consistent with my conclusion (Chapter 11) that the chromosomes extracted from South American varieties are not the product of teosinte introgression but may be the result of introgression from *Tripsacum*.

We have shown earlier in this chapter that chromosomes of teosinte introduced into an inbred strain of maize or chromosomes with tripsacoid effects extracted from open-pollinated varieties have a depressing effect upon vigor and yield when homozygous and a stimulating effect when heterozygous, thus exhibiting apparent overdominance. Since the blocks of genes of teosinte and *Tripsacum* introduced into maize during its evolution have become adapted to their new milieu through the accumulation of modifier complexes involving other genes on the same and other chromosomes, nonallelic interaction has undoubtedly also come to play a part in present-day heterosis in maize. In brief all the types of heterosis recognized by current theories: linked favorable dominants, overdominance, and epistasis, occur in maize, and the problem becomes one of determining which type or types are operating in any given set of circumstances.

Evidence of Overdominance

Both Crow (1948, 1952) and Brieger (1950) concluded on the basis of biometrical considerations that overdominance must be a factor in heterosis in maize. Some of the assumptions upon which their calculations were based we now know to be not completely valid. Crow, for example, assumed that there is no epistasis and no barriers to recombination that prevent each gene from reaching its own equilibrium frequency independently of other loci. He concluded that the dominance hypothesis is adequate to explain the deterioration that results from inbreeding and the recovery of vigor on outcrossing but that it is difficult to explain how the hybrid could greatly exceed in fitness the equilibrium populations from which their parents were derived. He recognized the fact that the overdominance hypothesis demands the assumption of a kind of gene action known to be rare but pointed out that if only a small proportion of the loci are of this type they may nevertheless represent a major factor in the population variance.

Brieger, on the basis of a different but comparable biometrical approach, concluded that (1) no breeding system would accumulate the number of undesirable recessives required by the dominance hypothesis and (2) that the dominance hypothesis is not adequate to account for the observed breeding results. Some of Brieger's observations on breeding results are not, however, completely accurate. On the basis of an inbred strain which he received from Rhoades and which he was able to improve in climatic adaptation by selection, though it appeared to be quite uniform initially, he concluded that many apparently uniform inbred strains are not homozygous. On this point he apparently overlooked Jones's (1939) data on long inbred strains which show that sib lines separated at various stages of continued inbreeding, although clearly differing in some cases, remained the same in others. When differences occurred in lines separated after many generations of inbreeding they could be attributed

to spontaneous variations and not to delayed segregation. Brieger assumed that Indian varieties should be highly inbred because of the small populations often grown but found that the reduction in vigor which they suffer with inbreeding is so pronounced that some are difficult to maintain after two or three generations of inbreeding. He predicted that methods of hybrid corn production followed in the United States would not succeed in Latin America because of large numbers of genes in Latin American races with a low survival index when homozygous. The methods are, however, contrary to his predictions, succeeding in some countries.

Neither Crow nor Brieger gave any consideration in their biometrical reasoning, nor at that time had any well-established reason to do so, to the possibility that modern maize is a complex hybrid which has been subjected to introgression by teosinte and *Tripsacum* or that the depression which occurs in maize when it is inbred is due in part to homozygosity for blocks of foreign genes. Crow, however, came close to describing the genetic situation which such blocks of genes might produce when he pointed out that only a small proportion of loci exhibiting overdominance might represent a major factor in variance. Brieger in an earlier paper (1944) had come even closer to such an explanation when he suggested that the extreme reduction in vigor following selfing in maize might be explained by recessive lethals becoming established as balanced lethals in an original interspecific hybrid owing to mutual incompatibility of the genes of either species when homozygous. More recently, however, Brieger et al. (1958) have questioned the whole concept of tripsacoid types of maize.

There has been a considerable number of experiments conducted in maize to distinguish between additive dominance and overdominance as a cause of heterosis, and there have been others which have some bearing on the problem.

Single-Gene Overdominance

There are few clear-cut cases of single-gene overdominance in maize or other organisms. The hybrid enzymes in maize mentioned earlier in this chapter are virtually unique. The case cited by Crow (1952) of overdominance at the *R* allele is a questionable example and probably represents a return to

normal anthocyanin formation of partially inhibited states. Now that the *R* locus has been shown to be compound it is at best an example of pseudo-overdominance. The mutants and inbred lines described by Jones, Schuler, and Schuler and Sprague, which exhibited heterosis when crossed with their mother lines, proved in critical tests to represent compound loci or to contain residual heterogenity. These are discussed in detail in the section on mutagenic effects of teosinte introgression.

The relative absence of evidence of single gene heterosis in maize speaks in itself against overdominance as a factor in heterosis in this species, since if loci exhibiting true overdominance are common some of them should probably be detectable individually. From the physiological standpoint it is somewhat strange that few instances of single-gene heterosis should have been discovered. Pontecorvo is correct in asserting that physiological genetics provides no basis for distinguishing between allelic and nonallelic interaction. Yet, at least in maize, there is no evidence of the former and abundant evidence, as we shall see later, of the latter.

Multiple-Gene Overdominance

Hull concluded that overdominance is a factor in yield heterosis although not necessarily the only one, because the dominance hypothesis alone seemed to him not to be adequate to explain, among other things, why: (1) mass selection and ear-to-row selection fails to improve yields; (2) recombinations of parent lines of elite hybrids yield little more than the original varieties; (3) hybrids of second and third cycle lines yield little more than those of the first cycle; (4) homozygous lines of corn yield about 30 percent as much as heterozygous varieties; (5) there is no evidence of epistasis in corn yield.

Hull has given particular attention to detecting overdominance by analyzing the regression of yields of F_1 crosses on the yields of the parent lines and has presented (1952) the results of regression analyses of a number of previously reported experiments. The estimates indicate overdominance in about half of the experiments. Hull was careful to point out, however, that the possibility of epistasis is not to be ruled out, although it was not revealed by his own graphic method of detecting it.

Comstock and Robinson (1948) described biometrical procedures for estimating the degree of dominance, and Robinson et al., (1949) and Comstock and Robinson (1952) presented data suggesting overdominance as a factor in yield. They pointed out, however, that close linkage in the repulsion phase can lead to biometrical estimates of overdominance even though the individual genes in the linked groups are no more than partially dominant.

Experiments by Gardner et al. (1953) involving F_2 plants backcrossed to each of their parental lines led to similar conclusions. The estimates of dominance were within the range of overdominance but could be equally well interpreted as pseudo-overdominance—the linkage of partially and completely dominant genes. The authors emphasize that the pseudo-overdominance effect resulting from close linkages has much the same significance for short-run breeding practice as true overdominance. Rumbaugh and Lonnquist (1959) made all possible crosses between eight inbred strains, of which four were of high and four were of low combining ability as shown by testcrosses with a single cross of WF9 × M14. The parental lines and their crosses through the F_1 to F_5 generations were compared in yield and other characteristics and the resulting data were analyzed by three different methods: (1) the regression of the means on levels of heterozygosity, (2) constant parent-progeny regression, (3) the graphic method of dialled analysis. The three methods were reasonably consistent in showing partial dominance to be the principal factor involved, but the second and third methods provided clear-cut indications of overdominance and the first showed almost none. There were also indications of epistasis.

In an extension of the preceding experiment, Gardner and Lonnquist (1959, 1961) crossed randomly chosen plants from several advanced generations of a cross to the two parental lines and made estimates of the additive and dominance variance in each generation. The estimates of dominance were in the overdominance range in the F_2 generation but declined with advances in generations. The authors suggested that the results could be explained by repulsion-phase linkages among genes no more than partially or completely dominant, these linkages being broken up in later generations of random mating. The authors also reviewed previous experiments concerned with estimating dominance in maize heterosis and these with their own include nine F_2's which have now been analyzed. All but one of these showed dominance within the range of overdominance. Nevertheless, the authors concluded that overdominance is probably not an important factor in yield heterosis in maize, although the possibility of overdominance at one or more loci is not ruled out.

Grafius (1960) has questioned the entire concept of overdominance in corn. He argues that with a uniform stand, yield must necessarily be the sum of its components: ear number per plant, kernels per row, rows per ear, and kernel weight. If, as in barley, the components are not correlated with each other, then different sets of genes must act upon each component and there are no genes for yield per se. Under these conditions there is no overdominance for yield, but there is also no heritability of yield per se and no additive of dominance effects. Grafius relies on the data of Hoen and Andrew (1959), which show almost no significant correlations between the components of yield and reaches the remarkable conclusion that the estimates of overdominance obtained by Robinson et al. (1949), Gardner et al. (1953), and Gardner and Lonnquist (1959) can only be the result of the "geometry" of the situation—that overdominance is in fact a "statistical artifact."

Evidence for Epistasis

A series of experiments which in retrospect suggest nonallelic gene action in yield heterosis in maize is that of Stringfield (1950), who compared the yields of inbred strains with several types of crosses, backcrosses, and advanced generations. The yields were generally in close agreement with the degree of heterozygosity with one exception. Backcrosses of single crosses to their inbred parents yielded significantly more than F_2's of single crosses, although theoretically they should be expected to yield the same. Lindstrom (1939) had earlier reported similar differences between mean yields of backcross and F_2 populations, 71.7 and 65.8 bushels per acre respectively. Stringfield suggested that the results might be explained by a more even distribution in the backcrosses of dominant genes or that lines which survive the selection procedures may tend to have favorable combinations of genes which may contribute more to vigor than the random assortment of genes equally good.

Similar experiments including an additional category of heterosis were conducted by Sentz et al. (1954). The data in general demonstrate curvilinear relationship between heterozygosis and performance indicating nonallelic gene interaction. The authors suggest that if nonallelic interactions are present between linked loci then favorable combinations entering into the backcrosses intact from the parental lines would tend to bias phenotypic expression upward, since the occurrence of certain especially favorable combinations would be greater than in a random mating of either an F_2 or F_3 population. The presence of linkage could therefore accentuate the effect of nonallelic interactions.

The explanations of both Stringfield and Sentz et al. come close to describing the situation which obtains if one or more of the inbred strains employed in the experiment contain blocks of genes from teosinte or *Tripsacum*. Every good inbred strain which carries such blocks of genes is the product of rigorous selection not only for these particular genes but for a genetic milieu which allows their favorable expression. Backcrosses furnish that milieu for the blocks of genes of the recurrent parent, while corresponding blocks in the nonrecurrent parent may remain heterozygous and without marked deleterious effects. F_2 populations on the other hand include many new combinations in which the milieu does not allow the most favorable expression of the blocks of genes involved. Of the inbred strains employed in Stringfield's experiments, at least one, Oh28, is known to be highly tripsacoid, another, I11 Hy, probably is also. I am not familiar with the inbred strains used by Sentz et al., but I do know that the majority of inbred strains isolated from southern dent varieties are tripsacoid.

A somewhat different method of identifying nonallelic interaction was employed by Jinks and Hayman (1953). Using Mather's (1949) method of distinguishing between the components of variation, these authors analyzed the previously published data on maize from Nilsson-Leissner (1927), Kinman and Sprague (1945), and Hull (MNL, 1946). In all three sets of data there was evidence of overdominance, but the analysis also showed nonallelic interaction to be present in all sets.

In a subsequent contribution, Jinks (1954), applying to the data of Kinman and Sprague a modified and more accurate form of the scaling test invented by Mather, found that the ten inbred strains employed fell into six groups with respect to their interaction relationships. Group A interacts with B, but neither interacts with any other. C interacts with D; E interacts with C and F. The inbred strains in each group are as follows:

A. Hy, C114
B. R46, 38–11
C. B2
D. WF9, Oh07
E. Oh04, Wv7
F. K159

Of special significance is the fact that the F_1 crosses showing genic interaction had a mean yield of 90.3 bushels per acre compared with 77.3 bushels for the noninteracting F_1's. Jinks concluded that although combining ability may be due to the operation of dominance in the F_1 families, genic interaction must be at the root of the special combining ability which leads to outstanding F_1 families. It should prove interesting to extract chromosomes with tripsacoid effects from these inbred strains, determine their linkage relations, and compare the results with the grouping suggested by estimates of nonallelic interaction. We are already aware that both 38–11 and WF9 are tripsacoid.

Hayman later (1957) estimated the percent components of the variance of specific combining ability in maize and arrived at the following percentages with respect to yield.

Detectable epistatic	25.17
Other genetic	61.83
Environmental	13.00

Thus yield in maize showed more epistatic variance than yield in cotton or flowering time or height in *Nicotiana rustica*.

The data of Rumbaugh and Lonnquist (1959) discussed in the section on overdominance also provided some indication of epistasis. Bauman (1959) conducted experiments to detect epistasis in maize heterosis by comparing three-way crosses (A × B) × C with the average of the two separate single crosses, A × C and B × C. Significant differences between the three-way and average of the two parents would indicate epistasis.

Eighteen sets of lines were tested in one year and sixteen in another. Significant epistatic deviations were found in each year, but the results were not significant when epistasis-year interaction was included in the error term. It is noteworthy that all except one of the epistatic deviations significant at the 0.01 level were in the direction of indicating complementary epistasis of genes in the repulsion phase, AAbb × aaBB and not AABB × aabb.

In an extension of Bauman's technique, Gorsline (1960) tested 6 single crosses, 6 double crosses, and 1 double-double cross for epistatic gene action involving 10 characters. Epistasis was established for all 10 characters including yield, although it and ear length exhibited less than the other 8 characters. The author quite properly raised the question regarding the validity of gene-action studies based on assumptions of no epistasis.

In a more general paper, mentioned here because of its bearing on heterosis in maize, Jinks and Jones (1958) state that heterosis can be partitioned into four components expressing the role of additivity, dominance, nonallelic interaction, and the degree of association or dispersion of the relevant genes in the homozygous parent. Heterosis does not indicate any particular type of gene action or interaction, *but it is more frequent and has a higher mean expression when nonallelic interactions are operative.*

That nonallelic interactions are usually, if not always, operative in heterosis in maize would be expected on the basis of what is known about the genetics of the plant. Unlike single-gene overdominance, of which there is but a single clear-cut example in maize, there are numerous cases of epistasis not only in the classical sense of the term—one locus concealing the action of another—but also in the broader sense in which the term epistasis is now commonly used. There is probably no "unit" character in corn affecting any part of the plant which has not been found to vary in its expression in different genetic backgrounds. Anyone who has attempted to transfer "single-gene" traits such as sugary endosperm controlled by a locus on chromosome 4 or yellow endosperm controlled by a locus on chromosome 6 to varieties of corn which have never had these characters will appreciate the importance of nonallelic interaction on the expression of inherited traits.

Our discovery that many varieties of corn carry blocks of genes from teosinte or *Tripsacum* also supports the assumption of nonallelic interaction in heterosis, for these blocks behave in a limited way like the different genomes of an allopolyploid; they contribute to vigor by the interaction of genes which are not now allelic, although they may, earlier in the evolutionary history of the genus, have been.

Agreement between Mathematical and Biological Models

It is remarkable how closely the mathematical and biological models have now come to agree in explaining heterosis in maize. The majority of recent experiments have shown heterosis in maize to conform in general to the biological model proposed by Jones in 1917 with particular emphasis on the role of linkage in explaining the results thus recognizing in the mathematical model what has always been an essential part of the biological model and what Jones accepted as an established fact, "that characters are inherited in groups and that it is these groups of factors which Mendelize." Yet the model of linked dominant favorable factors does not account completely for yield heterosis in maize or for the marked depression which follows inbreeding. It might do so if only popcorn varieties with virtually knobless chromosomes were considered. The majority of modern varieties, however, contain blocks of genes from teosinte or *Tripsacum* which exhibit both nonallelic interaction and an apparent overdominance. The latter is due in part to the fact that genotypes which are homozygous for blocks of such genes suffer disturbances in developmental homeostasis which have little relationship to the genetic worth of the genes in a heterozygous state. It is also these disturbances in homeostasis which account for the marked inbreeding depression suffered by the majority of modern races of maize.

The biological model for heterosis in maize then is the model of Jones modified by the concept of blocks of genes from teosinte or *Tripsacum* which (1) exhibit apparent overdominance; (2) interact epistatically with other genes; (3) exhibit repulsion phase linkage; (4) simulate multiple alleles when linkages are broken; (5) cause disturbances in developmental homeostasis when homozygous. What does this model suggest with respect to methods of exploiting heterosis in maize? To answer this question let us examine the various methods which have been proposed for developing inbred strains and hybrids and evaluate these in terms of the model described above. Lerner's comment (1958) that "a considerable degree of statistical sophistication is needed to follow the more advanced types of analyses—especially those used for plant material" is all too true and I, unfortunately, lack that degree of sophistication. The most that I can hope to do is to examine the assumptions and conclusions and to determine, so far as this can be done by verbal reasoning, how well these correspond to what is known about the biological model. I have been able to show by such reasoning that at least one proposed method of breeding, that known as convergent improvement, is not likely to be effective in view of the biological nature of the corn plant as revealed by my studies and those of my associates.

20 Modern Breeding Techniques

Convergent Improvement

Hybrid corn had not been long in production before various modifications designed to improve its techniques were suggested. One of the earliest of these was the method of convergent improvement proposed by Richey (1927). This involves backcrossing a hybrid to each of its parental lines for several generations, followed by selfing, selecting in each generation for desirable characteristics contributed by the other, the nonrecurrent line. The inbred strains should be improved, because each should have gained through selection some of the dominant favorable factors of the other while retaining, through repeated backcrossing, its own. Their hybrid should be improved, because some of its favorable loci will be homozygous, and since dominance for quantitative characters is often not complete, the homozygous loci might be expected to be superior to the heterozygous ones.

On the face of it, convergent improvement is a brilliantly conceived device for determining whether hybrid vigor is due to dominant favorable factors and, if it is, for also providing a virtually infallible method of improving inbred strains and their hybrids. If convergent improvement is practiced through a sufficient number of cycles it should be possible theoretically to produce true-breeding inbred strains as productive as their single crosses. If, however, heterosis is due in some instances, as we are now reasonably certain that it is, to blocks of genes originally from teosinte or *Tripsacum*, then convergent improvement will not in some cases be successful, because of the tendency of these blocks of genes to have deleterious effects when

homozygous. If the two inbreds of a single cross differ in their blocks of genes, then convergent improvement will tend to incorporate into each those of the other, and we know from experiments described in an earlier chapter on extracted teosinte and *Tripsacum* chromosomes that a good line can be virtually exterminated by incorporating into it too much foreign germ plasm. Experiments on convergent improvement may be expected, then, to show conflicting results, improvement in some cases and not in others depending upon the extent to which the heterosis is the product of blocks of genes from teosinte or *Tripsacum*.

The experiments of Richey and Sprague (1931) seem at first glance to fall into the first category, and they are cited by Sprague et al. (1959) and by Lonnquist (1960) as demonstrating the effectiveness of convergent improvement and supporting the theory of dominant favorable factors. Close examination, however, raises some question whether this is true, since the experiment actually employed only backcrossed lines which had not yet been selfed. These selected lines did yield 30 and 27 percent more respectively after three and four generations of backcrossing than would have been expected if no selection had been practiced. It was assumed that half of this excess would be retained following selfing, hence the figures 15 and 13 mentioned by Sprague et al. for the percentage improvement of the inbreds. No data have subsequently been presented, however, to show that this gain was actually obtained, that the recovered lines after selfing were superior to the original ones. The fact that the

lines were not employed for a subsequent experiment on a second cycle of selection because they were "poorly adapted" raises some question whether they were actually improved. Nor are data available on single crosses made with these lines after selfing had rendered them homozygous.

The results of Murphy (1942) are somewhat more convincing in demonstrating improvement through the method of convergent improvement. His experiments, in contrast to those of Richey and Sprague, did employ selfed recovered lines, and he stated that most of the recovered lines showed marked improvement in vigor, plant type, and yield over the original lines. Unfortunately no data were presented to support these observations or to show to what extent the lines were improved. Of 49 crosses between recovered lines and their original lines tested by Murphy, all but 6 were superior in yield to the original line by a difference equal to or greater than the standard error, and 34 were superior by a difference of two or more times S.E. But this does not prove that the lines themselves had been improved, it only shows that they had been modified to the extent of producing hybrid vigor when crossed with the original lines. It requires very little substitution in a line to accomplish this result.

Of 51 single crosses of the recovered lines to the nonrecurrent parent, 31 were lower than the corresponding original single crosses by an amount equal to or greater than S.E., 16 were essentially the same, and 6 were higher. Only one of the 51 crosses was superior to the original by an amount twice S.E. In another experiment reported by Murphy,

17 additional single crosses were made between recovered lines selected on the basis of their previous performance in crosses. Of these, four were superior in yield to the original single crosses by about twice S.E. or more.

Hayes et al. (1946) conducted extensive experiments on modifying inbred strains through convergent improvement by the methods of backcrossing to determine whether the inbred strains used in the double cross Minihybrid 403 could be improved. The pedigree of this hybrid is (11 × 14) (A374 × A375). The first two strains were derived from the variety Minnesota No. 13 and the last two as well as C23, which was also employed in the experiments, came from Reid's Yellow dent.

Rigorous selection for vigor and for desirable agronomic characteristics was practiced during the backcrossing and selfing generation, and the number of cultures grown and the ears pollinated in each generation was large. The published data and photographs showed that several of the inbred strains were improved both as lines and in their performance in various types of hybrids: topcrosses, single crosses, and double crosses. Unfortunately for our present purpose few of the data are concerned with convergent improvement in the strict sense, backcrossing the F_1 hybrid to both of its parents.

For the single cross, 11 × 14, the yield of only one recovered line (14 × 11) 14_3 is reported. This is lower than that of either parent. Three recovered lines of 11 were crossed with standard 14. All crosses were lower in yield than the original single cross, two probably significantly so. One recovered line of 14 was crossed with standard 11. The yield of the cross was lower than that of the original cross.

Line 14 was definitely improved by crossing with line 374. Of 8 selected recovered lines tested, 5 had higher yields than the original line. Line 374 crossed with 14 was not improved to the same extent. Of 5 selected recovered lines, 4 were lower in yield than the original. Unfortunately no data on single crosses of these lines were published, although such crosses were apparently made and tested for yield. Line 11 was clearly improved by crossing with C23. Of 10 selected recovered lines, 8 were more productive than the original, 6 of these probably significantly so. At

the same time line C23 was definitely improved by crossing with line 11. Of 8 selected recovered lines tested, all were distinctly higher in yield than the original. Of all of the experiments on convergent improvement reported in the published literature, this is the only one in which clear-cut improvement of both lines has been demonstrated. Even here, however, there are no data available to compare single crosses of recovered lines with the original single cross. That line 11 was improved through convergent improvement with line C23 is not surprising, since line 11 is probably not tripsacoid, being described as having its seeds attached loosely to the cob.

Two later experiments on convergent improvement have failed to show conspicuous improvement of either of the inbreds or their single crosses. Sprague et al. (1959) backcrossed an F_1 hybrid of two inbreds, B2 and K4, reciprocally to its parents for three generations. Certain sublines were then selfed for three generations, with selection for desired characters being practiced. Recovered lines were evaluated in single crosses, and the two best, designated as B2a and K4a, were employed in the second cycle of backcrossing and selfing. The two selected first-cycle lines were superior in yield to the original, B2a being 22.4 percent better than B2 and K4a 6.3 percent than K4. The latter gain is, however, not significant. But of 9 second-cycle recovered B2 lines only 3 were more productive than B2a, one significantly so and 6 lines were less productive, one significantly so. Of the 6 second-cycle K4 lines none were significantly better or poorer than K4a.

The results of comparing the single crosses in these second-cycle recovered lines are startling. Of the 54 single crosses not a single one was equal in yield to the original single cross, and only one to the second cycle cross, producing 92.0 bushels per acre compared to 91.6 bushels. At the 0.05 level of significance, 46 of the 54 second-cycle single crosses were inferior to the first-cycle single cross. The authors conclude that the data lend some support to the hypothesis of dominant favorable factors as an explanation for yield heterosis, but it is difficult to see how they reached this conclusion. What the data do support is the hypothesis that a significant part of the heterosis in this particular single cross is due to blocks of genes from teosinte or *Tripsacum* incorporated into the two in-

breds, B2 and K4. I am not familiar with these inbreds, but many widely used Corn Belt inbreds, 38–11 for example, are quite tripsacoid, and chromosomes with effects similar to those of teosinte chromosomes can be extracted from them. It would not be at all surprising to find that the second-cycle lines and their single crosses used in this experiment are more tripsacoid than the original inbreds and their single cross. If this proves to be true, then it can be predicted with some degree of confidence that a third cycle of convergent improvement will lead to still further deterioration of the lines.

Lonnquist (1960) backcrossed a single cross, WF9, × 38–11 twice to each of its parent lines and selfed the backcrossed lines for two generations. Selection was practiced during the backcrossing and selfing generations. No data on the yields of the recovered lines are reported, although it is stated that improvement of the parent lines was accomplished.

Of the crosses of the 21 recovered WF9 lines with the original 38–11 and of 19 recovered 38–11 lines with the original WF9, none was equal in yield to the original single cross. The average yield of the crosses involving recovered WF9 lines was 82.4 bushels compared to 94.3 bushels per acre for the original cross and for the crosses involving the recovered 38–11 lines, 80.4 bushels per acre compared to 90.2 bushels for the original. Crosses of the recovered lines with the recurrent parent were clearly more productive than the recurrent line itself, those involving WF9 averaging 45.3 bushels per acre against 38.4 bushels for the original and those of 38–11 averaging 44.5 against 19.4 for the original.

The author concluded that the reason why the single crosses were not improved was that the backcrossing was limited to two generations, implying that the lines had not been sufficiently modified to effect improvement in their crosses. But he also concluded that the relatively high yields of crosses of the recovered lines with their recurrent parents show that the recovered lines contain a considerable amount of the nonrecurrent parents. The two conclusions are somewhat conflicting; the second is probably more nearly correct than the first.

In another test the original WF9 and three recovered WF9's were each crossed with the original 38–11 and two recovered 38–11 lines. The crosses involving the recovered WF9 were distinctly lower in yield than the original, those involving recovered 38–11 were slightly but not significantly higher in yield, and those made up of recovered lines of both parents were almost identical with the original. Five double crosses in which single crosses of recovered lines were compared with the original single cross were all higher in yield than the original double crosses, the average for the recovered being 111.8 bushels per acre and for the standard 102.7 bushels. The author explained this improvement as the result of a single cross of recovered lines which were homozygous for some favorable genes or gene combinations for which they were previously heterozygous. This is undoubtedly the correct explanation, and it also accounts for the lower yields of single crosses of recovered lines; it is not in conflict with the concept of heterosis involving blocks of genes from teosinte or *Tripsacum*.

Using lines generously supplied to him by Dr. Lonnquist, one of my graduate students, Gordon Johnston (1965), repeated some of the comparisons that Lonnquist had made between the lines WF9 and 38–11 and their "improved" counterparts. Of the 15 single crosses included in this test involving the modified lines not one was equal in yield to the original cross.

Rarely in the history of agronomic experimentation have three experiments, those of Sprague et al. (1959), Lonnquist (1960), and Johnston (1965), involving a total of 109 comparisons, produced such consistent and conclusive results. There seems to be no doubt that the method of convergent improvement is not effective in developing improved single crosses. It does sometimes but by no means always result in improvement of inbreds and double crosses. Also it seems clear that the method of convergent improvement is not a critical test for the dominant, favorable factor theory of heterosis.

Earlier in this chapter I suggested that convergent improvement might be ineffective if it involves blocks of genes from teosinte or *Tripsacum* that are deleterious when homozygous. Some of the evidence for this was presented in the previous chapter. Johnston (1967) presented additional evidence in showing that the morphological effects of these blocks could be recognized in crosses with a multiple-gene linkage tester and that the highest-yielding crosses involved combinations of different chromosomes, while the lowest-yielding crosses tended to involve combinations of the same chromosomes. Johnston also showed a correlation between number of chromosomes knobs—which I regard as a measure of teosinte introgression—in the inbred strains and the yields of their crosses.

Considered as a whole, the data are in agreement with the statement made at the outset that experiments on convergent improvement may be expected to show conflicting results depending on the nature of the heterosis involved. Convergent improvement should be effective whenever the heterosis is primarily of the maize × maize type. It may be effective also when it is of the maize × teosinte or maize × *Tripsacum* type. It is not likely to be effective when the heterosis is predominantly of one of the remaining three types discussed in the preceding chapter. Agronomists planning experiments on convergent improvement would be well advised to study the morphological characteristics of the inbred strains in selecting those to be employed. Inbreds like 38–11, B10, I11R2, and Oh28 which are highly tripsacoid are not likely to be improved either as lines or as parents of single crosses by crossing with each other in a system of convergent improvement. They may, however, be improved as components of double crosses.

It should be obvious that when blocks of teosinte or *Tripsacum* genes are involved, convergent improvement is not a test to distinguish between dominance, overdominance, and epistasis as the cause of heterosis. Such blocks undoubtedly do contain loci which exert favorable effects that are dominant or partially dominant. Also they undoubtedly contain loci which interact epistatically with loci in other chromosomes or other parts of the same chromosome. The blocks themselves will exhibit overdominance in the sense that the heterozygote is superior to either homozygote.

The deleteriousness of these blocks of genes is not so much a matter of unfavorable recessive genes included in the blocks as it is of genes affecting the morphology of the plant and ear to the extent of causing derangements in developmental homeostasis. There is a limit to the amount of foreign germ plasm which can be introduced into maize in a homozygous state without causing marked disturbances in development which are accompanied by reductions in yield. In brief, heterosis in maize is far more complex than had previously been supposed. It is doubtful that it can be completely analyzed by the statistical techniques currently in use, although when carefully conducted and imaginatively interpreted these can provide some clue to what is happening. Gardner and Lonnquist, for example, (1959, 1961) crossed randomly chosen plants from the F_2, F_8, and F_{16} generations of a cross with the two parental lines. Estimates were made of additive variance and dominant variance in each generation tested. The estimate of dominance was in the overdominance range in F_2 but declined with advance in generations. In interpreting these results the authors suggest the possibility of repulsion phase linkage among genes no more than partially or completely dominant, these linkages being broken in later generations of random mating. This is a remarkably accurate description of what might be expected if the initial heterosis were due in part to blocks of teosinte or *Tripsacum* genes.

Recurrent Selection

The principle involved in the method of recurrent selection is one of alternating cycles of selection and crossing. The method, like many others, is the product of gradual evolution, and over a period of more than fifty years a number of workers have made significant contributions to it. The method had its beginning with the ear-to-row method of breeding proposed by Hopkins (1899) as a means of changing the chemical composition of corn through selection. Hayes and Garber (1919) and East and Jones (1920) introduced the idea of selection within self-fertilized lines, and the latter presented extensive data showing that the method is quite effective in raising the protein content of corn. The method as described by them is as follows: "The procedure was simply to cross different selected high-protein lines, to self-pollinate the first-generation plants, and to select again from the progenies which represent segregating generations." The rationale is explained in the following statement (1920), which in the ensuing 40 years has not yet been improved upon: "The plan is based upon the plausible assumption that since the various inbred high-protein strains differ in their morphological features, similar protein percentages may be due to different genetic constitutions."

Jenkins (1940), on the basis of extensive tests for segregation of genes affecting yield in top crosses, suggested intercrossing S_1 lines which had shown superior combining ability in their topcrosses to produce a synthetic and repeating the process at intervals. His suggestions added two important features to the method: (1) utilizing F_1 strains and (2) practicing recurrent selection for general combining ability. Hull (1945) contributed the idea of practicing recurrent selection for specific combining ability and is also responsible for giving the method its present name. The important feature of selecting reciprocally for combining ability was contributed by Comstock et al. (1949).

In considering the experimental results which have been obtained with recurrent selection it is convenient to divide, as Sprague (1955) has done the method into four categories with respect to the characteristics upon which selection is practiced: (1) general characteristics; (2) general combining ability; (3) specific combining ability; (4) reciprocal combining ability.

Recurrent Selection for General Characteristics

The effectiveness of recurrent selection in changing the chemical composition of the corn kernel is well illustrated by the experiments of East and Jones (1920). These involved selection for protein content in four varieties, but inbred strains from only two of these, Stadtmueller's Leaming and Illinois High Protein, were employed in the series on recurrent selection. In Stadtmueller's Leaming the protein content of 27 self-pollinated ears varied between 8.21 and 17.8, six having an average of 12.03 percent. Nine self-pollinated ears of Illinois High Protein varied in protein content from 14.97 to 16.64 percent, but a mixed sample of open-pollinated ears of this variety contained only 13.39 percent protein.

In 1914, thirty-seven crosses were made between lines which had been selected for high protein in these two varieties. In 1915, seven of these ears were planted and ten selfed ears in each progeny were analyzed for protein content. The average protein content of the F_1 families varied from 13.71 to 15.33, but

there was also considerable variation among the self-pollinated ears within each F_1 family. The selfed (S_1) ear having the highest protein content in each family was selected for producing the S_1 plant progenies grown in 1916. In this generation, as would have been expected, the variation within families was quite marked. A large number of intercrosses (101) was made between the individual plants of the S_1 progenies, and on the basis of protein content in both parents ten crosses were selected for growing in 1917.

In 1917, one family, C23 × B20, a cross between two S_1 plants having 18.16 and 18.35 percent protein was particularly outstanding. The average protein content of the individual plants was 16.31 percent, and the two highest self-pollinated ears had 17.59 and 17.25 percent protein. The progeny of these grown in 1918 averaged 17.68 and 16.39 percent respectively, the two highest ears in the two families having 20.14 and 18.54 percent of protein. In the first family eight of the fifteen ears analyzed had a protein content of more than 18 percent. A mixture of sib-pollinated ears of three ears highest in protein in 1917, the equivalent of a high-protein synthetic, averaged 15.84 percent in protein content compared to 12.03 and 13.39 percent for the two varieties with which the two cycles of recurrent selection had begun.

Significant experiments comparing the effectiveness of recurrent selection in increasing the oil percentage of the grain with inbreeding accompanied by selection were conducted by Sprague and Brimhall (1950) and Sprague et al. (1952). The results indicate that for the same land area, number of pollinations, and chemical analyses recurrent selection was consistently more effective than selection in selfed strains. In one population selection in selfed strains raised the oil content in five generations from 4.97 to 5.62 percent, or at the rate of 0.13 per year. In the recurrent series the oil content was raised in two cycles involving the same number of years from 4.97 to 8.2 percent, or at the rate of 0.65 percent a year. Thus recurrent selection was superior to standard inbreeding selection by a factor of five in this particular experiment. Recurrent selection had the additional advantage of retaining genetic variability, which in the inbred series was largely exhausted after five generations of selfing.

Comparable but somewhat different results

were obtained by Jenkins et al. (1954) in recurrent selection for resistance to the leaf blight of corn caused by the fungus *Helminthosporium turcicum*. Nine groups of progenies of two inbred strains, one susceptible to the disease and the other not, were subjected to three cycles of recurrent selection. No inbreeding was practiced in this experiment; pollen from open-pollinated resistant plants was mixed and applied to the plants contributing to the mixture. Significant gains were made in virtually all of the groups during the first and second cycles, but only one gain of significant size was made in the third cycle.

The results indicate that genetic diversity with respect to resistance had been greatly reduced by two cycles of selection. Jenkins and Robert (1952) and Jenkins et al. (1952) had previously shown resistance to this blight to be controlled by many genes, most of which appeared to have minor effects, but there was evidence that a few genes had major effects. Apparently two cycles of selection had created a state of relative homozygosity for the major genes. Additional selection subsequently operating upon the minor ones continued to be effective, but at a much reduced rate. Recurrent selection is obviously a very effective method of rapidly producing a relatively homogeneous population for characteristics controlled by small numbers of genes.

Practicing selection within five populations which were constantly intercrossing, Weaver and Thompson (1957) raised the popping expansion of white hulless popcorn from 22.25 volumes to 35.83 volumes. Selecting for increased tryptophan, Frey et al. (1949) raised the percentage in one cycle of selection from 0.106 to 0.119, an increase of 12.7 percent of the mean of the original population.

The presence of blocks of teosinte and *Tripsacum* genes should effect recurrent selection only to the extent that such blocks are associated with the characteristic for which selection is being practiced. If selection were for resistance to "stunt" disease, for example, the recurrently selected population would probably become progressively more tripsacoid, since Cervantes et al. (1958) have shown that there is a correlation between resistance to stunt and the estimates of Wellhausen et al. (1952) on teosinte introgression. If selection caused the population to become homozygous for teosinte introgression it would also cause a reduction in yield but might improve combining ability. The same result might occur if selection for methionine content were practiced, since Melhus et al. (1953) have found that teosinte contains greater amounts of this amino acid than maize although it does not appear to differ significantly from maize in tryptophan and lysine.

Recurrent Selection for General Combining Ability

This method appears to have been first suggested by Jenkins (1940), who outlined a procedure for producing synthetic varieties by intercrossing S_1 strains which had performed well as parents of topcrosses, repeating the process at intervals. "High" and "low" yield synthetics were produced by Lonnquist (1951) by combining inbreds from a single original source on the basis of their performance in topcrosses with Krug. Subsequently 152 S_0 plants in the high yield synthetic and 77 S_0 plants of the low yield synthetic were tested in crosses on the single cross WF9 × M14. Yields of the former group varied from 75 to 115 bushels per acre, with a mean of 99 compared to 101.7 for the check (WF9 × M14). The yields of the latter varied from 60 to 90, with a mean yield of 75 compared to 90.5 for the check. No strains from the low synthetic exceeded the tester, many strains from the high synthetic did so. It appears that a single cycle of selection has separated the original group population into two distinct groups with respect to combining ability.

Later Lonnquist and McGill (1956) compared four synthetics after one and two cycles of selection with a standard hybrid, US13, as the control. The mean yield of the first cycle synthetics was 82 percent of the hybrid but 13 percent greater than that of the original open-pollinated varieties. The mean yield of the second cycle synthetics was 96 percent of the control. One of the second cycle synthetics yielded slightly more than the control.

These results are quite promising not only in showing that synthetics, about as productive as commercial double crosses, can be developed but also in suggesting that a synthetic of this type, presumably still containing considerable genetic diversity, could be a source of inbred strains superior to those now generally in use.

The presence of blocks of teosinte or *Tripsacum* genes should not influence the effectiveness of this method except that additional cycles of selection might increase homozygosity for such blocks and reduce the yields of synthetics made up of strains carrying them. Also, to the extent that improvements in general combining ability effected by recurrent selection involve blocks of genes from teosinte, the corn breeder must exercise care not to lose this effective germ plasm by discarding too rigorously, without testing, inbred strains which are not in themselves productive and promising. Such strains, if they exhibit superior combining ability, can often be used as the pollinator parents of single crosses.

Recurrent Selection for Specific Combining Ability

Specific combining ability in maize is concerned with the ability of one genotype to combine with another specific genotype, either an inbred strain or a single cross. This method was first suggested by Hull (1945), who maintained on theoretical grounds that it is the best method of identifying new strains for improving a hybrid combination if heterosis involves overdominance and that it is a good method if it involves only simple dominance.

There are few experimental data on the effectiveness of the method. A brief report from the Florida Experiment Station (1958) states that three cycles of recurrent selection for specific combining ability had been practiced using an inbred as a tester. The method has been compared with one involving selection for general combining ability. Significant advances in yield have resulted from both methods, and there is yet no trend to indicate that one is appreciably more effective than the other. Lonnquist and Rumbaugh (1958) found that lines selected for specific combining ability using single cross WF9 × M14 as a tester represented a random sample with respect to general combining ability. When compared with an unselected group of lines from the same original source, the average yields of topcrosses to plants of the original variety involving the two groups was 83.5 and 82.5 bushels per acre respectively, but a synthetic comprising the lines showing superior general combining ability was more productive than one comprising approximately equal numbers of lines exhibiting superior specific combining ability. The former had an average yield of 98.2 bushels per acre, the latter 94.7 bushels. The data from this experiment lend support to the common procedure of testing new lines first for general combining ability and later for specific combining ability.

Sprague et al. (1959) tested strains of Lancaster Surecrop and Kolkmeier (an Indiana strain of Reid Yellow dent) for specific combining ability, first in crosses with WF9 × Hy and in later cycles in crosses with Hy. Synthetics produced by intercrossing the strains with specific combining ability were compared with the original variety in crosses with Hy. The crosses of Lancaster first and second cycles synthetics were better than those of the original by 3.9 and 6.5 bushels per acre, those of the Kolkmeier were better by 7.0 and 20.0 bushels. The synthetics themselves were not substantially different in yield from the original varieties, but the hybrids between them were conspicuously so. The hybrids of the original varieties and the first and second cycle synthetics yielded 70.6, 74.8, and 81.7 bushels per acre respectively. Selection for specific combining ability had in these particular experiments also improved general combining ability. These several experiments leave no doubt that selection for specific combining ability can be effective.

From the standpoint of interpreting the results in terms of the presence of blocks of genes from teosinte or *Tripsacum*, it is of interest that the tester strains used in these experiments, WF9, M14, and Hy, are all to some extent tripsacoid. Consequently selection for combining ability with these particular strains must involve selection either for different blocks of genes or for genes other than those in the teosinte segments. Thus selection for specific combining ability may in some instances also improve general combining ability.

Reciprocal Recurrent Selection

The method of reciprocal recurrent selection was designed by Comstock et al. (1949) to make maximum use of both general and specific combining ability, and on theoretical grounds it should be capable of doing so. It employs foundation material from two sources, A and B. These may be two varieties, two synthetics, two F_2 populations, or two single crosses involved in a successful double cross. Plants from source A are selfed and outcrossed to a number of plants of B. Likewise, plants from B are selfed and outcrossed to A. Selection is based on the testcrossed progenies, the selected S_1 strains being intercrossed to produce a synthetic with which a second cycle of selection is begun.

Compared with the method of selection for general and specific combining ability on the basis of theoretical biometrical considerations, reciprocal recurrent selection should theoretically be superior to both. It should be better than selection for general combining ability for loci at which there is overdominance and better than selection for specific combining ability at loci for which there is partial dominance. It appears to be generally agreed that the method is theoretically sound and does indeed have possibilities of making maximum use of both general and specific combining ability. Selection is initially for general combining ability, but with each cycle it approaches nearer to becoming selection for specific combining ability. On the basis of verbal reasoning it appears to combine the best features of both. It also includes to some extent the feature of early testing.

Since the method appears to hold so much promise it is strange that it has not been more widely employed or even tested.* The published data so far available on the method are still of a preliminary nature. Harvey et al.

*This was written in 1962 while I was at Cambridge University in England on sabbatical leave from Harvard. I have not been able to review all of the articles concerned with this method that have since been published and which are cited in a recent article by Moll and Robinson (1967). I have learned from Dr. Major Goodman that more recent results are generally similar to earlier ones cited here in showing the method to be effective in improving yields. The hybrids resulting from crossing the selected varieties do not, however, have the uniformity of plants and ears that the modern farmer, accustomed to growing hybrid corn, has come to demand, so that the method is not likely to be widely used except to provide populations from which superior inbred strains can be isolated. I have been informed that one of the commercial seed companies has produced an excellent hybrid by isolating inbred strains from populations resulting from practicing recurrent reciprocal selection in two of the best hybrids developed by two of its competitors. Practices such as this, although effective in improving yields, also tend to narrow the genetic base of the hybrid corn grown in the United States and so to contribute to widespread susceptibility to various diseases, such as the mutant strain of corn blight mentioned later in this chapter.

(1959) report a 14 percent increase in the yield of testcrossed progenies from the first and third cycles of selection in two synthetics, A and G. Estimates of genetic variance indicate that further improvement should be possible through selection.

Douglas et al. (1961) subjected two Texas varieties, Yellow Surcropper and Ferguson Yellow dent, to three cycles of reciprocal recurrent selection. Both varieties were improved in combining ability with the other. It is difficult to determine precisely how much improvement was effected, since the testcrosses of the different cycles were grown in different years. However, the testcrosses of Ferguson Yellow dent differed from the yield of Surcropper compared as a control by 1.08, 2.15, 2.15 times the standard deviation in successive generations and the mean yields of the Yellow Surcropper testcrosses differed from the yield of open-pollinated Ferguson Yellow dent by 1.36, 1.65, and 2.86 times the standard deviation in three cycles. Also a direct comparison of crosses of the original populations and the final synthetics were 50.2 and 59.9 bushels per acre respectively, a gain of approximately 14 percent.

Thomas and Grissom (1961) have presented data on two cycles of reciprocal recurrent selection for popping volume, resistance to root lodging, and yield in two populations of popcorn. Significant improvement was effected in all three characteristics, but selection in self-pollinated lines practiced during the same period proved to be almost as effective. However, in the recurrent series considerable genetic variation still remained after two cycles while in the selfing series it had been largely dissipated.

Although the results so far available from experiments on reciprocal recurrent selection are quite encouraging, it has not yet been shown that this method is superior to selection for either specific or general combining ability. Thompson and Harvey (1960) have presented preliminary data on a comparison of two types of selection. Two synthetics, A and B, were subjected to reciprocal recurrent selection for three cycles. The mean yields of the testcrosses in terms of percentage of its checks were 82.7, 89.1, and 94.6 for synthetic A and 82.2, 82.9, and 93.3 for synthetic B. One of the synthetics, A, was subjected to five cycles of recurrent selection, the open-pollinated variety, Jarvis, being the tester for the

first cycle and the single cross, NC7 × CI21, the tester for the remaining ones. The yields and the testcrosses in percentage of the checks in cycles 1 to 5 were 77.8, 86.0, 94.2, 96.7, and 98.8. To the extent that the data from the two series representing two types of selection are comparable they indicate that both types of selection are effective and that the gains are of about the same order in both for the first three cycles. The data of Sprague et al. (1959) mentioned above, although not directly comparable, are also of interest in this connection. Two cycles of selection improved the yield of test crosses in Lancaster Surecrop by 8 percent and in Kolkmeier by 29 percent, and the two synthetics when crossed yielded 16.5 percent more than the cross of the parental varieties.

The existence of blocks of genes from teosinte or *Tripsacum* should not affect the method of reciprocal recurrent selection to any great degree, favorably or unfavorably. As already pointed out, such blocks exhibit (1) overdominance, (2) nonallelic interaction, (3) similarity to multiple alleles, and (4) linkage in the repulsion phase. In all of these situations reciprocal recurrent selection should be more effective than selection for general combining ability. To the extent that these blocks of genes can ultimately be broken up by recombination and the action of their individual component genes are not overdominant, reciprocal recurrent selection should be more effective than selection for specific combining ability. As in other methods of recurrent selection, care must be taken not to discard, on the basis of appearance alone, inbred strains which are isolated from the improved gene pools resulting from recurrent selection.

To the extent that a limited number of blocks of teosinte and *Tripsacum* genes account for a substantial part of the combining ability in the two populations, A and B, it is possible that reciprocal recurrent selection may result in rather discouraging reductions in yield in one or both populations. This situation would be the counterpart of that described by Lerner (1958) using the single symbols X_1 and X_2 for the two populations undergoing selection:

Indeed, the effects of this method of selection on the gene pools of the population X_1 and X_2 may be expected to be disintegrative, since it tends to disregard intrapopulation fitness. For instance, it is possible that a certain allele A_1 produces in population X_1 a subvital effect not only as a homozygote but also when it is in combination with A_2 and A_3. Ordinary intrapopulation selection would tend to eliminate A_1. But now suppose that population X_2 contains allele A_4 (absent from the X_1 pool), which in combination with A_1 but not with A_2 or A_3 produces luxuriance. Selection for combining ability will, under these circumstances, encourage the rise in the frequency of A_1 perhaps to the point of danger to the continuation of X_1 as a selfed-reproducing entity. This process can thus produce a pool which has contents that combine well with the alleles of another pool but which is not coadapted in the sense of maximizing the fitness of a self-reproducing population.

The method of reciprocal recurrent selection would seem to offer its greatest promise in parts of the United States and other countries where hybrid corn produced by the conventional methods characteristic of the Corn Belt is not yet well established and an extensive hybrid seed corn industry does not yet exist. In these circumstances the method might produce substantial improvement in the breeding material before the isolation of inbred strains for the production of single and double crosses were undertaken. In tropical countries where a winter crop can be grown, the third step in each cycle, the intercrossing of selected F_1 strains, could be performed in a winter generation, thus reducing a cycle to two years. In some countries such as Colombia where two crops a year are regularly grown, the cycles can perhaps be reduced to a year and a half each.

Under some circumstances a combination of the method of reciprocal recurrent selection and the method of producing homozygous line from monoploids may be practical. It has not yet been shown that the homozygous lines derived in this way are superior in combining ability to those produced by successive generations of self-pollination, but they are apparently at least equal to these (Thompson, 1954). Thus the delay involved in employing two cycles of reciprocal recurrent selection might be compensated for by developing homozygous strains rapidly through the monoploid method. I have learned that the breeder for one commercial company has used this combination quite successfully.

Improving already established commercial double crosses by practicing reciprocal recurrent selection in the F_2 generation of two single crosses followed by the isolation of new inbred strains is a method which seems not yet to have been extensively tested. Theoretically, there is no reason why it should not succeed in improving double cross yields. However, if blocks of teosinte or *Tripsacum* genes are involved in both single crosses, recurrent selection like convergent improvement might increase homozygosity for these, with the result that inbred strains isolated from the recurrently selected populations as well as their single crosses would be inferior in productiveness to the original strains and crosses. The performance in double crosses might be improved. The data available on various types of recurrent selection indicate that all types are effective, but the tests are not adequate to show which is most effective. Other promising methods which have not yet been adequately tested are considered below.

Although the method of recurrent reciprocal selection seems to offer its greatest promise in regions not yet committed to established techniques of hybrid corn production, the possibility of using the method in the United States should not be overlooked. The data of Robinson et al. (1956) on dialled crosses of six southern varieties show the yields of the F_1 hybrids to vary from 4.6 to 46.2 percent above the average of the two parents and the F_1 means to be 19.9 percent more than the average of the parents and 11.5 percent more than the mean of the higher yielding parent. The two varieties which produced the highest F_1 yield should provide very promising material for a program of reciprocal recurrent selection.

Even in the Corn Belt, where in the opinion of Hull (1945) additive genetic variance has been largely dissipated by past selection, the possibility of reciprocal selection merits consideration. In the recent experiments of Lonnquist and Gardner (1961) on intervarietal crosses, an F_1 hybrid of Golden Republic and Barber Reid yielded 101.3 bushels per acre compared to 87.9 and 96.6 bushels respectively for its two parents and 104.6 bushels for U.S. 13, the commercial hybrid used as a check. Further improvement in intrapopulation combining ability of these varieties can undoubtedly be effected by selection. However, even more promising foundation stocks for this purpose were revealed by these experiments. An F_1 of two synthetics, Krug IIA and RII, which had previously been produced by combining lines subjected to recurrent selection, yielded 106.9 bushels per acre, 14 percent more than the mean of the parents, 96.2 and 90.4 bushels respectively, and slightly more then U.S. 13, the check. Since these two synthetics were developed by recurrent selection for general combining ability with different testers, they may still respond to selection for combining ability with each other. If so, there is a possibility of isolating from them, after one or more cycles of recurrent selection, inbred lines capable of producing double crosses far superior to those now in general use. There is also a real possibility that average yields in the Corn Belt can be raised to a new plateau as much higher than the present one as that was over prehybrid corn yields.

Gamete Selection

A method of gamete selection was proposed by Stadler (1944) as a means of improving inbred lines already in use by introducing selected germ plasm from open-pollinated varieties. He contended that the need for new breeding stocks makes the further sampling of open-pollinated varieties imperative. He assumed that the frequency of genotypes comparable to the elite lines in both yield potential and general agronomic value is probably too low to make their direct extraction feasible as a general practice.

The method is essentially as follows: If in the double cross $(A \times B) \times (C \times D)$ the inbred A has inferior combining ability with C and D and needs to be improved in this respect, it is crossed with a selected open-pollinated variety. The F_1 plants, all of which are A × variety topcrosses, differ from each other only in the gamete received from the variety. A number of these as well as the inbred A are crossed to the single cross, C × D, and the crosses are compared in a yield test. Any testcross, $(C \times D) \times (A \times$ variety$)$, which has a higher yield than the three-way cross used as check $(C \times D) \times A$ is assumed to represent a combination of A with a gamete superior to A in its ability to combine with C × D. Selfed seed of the F_1 plants (A × variety), having high testcross yields, furnish the basis for further inbreeding and the selection of lines superior to the original in its combining ability with C × D. One of the breeders for a well-known commercial seed corn company has employed this method quite successfully.

The theoretical advantage of the method lies in the simple fact inherent in Mendelian heredity that any particular type of gamete has a higher frequency than its corresponding homozygote by a ratio of a number to its square, $x : x^2$. Thus, if in an open-pollinated variety only one zygote per 10,000 is distinctly superior, then one gamete in 100 would be in this category. The difference becomes even more striking as the frequency of superior gametes decreases. If these occur in a frequency of 1 per 1,000 then the corresponding zygotes have a frequency of 1 per 1,000,000. To produce the zygote in this case is an impossible task; to identify and preserve the gamete is within the realm of possibility. The method thus offers its greatest promise in identifying and preserving superior combinations which are rare.

Richey (1947), although agreeing with Stadler on the need for further sampling of open-pollinated varieties and on the greater efficiency of sampling gametes versus sampling zygotes, objected to the feature of early testing which Stadler's method involves. He considered it preferable to employ the conventional method of selection in inbred lines following the crossing of an inbred line with an open-pollinated variety. Richey also pointed out, as had Burnham (1946) earlier, that gametes which are themselves rare may still be perpetuated in zygotes and can be recovered from these through selfing. Consequently the ratio $x : x^2$ does not tell the whole story so far as the breeding possibilities are concerned. It should be pointed out, however, that if the frequency of the superior gamete is quite low then its frequency in zygotes will also be quite low and the possibility of identifying the satisfactory zygotes and recovering the superior gametes from them becomes remote. The method of gamete selection obviously offers its greatest promise, as Stadler saw clearly, when the superior gametes have a low frequency. Nei (1963) has reached a similar conclusion.

The few experiments which have been conducted on gamete selection have all shown the method to be promising in improving

elite lines by combining them with superior gametes extracted from open-pollinated varieties.

Pinnell et al. (1952) crossed three inbred strains employed in commercial hybrids with open-pollinated varieties. A344 was crossed to Minnesota 13 and for comparison to 8 inbreds of diverse origin. Of 35 gametes tested, 16 gave significant increases over the testcrosses and 5 additional ones, although not producing significant yield increases, had testcrosses which were earlier. Three of the selected inbreds produced increases in yield. The inbred A25 was crossed to the variety Golden King. None of its 32 gamete testcrosses were superior in yield to the checks, but 8 had about the same yields and were earlier. Inbred A73 was crossed to the variety Murdock. Of 38 gametes tested, 8 produced testcrosses significantly higher in yield than the checks and 14 testcrosses with about the same yield but earlier in maturity.

Altogether about one-fourth of the 105 gametes tested produced significant increases in yield in the testcrosses and another fourth increases in earliness. A random sample of gametes from the open-pollinated varieties performed about as well as the gametes from a selected sample of inbred strains. Similar but somewhat different and more extensive experiments on gamete improvement were conducted by Lonnquist and McGill (1954) with two three-way crosses, one the popcorn hybrid K4 and the other a dent hybrid, Nebraska 501.

In an experiment involving one generation of mass selection, one of selection for specific combining ability, followed by gamete selection, the popcorn inbred SA24 was crossed with the open-pollinated variety South American pop. Ninety of the best-appearing plants of the variety were selected for yield-testing of their topcrosses with SA24. On the basis of topcross yields, six families, three high and three low, were tested in crosses with the single-cross parent of the commercial hybrid. Of 70 gametes tested from three F_1 plants in the high series, eleven produced better testcrosses than the checks by one or two times the standard error, while in the low series only one gamete of 71 tested gave a similar performance.

Similar procedures were employed to improve the dent hybrid Nebraska 501. The inbred, N6, was crossed on the open-pollinated

variety, Hayes Golden, and 118 topcrossed plants were chosen for yield tests. Remnant seed of 6 of these selected over the range were subjected to gamete sampling, about 20 plants in each family being selfed and outcrossed to the single cross WF9 × Hy. Of 123 gametes tested, 23 were equal or superior to those derived from the inbred N6 in yield and were as early or earlier in maturity. As the next step, five plants from the best progenies within each of the six families were selfed and outcrossed to the single-cross tester. The average yields of the families continued to be correlated with those of previous generations and of the ten gametes tested in the two higher yielding families, nine were equal or superior in their testcross yields to those of N6. Taking maturity into consideration, the authors concluded that 80 percent of the Hayes Golden gametes included in this test were worth further testing compared to the 25 percent of those from elite strains.

Additional experiments on gamete-testing at the Nebraska Station were carried out by Helmerick (1959) to improve the hybrid Nebraska 503, (WF9 × N6) × (Oh7 × 187–2). A population of 148 S_1 plants of the Stiff Stalk Synthetic were evaluated in topcrosses with a broad gene base and partitioned according to intervals of one standard deviation on either side of the mean. Four S_1 lines from each of the four resulting intervals were selected for gamete-sampling and were crossed to the two elite inbreds Oh7 and 187–2. Twelve gametes from each of the resulting F_1 families were then evaluated in crosses with WF9 and N6. About 48 percent of the 192 gametes tested were superior to those of the elite inbreds in the testcrosses, and, as might be expected, the greater percentage of superior gametes originated from the upper half of the frequency distribution of the original 148 plants of Stiff Stalk Synthetic.

Gametes giving the better performance were selected from each inbred series regardless of the interval with which they were associated and combined into 35 single crosses which replaced the original single cross, Oh7 × 187–2 in the double cross Nebraska 503. Predicted values of the double crosses based upon the gamete testcrosses indicated that 31 of the 35 selected substitute double crosses would exceed the control in yield. At the 0.05 level of probability, 22 of them were equal to the control. The single crosses themselves were no better on the average than the original. The latter result is not too surprising, since all of the inbreds employed had been modified by the introduction of germ plasm from the same source, the Stiff Stalk Synthetic.

Considered as a whole, the experiments on gamete selection have yielded promising results and the method appears to merit wider use than it has had so far. It offers its greatest promise in sampling exotic varieties which are poorly adapted to the region in which it is desired to employ them but which are known from tests in other regions or from their past history to contain superior germ plasm. In such varieties desirable zygotes may be too rare even to be sought for but desirable gametes are not beyond reach. A minor modification of Stadler's method of gamete selection to make use of the germ plasm of unadapted exotics is discussed in a later section of this chapter.

Homozygous Lines from Monoploids and Parthenogenetic Diploids

Another method which involves the sampling of gametes instead of zygotes is that of developing homozygous diploids from monoploids (haploids). The procedure is to pollinate the population to be sampled with pollen from a stock carrying dominant marker genes which will show up in the seedling progeny. Seedlings which do not exhibit the marker genes are putative monoploids. Because of meiotic irregularities these are usually sterile. However, a varying proportion of them develop sectors which are diploid and fertile. When self-pollinated, such sectors produce seeds which are completely homozygous because they have resulted from the doubling of a haploid complement. Thus the method can in one generation produce lines which are comparable in their degree of homozygosity to those resulting from five, six, or more generations of self-pollination. Chase (1952a, b, c, 1958, 1959), who has given particular attention to employing this method, has now produced several thousand monoploids from which several hundred different homozygous diploids have been developed. A high percentage of these have given good performance in hybrids, and some are now in commercial production.

There is great variation in the frequency of monoploids in different populations, the

average being of the order of 1 per 1,000 (Chase, 1952a; Gerrish, 1956; Seaney, 1955), but frequencies much higher than this have been encountered. Coe (1959) has described an inbred stock with a frequency of 32 monoploids per 1,000. The tendency to produce monoploids obviously has a hereditary basis, and the trait is one which responds to selection. The Stiff Stalk Synthetic sampled by Chase (1952c) yielded monoploids at a rate of 1.3 per 1,000, but single crosses involving homozygous lines derived from these monoploids produced new monoploids at the rate of 4.3 per 1,000. The frequency of monoploids also varies with the cytoplasm, the pollinator, the time of pollination, and treatment of the silks.

Mazoti and Muhlenberg (1958) found that significantly more haploids were produced in teosinte than in maize cytoplasm. A number of investigators have reported variation in frequencies resulting from differences in the pollinator (Chase, 1952; Gerrish, 1956; Seaney, 1955). Delayed pollination resulted in higher frequencies of monoploids in the experiments of Gerrish (1956) and Seaney (1955). Deanon (1957) reported that the frequency of monoploids was increased significantly by treating the silks with maleic hydrazide.

Also variable is the frequency of monoploids developing fertile diploid sectors which make self-pollination possible. On the average this is of the order of one per ten monoploids, but differences between the stocks may be tenfold. Of the monoploids (1.3 per 1,000) isolated from Stiff Stalk Synthetic by Chase (1952c), 9.4 percent set seed. But of those isolated (4.3 per 1,000) from single crosses involving homozygous lines derived from monoploids, 33 percent set seed. The ratio of fertile monoploids to seedlings tested is 1.3 per 10,000 in the first population and 14.2 in the second, which is a sample of the first.

Even more important than the frequencies of fertile monoploids is the frequency of homozygous lines which can be successfully employed in the production of commercial hybrid seed. On this point the data are still inadequate or not yet published. Thompson (1954) compared 23 homozygous diploid lines with 37 randomly selected S_1 lines of Golden Cross Bantam in three-way crosses. The two groups had mean yields of 61.8 and 62.2

bushels per acre respectively. Similar comparisons were made between three-way crosses involving 31 homozygous diploid lines, 35 unselected S_5 lines, and 5 selected inbreds. The mean yields of the three groups of crosses testing these lines were 58.4, 58.1, and 60.1 bushels per acre respectively. In another series of tests comparing 43 homozygous diploids with 38 unselected S_5 lines and 5 selected inbreds, the testcross yields were 103.8, 103.8, and 99.5 bushels per acre respectively. Thompson concluded from these several sets of data that survival of individuals in the haploid state did not constitute a selective advantage as measured by combining ability in the populations studied. Perhaps a more important conclusion which may be drawn is that it also did not constitute a selective disadvantage. Apparently lines derived by the monoploid method represent approximately a random sample so far as combining ability of the original population is concerned. Consequently the merit of the method lies in the rapidity with which useful homozygous lines can be established. We have already suggested that the method might be used in combination with the method of reciprocal recurrent selection. Here it would offer two advantages: (1) it would compensate for the time consumed in subjecting the open-pollinated populations to two or more cycles of recurring selection; (2) the proportion of superior gametes—those imparting good combining ability, both general and specific—should be high and the percentage of derived homozygous lines which can be employed in hybrid seed production should also be high.

The time required for the development and evaluation of a new hybrid combination is usually estimated at a minimum of ten years (Sprague, 1955) and is often more than this. In about that same length of time, two selected open-pollinated varieties which combine well in the varietal cross can be put through two cycles of recurrent reciprocal selection, monoploids isolated from them and selfed, crosses made and tested, and double-cross combinations tentatively chosen. Dr. Sherret S. Chase, the chief corn breeder for one of the larger seed companies tells me that employing the monoploid method, winter nurseries, and intensive yield-testing he has produced well-tested double crosses in six years. The inbred strains derived from such a program may not be better than those now in common use, but the hybrid combinations should be markedly superior, since the stocks from which the inbreds have been derived have been rigorously selected for both general and specific combining ability.

For beginning a program of corn improvement in a country where hybrid corn is not already well established, the combination of reciprocal recurrent selection with gamete selection through monoploids would appear to be one of the most promising lines of approach. To what extent blocks of genes from teosinte or *Tripsacum* affect the frequency of monoploids or the occurrence of fertile diploid sectors among them is completely unknown and represents one of the numerous gaps in our present knowledge. The fact that a Mexican variety which yields a high frequency of monoploids is quite tripsacoid in some of its characteristics may be suggestive; it is no more than this.

Paternal Monoploids

Monoploids arising from sperm nuclei also occur and may be identified in crosses with appropriate marker stocks, but their frequency is quite low. Seaney (1955) reported a frequency of 1 to 187,500. Goodsell (1961) states that he has found approximately 20 paternal monoploids in a period of ten years and in a recent experiment identified 5 paternal monoploids among approximately 400,000 seedlings. This frequency is obviously too low to be employed usefully in the production of homozygous lines except for special purposes such as the rapid conversion of an important commercial inbred into a male-sterile line. Goodsell has successfully converted an inbred into a male-sterile line in two generations. A cross of a male-sterile stock by the inbred N6 yielded 2 paternal monoploids among approximately 150,000 seedlings. One of these pollinated by the fertile N6 bore several seeds which gave rise to plants resembling the original line in all respects except that they were male sterile.

Chase (1963) in similar experiments found the same frequency of paternal monoploids as Goodsell and also succeeded in incorporating cytoplasmic male sterility in an inbred strain in one generation. In a more recent publication (1969) he has summarized the experimental data involving the use of paternal monoploids. Now that the T-type of

cytoplasmic sterility described later in this chapter has proved to be susceptible to infection by a strain of *Helminthosporium maydis*, the method of utilizing monoploids may be feasible in quickly converting standard inbred strains to other types of cytoplasmic male sterility or in converting to the T-type any inbred strains that may prove to be capable of overcoming the susceptibility. It is possible that this procedure may be facilitated by the important discovery of Kermicle (1969) that the frequency of paternal monoploids is greatly increased in the presence of a mutant gene affecting the development of the gametophyte.

The observations on paternal monoploids are also of interest in providing confirmation for a report made many years ago by Collins and Kempton (1916) on a cross of *Tripsacum* and teosinte. A single plant resulting from this cross resembled the pollen parent, teosinte, in all of its characteristics, a phenomenon to which Collins and Kempton applied the term "patrogenesis." Since these results have never been repeated there has sometimes been some question about their authenticity.

Parthenogenetic Diploids

It may be possible in some cases to produce homozygous lines directly from parthenogenetic diploids without going through the monoploid stage. Yarnell and Hills (1959) described a case in which a bagged ear, never artificially pollinated, produced seeds containing both endosperm and embryo but weighing only about one-fourth as much as normal seeds. Data on the segregation and vigor of the plants grown from these indicate that a high proportion of them had originated parthenogenetically and the remainder through accidental crossing. The authors suggest that parthenogenetic development may occur in the absence of pollination in the vicinity of fertilized kernels under special circumstances which may include limited amounts of pollen applied during a period of high temperatures. The possibility of producing parthenogenetic diploids directly obviously is one which merits much further investigation.

The *Oenethera* Method of Establishing Homozygous Lines

An ingenious method of establishing homozygous lines directly from gametes has been proposed by Burnham (1946). This would

employ a stock in which all of the chromosomes of the haploid set are involved in translocations in such a way that the F_1 crosses with normal stocks will, like certain crosses in *Oenethera*, have at meiosis a ring containing the entire diploid number of chromosomes. Such a plant would produce principally two kinds of functional spores corresponding to the parental gametic combinations. Selfing the F_1 would produce three kinds of plants, one with high pollen abortion and two with normal pollen. One of the latter would be a new homozygous line derived by the doubling of the haploid set of chromosomes contributed by the gamete. Crossing over within the ring would produce some variability in the gametes and some heterozygosity in the lines initially developed from them.

The method has not yet been tested, primarily because a stock combining appropriate translocations in all of the ten chromosomes has not yet been developed. If and when this is accomplished the success of the method will depend largely on the amount of crossing over which occurs within the ring and the extent to which the germ plasm from the multiple translocation stock, which finds its way into the homozygous lines through crossing over in the F_1 hybrid, contributes to or detracts from the genetic worth of the lines. Since some crossing over will undoubtedly occur, it would appear to be important to develop such a stock from translocations derived from inbreds of high general combining ability.

The Use of Exotic Germ Plasm

The degree of heterosis exhibited by crosses of maize varieties or inbred strains is highly correlated with the distance of the relationship between them. The importance of diversity of origin in making up commercial hybrids is generally recognized, and it is a common practice to include in such hybrids, lines derived from two or more open-pollinated varieties. There is adequate experimental evidence to support this practice. Wu (1939) has shown that single crosses of lines of related origin are consistently lower yielding than single crosses of unrelated lines. Hayes and Johnson (1939) showed that of 43 single crosses between lines unrelated in origin, 28 were equal or superior to standard double crosses used as checks, while of the 15 single crosses of related lines only 6 were equal or superior to the double crosses. The experimental results of Eckhardt and Bryan (1940a) and Cowan (1943) also emphasize the importance of genetic diversity in the production of high-yielding hybrids.

Since the importance of genetic diversity is so generally recognized it is somewhat surprising that so little attention has been given to employing exotic germ plasm—derived from varieties from other countries—in the production of hybrid corn in the United States. One reason for this is the belief, by no means an erroneous one, that the possibilities of improvement inherent in native varieties are still far from exhausted. It is certainly true that inbred strains and their hybrids can be and have been improved by the methods of recurrent selection for general and specific combining ability, gamete selection, and other methods without introducing exotic germ plasm.

A second reason that the exotic varieties have not appealed to the majority of American corn breeders is that these varieties are not as such usually promising. Although corn is classified by plant physiologists as a "day neutral" plant, many tropical varieties are responsive in varying degrees to photoperiod when grown in northern latitudes during the summer months. They are vegetatively luxuriant, quite late in flowering, fail often to produce ears, develop high concentrations of sugar in their sap and become susceptible to corn smut, and in general appear to have little to offer to the practical plant breeder. F_1 crosses of exotics with native varieties or inbreds are often not much more promising. It is not until at least one back cross to adapted native corn has been made that the merit of the exotic germ plasm becomes apparent. Finally the general indifference to the use of exotic varieties lies in the fact that not much has been known about them and very little critical experimental work has been done with them.

Griffing and Lindstrom (1954) compared the single crosses of three kinds of lines: (C), inbreds cycled with other inbreds; (B), inbreds isolated from partially inbred Brazilian material; and (E), Corn Belt lines having 50 or 25 percent Mexican germ plasm. Inbreds in each of these three groups were crossed with inbreds in the same and other groups. The results were striking. Of 36 single crosses compared, the first 10 had for one or both

parents the inbreds containing Mexican germ plasm. The average yields of the E × E single crosses were 13 percent higher than those of the C × C crosses. Two of the three E lines were superior to the three C lines in general combining ability. Unfortunately the source of the Mexican germ plasm was not reported; it is unlikely that it was the best that is available. Melhus et al. (1949) conducted trials of crosses of U.S. inbreds with Guatemalan varieties in Guatemala and Iowa. The preliminary trials in 1946 showed some of the hybrids to be promising, but no data on yields were reported. In replicated tests conducted at Ames, Iowa, in 1947 and 1948 the data are unfortunately reported only in average weights per ear, and to the extent that the hybrids were single eared and the stands comparable, such data give some indication of productiveness. Several three-way crosses involving Guatemala germ plasm had average ear weights substantially higher than those of the original single crosses. These represent increases of more than 40 percent over the original single crosses.

Kramer and Ullstrup (1959) made a study of 1,066 introduced maize varieties. These were first screened as seedlings for resistance to two rusts, *Puccinia sorghi* and *P. polysora*, and later for two strains of *Helminthosporium*, *H. turcicum* and *H. maydis*. About 600 of these were finally used as pollen parents in crosses with the single cross WF9 × Hy, and 572 of these were grown in a yield test. Of these, only 8 yielded more than the single cross itself, and these were all later in maturity than the single cross. At first glance these results might appear to be somewhat discouraging with respect to the possibility of making profitable use of exotic germ plasm, but actually the results are in some respects quite encouraging. Of the 572 crosses tested, 103 did not differ significantly in yield from the single cross WF9 × Hy. Almost all of these might have among their best gametes some which would have raised the level of performance quite substantially. Also the germ plasm in this particular experiment was restricted to a rather narrow range. Only introductions which matured at Ames, Iowa, were used. This preliminary screening of the exotics would have eliminated the great majority of tropical varieties which, under the climatic conditions of the Corn Belt, grow luxuriantly but do not flower until quite late.

We have had tropical varieties from Mexico, Guatemala, Honduras, Ecuador, and other countries of the hemisphere which without short-day treatment do not flower until almost the time of the first frost. If evaluated on the basis of this performance, these varieties would be arbitrarily eliminated. By subjecting them to short-day treatment they can be induced to produce pollen in time for crossing with standard inbreds. Even their crosses, however, may be so late in maturity as to be quite unpromising. It is not until they have been backcrossed at least once to an adapted native inbred that whatever good qualities they have to transmit can begin to be appreciated.

The question may be raised whether modifying one-fourth of the germ plasm of an inbred strain which itself will contribute only one-fourth of the parentage of a commercial double cross can have any appreciable effect upon the yield of the double cross. A simple calculation will show that it can. An inbred strain that has been modified by crossing with an exotic followed by backcrossing to the inbred will on the average carry the equivalent of about 5 chromosomes from the exotic and will transmit about $2\frac{1}{2}$. If there has been selection for desirable characteristics from the exotic the modified strain may well transmit as many as three chromosomes derived from the exotic. A single cross having the modified inbred as one parent will be heterozygous for the three exotic chromosomes. Employed as one parent of a double cross, the single cross, its chromosomes segregating at random, will transmit all three exotic chromosomes to 1/8 of the double-cross plants, two exotic chromosomes to 3/8, one exotic chromosome to another 3/8, and none to 1/8. In other words, 7/8 of the double-cross plants will be affected to some degree by the exotic germ plasm.

In some cases the desirable characteristics contributed by the exotic may involve not more than one chromosome. In this case half of the double-cross plants will receive it. Since single crosses have now replaced double crosses to a large extent the problem becomes even more simple. No matter how many exotic chromosomes are involved in the modified strain all of the single-cross plants will receive all of them. The single crosses will of course be heterozygous for the exotic chromosomes, but if these involve blocks of genes from teosinte or *Tripsacum* they may exert their desirable traits most effectively in this state.

Corn breeders of the other countries of this hemisphere, concerned only with improving the productiveness of corn and usually having no provincial ideas about ideal types, are much less restricted in their choice of breeding materials than are United States corn breeders. As a consequence they have also made wider use of exotic varieties in developing hybrid combinations and have done so with notable success. There are few commercial hybrids in any of the countries of Latin America which do not employ inbreds derived from varieties from other countries. One of the best hybrids in Colombia, for example, contains inbreds from Cuban and Peruvian varieties.

Inventory of Exotic Races

The use of exotic germ plasm has become all the more promising now that the collection and classification of the races of the countries of this hemisphere has been virtually completed. It is no longer necessary for the corn-breeders to experiment with exotic varieties at random. They now have at their disposal a useful inventory of the germ plasm of this hemisphere and can approach new breeding problems with some degree of confidence in their choice of stocks. In this connection a study of the origins and relationships of exotic races may prove useful. The genealogies which have been proposed for some of the modern races of maize may seem highly speculative to those who have not worked intimately with exotic races, and it is probable that some of them will require modification as new evidence is brought to bear upon the problem. Still, certain important features stand out as shown by the genealogies published in Wellhausen et al. (1952) and Grobman et al. (1961) of Corn Belt dent, the world's most productive corn, of Jala, the world's largest-eared corn, and of Cuzco Gigante, the world's largest-kerneled corn.

All of these highly evolved races have in common an eight-rowed corn: Northern flint in the case of Corn Belt dent, Tabloncillo in the case of Jala, and Cuzco Gigante, both of which are thought to trace back to a South American eight-rowed corn, Cabuya.

A second feature which all of these outstanding races have in common is that they contain teosinte or *Tripsacum* germ plasm from more than one source. Corn Belt dent may have South American *Tripsacum* in both of its principal ancestral lines through Sabanero and has teosinte in its Southern dent line through Olotillo and Tepecintle. Jala has teosinte through Reventador and Oloton; Cuzco Gigante appears to have *Tripsacum* in one principal ancestral line and teosinte in the other.

All of the races which have participated in the ancestry of these three remarkable modern races offer some promise to corn-breeders in the United States, and it is difficult to say which offers most. High priority should undoubtedly be given to the race Tabloncillo, not only because it is the immediate ancestor of Jala and Cuzco Gigante, as well as the productive Mexican race, Celaya, but also because it apparently has previously played no part in the ancestry of Corn Belt corn. Consequently it offers a new source of genetic diversity as well as one of proved merit. Another group of races which seem to offer particular promise are Tepecintle and its descendants, Tuxpeño, Zapalote Chico, and Zapalote Grande. These four races are especially promising as a source of teosinte germ plasm, a subject discussed in more detail later. Other exotic races will be chosen for specific characteristics which they can provide. A number of tropical races are resistant to *Helminthosporium*. Many high altitude races are resistant to rust. A high altitude maize from the slopes of Lake Titicaca in Peru and Bolivia is quite resistant to cold but unfortunately quite susceptible to smut.

Using Exotic Races

It is very unlikely that inbred strains isolated directly from exotic races will be found generally useful in the United States. The most promising method of utilizing such races, at least initially, would seem to be in introducing new genetic diversity into established hybrid combinations by modifying one or more of the inbred strains through the incorporation of exotic germ plasm. This procedure is suggested by the fact that the exotic races are themselves usually poorly adapted to the climatic conditions of the United States. Even their first generation hybrids are often not too promising. But hybrids

between well-adapted inbreds and exotic races backcrossed once or twice to their inbred parent are often remarkably vigorous and produce handsome, well-formed ears. Their undesirable characteristics seem, in these combinations, to be almost completely masked by the effects of the native germ plasm. This may be an example of the phenomenon of antithetical dominance described by Anderson and Erickson (1941). More probably it is a result of the fact that the strong photoperiodic response of some tropical varieties may be governed by a relatively small number of major genes.

Although I cannot present actual data to support these observations, I have, in the past twenty-five years in connection with studies of chromosome-knob numbers and various gene frequencies, grown exotic varieties from virtually all of the countries of Latin America. In 1950 alone we had 525 Latin American varieties in our experimental fields. Seldom has one of these exotic races as such seemed promising as a source of germ plasm for corn improvement. Hundreds of these varieties have been crossed with one or more inbreds, especially A158, P39, and a strain of Wilbur's flint. The F_1's among them which seemed promising are decidedly in the minority, but when backcrossed to A158, as many of them have been for the purpose of extracting from them certain chromosomes, both the three-fourths A158 and seven-eighths A158 stocks have included lines which would attract the attention of the most critical corn breeder.

How best to incorporate the exotic germ plasm into well-established inbreds is a problem which deserves both thought and experimentation. My own inclination at this time would be to rely heavily upon a modification of Stadler's method of gamete selection described earlier in this chapter. The method is particularly promising in sampling exotic varieties in which desirable zygotes are too rare to be even sought for but desirable gametes are not beyond reach. There is little possibility of isolating superior zygotes directly from the majority of exotic races and not much to be gained by doing so.

The only modification which I would suggest to the method proposed by Stadler is that the gamete-testing begin not with the F_1 but with the first backcross to the elite strain. This suggestion is made for two practical reasons:

(1) many F_1's containing excellent exotic germ plasm are, because of their tropical origin or photoperiodic response, too luxuriant vegetatively to produce selfed ears; (2) since the modified strains which will eventually be employed will seldom have more than one-fourth of their germ plasm from exotic sources, it seems sensible to concentrate, although not necessarily to confine, time and space-consuming testing to gametes of the first backcross. These will on the average have one-fourth exotic germ plasm but will vary theoretically between gametes carrying one-half exotic germ plasm and gametes carrying none.

An important fact to be remembered in introducing the exotic germ plasm into elite inbreds is that the inbreds themselves may not be improved by this procedure. This will be especially true if the introduced germ plasm involves blocks of genes from teosinte or *Tripsacum* and the elite strains already contain an appreciable amount of such germ plasm. In such circumstances the modified inbred strains will, like those produced by convergent improvement, be generally less productive than the original ones, but unlike those resulting from convergent improvement they may produce much better single crosses. A slight reduction in yield of an inbred can be tolerated if its combining ability has been substantially improved, and even a marked reduction may not be of serious consequences if the modified inbred can be used as a pollen parent. Actually the pollen-shedding ability of many United States inbreds can be improved by introducing germ plasm from exotic varieties.

Using Exotic Germ Plasm from Diverse Sources

There is no doubt that many of the commercial hybrids now in use in the United States can be improved by introducing exotic germ plasm into one or even two of the inbred strains which enter into their parentage. Whether they can be improved still further by modifying all of them is a question which only experimentation can answer. The fact that the most productive races of maize of Latin America are complex hybrids usually containing germ plasm from several primitive races of maize as well as admixture with teosinte and *Tripsacum* from several sources may

suggest that attempts to create hybrids with pedigrees as complex as those of these races may meet with some success.

The Use of Cytoplasmic Male Sterility

A major problem in the production of hybrid seed corn is that of removing the tassels of one of the parents in a crossing field. In the early years of the hybrid seed corn enterprise it appeared that the labor and expense involved in detasseling might prove to be the principal obstacle to the large-scale production and widespread use of hybrid corn. This fortunately has not proved to be the case. Nevertheless tassel removal on a scale sufficient to produce the 15 million bushels of hybrid seed needed to plant some 65 million acres of corn annually is an operation of considerable magnitude.

Before the use of cytoplasmic sterility was introduced the hybrid seed corn industry had to find and to train each summer thousands of casual laborers for this purpose. One seed firm alone employed 20,000 laborers during the tassel-pulling season. It has been estimated that on the peak day of the season some 125,000 persons in the United States were engaged in removing tassels from corn plants. When there were labor shortages or when rainy weather prevailed during the detasseling season producers of hybrid seed corn had serious problems on their hands and the quality of their product sometimes suffered from unavoidable self-pollination.

When I joined the staff of the Connecticut Agricultural Experiment Station in 1921 as a graduate assistant to Donald F. Jones, one of the originators of hybrid corn, I found that, convinced though he was of the feasibility of producing hybrid seed on a large scale, he was nonetheless concerned about the possible magnitude of the problem of detasseling, and he was giving much thought to the possibility of developing other methods of producing crossed seed. Together we conducted a number of experiments aimed at finding methods of avoiding detasseling or reducing the amount required. The results of these are reported in Connecticut Station Bulletin 550, published in 1951. Only one of our early methods proved to be at all promising. This involved a linkage with about 6 percent of crossing over between white endosperm color and a recessive gene for male sterility that we

had discovered in one of our testcrosses. By sorting out the white seeds in the backcross of the heterozygote by the double recessive we produced populations in which 94 percent of the plants were male sterile. The chief objection to this method was that the hybrid seed produced by it would either be all white kernels or mixed white and yellow. By the middle twenties farmers generally, even in the South, were preferring yellow corn. In the hope of developing stocks in which the linkage phase was between male sterility and yellow endosperm color instead of white, we continued to experiment with this method, Jones in Connecticut and I, beginning in 1927, in Texas, neither of us with notable success.

It was in 1938 that I discovered a form of male sterility that was to become the basis of a method of completely avoiding the operation of detasseling in the production of hybrid seed corn. This occurred in Honey June, a sweet corn variety that I had developed by introducing the gene *su* for sugary endosperm into Mexican June, a well-adapted and somewhat drought-resistant Texas field-corn variety. A planting in 1939 of the seeds from a sterile plant produced a row in which all the plants were male sterile. I concluded at once that this was a case of cytoplasmically inherited male sterility similar to the one Marcus Rhoades had described in 1933. It also occurred to me almost immediately that here might be the ideal way of avoiding detasseling in the production of hybrid seed corn, but I failed to see how fertility was to be restored in the farmer's crop. It was not until some years later, in the spring of 1944, that Dr. Jones suggested that the problem could be solved by producing two versions of any particular hybrid, one by employing cytoplasmic male sterility to avoid detasseling and the other by the conventional method of removing the tassels. A blend of the two in proportions of 2:1 or 3:1 planted by the farmer would provide sufficient pollen for the entire planting.

At Dr. Jones's request I asked John Rogers, who had succeeded me in charge of the corn-breeding program in Texas, to send seed of the male-sterile line to Jones and at the same time suggested to him that he should apply the method to the production of hybrid corn in Texas. Rogers could find no seed of the male sterile that I had discovered in Honey June, but he sent Jones seed of a sterile that had appeared in Golden June, a variety that I had developed by introducing yellow endosperm into the Mexican June variety.

Jones found that the Texas male sterile, later known as the "T" type, could be readily introduced into a number of commonly used inbred strains through repeated backcrossing, and by 1947 when Rogers, then my graduate student at Harvard, and I paid him a visit he had a number of excellent inbred strains that were almost identical in their characteristics to the original ones except that they were male sterile. He also had some lines in which the plants were fertile, indicating that these lines carried genes capable of overcoming the cytoplasmic male sterility. Such genes later came to be known as fertility-restoring genes and were employed as a substitute for blending in insuring fertility in the farmer's crop.

In September 1948, Jones and I applied jointly for a patent on the method of employing cytoplasmic male sterility and restoring fertility by blending, and in April 1950, Jones applied for a patent on the method of employing restoring genes to insure fertility in the farmer's crop. The first application was rejected by the Patent Office, but the second was granted and a patent issued on July 10, 1956. By previous agreement I had equal rights with Jones in this invention and we had both, in 1949, assigned our rights to Research Corporation, which agreed to pay each inventor royalties equivalent to those usually paid authors of books. Research Corporation is a foundation for the advancement of science one of whose functions is to contribute patent assistance services without cost to educational, scientific, and other nonprofit institutions.

Both Jones and I knew that our patent applications might not meet with favor in certain circles, but we greatly underestimated the amount of animosity that they would create. Resolutions against us and the patent were passed by the Southern Corn Conference and the American Society of Agronomy. The latter, it turned out later, was influenced in part by the American Seed Trade Association. Despite the oppostion to our patent, the hybrid seed corn producers quickly adopted the method of employing cytoplasmic male sterility and restorer genes to avoid detasseling. Even the small producers, whom certain agronomists considered lacking in adequate

training to employ this sophisticated new method, seemed to have no difficulty in mastering it. Soon a substantial portion of the corn grown in the United States contained the Texas or "T" type of cytoplasm and much of it was being used in combination with restorer genes. By 1969 it was estimated that 70 to 90 percent of the hybrid corn grown in the United States carried the T-cytoplasm.

This was the year when it first became apparent that corn with T-cytoplasm might be more susceptible to the southern blight than corn with normal cytoplasm. For 25 years hybrids containing this cytoplasm had undergone, to quote one writer, "the tests of time against the challenges of soil, weather, climate and diseases such as the southern corn leaf blight." Then a new mutant strain of the fungus *Helminthosporium maydis* made its appearance and attacked many corn hybrids. Many observations showed that the hybrids with the T-cytoplasm were more susceptible to infection than those with the normal cytoplasm.

In 1970 the blight was much worse than in the previous year. Appearing early on winter crops of sweet corn grown in Florida, it soon infected field-corn crops in the southern states. Through wind-blown spores the blight spread westward and northward, until by the end of the summer it occurred in every state east of the Rockies from Texas to Minnesota, from Florida to New England. In many areas spells of unusually hot humid weather contributed to the development of the disease. Estimates of the damage that it was causing varied greatly, partly because there was little past experience to serve as a guide. The final figures for the 1970 crop showed a loss of 13 percent from the earlier estimates made by the Department of Agriculture.

In the years before the new strain of the blight fungus appeared on the scene the use of cytoplasmic sterility was remarkably successful. It not only drastically reduced the labor required in producing hybrid seed but it also eliminated the loss of production of hybrid seed caused by the removal of one or more leaves in the detasseling operation. I believe it was also a factor in making possible on an extensive scale the replacement of double crosses by higher-yielding single and three-way crosses, and although as I pointed out in an earlier chapter there is a danger inherent in growing these genetically more uniform

hybrids they have undoubtedly contributed to the spectacular increase in average yields in the past decade which the following tabulation shows.

Year	Percent hybrid corn	Bushels per acre	Percent increase over previous decade
1929	0	25.7	—
1939	22.9	29.7	16
1949	78.3	37.8	27
1959	94.8	51.5	36
1969	99+	80.0	55

Susceptibility to the corn blight of inbred strains and hybrids carrying the T-cytoplasm irrespective of the diversity of their residual heredity has focused attention on the possibility that the new dwarf wheats and rices which are revolutionizing the agriculture of underdeveloped countries may, because each crop has genes for dwarfness in common, become susceptible to new strains of pathogens despite the heterogeneity of their residual germ plasm. Recognizing this possibility the National Academy of Sciences in late 1970 initiated a study financed in large part by Research Corporation from proceeds derived from the Jones patent—on the genetic vulnerability of major food crops. If this study should help to foresee and to forestall truly disastrous epidemics in the world's major food crops it may turn out that the 1970 corn blight was in some respects, at least, a blessing in disguise. In the meantime the corn breeders of the United States, faced with a situation of unprecedented urgency, seeking resistance to the new fungus and incorporating new forms of cytoplasmic sterility in established inbreds, may find themselves employing extensively some of the methods described in this chapter: the use of exotic germ plasm, of gamete selection, and of maternal and paternal monoploids.

Breeding Corn for Improved Protein Quality

One of the most interesting, exciting, and significant developments in corn breeding in recent years has been the discovery that protein quality in corn can be substantially improved by employing certain mutant genes which affect especially the lysine content. This chapter would not complete without at least a brief mention of these developments.

Corn has long been recognized by nutritionists as a far-from-perfect source of protein in the diet. Thomas B. Osborne and his associates at the Connecticut Agricultural Experiment Station found many years ago that rats suffered from malnutrition on a diet containing corn as the sole source of protein. There is a counterpart of this situation in human nutrition; among children in countries where corn is the principal food there occurs a high frequency of a protein-deficiency syndrome known as "kwashiorkor." In both rats and humans the malnutrition can be corrected by supplying the diet with small amounts of lysine and tryptophane, two amino acids in which corn is usually deficient.

Numerous attempts have been made by corn breeders to improve the lysine and tryptophan content through selection but without notable success until what has now come to be regarded as a "break through" occurred in the middle sixties, when Mertz et al. (1964) reported that a single recessive gene, "opaque-2," introduced into an ordinary strain of corn could increase the lysine content of the kernels by as much as 69 percent. Opaque-2 is a recessive mutant discovered in the early thirties by Ralph Singleton, then associated with Donald F. Jones at the Connecticut Agricultural Experiment Station. It was maintained for many years by Maize Genetics Cooperation as a useful marker gene for corn's chromosome 7, no one suspecting that it had unusual nutritional qualities and would one day become famous.

When Mertz and his associates at Purdue University, who had long been searching for variants in corn rich in protein, were joined by Oliver E. Nelson, a maize geneticist, they began to investigate certain soft-kernel mutants including opaque-2 and floury-2. Both proved to be capable of increasing the lysine content of corn kernels. The discovery has been described by one writer (Harpstead, 1971) as "a breath of fresh air to plant breeders and nutritionists. It stirred the imagination of many people who were concerned with the worldwide problem of protein need. The possibility that corn might become an agent for relief of the world hunger for protein appeared to be as important a find as the historic discovery that pellagra could be prevented by niacin."

The nutritional value of varieties of corn in which the opaque gene had been introduced was soon established. In Colombia children suffering from third-degree malnutrition were brought back to health on a diet consisting mainly of high-lysine corn. Pigs on a ration of high-lysine corn were grown to marketable weight without the usual supplement of soybean meal or other high-protein food. This fact is of special importance in the United States where a substantial part of the crop is fed to pigs. Thus high-lysine corn may become as important to the United States as it is to the developing countries.

The gene opaque-2 is easily introduced into varieties and inbred strains through crossing followed by repeated backcrossing to the non-opaque parent, the recessive gene being finally recovered through self-pollination. But the successful exploitation of the gene's potentialities has met with certain problems. Many of the hybrids into which the opaque gene has been incorporated are less productive than their normal counterparts. In experiments conducted in Illinois, the opaque-2 hybrids averaged 8 percent less in grain yield than their normal counterparts (Alexander et al., 1971). This is not surprising. Opaque-2 is essentially a defect in normal metabolism and its deleterious effects can be cured only by assembling a complex of modifying factors that will permit it to function effectively in the new niche that plant breeders have created for it as has been done with both sugary and waxy endosperms (see Chapter 12). Fortunately several of the opaque-2 hybrids tested in Illinois produced as well as their normal counterparts; this suggests that the problem of producing high-yielding, high-lysine hybrid corn can be solved.

Another problem in exploiting opaque-2 involves the texture of its endosperm. This is soft and floury and lacks the vitreous areas that are characteristic of dent corns and that make up almost the entire endosperm of flint corns. Thus opaque-2 varieties are difficult to mill by conventional methods and the meal is difficult for the consumer to use in the usual ways of preparing traditional foods. Attempts are being made in Mexico and elsewhere to develop new types of corn that have the dent or flint type endosperm texture combined with the high-lysine content of opaque-2 endosperm. Some progress has already been made in this direction and if the magnitude of the effort is any guarantee of eventual success the present prospects are good; each year thousands of samples of corn carrying the opaque-2 gene but having "normal" endosperm are tested for lysine content in the United States, Mexico and European maize-growing countries. The method of recurrent selection described in this chapter should be effective in achieving the desired objective.

A recent discovery by Zuber et al. (1972) may also prove to be significant from a nutritional standpoint. They have found that a variety of floury corn of Peru, known as Coroico, is quite unusual in having tissues of the aleurone that are multi-layered instead of single layered as is the usual condition in corn (see Chapter 1). This corn, which was discovered by Hugh Cutler in his collection of corn varieties in South America in the early forties is also peculiar in that its ears often have an odd number of rows of grain. Zuber and his associates found that Coroico has a higher protein and higher lysine content than dent corn, probably because of its multilayer of aleurone in which part of the protein content resides. This characteristic may prove to be especially useful when combined with opaque-2. This would also introduce into modern corn some exotic germ plasm that has so far been little used.

These recent developments make it clear that the improvement of corn, especially with respect to its nutritive value has a bright future. Of the world's major food plants, corn is by all odds the most versatile. As I contemplate the many possibilities that lie ahead, I could almost wish that my career were only beginning instead of coming to an end. Still, I doubt that the next fifty years can be any more rewarding to a student of *Zea Mays* than the past fifty years have been for me.

Bibliography   Index

Bibliography

Alexander, D. E., J. W. Dudley, and R. J. Lambert. 1971. The modification of protein quality in maize by breeding. *In* Proc. Fifth Meeting Maize and Sorghum Section of Eucarpia. Akadémiai Kiadō, Budapest.

Anderson, E. 1943. A variety of maize from the Rio Loa. Ann. Missouri Bot. Gard. 30: 469–476.

————. 1944a. Cytological observations on *Tripsacum dactyloides*. Ann. Missouri Bot. Gard. 31: 317–323.

————. 1944b. Maize reventador. Ann. Missouri Bot. Gard. 31: 301–314.

————. 1945. What is *Zea Mays*? A report of progress. Chron. Bot. 9: 88–92.

————. 1947a. Field studies of Guatemalan maize. Ann. Missouri Bot. Gard. 34: 433–467.

————. 1947b. Corn before Columbus. Des Moines: Pioneer Hi-Bred Corn Co.

————. 1953. Introgressive hybridization. Biol. Rev. 28: 280–307.

————. 1959. Zapalote Chico: An important chapter in the history of maize and man. Actas 33 Congreso Internacional de Americanistas, pp. 230–237.

————, and F. D. Blanchard. 1942. Prehistoric maize from Cañon del Muerto. Amer. J. Bot. 29: 832–835.

————, and W. L. Brown. 1950. The history of common maize varieties in the United States corn belt. J. New York Bot. Gard. 51: 242–267.

————, ————. 1952a. The history of the common maize varieties of the United States corn belt. Agric. History 26: 2–8.

————, ————. 1952b. Origin of corn belt maize and its genetic significance. *In* Heterosis: 124–148. Iowa State College Press.

————, ————. 1953. The popcorns of Turkey. Ann. Missouri Bot. Gard. 40: 33–48.

————, and H. C. Cutler. 1942. Races of *Zea Mays*: I. Their recognition and classification. Ann. Missouri Bot. Gard. 29: 69–89.

————, and R. O. Erickson. 1941. Antithetical dominance in North American maize. Proc. Nat. Acad. Sci. 27: 436–440.

Andres, J. M. 1950. Granos semivestidos, restos de un carácter ancestral del maíz. Revista Argentina de Agron. 17: 252–256.

Anon. 1958. Ann. Rep. Univ. Florida Agric. Exp. Sta.

Arber, Agnes. 1934. The Gramineae. Cambridge University Press.

Arnason, T. J. 1936. Cytogenetics of hybrids between *Zea mays* and *Euchlaena mexicana*. Genetics 21: 40–60.

Ascherson, P. 1875. Ueber *Euchlaena mexicana* Schrad. Bot. Vereins Prov. (Brandenburg) 17: 76–80.

————. 1877. Proc. Bull. Linn. Soc. (Paris), pp. 105–108.

————. 1880. Bemerkungen über ästigen Maiskolben. Bot. Vereins Prov. (Brandenburg) 21: 133–138.

Azara, F. de. 1809. Voyages dans l'Amérique méridionale, vol. 1, pp. 146–148. Paris.

Barghoorn, E. S., M. K. Wolfe, and K. H. Clisby. 1954. Fossil maize from the Valley of Mexico. Bot. Mus. Leafl. Harvard Univ. 16: 229–240.

Bartlett, A. S., E. S. Barghoorn, and R. Berger. 1969. Fossil maize from Panama. Science 165: 389–390.

Bauman, L. F. 1959. Evidence of non-allelic gene interaction in determining yield, ear height, and kernel row number in corn. Agron. J. 51: 531–534.

Beadle, G. W. 1930. Genetical and cytological studies on Mendelian asynapsis in *Zea mays*. Cornell Univ. Agric. Exp. Sta. Mem. 129: 1–23.

————. 1931. A gene in maize for supernumerary cell divisions following meiosis. Cornell Univ. Agric. Exp. Sta. Mem. 135: 1–12.

————. 1932a. The relation of crossing over to chromosome association in *Zea-Euchlaena* hybrids. Genetics 17: 481–501.

————. 1932b. Studies of *Euchlaena* and its hybrids with *Zea* I. Chromosome behavior in *E. mexicana* and its hybrids with *Zea mays*. Zeitschr. Abst. Vererb. 62: 291–304.

————. 1939. Teosinte and the origin of maize. J. Hered. 30: 245–247.

————. 1972. Corn—gift of the gods. Garden Talk, Chicago Hort. Soc. Jan.–Feb., pp. 12–15.

————. 1972. The mystery of maize. Field Mus. Nat. Hist. Bull. 43: 2–11.

Bear, R. P. 1944. Mutations for waxy and sugary endosperm in inbred lines of dent corn. J. Am. Soc. Agron. 36: 89–91.

Benzer, S. 1961. Genetic fine structure. Harvey Lectures 56. New York: Academic Press.

Bhagwat, S. D., and G. B. Deodikar. 1961. *Trilobachne*, an imperfectly known genus. Agharkar Commemoration Volume, pp. 139–144.

Bianchi, A. 1957. Defective caryopsis factors from maize-teosinte derivatives. I. Origin, description and segregation. Soc. Ital./Rom./*Genetica agraria* 7: 1–38.

————. 1958. Defective endosperm factors from maize-teosinte derivatives. Maize Gen. Coöp. News Letter 32: 11–12.

————. 1959. Fattori genetici determinanti cariossidi difettose in descendenti di incroci fra mais e teosinte. II. Curve di variabilità dei pesi. Supplemento La Ricerca Scientifica 29: 3–21.

————, and F. Salamini. 1963. Detection of linkage in maize by means of balanced lethal systems. Estratto da Maydica 8: 52–58.

Birket-Smith, K. 1943. The origin of maize cultivation. Kgl. Danske Videnskabernes Selskab; Hist.—Filol. Meddel 29: 1–49.

Blaringhem, L. 1924. Note sur l'origine du maïs. Métamorphose de l'Euchlaena en Zea, obtenue au Brésil par Bento de Toledo. Ann. Sci. Nat. Bot. Series 10, 6: 245–263.

Bonafous, M. 1836. Histoire naturelle, agricole et economique du maïs. Paris.

Bonnett, O. T. 1940. Development of the staminate and pistillate inflorescences of sweet corn. J. Agric. Res. 60: 25–37.

————. 1953. Developmental morphology of the vegetative and floral shoots of maize. Ill. Agric. Exp. Sta. Bull. 568.

————. 1966. Inflorescences of maize, wheat, rye, barley, and oats: Their initiation and development. Ill. Agric. Exp. Sta. Bull. 721: 1–105.

Bonvicini, M. 1932. Sulla ereditarietà di una anomalia nel mais. L'Italia Agricola 69: 3–9.

Bor, N. L. 1960. The grasses of Burma, Ceylon, India and Pakistan. Oxford: Pergamon Press.

Brandolina, A. 1970. Maize. In Genetic Resources in Plants. Philadelphia: F. A. Davis, pp. 273–309.

Breggar, T. 1928. Waxy endosperm in Argentine maize. J. Hered. 19: 111.

Brieger, F. G. 1944. Estudos experimentais sôbre a origen do milho. Anais esc. sup. agric. 2: 225–278.

———. 1950. The genetic basis of heterosis in maize. Genetics 35: 420–445.

———, J. T. A. Gurgel, E. Paterniani, A. Blumenschein, and M. R. Alleoni. 1958. Races of maize in Brazil and other eastern South American countries. Nat. Acad. Sci.-Nat. Res. Council Publ. 593.

Brink, R. A. 1954. Very light variegated pericarp in maize. Genetics 39: 724–740.

———, and D. C. Cooper. 1947. The endosperm in seed development. Bot. Rev. 13: 423–477, 479–541.

———, and R. A. Nilan. 1952. The relation between light variegated and medium variegated pericarp in maize. Genetics 37: 519–544.

Brown, W. L. 1949. Numbers and distribution of chromosome knobs in United States maize. Genetics 34: 524–536.

———, and E. Anderson. 1947. The northern flint corns. Ann. Missouri Bot. Gard. 34: 1–28.

Bruce, A. B. 1910. The Mendelian theory of heredity and the augmentation of vigor. Science 32: 627–628.

Brunson, A. M. 1937. Popcorn breeding. U.S.D.A. Yearbook, pp. 395–404.

———, and G. M. Smith. 1945. Hybrid popcorn. J. Am. Soc. Agron. 37: 176–183.

Burdick, A. B. 1951. Dominance as function of within organism environments in kernel-row number in maize (Zea mays L.) Genetics 36: 652–666.

Burnham, C. R. 1946. An "Oenothera" or multiple translocation method of establishing homozygous lines. J. Am. Soc. Agron. 38: 702–707.

Candolle, A. de. 1886. Origin of cultivated plants. Reprint of 2d edition, 1959. New York: Hafner Publishing Co.

Carter, G. F. 1945. Plant geography and culture history in the American Southwest. Viking Fund Pub. Anthrop. no. 5.

———. 1948. Sweet corn among the Indians. Geog. Rev. 38: 206–221.

———. 1950. Plant evidence for early contacts with America. Southwestern J. Anthrop. 6: 161–182.

———. 1953. Plants across the Pacific. Mem. Soc. Am. Arch. 9: 62–71.

Celarier, R. P. 1957. Cytotaxonomy of the Andropogoneae. II: Subtribes Ischaeminae, Rottboellinae and the Maydeae. Cytologia 22: 160–183.

Cervantes, R. J., A. Rodriguez V., and J. S. Niederhauser. 1958. Resistencia al virus causante del achaparramiento del maíz. Secretaria de Agricultura y Ganaderia, Mexico, Folleto Tecnico 29: 1–18.

Chaganti, R. S. K. 1965. Cytogenetic studies of maize-Tripsacum hybrids and their derivatives. Bussey Inst. Harvard Univ.

Chase, S. S. 1952a. Monoploids in maize. In Heterosis, pp. 389–399. Ames: Iowa State College Press.

———. 1952b. Production of homozygous diploids of maize from monoploids. Agron. J. 44: 263–267.

———. 1952c. Selection for parthenogenesis and monoploid fertility in maize. Genetics 37: 573–574.

———. 1958. The utilization of parthenogenesis in maize breeding. Proc. Xth Inter. Cong. Gen. 2: 48.

———. 1963. Androgenesis—its use for transfer of maize cytoplasm. J. Hered. 54: 152–158.

Clark, G. 1970. Aspects of prehistory. Berkeley: Univ. California Press.

Coe, E. H. 1959. A line of maize with high haploid frequency. Am. Nat. 93: 381–382.

Collins, G. N. 1909. A new type of Indian corn from China. U.S.D.A. Bur. Pl. Ind. Bull. 161.

———. 1912. The origin of maize. J. Wash. Acad. Sci. 2: 520–530.

———. 1914. A drought-resisting adaptation in seedlings of Hopi maize. J. Agric. Res. 1: 293–302.

———. 1917. Hybrids of Zea tunicata and Zea ramosa. J. Agric. Res. 9: 383–396.

———. 1918. Maize, its origin and relationships. J. Wash. Acad. Sci. 8: 42–43.

———. 1919a. A fossil ear of maize. J. Hered. 10: 170–172.

———. 1919b. Notes on the agricultural history of maize. Am. Hist. Assoc. Ann. Rep. 1: 409–429.

———. 1920. Waxy maize from upper Burma. Science 52: 48–51.

———. 1921a. Teosinte in Mexico. J. Hered. 12: 339–350.

———. 1921b. Dominance and the vigor of first generation hybrids. Am. Nat. 55: 116–133.

———. 1931. The phylogeny of maize. Bull. Torrey Bot. Club 57: 199–210.

———, and J. H. Kempton. 1914. A hybrid between Tripsacum and Euchlaena. J. Wash. Acad. Sci. 4: 114–117.

———, ———. 1916. Patrogenesis. J. Hered. 7: 106–118.

———, ———. 1920. A teosinte-maize hybrid. J. Agric. Res. 19: 1–38.

———, ———, and R. Stadelman. 1937. Maize investigations. Carnegie Inst. Washington Year Book, pp. 149–150.

———, and A. E. Longley. 1935. A tetraploid hybrid of maize and perennial teosinte. J. Agric. Res. 50: 123–133.

Comstock, R. E., and H. F. Robinson. 1948. The components of genetic variance in populations of biparental progenies and their use in estimating the average degree of dominance. Biometrics 4: 254–266.

———, ———. 1952. Estimation of average dominance of genes. In Heterosis, pp. 494–516. Ames: Iowa State College Press.

———, ———, and P. H. Harvey. 1949. A breeding procedure designed to make maximum use of both general and specific combining ability. Agron. J. 41: 360–367.

Cowan, J. R. 1943. The value of double cross hybrids involving inbreds of similar and diverse genetic origin. Sci. Agric. 23: 287–296.

Crow, J. F. 1948. Alternative hypotheses of hybrid vigor. Genetics 33: 477–487.

———. 1952. Dominance and overdominance. In Heterosis, pp. 282–297. Ames: Iowa State College Press.

Cutler, H. C. 1946. Races of maize in South America. Bot. Mus. Leafl. Harvard Univ. 12: 257–291.

———. 1952. A preliminary survey of plant remains of Tularosa Cave. Fieldiana: Anthropology (Chicago Nat. Hist. Mus.) 40: 461–479.

———, and E. Anderson. 1941. A preliminary survey of the genus Tripsacum. Ann. Missouri Bot. Gard. 28: 249–269.

———, and M. C. Cutler. 1948. Studies on the structure of the maize plant. Ann. Missouri Bot. Gard. 35: 301–316.

———, and L. W. Blake. 1971. Travels of corn and squash. In Man across the Sea: Austin: Univ. Texas Press, pp. 367–375.

Darlington, C. D. 1956. Natural populations and the breakdown of classical genetics. Proc. Roy. Soc. London B 145: 350–364.

———, and K. Mather. 1949. The elements of genetics. London: Allen and Unwin.

Darwin, C. 1859. On the Origin of Species. A Facsimile. New York: Atheneum, 1967.

———. 1868. The variation of animals and plants under domestication. London.

———. 1877. The effects of cross- and self-fertilization in the vegetable kingdom. London.

Deanon, J. R. 1957. Treatment of sweet corn silks with maleic hydrazide and colchicine as means of increasing the frequency of monoploids. Philipp. Agric. 41: 364–377.

de Wet, J. M. J., J. R. Harlan, and C. A. Grant. 1971. Origin and evolution of teosinte (Zea mexicana (Schrad.) Kuntze). Euphytica 20: 255–265.

Dhawan, N. L. 1964. Primitive maize in Sikkim. Maize Gen. Coöp. News Letter 38: 69–70.

Dobrizhoffer, M. 1822. An account of the Abipones. London.

Dobzhansky, T. 1949. Observations and experiments on natural selection in Drosophila. Proc. 8th Int. Cong. Genetics, pp. 210–224.

———. 1952. Nature and origin of heterosis. In Heterosis, pp. 218–223. Ames: Iowa State College Press.

———. 1955. Evolution, Genetics, and Man. New York: John Wiley & Sons.

Dodds, K. S., and N. W. Simmonds. 1946. A cytological basis for sterility in Tripsacum laxum. Ann. Bot. n.s. 9: 109–116.

Doggett, H., and B. N. Najisu. 1968. Disruptive selection in crop development. Heredity 23: 1–23.

Douglas, A. G., J. W. Collier, M. F. El-Ebrashy, and J. S. Rogers. 1961. An evaluation of three cycles of reciprocal recurrent selection in a corn improvement program. Crop Sci. 1: 157–161.

Dupree, A. H. 1959. Asa Gray. Cambridge: Harvard Univ. Press.

Duvick, D. N. 1965. Cytoplasmic pollen sterility in corn. In Advances in genetics 13: 1–56. New York: Academic Press.

East, E. M. 1913. A chronicle of the tribe of corn. Pop. Sci. Monthly 82: 225–236.

———. 1936. Heterosis. Genetics 21: 375–397.

———, and H. K. Hayes. 1911. Inheritance in maize. Connecticut Agric. Exp. Sta. Bull. 167: 1–142.

———, ———. 1912. Heterozygosis in evolution and in plant breeding. U.S.D.A. Bur. Plant Indust. Bull. 243.

———, and D. F. Jones. 1920. Genetic studies on the protein content of maize. Genetics 5: 543–610.

Eckhardt, R. C., and A. A. Bryan. 1940. Effect of method of combining the four inbred lines of a double cross of maize upon the yield and variability of the resulting double crosses. J. Am. Soc. Agron. 32: 347–353.

Edwardson, J. R. 1956. Cytoplasmic sterility. Bot. Rev. 22: 696–738.

Emerson, R. A. 1924. Control of flowering in teosinte. J. Hered. 15: 41–48.

———. 1929. Genetic notes on hybrids of perennial teosinte and maize. Amer. Nat. 63: 289–300.

———, and G. W. Beadle. 1930. A fertile tetraploid hybrid between Euchlaena perennis and Zea mays. Amer. Nat. 64: 190–193.

———, ———. 1932. Studies of Euchlaena and its hybrids with Zea. II: Crossing over between the chromosomes of Euchlaena and those of Zea. Zeitschr. Ind. Abstamm. Vererb. 62: 305–315.

———, ———, and A. C. Fraser. 1935. A summary of linkage studies in maize. Cornell Agric. Exp. Sta. Memoir 180: 1–83.

———, and E. M. East. 1913. The inheritance of quantitative characters in maize. Nebraska Agric. Exp. Sta. Res. Bull. 2.

Erdtman, G. 1943. An introduction to pollen analysis. Waltham, Mass.: Chronica Botanica.

Erwin, A. T. 1934. Sweet corn—its origin and importance as an Indian food plant in the United States. Iowa State College J. Sci. 8: 385–389.

———. 1942. Anent the origin of sweet corn. Iowa State College J. Sci. 16: 481–485.

Farquharson, L. I. 1957. Hybridization in Tripsacum and Zea. J. Hered. 48: 295–299.

Finan, J. J. 1950. Maize in the great herbals. Waltham, Mass.: Chronica Botanica.

Fisher, R. A. 1949. The theory of inbreeding. Edinburgh: Oliver and Boyd.

Frey, K. J., B. Brimhall, and G. F. Sprague. 1949. The effects of selection upon protein quality in the corn kernel. Agron. J. 41: 399–403.

Galinat, W. C. 1954a. Corn grass. I: Corn grass as a possible prototype or a false progenitor of maize. Am. Nat. 88: 101–104.

———. 1954b. Corn grass. II: Effect of the corn grass gene on the development of the maize inflorescence. Am. J. Bot. 41: 803–806.

———. 1954c. Argentine popcorn as a modern relic of prehistoric corn. Maize Gen. Coöp. News Letter 28: 26.

———. 1956. Evolution leading to the formation of the cupulate fruit case in the American Maydeae. Bot. Mus. Leafl. Harvard Univ. 17: 217–239.

———. 1957. The effects of certain genes on the outer pistillate glume of maize. Bot. Mus. Leafl. Harvard Univ. 18: 57–76.

———. 1959. The phytomer in relation to floral homologies in the American Maydeae. Bot. Mus. Leafl. Harvard Univ. 19: 1–32.

———. 1963a. Form and function of plant structures in the American Maydeae and their significance for breeding. Econ. Bot. 17: 51–59.

———. 1964. Tripsacum a possible amphidiploid of Manisuris and wild maize. Maize Gen. Coöp. News Letter 38: 50.

———. 1967. Plant habit and the adaptation of corn. Univ. Mass. Exp. Sta. Bull. 565.

———. 1970. The cupule and its role in the origin and evolution of maize. Mass. Agric. Exp. Sta. Bull. 585.

———. 1971. The origin of sweet corn. Mass. Agric. Exp. Sta. Res. Bull. 591.

———. 1971. The origin of maize. Ann. Rev. Genetics 5: 447–478.

———, and R. G. Campbell. 1967. The diffusion of eight-rowed maize from the Southwest to the Central Plains. Mass. Agric. Exp. Sta. Monograph series 1.

———, R. S. K. Chaganti, and F. D. Hager. 1964. Tripsacum as a possible amphidiploid of wild maize and Manisuris. Bot. Mus. Leafl. Harvard Univ. 20: 289–316.

———, and J. H. Gunnerson. 1963. Spread of eight-rowed maize from the prehistoric Southwest. Bot. Mus. Leafl. Harvard Univ. 20: 117–160.

———, and P. C. Mangelsdorf. 1966. Genetic correspondence of Tripsacum chromosomes to their homeologs from corn. Maize Gen. Coöp. News Letter 40: 99–101.

———, ———, and L. Pierson. 1956. Estimates of teosinte introgression in archaeological maize. Bot. Mus. Leafl. Harvard Univ. 17: 101–124.

———, and R. J. Ruppé. 1961. Further archaeological evidence on the effects of teosinte introgression in the evolution of modern maize. Bot. Mus. Leafl. Harvard Univ. 19: 163–181.

Gardner, C. O., P. H. Harvey, R. E. Comstock, and H. F. Robinson. 1953. Dominance of genes controlling quantitative characters in maize. Agron. J. 45: 186–191.

———, and J. H. Lonnquist. 1959. Linkage and the degree of dominance of genes controlling quantitative characters in maize. Agron. J. 51: 524–528.

———, ———. 1961. Effect of linkage on genetic variances and estimates of average degree of dominance in corn. Agron. Abst.: 50–51.

Gerrish, E. E. 1956. Studies of the monoploid method of producing homozygous diploids in Zea mays. Diss. Abst. 16: 2285–2286.

———. 1967. Survey of attempts to hybridize maize with sorghum. Maize Gen. Coöp. News Letter 41: 26–28.

Gerstel, D. U. 1953. Chromosomal translocations in interspecific hybrids of the genus Gossypium. Evolution 7: 234–244.

Gilmore, M. R. 1931. Vegetal remains of the Ozark Bluff-Dweller culture. Mich. Acad. Sci. Arts and Letters 14: 83–102.

Gini, E. 1939. Estudios sobre esterilidad en maíces regionales de la Argentina. An. Inst. Fito. Santa Catalina 1: 135–158.

Goebel, K. 1910. Über sexuellen Dimorphismus bei Pflanzen. Biol. Centralbl. 30: 692–718.

Goodman, M. M. 1965. The history and origin of maize. N. C. Agric. Exp. Sta. Bull. 170.

Goodrich, L. C. 1938. China's first knowledge of the Americas. Geog. Rev. 27: 400–411.

Goodsell, S. F. 1961. Male sterility in corn by androgenesis. Crop. Sci. 1: 227–228.

Goodwin, A. J. H. 1953. The origin of maize. S. African Arch. Bull. 29: 13–14.

Gorsline, G. W. 1960. Quantitative epistatic gene action in maize (Zea mays L.). Diss. Abst. 20: 3915.

Gowen, J. W., Editor. 1952. Heterosis. Ames: Iowa State College Press.

Grafius, J. E. 1960. Does overdominance exist for yield in corn? Agron. J. 52: 361.

Graner, E. A. 1950. Genética da coloracão amarela da semente de milho. Escola Super. Agric. "Luiz de Queiroz" Univ. São Paulo.

———, and G. Addison. 1944. Meiose em Tripsacum australe. Cutler e Anderson. Anais Escola Super. Agric. "Luiz de Queiroz," Univ. São Paulo. 9: 213–224.

Grant, U. J., W. H. Hatheway, D. H. Timothy, C. Cassalett D., and L. M. Roberts. 1963. Races of maize in Venezuela. Nat. Acad. Sci.-Nat. Res. Council Publ. 1136.

Gregory, R. A. 1916. Discovery, or the spirit and service of science. London: Macmillan.

Griffing, B., and E. W. Lindstrom. 1954. A study of the combining abilities of corn inbreds having varying proportions of corn belt and noncorn belt germ plasm. Agron. J. 46: 545–552.

Grobman, A. 1967. Tripsacum in Peru. Bot. Mus. Leafl. Harvard Univ. 21: 285–287.

———, W. Salhuana, and R. Sevilla, in collaboration with P. C. Mangelsdorf. 1961. Races of maize in Peru. Nat. Acad. Sci.-Nat. Res. Council Publ. 915.

Grohne, U. 1957. The importance of phase-contrast microscopy in pollen analysis, demonstrated on corn-type Gramineae pollen. Photograph. Forsch. 7: 237–249.

Guignard, L. 1901. La double fécondation dans le maïs. J. de Bot. 15: 37–50.

Hackel, E. 1890. The true grasses. New York: Henry Holt.

Haines, H. H. 1924. The botany of Bihar and Orissa. London.

Haldane, J. B. S. 1955. On the biochemistry of heterosis, and the stabilization of polymorphism. Proc. Roy. Soc. B 144: 217–220.

Harada, K., M. Murakami, A. Fukushima, and M. Nakazima. 1954. Breeding study on the forage crops: Studies on the intergeneric hybridization between the genus Zea and Coix (Maydeae). I: Kyoto Prefectural Univ. Faculty

Agric. Sci. Rep. 6: 139–145.

———, O. Umekage, and M. Nakazima. 1955. Studies on the intergeneric hybridization between the genus *Zea* and *Coix* (*Maydeae*). Japanese J. Breed. 4: 288.

Harpstead, D. D. 1971. High-lysine corn. Sci. Amer. August, 1971.

Harshberger, J. W. 1893. Maize, a botanical and economic study. Contr. Bot. Lab. Univ. Penna. 1: 75–202.

———. 1896. Fertile crosses of teosinthe and maize. Gard. and Forest 9: 522–523.

———. 1900. A study of the fertile hybrids produced by crossing teosinte and maize. Contrib. Bot. Lab. Univ. Penna. 2: 231–235.

———. 1911. An unusual form of maize. Proc. Delaware Co. Inst. Sci. 6: 49–53.

Harvey, P. H., D. L. Thompson, and R. H. Moll. 1959. Three cycles of reciprocal recurrent selection in corn. Proc. Assn. South. Agric. Workers: 64.

Hatheway, W. H. 1957. Races of maize in Cuba. Nat. Acad. Sci.-Nat. Res. Council Publ. 453.

Hatt, G. 1951. The corn mother in America and in Indonesia. Anthropos 46: 853–914.

Hayes, H. K. 1963. A professor's story of hybrid corn. Minneapolis: Burgess Publishing Co.

———, and R. J. Garber. 1919. Synthetic production of high-protein corn in relation to breeding. J. Am. Soc. Agron. 11: 309–318.

———, and I. J. Johnson. 1939. The breeding of improved selfed lines of corn. J. Am. Soc. Agron. 31: 710–724.

———, E. H. Rinke, and Y. S. Tsiang. 1946. Experimental study of convergent improvement and backcrossing in corn. Minn. Agric. Exp. Sta. Tech. Bull. 172: 1–40.

Hayman, B. I. 1957. Interaction, heterosis and diallel crosses. Genetics 42: 336–355.

Heine-Geldern, R. 1958. Kulturpflanzengeographie und das Problem vorkolumbisher Kulturbeziehungen zwischen Alter und Neuer Welt. Anthropos 53: 361–402. English translation by Goodman et al., 1964.

Helbaek, H. 1959. Domestication of food plants in the Old World. Science 130: 365–372.

Helmerick, R. H. 1959. Gamete sampling from a seriated sample of S_1 lines of corn. Diss. Abst. 19: 1504.

Henrard, J. T. 1931. A contribution to the knowledge of the Indian Maydeae. Meded. Rijks Herb. Leiden 67: 1–17.

Hepperly, I. W. 1949. A corn with odd-rowed ears. J. Hered. 40: 62–64.

Hernández-Xolocotzi, E., and L. F. Randolph. 1950. Descripción de los *Tripsacum* diploides de Mexico: *Tripsacum maizar* y *Tripsacum zopilotense* spp. nov. Ofic. Estud. Esp. Sec. Agric. y Ganad. Fol. tec. 4: 1–28.

Hitchcock, A. S. 1922. A perennial species of teosinte. J. Wash. Acad. Sci. 12: 205–208.

———. 1935. Manual of the grasses of the United States. U.S.D.A. Misc. Pub. no. 200.

Ho, P. T. 1956. The introduction of American food plants into China. Am. Anthrop. 57: 191–201.

———. 1969. The loess and the origin of Chinese

agriculture. Am. Hist. Rev. 75: 1–36.

Hoen, K., and R. H. Andrew. 1959. Performance of corn hybrids with various ratios of flint-dent germplasm. Agron. J. 51: 451–454.

Hopkins, C. G. 1899. Improvement in the chemical composition of the corn kernel. Ill. Agric. Exp. Sta. Bull. 55: 205–240.

Horowitz, S., and A. H. Marchioni. 1940. Herencia de la resistencia a la langosta en el maiz "amargo." An. Inst. Fito. Santa Catalina 2: 27–52.

Hull, F. H. 1945. Recurrent selection for specific combining ability in corn. J. Am. Soc. Agron. 37: 134–145.

———. 1946. Regression analyses of yields of hybrid corn and inbred parent lines. Maize Gen. Coöp. News Letter 20: 9–13.

———. 1946. Overdominance and corn breeding where hybrid seed is not feasible. J. Am. Soc. Agron. 38: 1100–1103.

———. 1952. Recurrent selection and overdominance. *In* Heterosis, pp. 451–473. Ames: Iowa State College Press.

Hurst, C. T., and E. Anderson. 1949. A corn cache from western Colorado. Am. Antiquity 14: 161–167.

Hutchinson, J. B. 1962. The history and relationships of the world's cottons. Endeavor 21: 5–15.

———, R. A. Silow, and S. G. Stephens. 1947. The evolution of Gossypium and the differentiation of the cultivated cottons. London: Oxford University Press.

Irwin, H., and E. S. Barghoorn. 1965. Identification of the pollen of maize, teosinte and Tripsacum by phase contrast microscopy. Bot. Mus. Leafl. Harvard Univ. 21: 37–56.

Janaki-Ammal, E. K. 1938. A Saccharum-Zea cross. Nature 142: 618–619.

Jeffreys, M. D. W. 1953a. Pre-Columbian negroes in America. Scientia (July-Aug.): 1–18.

———, 1953b. Pre-Columbian maize in Africa. Nature 172: 965–966.

———. 1953c. The history of maize in Africa. Eastern Anthrop. 7: 138–147.

———. 1965. Pre-Columbian maize in the Phillipines. South African J. Sci. 61: 5–10.

———. 1967. Who introduced maize into southern Africa? South African J. Sci. 63: 24–40.

———. 1971. Pre-Columbian maize in Asia. *In* Man across the Sea. Austin: Univ. Texas Press, pp. 376–400.

Jenkins, M. T. 1929. Correlation studies with inbred and crossbred strains of maize. J. Agric. Res. 39: 677–721.

———. 1940. The segregation of genes affecting yield of grain in maize. J. Am. Soc. Agron. 32: 55–63.

———. 1943. A new locality for teosinte in Mexico. J. Hered. 34: 206.

———, and A. L. Robert. 1952. Inheritance of resistance to the leaf blight of corn caused by *Helminthosporium turcicum*. Agron. J. 44: 136–140.

———, ———, and W. R. Findley, Jr. 1954. Recurrent selection as a method for concentrating genes for resistance to *Helminthosporium tur-*

cicum leaf blight in corn. Agron. J. 46: 89–94.

Jinks, J. L. 1954. The genetical basis of heterosis. Maize Gen. Coöp. News Letter 28: 47–50.

———, and B. I. Hayman. 1953. The analysis of diallel crosses. Maize Gen. Coöp. News Letter 27: 48–54.

———, and R. M. Jones. 1958. Estimation of the components of heterosis. Genetics 43: 223–234.

———, and K. Mather. 1955. Stability in development of heterozygotes and homozygotes. Proc. Roy. Soc. B 143: 561–578.

Johnston, G. S. 1966. Manifestations of teosinte and "Tripsacum" introgression in Corn Belt maize. Bussey Inst. Harvard Univ.

Jones, D. F. 1918. The effects of inbreeding and crossbreeding upon development. Conn. Agric. Exp. Sta. Bull. 207.

———. 1939. Continued inbreeding in maize. Genetics 24: 462–473.

———. 1944. Equilibrium in genic materials. Proc. Nat. Acad. Sci. 30: 82–87.

———. 1945. Heterosis resulting from degenerative changes. Genetics 30: 527–542.

———. 1952. Plasmagenes and chromogenes in heterosis. *In* Heterosis, pp. 224–235. Ames: Iowa State College Press.

———. 1957. Gene action in heterosis. Genetics 42: 93–103.

———. 1958. Heterosis and homeostasis in evolution and in applied genetics. Am. Nat. 92: 321–328.

———, and P. C. Mangelsdorf. 1951. The production of hybrid corn seed without detasseling. Conn. Agric. Exp. Sta. Bull. 550.

Jorgenson, L., and H. E. Brewbaker. 1927. A comparison of selfed lines of corn and first generation crosses between them. J. Am. Soc. Agron. 19: 819–830.

Kellerman, W. A. 1895. Primitive corn. Meehan's Monthly 5: 44.

Keeble, F., and C. Pellew. 1910. The mode of inheritance of stature and time of flowering in peas (*Pisum sativum*). J. Genet. 1: 47–56.

Kelly, I., and E. Anderson. 1943. Sweet corn in Jalisco. Ann. Missouri Bot. Gard. 30: 405–412.

Kempton, J. H. 1924. Inheritance of the crinkly ramose and brachytic characters of maize in hybrids with teosinte. J. Agric. Res. 27: 537–596.

———. 1931. Maize, the plant-breeding achievement of the American Indian. Smiths. Sci. Series 11: 319–349.

———. 1936. Maize as a measure of Indian skill. Univ. N.M. Bull. 296: 19–28.

———, and W. Popenoe, 1937. Teosinte in Guatemala. Carnegie Inst. Washington Publ. 483: 199–218.

Kermicle, J. L. 1969. Androgenesis conditioned by a mutation in maize. Science 166: 1422–1424.

Kiesselbach, T. A. 1926. The immediate effect of gametic relationship and of parental type upon the kernel weight of corn. Neb. Agric. Exp. Sta. Res. Bull. 33: 1–69.

———. 1949. The structure and reproduction of corn. Neb. Agric. Exp. Sta. Res. Bull. 161: 1–96.

———, and F. D. Keim. 1921. The regional

248 Bibliography

adaptation of corn in Nebraska. Neb. Agric. Exp. Sta. Res. Bull. 19.

Kinman, M. L., and G. F. Sprague. 1945. Relation between number of parental lines and theoretical performance of synthetic varieties of corn. J. Am. Soc. Agron. 37: 341–351.

Kramer, H. H., and A. J. Ullstrup. 1959. Preliminary evaluation of exotic maize germplasm. Agron. J. 51: 687–689.

Kuleshov, N. N. 1928. Some peculiarities in the maize of Asia. Bull. Appl. Bot. and Pl. Breed. 19: 325–374.

———. 1929. The geographical distribution of the varietal diversity of maize in the world. Bull. Appl. Bot. and Pl. Breed. 20: 506–510.

Kurtz, E. B., J. L. Liverman, and H. Tucker. 1960. Some problems concerning fossil and modern corn pollen. Bull. Torr. Bot. Club 87: 85–94.

Kuwada, Y. 1911. Meiosis in the pollen mother cells of *Zea mays* L. Bot. Mag. Tokyo 25: 163–181.

———. 1915. Über die Chromosomenzahl von *Zea mays* L. Bot. Mag. Tokyo 29: 83–89, 171–184.

———. 1919. Die Chromosomenzahl von *Zea mays* L. J. Col. Sci. Imp. Univ. Tokyo 39: 1–148.

Lambert, R. J. 1964. Analysis of certain backcross populations derived from *Zea mays* × *Zea mexicana* (Schrad., M.R.) hybrids. Thesis, Univ. Illinois.

Landstrom, B. 1967. Columbus. New York: Macmillan.

Langham, D. G. 1940. The inheritance of intergeneric differences in *Zea-Euchlaena* hybrids. Genetics 25: 88–107.

———, O. Gorbea, O. Villanueva, and C. Rojas. 1945. El maiz en Venezuela y su mejoramiento. Third Inter. Conf. Agric. Caracas: 54–55.

Laubengayer, R. A. 1949. The vascular anatomy of the eight-rowed ear and tassel of Golden Bantam sweet corn. Am. J. Bot. 36: 236–244.

Laughnan, J. R. 1952. The action of allelic forms of the gene *A* in maize. IV: On the compound nature of A^b and the occurrence and action of its A^d derivatives. Genetics 37: 375–395.

Lehman, W., and H. Doering. 1929. The Art of Old Peru. London: Ernest Benn.

Lerner, I. M. 1954. Genetic homeostasis. New York: John Wiley.

———. 1958. The genetic basis of selection. New York: John Wiley.

Lewis, D. 1941. Male sterility in natural populations of hermaphrodite plants. New Phytologist 40: 56–63.

Lewis, E. B. 1950. The phenomenon of position effect. *In* Advances in genetics 3: 73–115.

———. 1951. Genes and mutations. Cold Spring Harbor Symp. Quant. Biol. 16: 159–174.

Li, H. L. 1961. A case for pre-Columbian transatlantic travel by Arab ships. Harvard J. Asiatic Studies 23: 114–126.

Li, H. W., T. H. Ma, and K. C. Shang. 1954. Cytological studies of sugarcane and its relatives. XI: Hybrids of sugarcane and corn. Taiwan-Sugar 1: 13–24.

Libby, W. F. 1952. Chicago radiocarbon dates, III: Science 116: 673–681.

Lindstrom, E. W. 1935. Some new mutants in maize. Iowa State College J. Sci. 9: 237–245.

———. 1939. Analysis of modern maize breeding principles and methods. Proc. Seventh Int. Genetics Cong., pp. 191–196.

Longley, A. E. 1924. Chromosomes in maize and maize relatives. J. Agric. Res. 28: 673–682.

———. 1934. Chromosomes in hybrids between *Euchlaena perennis* and *Zea mays*. J. Agric. Res. 48: 789–806.

———. 1937. Morphological characters of teosinte chromosomes. J. Agric. Res. 54: 836–862.

———. 1939. Knob positions on corn chromosomes. J. Agric. Res. 59: 475–490.

———. 1941a. Knob positions on teosinte chromosomes. J. Agric. Res. 62: 401–413.

———. 1941b. Chromosome morphology in maize and its relatives. Bot. Rev. 7: 262–289.

———, and T. A. Kato Y. 1965. Chromosome morphology of certain races of maize in Latin America. Internat. Center Improvement Maize and Wheat Res. Bull. 1. Chapingo, Mexico.

Lonnquist, J. H. 1951. Recurrent selection as a means of modifying combining ability in corn. Agron. J. 43: 311–315.

———. 1960. Modifying double-cross hybrid corn performance through convergent improvement. Agron. J. 52: 226–228.

———, and C. O. Gardner. 1961. Heterosis in intervarietal crosses in maize and its implication in breeding procedures. Crop Sci. 1: 179–183.

———, and D. P. McGill. 1954. Genetic sampling from selected zygotes in corn breeding. Agron. J. 46: 147–150.

———, ———. 1956. Performance of corn synthetics in advanced generations of synthesis and after two cycles of recurrent selection. Agron. J. 48: 249–253.

———, and M. D. Rumbaugh. 1958. Relative importance of test sequence for general and specific combining ability in corn breeding. Agron. J. 50: 541–544.

Lumholtz, C. 1902. Unknown Mexico. New York: Charles Scribner's Sons.

Lyte, H. 1619. A new herbal, or historie of plants. [A translation of Dodoens] London.

MacNeish, R. S. 1964. Ancient Mesoamerican civilization. Science 143: 531–537.

———. 1964. The origins of New World civilizations. Sci. Amer. 211: 29–37.

Maguire, M. P. 1957. Cytogenetic studies of *Zea* hyperploid for a chromosome derived from *Tripsacum*. Genetics 42: 473–486.

———. 1961. Divergence in *Tripsacum* and *Zea* chromosomes. Evolution 15: 394–400.

———. 1962. Common loci in corn and *Tripsacum*. J. Hered. 53: 87–88.

Mangelsdorf, A. J. 1952. Gene interaction in heterosis. *In* Heterosis, pp. 320–329. Ames: Iowa State College Press.

Mangelsdorf, P. C. 1924. Waxy endosperm in New England maize. Science 60: 222–223.

———. 1926. The genetics and morphology of some endosperm characters in maize. Conn. Agric. Exp. Sta. Bull. 279: 513–612.

———. 1945. The origin and nature of the ear of maize. Bot. Mus. Leafl. Harvard Univ. 12: 33–88.

———. 1947. The origin and evolution of maize. *In* Advances in genetics I: 161–207.

———. 1952. Hybridization in the evolution of maize. *In* Heterosis, pp. 175–198. Ames: Iowa State College Press.

———. 1958a. Reconstructing the ancestor of corn. Proc. Am. Philos. Soc. 102: 454–463.

———. 1958b. The mutagenic effect of hybridizing maize and teosinte. Cold Spring Harbor Symp. Quant. Biol. 23: 409–421.

———. 1961. The genotypes of two primitive races of maize. Maize Gen. Coöp. News Letter 35: 35.

———. 1961. Introgression in maize. Euphytica 10: 157–168.

———. 1965. The evolution of maize. *In* Crop Plant Evolution, pp. 23–49. Cambridge: Cambridge Univ. Press.

———, and J. W. Cameron. 1942. Western Guatemala, a secondary center of origin of cultivated maize varieties. Bot. Mus. Leafl. Harvard Univ. 10: 217–252.

———, H. W. Dick, and J. Cámara-Hernández. 1967. Bat Cave revisited. Bot. Mus. Leafl. Harvard Univ. 22: 1–31.

———, and W. C. Galinat. 1964. The tunicate locus in maize dissected and reconstituted. Proc. Nat. Acad. Sci. 51: 147–150.

———, and D. F. Jones. 1926. The expression of Mendelian factors in the gametophyte of maize. Genetics 11: 423–455.

———, and R. H. Lister. 1956. Archaeological evidence on the evolution of maize in northwestern Mexico. Bot. Mus. Leafl. Harvard Univ. 17: 151–178.

———, R. S. MacNeish, and W. C. Galinat. 1956. Archaeological evidence on the diffusion and evolution of maize in northeastern Mexico. Bot. Mus. Leafl. Harvard Univ. 17: 125–150.

———, ———, ———. 1964. Domestication of corn. Science 143: 538–545.

———, ———, ———. 1967. Prehistoric maize, teosinte and Tripsacum from Tamaulipas, Mexico. Bot. Mus. Leafl. Harvard Univ. 22: 33–62.

———, and H. P. Mangelsdorf. 1957. Genotypes involving the *Tu-tu* locus compared in isogenic stocks. Maize Gen. Coöp. News Letter 31: 65–66.

———, and D. L. Oliver. 1951. Whence came maize to Asia? Bot. Mus. Leafl. Harvard Univ. 14: 263–291.

———, and R. G. Reeves. 1931. Hybridization of maize, Tripsacum and Euchlaena. J. Hered. 22: 329–343.

———, ———. 1939. The origin of Indian corn and its relatives. Texas Agric. Exp. Sta. Bull. 574: 1–315.

———, ———. 1959. The origin of corn. I. Pod corn, the ancestral form. Bot. Mus. Leafl. Harvard Univ. 18: 329–356.

———, ———. 1959. The origin of corn. IV. Place and time of origin. Bot. Mus. Leafl. Harvard Univ. 18: 413–427.

249 Bibliography

————, and C. E. Smith, Jr. 1949. New archaeological evidence on evolution in maize. Bot. Mus. Leafl. Harvard Univ. 13: 213–247.

Manwell, C., C. M. A. Baker, and W. Childers. 1963. The genetics of hemoglobin in hybrids. I. A molecular basis for hybrid vigor. Comp. Biochem. Physiol. 10: 103–120.

Mather, K. 1949. Biometrical genetics. London: Methuen.

————. 1955. The genetical basis of heterosis. Proc. Roy. Soc. B 144: 143–150.

Mazoti, L. B. 1958. Estudio sobre differencias citoplasmaticas heredables entre "Zea Mays" y "Euchlaena Mexicana." Rev. Argent. Agron. 25: 12–44.

————, and C. E. Muhlenberg. 1958. Haploides naturales en maíz. Rev. Argent. Agron. 25: 171–178.

McClintock, B. 1933. The association of non-homologous parts of chromosomes in a midprophase of meiosis in Zea mays. Zeitschr. Zellf. mikro. Anat. 19: 191–237.

————. 1959. Genetic and cytological studies of maize. Carnegie Inst. Washington Year Book 58: 452–456.

————. 1960. Chromosome constitutions of Mexican and Guatemalan races of maize. Carnegie Inst. Washington Year Book 59: 461–472.

————. 1961. Some parallels between gene control systems in maize and in bacteria. Am. Nat. 95: 265–277.

————. 1965. The control of gene action in maize. Brookhaven Sympos. Biol. 18: 162–184.

Melhus, I. E., F. Aguirre, and N. S. Scrimshaw. 1953. Observations on the nutritive value of teosinte. Science 117: 34–35.

————, and I. M. Chamberlain. 1953. A preliminary study of teosinte in its region of origin. Iowa St. Coll. J. Sci. 28: 139–164.

————, G. Semeniuk, and J. R. Wallin. 1949. A preliminary study of the growth response in Iowa of hybrids between Guatemalan and United States corn. Iowa Agric. Exp. Sta. Res. Bull. 371: 625–638.

Merrill, E. D. 1950. Observations on cultivated plants with reference to certain American problems. Ceiba 1: 3–36.

————. 1954. The botany of Cook's voyages. Waltham, Mass.: Chronica Botanica.

Mertz, E. T., L. S. Bates, and O. E. Nelson. 1964. Mutant gene that changes protein composition and increases lysine content of maize endosperm. Science 148: 1741–1742.

Metcalfe, C. R. 1960. Anatomy of the Monocotyledons. Oxford: Clarendon Press.

Miller, E. C. 1919. Development of the pistillate spikelet, and fertilization in Zea mays. J. Agric. Res. 18: 255–266.

Miracle, M. P. 1963. Interpretation of evidence on the introduction of maize into West Africa. Africa 33: 132–135.

Miranda, C. S. 1966. Discusion sobre el origen y la evolución del maíz. Memorias Segunda Congreso Nacional Fitogenetica, Monterey, N.L., Mexico.

Moll, R. H., and H. F. Robinson. 1967. Quanti-

tative genetic investigations of yield of maize. Der Züchter 37: 192–199.

Montgomery, E. G. 1906. What is an ear of corn? Pop. Sci. Monthly 68: 55–62.

Murphy, R. P. 1942. Convergent improvement with four inbred lines of corn. J. Am. Soc. Agron. 34: 138–150.

Murray, J. 1970. The First European Agriculture University Press, Edinburgh.

Nei, M. 1963. The efficiency of haploid methods of plant breeding. Heredity 18: 95–100.

Nelson, O. E. 1959. Intracistron recombination in the Wx-wx region in maize. Science 130: 794–795.

————. 1969. The modification by mutation of protein quality in maize. In New Approaches to breeding for Improved Plant Protein. Vienna: Internat. Atomic Energy Agency.

Nickerson, N. H. 1954. Morphological analysis of the maize ear. Am. J. Bot. 41: 87–92.

Nilsson-Leissner, G. 1927. Relation of selfed strains of corn to F_1 crosses between them. J. Am. Soc. Agron. 19: 440–454.

O'Mara, J. G. 1942. A cytogenetic study of Zea and Euchlaena. Res. Bull. Missouri Agric. Exp. Sta. 341: 1–16.

Painter, R. H. 1955. Insects on corn and teosinte in Guatemala. J. Econ. Entom. 48: 36–42.

Parodi, L. R. 1935. Relaciones de la agricultura prehispanica con la agricultura Argentina actual. Anales Acad. Nac. Agron. y Vet. de Buenos Aires 1: 115–167.

Parsons, P. A., and W. F. Bodmer. 1961. The evolution of overdominance: natural selection and heterozygote advantage. Nature 190: 7–12.

Paterniani, E., and J. H. Dennquist. 1963. Heterosis in inter-racial crosses of corn (Zea mays L.). Crop Science 3: 504–507.

Payne, E. J. 1892. History of the New World Called America, vol. I. Oxford.

Paxson, J. B. 1953. Pilosity and hispidulousness of the leaf sheath. Maize Gen. Coöp. News Letter 27: 36–38.

Phillips, L. L. 1963. The cytogenetics of Gossypium and the origin of New World cottons. Evolution 17: 460–469.

Pinnell, E. L., E. H. Rinke, and H. K. Hayes. 1952. Gamete selection for specific combining ability. In Heterosis, pp. 378–388. Ames: Iowa State College Press.

Plumb, C. S. 1898. The geographic distribution of cereals in North America. U.S.D.A. Div. Biol. Surv. Bull. 11: 1–24.

Pontecorvo, G. 1955. Gene structure and action in relation to heterosis. Proc. Roy. Soc. B 144: 171–177.

Post, T. von, and O. Kuntze. 1904. Lexicon Generum Phanerogamarum. Stuttgart: Deutsche Verlags-Anstalt.

Prywer, C. 1954. Meiosis en Tripsacum maizar. H. & R. Revist. Soc. Mex. Hist. Nat. 15: 59–64.

————. 1963. Meiosis en Tripsacum zopilotense. H. & R. Bol. Soc. Bot. 28: 11–18.

Ramírez, E. R., D. H. Timothy, E. Díaz B., and U. J. Grant, in collaboration with G. E. Nicholson, C. E. Anderson, and W. L. Brown. 1960.

Races of maize in Bolivia. Nat. Acad. Sci.-Nat. Res. Council Publ. 747.

Randolph, L. F. 1952. New evidence on the origin of maize. Am. Nat. 86: 193–202.

————. 1955. Cytogenetic aspects of the origin and evolutionary history of corn. In Corn and corn improvement, pp. 16–61. New York: Academic Press.

————. 1959. The origin of maize. Indian J. Gen. Plant Breed. 19: 1–12.

————. 1970. Variation among Tripsacum populations of Mexico and Guatemala. Brittonia 22: 305–337.

Reeves, R. G. 1944. Chromosome knobs in relation to the origin of maize. Genetics 29: 141–147.

————. 1950. Morphology of the ear and tassel of maize. Am. J. Bot. 37: 697–704.

————. 1953. Morphology of plant organs related to the maize ear. Am. J. Bot. 40: 266–271.

————. 1953. Comparative anatomy of the American Maydeae. Texas Agric. Exp. Sta. Bull. 761.

————, and P. C. Mangelsdorf. 1935. Chromosome numbers in relatives of Zea mays. L. Am. Nat. 69: 633–635.

————, ————. 1942. A proposed taxonomic change in the tribe Maydeae (family Gramineae). Am. J. Bot. 29: 815–817.

————, ————. 1959. The origin of corn. II. Teosinte, a hybrid of corn and Tripsacum. Bot. Mus. Leafl. Harvard Univ. 18: 357–387.

————, ————. 1959. The origin of corn. V. A critique of current theories. Bot. Mus. Leafl. Harvard Univ. 18: 428–440.

Renfrew, J. M. 1969. The archaeological evidence for the domestication of plants: methods and problems. In The Domestication and Exploitation of Plants and Animals, pp. 149–152. London: Gerald Duckworth.

Rhoades, M. M. 1955. The cytogenetics of maize. In Corn and corn improvement, pp. 123–319. New York: Academic Press.

Richey, F. D. 1927. The convergent improvement of selfed lines of corn. Am. Nat. 61: 430–449.

————. 1947. Corn breeding: gamete selection, the Oenothera method, and related miscellany. J. Am. Soc. Agron. 39: 403–411.

————. 1950. Corn breeding. In Advances in genetics 3: 159–192.

————, and G. F. Sprague. 1931. Experiments on hybrid vigor and convergent improvement in corn. U.S.D.A. Tech. Bull. 267, pp. 1–22.

Riley, R. 1965. Cytogenetics and the evolution of wheat. In Crop plant evolution, pp. 103–122. Cambridge: Cambridge Univ. Press.

Roberts, L. M., U. J. Grant, R. Ramírez E., W. H. Hatheway, and D. L. Smith, in collaboration with P. C. Mangelsdorf. 1957. Races of maize of Colombia. Nat. Acad. Sci.-Nat. Res. Council Publ. 510.

Robinson, H. F., R. E. Comstock, and P. H. Harvey. 1949. Estimates of heritability and the degree of dominance in corn. Agron. J. 41: 353–359.

Rogers, J. S. 1950a. Inheritance of photoperiodic response and tillering in maize-teosinte hybrids. Genetics 35: 513–540.

———. 1950b. Inheritance of inflorescence characters in maize-teosinte hybrids. Genetics 35: 541–558.

———. 1950c. Fertility relationships in maize-teosinte hybrids. Texas Agric. Exp. Sta. Bull. 730.

Rollins, R. C. 1953. Cytogenetical approaches to the study of genera. Chronica Botanica 14: 133–139.

Rosbaco, U. F. 1951. Consideraciones sobre maíces "amargos" con especial referencia a su cultivo en la provincia de Entre Ríos. Idia no. 46: 1–12.

Rumbaugh, M. D., and J. H. Lonnquist. 1959. Inbreeding depression of diallel crosses of selected lines of corn. Agron. J. 51: 407–412.

Saint-Hilaire, A. de. 1829. Lettre sur une variété remarquable de maïs du Brésil. Ann. Sci. Nat. 16: 143–145.

Salmon, S. C., and A. A. Hanson. 1964. The principles and practice of agricultural research. London: Leonard Hill.

Sass, J. E. 1955. Vegetative morphology. In Corn and corn improvement, pp. 63–97. New York: Academic Press.

Sauer, C. O. 1952. Agricultural origins and dispersals. Am. Geog. Soc. Bowman Memorial Lectures, Ser. 2. New York.

———. 1960. Maize into Europe. Acts Internat. Americanist Cong. (Vienna) 34: 777–787.

Schiemann, E. 1932. Entstehung der Kulturpflanzen. Berlin.

Schuler, J. F. 1954. Natural mutations in inbred lines of maize and their heterotic effect. I. Comparison of parent, mutant, and their F_1 hybrid in a highly inbred background. Genetics 39: 908–922.

———, and G. F. Sprague. 1956. Natural mutations in inbred lines of maize and their heterotic effects. II. Comparison of mother line vs. mutant when outcrossed to unrelated inbred. Genetics 41: 281–291.

Schuman, K. M. 1904. Mais und Teosinte. In P. Ascherson, Festschrift. Leipzig.

Schwartz, D., and W. J. Laughner. 1969. A molecular basis for heterosis. Science 166: 626–627.

Seaney, R. R. 1955. Studies on monoploidy in maize. Diss. Abst. 15: 187–188.

Sehgal, S. M. 1963. Effects of teosinte and "Tripsacum" introgression in maize. Bussey Inst. Harvard Univ.

———, and W. L. Brown. 1965. Introgression in Corn Belt maize. Econ. Bot. 19: 83–88.

Sentz, J. C., H. F. Robinson, and R. E. Comstock. 1954. Relation between heterozygosis and performance in maize. Agron. J. 46: 514–520.

Shaver, D. L. 1962. Cytogenetic studies of allotetraploid hybrids of maize and perennial teosinte. Am. J. Bot. 49: 348–354.

Shull, G. H. 1908. The composition of a field of maize. Rep. Am. Breeder's Assn. 4: 296–301.

———. 1909. A pure-line method of corn breeding. Rep. Am. Breeder's Assn. 5: 51–59.

———. 1911. The genotypes of maize. Am. Nat. 45: 234–252.

———. 1914. Duplicate genes for capsule form in Bursa bursa-pastoris. Zeitschr. Abst. Vererb. 12: 97–149.

Singleton, W. R. 1939. Opaque endosperm-2 (O_2), recent linkage studies in maize. Genetics 24: 61.

———. 1943a. Hybrid vigor in the intra-inbred cross P39 × C30 in maize. Records Genetics Soc. Am. 12: 52–53.

———. 1943b. Breeding behavior of C30 diminutive mutant whose hybrids show increased vigor. Genetics 28: 89.

———. 1946. Inheritance of indeterminate growth in maize. J. Hered. 37: 61–64.

———. 1951. Inheritance of corn grass, a macromutation in maize, and its possible significance as ancestral type. Am. Nat. 85: 81–86.

———. 1954. The effect of chronic gamma radiation on endosperm mutations in maize. Genetics 39: 587–603.

Sinnott, E. W., L. C. Dunn, and Th. Dobzhansky. 1958. Principles of genetics. 5th ed. New York: McGraw-Hill.

Sprague, G. F. 1939. An estimation of the number of top-crossed plants required for adequate representation of a corn variety. J. Am. Soc. Agron. 31: 11–16.

———. 1955. Problems in the estimation and utilization of genetic variability. Cold Spring Harbor Symp. Quant. Biol. 20: 87–92.

———. 1955. Corn breeding. In Corn and corn improvement. New York: Academic Press.

———, and B. Brimhall. 1950. Relative effectiveness of two systems of selection for oil content of the corn kernel. Agron. J. 42: 83–88.

———, P. A. Miller, and B. Brimhall. 1952. Additional studies of the relative effectiveness of two systems of selection for oil content of the corn kernel. Agron. J. 44: 329–331.

———, W. A. Russell, and L. H. Penny. 1959. Further studies on convergent improvement in corn. Genetics 44: 341–346.

Stadler, L. J. 1942. Some observations on gene variability and spontaneous mutations. Spragg Memorial Lectures, Michigan State College.

———. 1944. Gamete selection in corn breeding. J. Am. Soc. Agron. 36: 988–989.

———. 1951. Problems of gene structure. I. The inter-dependence of the elements (S) and (P) in the gene R^r of maize. Science 114: 488.

———, and M. H. Emmerling. 1955. Relation of unequal crossing over to the interdependence of R^r elements (P) and (S). Genetics 41: 124–137.

———, and M. G. Nuffer. 1953. Problems of gene structure. II. Separation of R^r elements (S) and (P) by unequal crossing over. Science 117: 471–472.

Standley, P. C. 1950. Teosinte in Honduras. Ceiba 1: 58–61.

Stanton, W. R. 1963. Archaeological evidence for changes in maize types in West Africa: An experiment in technique. Man 150: 117–123.

Stebbins, G. L., Jr. 1950. Variation and evolution in plants. New York: Columbia Univ. Press.

———. 1956. Cytogenetics and evolution of the grass family. Am. J. Bot. 43: 890–905.

Stephens, S. G. 1963. Polynesian cottons. Ann. Missouri Bot. Gard. 50: 1–22.

Stonor, C. R., and E. Anderson. 1949. Maize among the hill peoples of Assam. Ann. Missouri Bot. Gard. 36: 355–404.

Stringfield, G. H. 1950. Heterozygosis and hybrid vigor in maize. Agron. J. 42: 145–152.

Sturtevant, A. H. 1965. The early Mendelians. Proc. Am. Philos. Soc. 109: 199–204.

Sturtevant, E. L. 1879. Indian corn. Trans. N.Y. State Agric. Soc. 33: 37–74.

———. 1881. The superabundance of pollen in Indian corn. Am. Nat. 15: 1000.

———. 1894. Notes on Maize. Bull. Torrey Bot. Club 21: 319–343, 503–523.

———. 1899. Varieties of corn. U.S.D.A. Off. Exp. Sta. Bull. 57.

Sutô, T., and Y. Yoshida. 1956. Characteristics of the Oriental maize. In Land and crops of Nepal Himalaya. II: 373–529. Kyoto, Japan: Fauna and Flora Research Society.

Tantravahi, R. V. 1968. Cytology and crossability relationships of Tripsacum. Bussey Inst. Harvard Univ.

———. 1971. Multiple character analysis and chromosome studies in Tripsacum lanceolatum complex. Evolution 25: 38–50.

Tapley, W. T., W. D. Enzie, and G. P. van Eseltine. 1934. Vegetables of New York. I. Part III. Sweet Corn. N.Y. State Agric. Exp. St. Report, 1934.

Tavcar, A. 1935. Beitrag zur Vererbung der Kornreihenanzahl an Maiskolben. Zeits. Züchtung, A. 20: 364–376.

Thapa, J. K. 1966. Primitive maize with the Lepchas. Bull. Tibetology 3: 29–31.

Thomas, W. I., and D. B. Grissom. 1961. Cycle evaluation of reciprocal recurrent selection for popping volume, grain yield, and resistance to root lodging in popcorn. Crop Sci. 1: 197–200.

Thompson, D. L. 1954. Combining ability of homozygous diploids of corn relative to lines derived by inbreeding. Agron. J. 46: 133–136.

———, and P. H. Harvey. 1960. Progress from recurrent and reciprocal recurrent selection for yield of corn. Agron. Abst. 1960: 54–55.

Timothy, D. H., B. Peña V., and R. Ramírez E. in collaboration with W. L. Brown and E. Anderson. 1961. Races of maize in Chile. Nat. Acad. Sci.-Nat. Res. Council Publ. 847.

———, W. H. Hatheway, U. J. Grant, M. Torregroza C., D. Sarria V., and D. Varela A. 1963. Races of maize in Ecuador. Nat. Acad. Sci.-Nat. Res. Council Publ. 975.

Ting, Y. C. 1956. Cytological studies of maize-teosinte derivatives. Maize Gen. Coöp. News Letter 30: 36–37.

———. 1957. Further cytological studies of maize-teosinte derivatives. Maize Gen. Coöp. News Letter 31: 70–72.

———. 1959. Chromosomes in three teosinte varieties. Maize Gen. Coöp. News Letter 33: 35–36.

———. 1960. Cytological observations on two tropical forms of Tripsacum. Bot. Mus. Leafl. Harvard Univ. 19: 97–108.

———. 1964. Chromosomes of maize-teosinte

hybrids. Bussey Inst. Harvard Univ.

————. 1967. Common inversion in maize and teosinte. Am. Nat. 101: 87–89.

Trattner, E. R. 1938. Architects of ideas. New York: Carrick & Evans.

Vargas, C. 1962. Phytometric representations of ancient Peruvians. Econ. Bot. 16: 106–115.

Venkateswarlu, J. 1962. Origin of maize. Proc. Summer School Bot.: 494–504. Darjeeling.

————. 1963. Cytogenetic evolution in angiosperms—Maydeae. Indian Bot. Soc. Mem. 4: 65–73.

Venkatraman, T. S., and R. Thomas. 1932. Sugarcane-sorghum hybrids. Indian J. Agric. 2: 19–27.

Vetukhiv, M. 1954. Integration of the genotype in local populations of three species of Drosophila. Evolution 8: 241–251.

Vishnu-Mittre and H. P. Gupta. 1966. Pollen morphological studies of some primitive varieties of maize (Zea Mays L.) with remarks on the history of maize in India. Palaeobotanist 15: 176–184.

Wallace, H. A., and W. L. Brown. 1956. Corn and its early fathers. East Lansing: Michigan State Univ. Press.

Watson, S. 1891. Contributions to American botany. 3. Upon a wild species of Zea from Mexico. Proc. Am. Acad. Arts and Sci. 26: 108–161.

Weatherwax, P. 1918. The evolution of maize. Bull. Torrey Bot. Club 45: 309–342.

————. 1919a. Gametogenesis and fecundation in Zea mays as the basis of xenia and heredity in the endosperm. Bull. Torrey Bot. Club 46: 73–90.

————. 1919b. The ancestry of maize—a reply to criticism. Bull. Torrey Bot. Club 46: 275–278.

————. 1920. A misconception as to the structure of the ear of maize. Bull. Torrey Bot. Club 47: 359–362.

————. 1926. Comparative morphology of the oriental Maydeae. Indiana Univ. Studies 73: 3–18.

————. 1935. The phylogeny of Zea mays. Am. Midland Nat. 16: 1–71.

————. 1942. The Indian as a corn breeder. Proc. Indiana Acad. Sci. 51: 13–21.

————. 1948. Right-handed and left-handed corn embryos. Ann. Missouri Bot. Gard. 35: 317–321.

————. 1950. The history of corn. Sci. Month. 71: 50–60.

————. 1954. Indian corn in old America. New York: Macmillan.

————. 1955. History and origin of corn. I. Early history of corn and theories as to its origin. In Corn and corn improvement, pp. 1–16. New York: Academic Press.

Weaver, B. L., and A. E. Thompson. 1957. White hulless popcorn: Fifteen generations of selection for improved popping expansion. Bull. Ill. Agric. Exp. Sta. 616: 1–19.

Wellhausen, E. J., L. M. Roberts, and E. Hernández X. in collaboration with P. C. Mangelsdorf. 1951. Razas de maiz en Mexico. Secretaria de Agricultura y Granaderia Folleto Tecnico 5, Mexico.

————, ————, ————. 1952. Races of maize in Mexico. Bussey Inst. Harvard Univ.

————, A. Fuentes O., and A. Hernández C. in collaboration with P. C. Mangelsdorf. 1957. Races of maize in Central America. Nat. Acad. Sci.-Nat. Res. Council Publ. 511.

Werth, E. 1922. Zur experimentellen Erzeugung

eingeschlechtiger Maispflanzen und zur Frage: wo entwickeln sich gemischte (androgyne) Blütenstände am Mais? Ber. Deut. Bot. Gesell. 40: 69–77.

Whaley, W. G. 1944. Heterosis. Bot. Rev. 10: 461–498.

Wiener, L. 1920–22. Africa and the discovery of America. 3 vols.

Wilkes, H. G. 1967. Teosinte: the closest relative of maize. Bussey Inst. Harvard Univ.

————. 1972. Maize and its wild relatives. Science 177: 1071–1077.

Will, G. F., and G. E. Hyde. 1917. Corn among the Indians of the upper Missouri. St. Louis: W. H. Miner.

Willett, F. 1962. The introduction of maize into West Africa: An assessment of recent evidence. Africa 32: 1–13.

Willey, G. R. 1966. An introduction to American archaeology. I. North and Middle America. Englewood Cliffs, N.J.: Prentice-Hall.

Wolf, M. J., H. C. Cutler, M. S. Zuber, and U. Khoo. 1972. Maize with multilayer aleurone of high protein content. Crop Sci. 12: 440–442.

Woodworth, C. M., E. R. Leng, and R. W. Jugenheimer. 1952. Fifty generations of selection for protein and oil in corn. Agron. J. 44: 60–65.

Worsdell, W. C. 1916. The principles of plant-teratology. London.

Wu, S. K. 1937. The relationship between the origin of selfed lines of corn and their value in hybrid combinations. J. Am. Soc. Agron. 31: 131–140.

Yarnell, S. H., and W. A. Hills. 1959. Is parthenogenesis possible in sweet corn? Proc. Am. Soc. Hort. Sci. 73: 407–414.

Zelinsky, W. 1950. A letter in Scientific American 183 (3): 2.

Index

Abejas phase, 167, 169, 171
Acoma, the, 110
Addison, G., 56
Aegean type of corn, 206
Africa: species of cotton from, 203; pre-Columbian corn in, 205
Agriculture, early evidence of, 169
Agriculture, U.S. Department of, 17
Aguirre, F., 24, 156, 231
Ajalpan phase, 126, 170
Aleurone, 8; color of, 144; multi-layered, 241
Alexander, D. E., 241
Allard, H. A., 23
Allelic interaction. *See* Overdominance
Alleoni, M. R., 78, 105, 122, 128
Allopolypoid: *Tripsacum* as, 58, 68; modern corn as, 135; nonallelic genic interaction in, 217
Altitude: and growing of corn, 95, 106, 114, 119, 145; high, 202, 204
Amaranths, 169
Amarillo de Ocho, 114A
Amecameca, 123
American Indians. *See* Indians, American
Amino acids, 1
Amylopectin, 8
Ancachino, 141
Ancestral form of maize, model of, 81–84; in simulated wild habitat, 84–85; later models, 93–97; pedigree of, 97–98; wild prehistoric corn similar to, 168
Ancient Indigenous maize, 101, 102, 117, 121; Palomero Toluqueño as, 106, 156; Chapalote as, 151; Nal-Tel as, 153
Andean region: maize with knobless chromosomes in, 118–119; gene frequencies in, 144
Anderson, Edgar, ix, 48, 108, 141, 145, 160, 163; on *Tripsacum*, 53, 54, 55, 170, 220; on Tripsacum as amphidiploid, 58; on *Manisuris*, 68, 73; on Asian origin of maize, 72, 201, 202, 203; on races of maize, 101, 102; on sweet corn, 109, 110, 113; on shape of ears, 117; on Corn Belt dent, 121, 122; on waxy maize, 143; on Norsemen as bearers of corn, 206; on Indian cultivation of corn, 208; on antithetical dominance, 238

Andres, J. M., 12
Andrew, R. H., 223
Andropogoneae: resemblance of maize to, 12, 34, 158; and teosinte, 19, 21; as remote ancestor of maize, 52, 66, 71, 73
Antevs, Ernst, 148, 149
Antibes, teosinte grown in, 16
Apomixis, ix, 72
Arabs, as bearers of corn, 205
Araguito, 117
Arapaho Indians, 116
Araucano, 114
Arber, Agnes, 5, 175
Archaeology: as evidence in origin of plants, 3, 147, 158; and teosinte introgression, 125; and *Tripsacum* introgression, 125; and prehistoric art, 187, 188; problems of forgeries in, 188
Argentina, 106, 128, 130; papyrescent corn in, 12; sweet corn in, 107, 110; eight-rowed corn in, 114; Culli, 115; Cateto, 118
Argentine pop, 98, 220
Argentino, 128
Arizona, 155; *Tripsacum* in, 56; cave sites in, 126, 150, 160, 163
Arnason, T. J., 24, 25
Arrocillo Amarillo, 102, 178
Art, prehistoric: representation of corn in, 187, 188, 189–199; idols, 189, 191; funerary urns, 190, 191, 192; replicas of corn, 194
Ascherson, P., on teosinte theory, 11, 17, 19, 34, 35, 201; on relation of teosinte to *Tripsacum*, 48
Asia: waxy maize in, 143–144, 145; as site of origin of maize, 169, 181, 201–204; theory of cotton from, 203
Assam, 143, 202
Avocados, 169
Azara, F. de, 75–76, 78
Aztecs, 3, 187; Sahagún on, 15; Hernández on, 16

Baby Golden popcorn, 81, 98, 220
Backcrossing, 38–39, 223; in convergent improvement, 227, 228
Bagnal, Richard, 110, 111
Bailey, Liberty Hyde, 18
Baker, M. A., 217
Baker, Raymond, 54

Bal-baeck, Lebanon, 188
Balsas, 26, 31
Banerjee, Umesh, C., ix, 49, 184
Barghoorn, Elso S., ix, xi, 49; on fossil pollen, 12, 181–183; re-examination of pollen by, 183–184
Barber Reid, 233
Barlow, R. H., 12n
Bartlett, Alexandria, 184
Bat Cave: corn from, x, 81, 126, 141, 158, 208; excavations in, 147–148; maize remains from, 148–149, 160; dating of remains from, 148, 149–150; second visit to, 149–152; teosinte contamination shown in, 151; significance of, 152; archaeological pollen from, 184; pore-axis ratio of maize from, 185
Bates, L. S., 241
Bateson, W., 216
Bauman, L. F., 224
Beadle, G. W., viii, xi, 12, 24, 25, 139, 156; on fertility of hybrids, 26; on fruit cases, 35; on chromosomes, 38, 66; on genes, 43, 123; on teosinte as ancestral form of maize, 49, 98, 180n; teosinte mutation hunt by, 51, 100
Beal, William, 211–212
Beans: planted with corn, 1, 148; dietary quality of, 2; in archaeological sites, 169; Johannsen's work on, 212
Bear, R. P., 108, 143, 202
Belles Artes (Mexico City) fossil pollen, 181–182; large grains from, 183; re-examination of, 184
Bentham, Jeremy, 20
Bento de Toledo, 18
Benzer, S., 217
Berger, R., 184
Bhagrat, S. D., 72
Bianchi, Angelo, xi, 134, 137, 138
Bible, the, 3, 204
Biological model for heterosis in maize, 224–225
Birket-Smith, K., 14
Bisingallo, 76, 78
Blaringhem, L., 18
Blight of 1970, 240. *See also* Diseases of corn
Blue-kerneled corn, 116
Blumenschein, A., 78, 105, 122, 128
Bock, Jerome, 201, 206